A TEXTBOOK OF INORGANIC CHEMISTRY

Volume I

Mandeep Dalal

DALAL INSTITUTE

Publisher:

Dalal Institute, Main Market, Sector 14, Rohtak, Haryana 124001, India

(info@dalalinstitute.com, +91-9802825820)

www.dalalinstitute.com

Disclaimer:

Although every precaution has been taken to verify the accuracy of the information contained herein, the author and publisher assume no responsibility for any errors or omissions. No liability is assumed for damages that may result from the use of the information contained within.

Credits:

The dedication image is a derivative work of the famous painting "Innocence" by William-Adolphe Bouguereau, whereas the image on the front cover is a derivative work of Dimitris Christou's "Precious Diamond".

Dedicated to my mother "Darshana Devi"

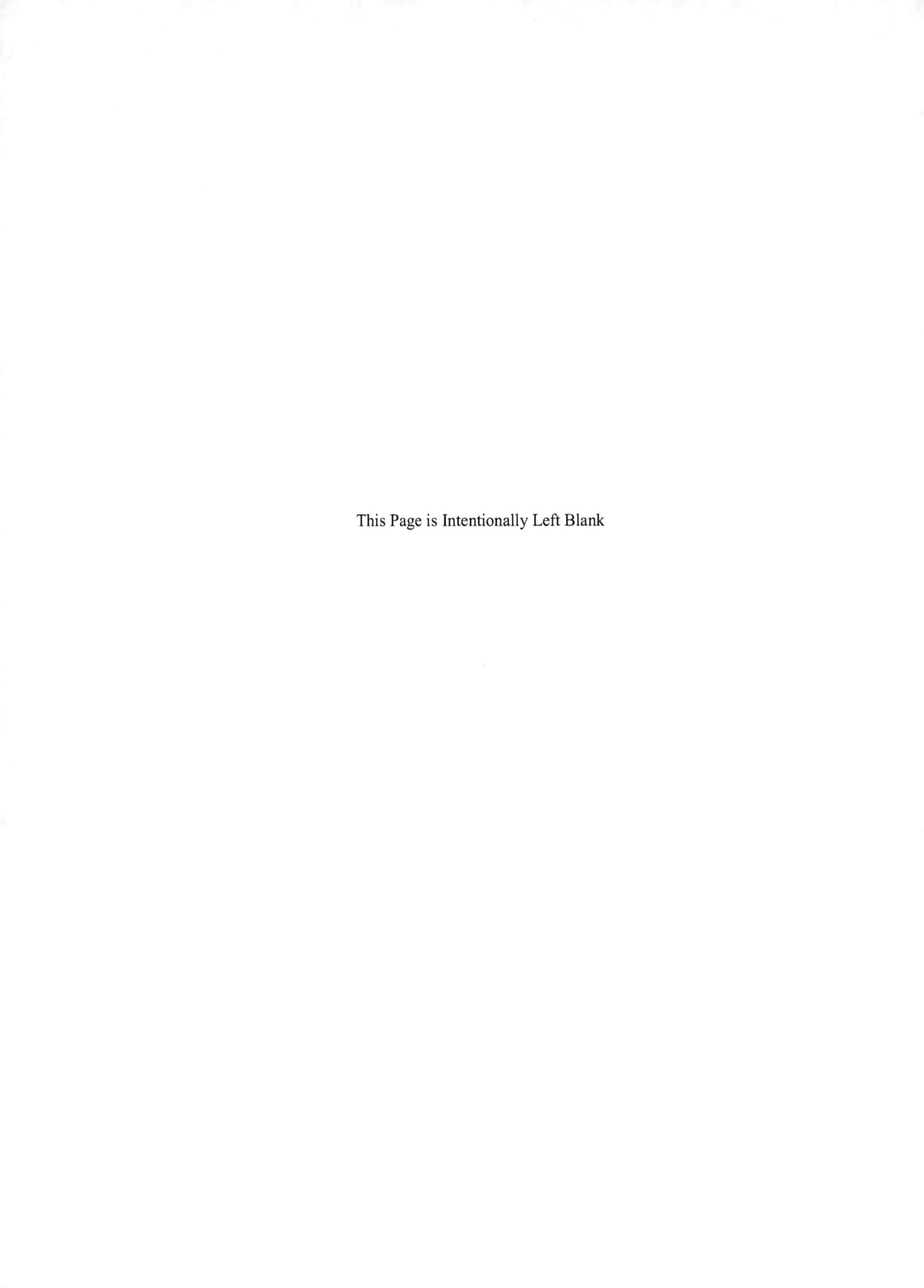

This Page is Intentionally Left Blank

PREFACE

The preface writing has always been a wonderful feeling that cannot be expressed in words as it relates you to your audience through your work. I conceived the idea of writing a new advanced-level textbook in inorganic chemistry during my Ph.D pursuit when I saw post-graduate chemistry students who were tired in search of the syllabus topics because of their ill-resourced university or college library. I also decided to write the textbooks of physical and organic chemistry because I think that someone who wants to teach or text one stream must have the core conceptual understanding of all the three streams of chemical science otherwise one would not be able to connect and explain the interdisciplinary topics in a comprehensive manner.

Out of the series of three textbooks, the present book, entitled "A Textbook of Inorganic Chemistry – Volume 1", is the first installment of "A Textbook of Inorganic Chemistry" which is a four-volume set in all. All the students and teachers are advised to read and consult all the four volumes in a subsequent pattern for a more efficient understanding of the subject of inorganic chemistry.

I also celebrate this opportunity for expressing the bottom hearted gratitude towards the people who supported me at all stages of my work. First of all, I would like to express my sincere gratitude to my doctoral supervisors, Prof. S. P. Khatkar and Prof. V.B. Taxak for their continuous support and guidance from day one. Then I would like to record appreciation to my lovely sister, Jyoti Dalal, for her unconditional love, support and for being the guiding light when life threw me in the darkest of corners. I am very much thankful to my beautiful wife, Anita Sangwan, who always stands shoulder to shoulder with me in my good and bad times. I especially want to thank my brother Sandeep Dalal for his positive criticism, encouragement, motivation and truly selfless support. A special thanks to my dearest sister Garima Sheoran for her love, care, and all-time encouragement. I also wish to thank my entire family, friends and teachers for providing a loving environment for me.

Lastly, and most importantly, I wish to thank my mother, Darshana Devi, who bore me, raised me, supported me, taught me, and loved me.

Mandeep Dalal

This Page is Intentionally Left Blank

Table of Contents

CHAPTER 1

Stereochemistry and Bonding in Main Group Compounds:

❖ VSEPR Theory

One of the most important discoveries of the 20th century was Lewis's description of the chemical bond as a shared pair of electrons. This remarkably brilliant idea connected some of the most important inventions in chemistry from the 19th century; like the Mendeleev's periodic table of elements and the van 't Hoff's formulation of the tetrahedral carbon. Lewis's idea also laid down the foundation of some advanced theoretical models for chemical bonding used today. The cubical atoms and the concept of shared electron pairs proposed by Lewis in 1919 can be illustrated as shown below.

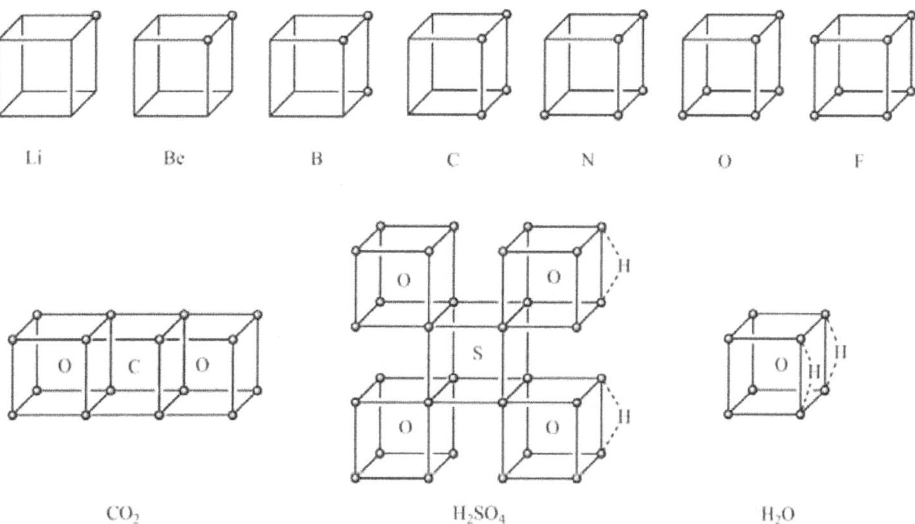

Figure 1. The Lewis concept of electron-pair sharing between cubical atoms.

The valence-shell-electron-pair-repulsion (VSEPR) theory is actually the successor of Lewis's idea which also says that the covalent bond can be portrayed as a shared electron-pair. Now, although Lewis's model explained the correlation between valence and bonding in an extremely beautiful manner, it had no theoretical basis at that time. Later in 1924, Wolfgang Pauli rationalized the electron pairing by proposing the Pauli exclusion principle. Now being an extension of the Lewis idea, the VSEPR theory also finds its roots in the Pauli exclusion principle. The initial formulation of the VSEPR model was actually carried out by two British chemists, Nevil Sidgwick and Herbert Powell, who correlated the number of valence shell electron pairs of the central atom in a molecule to the bonding profile around.

The basic proposal of Sidgwick and Powell was that all the electron-clouds in the outermost shell of an atom (valence shell) must be taken into consideration before any geometry profiling is carried out. In other words, all electron pairs of the valence shell, whether they participate in bonding or not (lone pair as well as bond pairs), have their space requirement; and therefore govern the bonding profile around the central atom. The initial version of the VESPR model also postulated that electron pairs in a Lewis description of a molecule can be represented by points which are arranged on the surface of a hypothetical sphere as far apart as possible. However, later it was thought that the more realistic representation of valence shell electron-pair is a negatively charged cloud which is comprised of two opposite-spin electrons; and this cloud is trying to occupy as much space as possible while eliminating its other counterparts from this space. Therefore, in addition to the "points on the sphere" model; an alternative model can also be given which is based on the different numbers of circles of equal radii arranged in such a way that they occupy maximum possible surface of a sphere without any mutual overlap; though both of these models lead to the same geometrical profile.

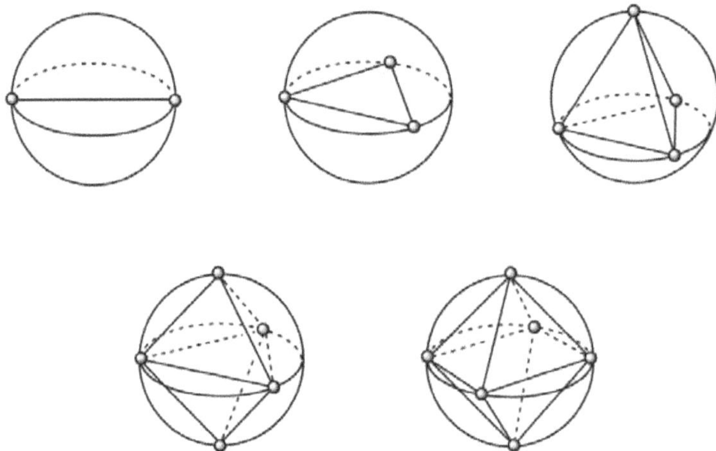

Figure 2. The points-on-the-sphere model of molecular geometries.

A third version, the tangent-sphere model, was also developed by Kimball and Bent that considers all electron-pairs as the spherical entities of the same size. These spherical domains are expected to get packed around the central atom as efficiently as possible. Sidgwick and Powell, in 1940, proposed these most primitive coordination profiles of two to six electron pair domains which are actually fundamental to the VSEPR model and they set the stage for the prediction of the molecular geometries. Now although these predictions explained a wide range of molecular geometries, the distortion from these ideal structures was still a challenge to solve. The most decisive step towards the development of modern VSEPR theory was made Gillespie and Nyholm in 1957 when they published their revolutionary paper entitled "Inorganic Stereochemistry". They treated bond pair and lone pair distinctly and incorporated the necessary allowances. They not only coined the term "VSPER theory", but also worked a lot to popularize the same.

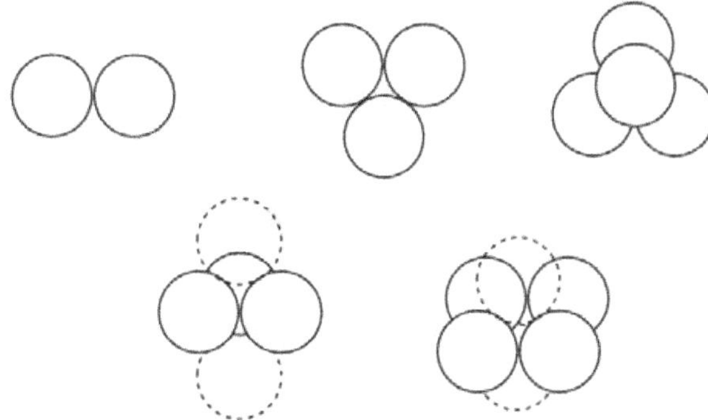

Figure 3. The spherical electron-pair domain model of molecular geometries.

The valence-shell-electron-pair-repulsion or simply the VSEPR theory is a theoretical model that is used to predict the geometry of individual molecules or complexes from the number of electron pairs surrounding their central atoms or ions.

It is also worthy to mention that the VSEPR theory is based purely upon observable electron-density rather than single electron wave functions or orbitals, and therefore is not related to the hybridization in any sense.

> *Basic Postulates of VSEPR Theory*

The modern valence-shell-electron-pair-repulsion or the VSEPR theory is founded upon five basic postulates as given below.

1. All of the electron pairs of the valence shell, whether they participate in bonding or not i.e. lone pairs as well as bond pairs, have space requirements; and therefore govern the bonding profile around the central atom.

2. These valence electron pairs domains, surrounding an atom, tend to repel each other, and will thus prefer to adopt an arrangement that minimizes this repulsion, thus determining the molecule's geometry.

3. Lone pairs of electrons (nonbonding domains) are bigger in size than their single-bond counterparts; which in turn implies that they require more space in the valence shell comparatively. This is simply because the non-bonding electron-pair domain is influenced by only one positive core while the bonding one is held by two positively charged centers. This rationalization, therefore, predicts the following order of domain repulsion:

Lone pair − Lone pair > Lone Pair − Bond Pair > Bond Pair − Bond Pair

4. The size of the valence shell electron pair domain participating in a single bond decreases with rising electronegativity strength of the attached group.

5. The double and triple bonds should be considered as two- and three-electron-pair-domains, respectively; in which the individual electron pairs are not distinguished. Owing to the greater electron density, electron-pair-domain size increases as we move from a single to the triply bonded system.

> ➤ *Application of the VSEPR Theory to Predict Molecular Geometries*

The VSEPR theory can successfully be used to explain the qualitative geometrical profile of molecular species with coordination numbers ranging from two to seven. Some of the most common illustrative examples are given below.

1. Two electron-pair domains: *i) BeCl₂:* The central atom in $BeCl_2$ molecule is Be which has two valence electrons (2, 2). Now because each chlorine atom needs one electron to complete its octet (2, 8, 7), the Be atom uses its both valence electrons to create two bond pair domains only. Hence the geometry for minimum repulsion will be linear and the normal bond angle will be 180°.

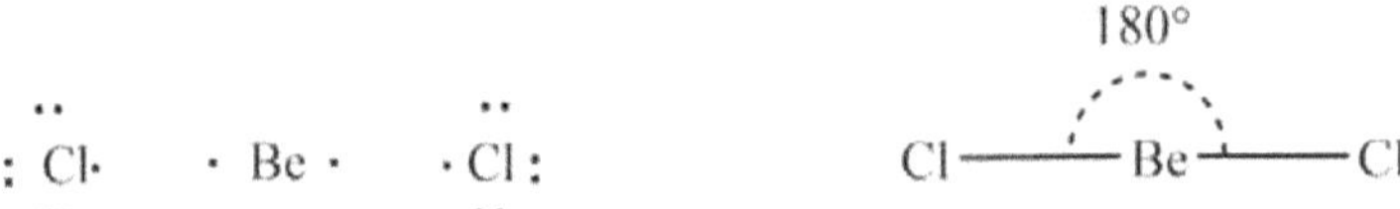

Figure 4. Structure of $BeCl_2$ molecule from the VSEPR model.

2. Three electron-pair domains: *i) BF₃:* The central atom in BF_3 molecule is B which has three valence electrons (2, 3). Now because each fluorine atom needs one electron to complete its octet (2, 7), the B atom uses its three valence electrons to create three bond pair domains only. Hence the geometry for minimum repulsion will be trigonal and the normal bond angle will be 120°.

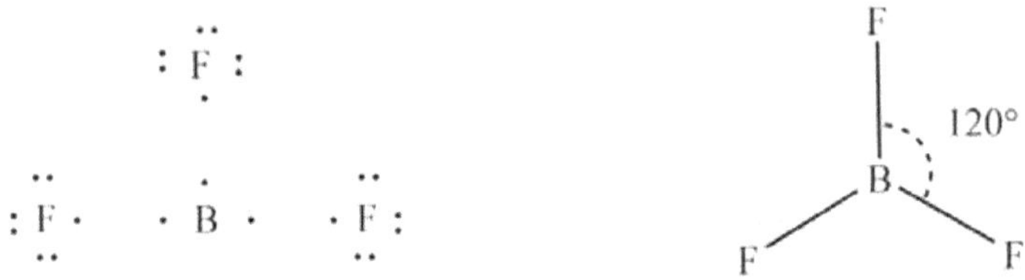

Figure 5. Structure of BF_3 molecule from the VSEPR model.

ii) SO₂: The central atom in the SO_2 molecule is S which has six valence electrons (2, 8, 6). Now because each oxygen atom needs two electrons to complete its octet (2, 6), the S atom uses its four valence electrons to create two bonding two-electron-pair domains, while two electrons are left as a lone pair domain. Now though the geometry for three electron pair domains is trigonal planar with 120°, in this case, the non-bonding one-electron-pair domain would require more space than the bonding domains. This, in turn, would result in a greater lone-pair–bond-pair repulsion yielding a V-shapes geometry with an actual bond angle slightly less than the normal 120° of a perfectly trigonal planar system. However, the actual O−S−O bond angle is 119.3° which is still not very much less than ideal 120° as we expected it to be; this can be attributed to the larger electron density from the two-electron-pair nature of each bond pair domains.

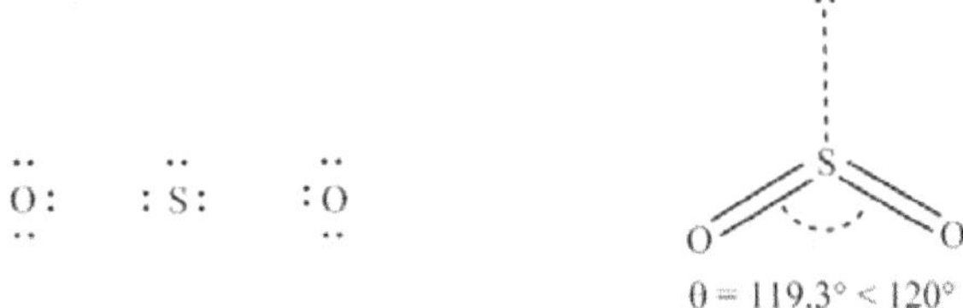

Figure 6. Structure of SO_2 molecule from the VSEPR model.

3. Four electron-pair domains: *i) CH_4:* The central atom in CH_4 molecule is C which has four valence electrons (2, 4). Now because each hydrogen atom needs one electron to complete its duplet (1), the C atom uses its all four valence electrons in sharing to create four bonding one-electron-pair domains only. Hence the geometry for minimum repulsion will be perfect tetrahedral and the normal bond angle will be 109°28′.

Figure 7. Structure of CH_4 molecule from the VSEPR model.

ii) NH_3: The central atom in NH_3 molecule is N which has five valence electrons (2, 5). Now because each hydrogen atom needs one electron to complete its duplet (1), the N atom uses its three valence electrons to create three bonding one-electron-pair domains, while two electrons are left as lone pair domain. Now though the geometry for four electron-pair domains is tetrahedral with 109°28′, in this case, the non-bonding domain would require more space than the bonding domains. This, in turn, would result in greater lone-pair–bond-pair repulsion yielding a pyramidal-shaped geometry with H−N−H bond angle (107.8°) slightly less than the normal 109°28′ of a perfectly tetrahedral system.

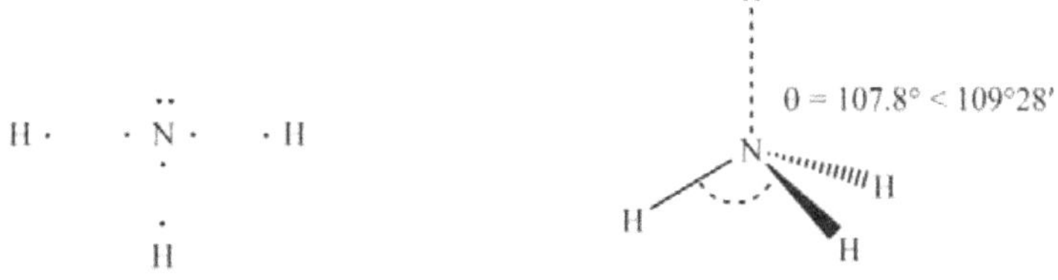

Figure 8. Structure of NH_3 molecule from the VSEPR model.

iii) H_2O: The central atom in H_2O molecule is O which has six valence electrons (2, 6). Now because each hydrogen atom needs one electron to complete its duplet (1), the O atom uses its two valence electrons to create two bond pair domains, while four electrons are left as two lone pair domains. Now though the geometry for four electron-pair domains is tetrahedral with 109°28′, in this case, the two non-bonding domains would require more space than the bonding domains. This, in turn, would result in a greater lone-pair–lone-pair repulsion yielding a V-shaped geometry with H−O−H bond angle (104.5°) less than the normal 109°28′ of a perfectly tetrahedral system. It is also worthy to note that the bond angle in water is also less than NH_3 which is obviously due to the presence of one extra lone pair in H_2O.

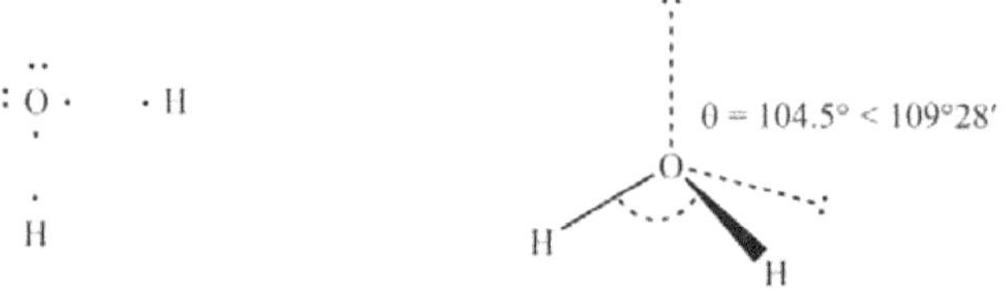

Figure 9. Structure of H_2O molecule from the VSEPR model.

4. Five electron-pair domains: *i) PF_5:* The central atom in PF_5 molecule is P which has five valence electrons (2, 8, 5). Now because each fluorine atom needs one electron to complete its octet (2, 7), the P atom uses its all five valence electrons in sharing to create five bond pair domains only. Hence the geometry for minimum repulsion will be perfect trigonal bipyramidal and the normal bond angle between equatorial groups should be 120°, while the normal bond angle between axial fluorine should be 180°. Now because each axial position has three neighbors at 90° while every equatorial position has only two 90° neighbors, we can conclude that the more crowding at the axial position would lead to longer axial bonds comparatively.

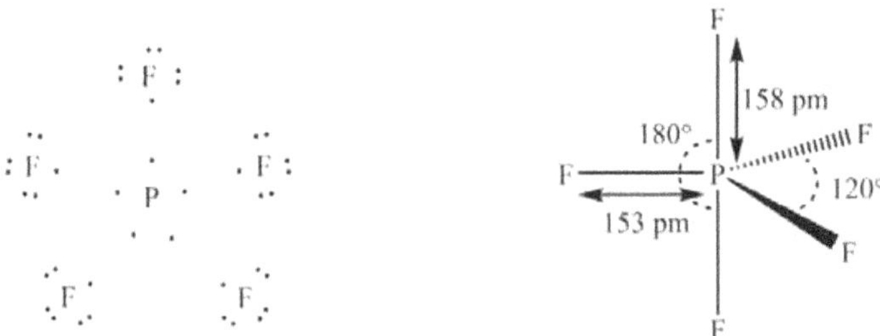

Figure 10. Structure of PF_5 molecule from the VSEPR model.

ii) SF_4: The central atom in SF_4 molecule is S which has six valence electrons (2, 8, 6). Now because each fluorine atom needs one electron to complete its octet (2, 7), the S atom uses its four valence electrons to create four bond pair domains, while two electrons are left as lone pair domain. Now, although the geometry for five

electron-pair domains is trigonal bipyramidal with equatorial and axial bond angles of 120° and 180°, respectively; in this case, the non-bonding domain would require more space than the bonding domains. Now the question arises here is what position this non-bonding electron-pair domain should occupy in a trigonal-bipyramidal frame. After seeing that each axial position has three neighbors at 90° while every equatorial position has only two 90° neighbors, we can conclude that the equatorial position is more suitable for the placement of lone pair. This, in turn, would result in a greater lone-pair–bond-pair repulsion yielding the distortion of a perfect seesaw-shaped derivative with axial and equatorial bond angles slightly less than their normal of 180° and 120°.

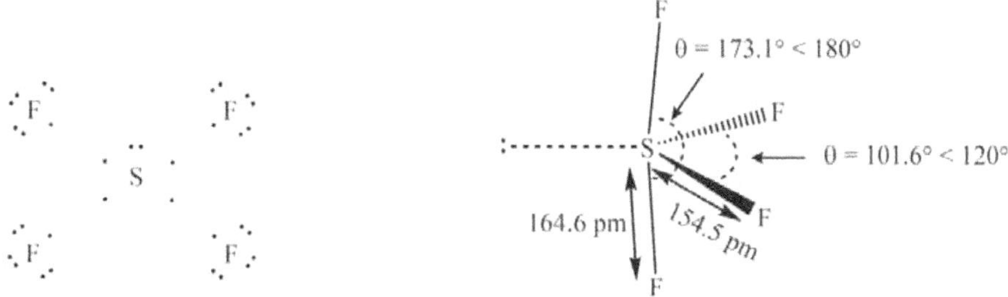

Figure 11. Structure of SF_4 molecule from the VSEPR model.

iii) ClF₃: The central atom in ClF_3 molecule is Cl which has seven valence electrons (2, 8, 7). Now because each fluorine atom needs one electron to complete its octet (2, 7), the Cl atom uses its three valence electrons to create three bond pair domains, while four electrons are left as two lone pair domain. Now though the geometry for five electron pair domains is trigonal bipyramidal equatorial and axial bond angles of 120° and 180°, respectively; but in this case, the non-bonding domain would require more space than the bonding domains, and therefore need to be placed on the equatorial position. This, in turn, would result in greater lone-pair–lone-pair repulsion yielding the distortion of perfect T-shaped derivative with axial equatorial bond angles less than their normal of 180°.

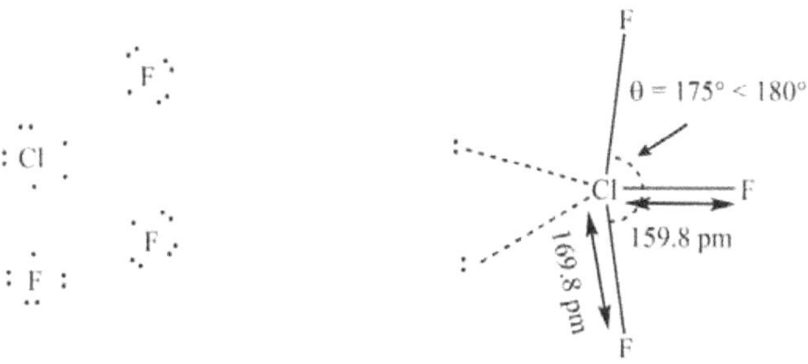

Figure 12. Structure of ClF_3 molecule from the VSEPR model.

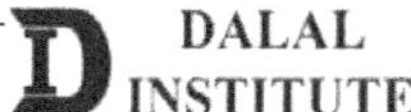

iv) I_3^-: The central atom in I_3^- ion is I which has seven valence electrons (2, 8, 18, 18, 7). Now because one iodine atom needs one electron to complete its octet (2, 8, 18, 18, 7) while the iodide ion needs zero electrons (2, 8, 18, 18, 8), the I atom uses its one valence electrons to create two bond pair domains, while six electrons are left as three lone pair domains. Now though the geometry for five electron pair domains is trigonal bipyramidal with equatorial and axial bond angles of 120° and 180°, respectively; but in this case, the non-bonding domain would require more space than the bonding domains, and therefore need to be placed on the equatorial position. This, in turn, would result in a perfectly linear geometry with a normal bond angle of 180°.

Figure 13. Structure of I_3^- from the VSEPR model.

5. Six electron-pair domains: *i) SF_6:* The central atom in SF_6 molecule is S which has six valence electrons (2, 8, 6). Now because each fluorine atom needs one electron to complete its octet (2, 7), the S atom uses its all six valence electrons in sharing to create six bond pair domains only. Hence the geometry for minimum repulsion will be perfect octahedral and the normal angle between two any adjacent bonds will be 90°.

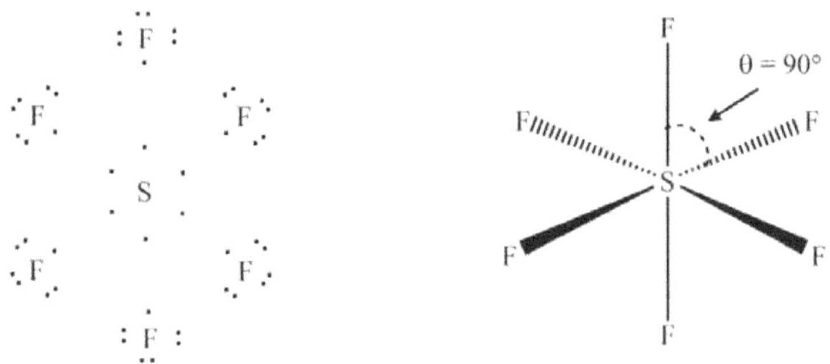

Figure 14. Structure of SF_6 molecule from the VSEPR model.

ii) BrF₅: The central atom in BrF$_5$ molecule is Br which has seven valence electrons (2, 8, 18, 7). Now because each fluorine atom needs one electron to complete its octet (2, 7), the Br atom uses its five valence electrons to create five bond pair domains, while two electrons are left as a lone pair domain. Now though the geometry for six electron pair domains is perfectly octahedral; but in this case, the non-bonding domain would require more space than the bonding domains. The question arises here is what position this non-bonding electron pair domain should occupy in an octahedral frame. Now owing to the fact that all positions in an octahedral frame are equivalent, we can conclude that the lone pair can be placed at any of the six site. This, in turn, would result in greater lone-pair–bond-pair repulsion yielding the distortion of perfect octahedral geometry to a square-pyramidal one.

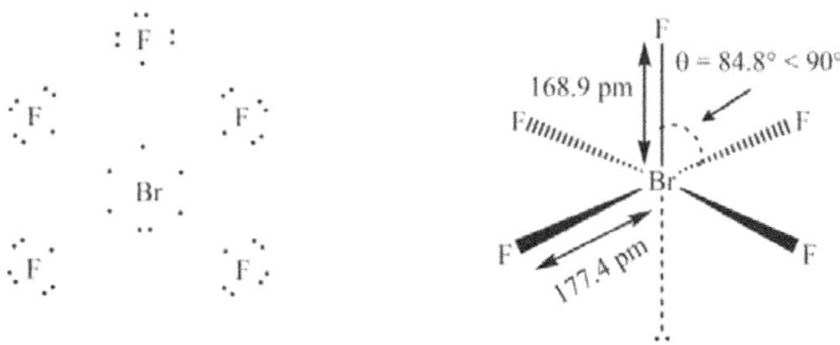

Figure 15. Structure of BrF$_5$ molecule from the VSEPR model.

iii) XeF₄: The central atom in XeF$_4$ molecule is Xe which has eight valence electrons (2, 8, 18, 18, 8). Now because each fluorine atom needs one electron to complete its octet (2, 7), the Xe atom uses its four valence electrons to create four bond pair domains, while the four electrons are left as two lone pair domains. Now though the geometry for six electron pair domains is perfectly octahedral; but in this case, the non-bonding domain would require more space than the bonding domains. The question arises here is what position this non-bonding electron pair domain should occupy in an octahedral frame. Now owing to the fact that lone-pair–lone-pair repulsion is highest, both non-bonding domains should be placed trans to each other in an octahedral frame to give a perfect square-planar geometry.

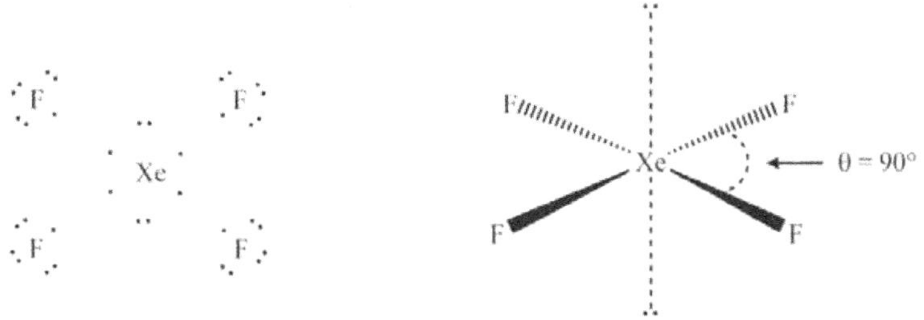

Figure 16. Structure of XeF$_4$ molecule from the VSEPR model.

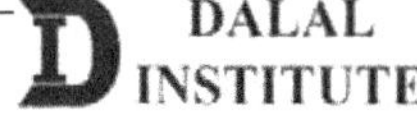

6. Seven electron-pair domains: *i) IF₇:* The central atom in IF₇ molecule is I which has seven valence electrons (2, 8, 18, 18, 7). Now because each fluorine atom needs one electron to complete its octet (2, 7), the I atom uses its all seven valence electrons in sharing to create seven bond pair domains only. Hence the geometry for minimum repulsion will be pentagonal bipyramidal; and the normal bond angle between equatorial groups will be 72°, while the normal bond angle between axial fluorine will be 180°. Now because each axial position has nearest neighbors at 90° while every equatorial position has nearest neighbors at 72°, we can conclude that the less crowding at the axial position would lead to shorter axial bonds comparatively.

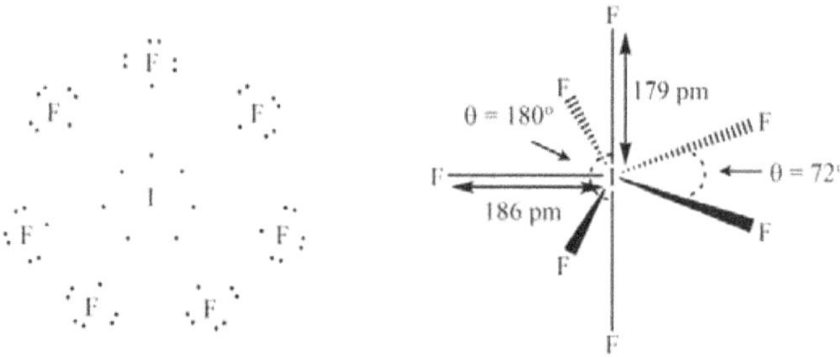

Figure 17. Structure of IF₇ molecule from the VSEPR model.

ii) XeOF₅⁻: The central atom in XeOF₅⁻ molecule is Xe which has eight valence electrons (2, 8, 18, 18, 8). Now because each fluorine atom, as well as O⁻, needs one electron to complete its octet (2, 7), the Xe atom uses its six valence electrons to create six bond pair domains, while two electrons are left as a lone pair domain. Now though the geometry for seven electron pair domains is pentagonal bipyramidal; but in this case, a non-bonding domain is present which would require more space than the bonding domains. Now because each axial position has the nearest neighbors at 90° while every equatorial position has nearest neighbors at 72°, we can conclude that the axial position is more suitable for the placement of lone pair. This, in turn, would result in greater lone-pair–bond-pair repulsion yielding the distortion of perfect pentagonal pyramidal-shaped geometry with equatorial bond angles slightly less than their normal of 72°.

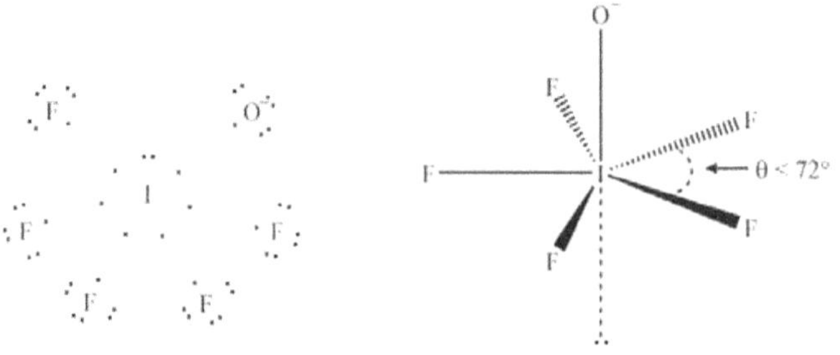

Figure 18. Structure of XeOF₅⁻ molecule from VSEPR model.

iii) XeF$_5^-$: The central atom in the XeF$_5^-$ molecule is Xe which has eight valence electrons (2, 8, 18, 18, 8). Now because four fluorine atom needs one electron to complete its octet (2, 7) while the fifth group F$^-$ needs zero electrons, the Xe atom uses its four valence electrons to create five bond-pair-domains, while the four electrons are left as two lone pair domains. Now though the geometry for seven electron pair domains is perfectly pentagonal bipyramidal; but in this case, the non-bonding domain would require more space than the bonding domains. The question arises here is what position these non-bonding electron pair domains should occupy in a pentagonal bipyramidal frame. Out of the axial and equatorial sites, the axial site is more suitable for the placement of lone pair as explained earlier. Now owing to the fact that lone-pair–lone-pair repulsion is highest, the second non-bonding domains should be placed trans to the first one in a pentagonal bipyramidal frame to give a perfect pentagonal-planar geometry.

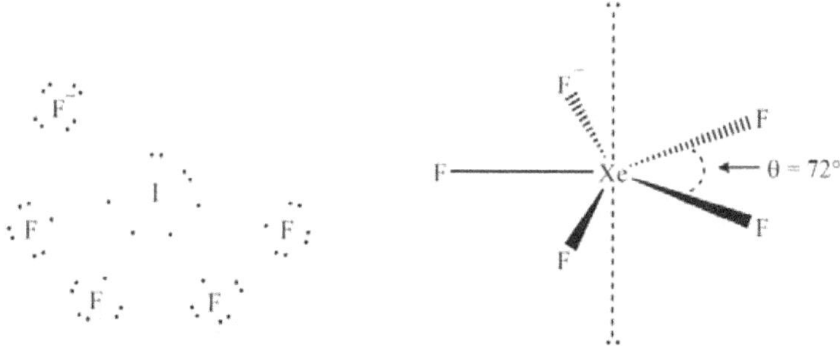

Figure 19. Structure of XeF$_5^-$ molecule from the VSEPR model.

➢ *The AXE Method*

The commonly used electron counting scheme when applying the VSEPR theory is generally called as the "AXE method". The A represents the central atom while the X represents each of the ligands attached. The latter E symbolizes the number of lone electron pair domains around the central atom. The value of X and E is collectively labeled as the steric number. The VSEPR uses this steric number to predict the molecular geometries according to the atomic positions only. For instance, the AX$_2$E$_1$ infers a bent molecule that has three atoms AX$_2$ which are not in one straight line because of the presence of a lone pair. The deviations from idealized geometries can be rationalized from the lone pairs, multi-electron-electron-pair domains and the presence of electronegative groups.

Steric Number	Molecular geometry 0 lone pairs	Molecular geometry 1 lone pair	Molecular geometry 2 lone pairs	Molecular geometry 3 lone pairs
2	X—A—X Linear (CO$_2$)			

Figure 20. Continued on the next page…

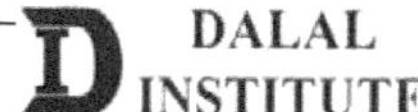

 A Textbook of Inorganic Chemistry – **Volume I**

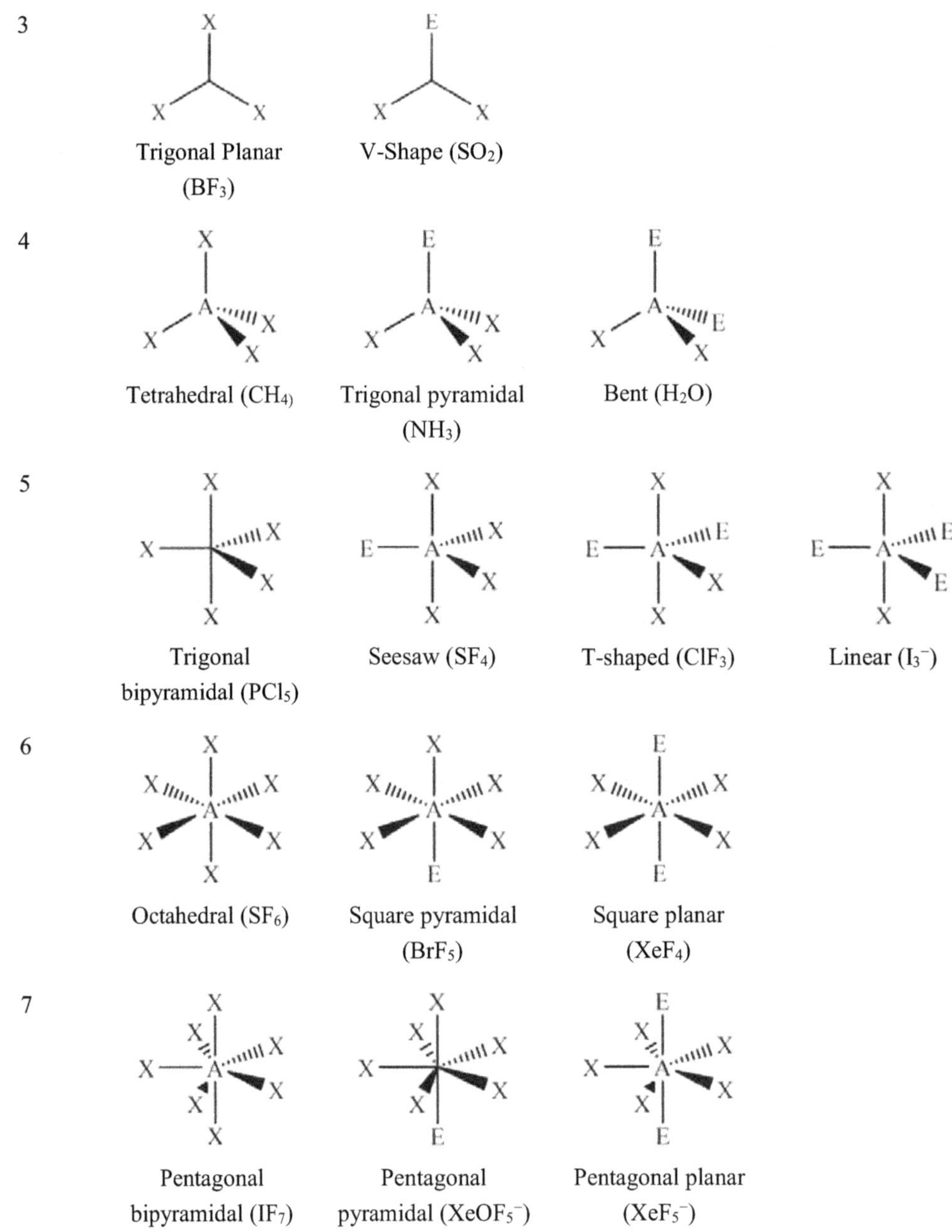

Figure 20. The overall geometry of the compound is predicted by the total number of lone pairs and bond pairs of the electrons i.e. the steric number of the system.

➤ *Limitations of VSEPR Theory*

The VSEPR theory is pretty much successful in explaining a wide range of geometries in main group chemistry but there are some serious limitations also for which it offers no rational solution. Some of the most common limitations of the VSEPR theory are given below.

1. The valence-shell-electron-pair-repulsion theory has no sound explanation for molecular geometries having very polar bonds; like Li_2O is linear whereas its counterpart H_2O is bent is nature.

2. The VSEPR theory offers no solution for molecular geometries with a high magnitude of π-cloud delocalization.

3. It does not consider the "inert pair effect", and thus fails to rationalize the bonding characteristic in molecular geometries possessing the same. For example, $[SeCl_6]^{2-}$, $[TeCl_6]^{2-}$ and $[BrF_6]^-$ are expected to adopt a pentagonal bipyramidal structure due to seven electron pair domains but are found to be octahedral systems.

4. It fails to predict the structure of transition metal complexes. For example, many bivalent complexes of nickel are square-planar and not tetrahedral.

5. The VSEPR theory does not consider the wave function treatment of chemical bonds and therefore fails to explain many bond length and bond angle variations. For example, the bond C−Cl bond length decreases as we move from CH_3Cl to CCl_4, which is unexpected because of the larger size of Cl in comparison to H.

❖ *dπ–pπ* Bonds

This is a special type of bonding found in the molecular species having a central atom with *d* or *p* valence shell and surrounding groups with empty, partially or completely filled *p* or *d* orbitals. In addition to direct overlap resulting in the σ-bonding, *dπ–pπ* bonds are formed by the sidewise overlap. The exact nature of the orbitals from central atom participating in the formation of *dπ–pπ* bond can be obtained by resolving the irreducible components of the reducible representation based upon the vectors set perpendicular to the σ bonds for a particular geometry.

➤ *Molecules with Central Atom Having d-Valence Shell for Sidewise Overlap*

Some of the most well-documented cases in main-group chemistry are AB_4 type molecules like SiO_4^{4-}, SO_4^{2-}, PO_4^{3-}, ClO_4^-, SiF_4; which are found to have A–O bond lengths too short for the single bond confirming a *dπ–pπ* overlap responsible for this anomaly.

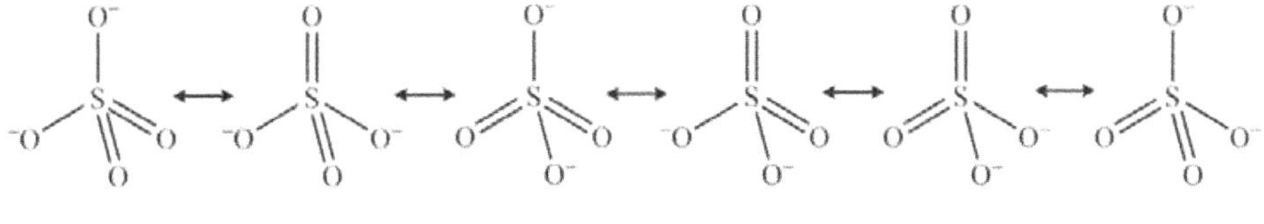

Figure 21. The *dπ–pπ* bonding in SO_4^{2-} ion.

The character table for T_d point group is given below.

T_d	E	$8C_3$	$3C_2$	$6S_4$	$6\sigma_d$		
A_1	1	1	1	1	1		$x^2+y^2+z^2$
A_2	1	1	1	−1	−1		
E	2	−1	2	0	0		$(2z^2-x^2+y^2,\ x^2-y^2)$
T_1	3	0	−1	1	−1	(R_x, R_y, R_z)	
T_2	3	0	−1	−1	1	(x, y, z)	(xy, xz, yz)

The reducible representation based upon the s, p and d orbitals of the central atom in tetrahedral geometry is:

Table 1. Reducible representation based on s, p and d orbitals.

T_d	E	$8C_3$	$3C_2$	$6S_4$	$6\sigma_d$
Γ_π	9	0	1	−1	3

Resolving the reducible into irreducible components, the symmetry designations of different orbitals of the central atom taking part in tetrahedral overlap are given below:

s	—	a_1
p_x, p_y, p_z	—	t_2
d_{xy}, d_{xz}, d_{yz}	—	t_2
$d_{z^2}, d_{x^2-y^2}$	—	e

The symmetry adapted linear combinations of atomic orbitals (SALCs) of surrounding groups for sidewise overlap can be obtained just by resolving the reducible representation based on the displacement vectors perpendicular to the axis of σ overlap.

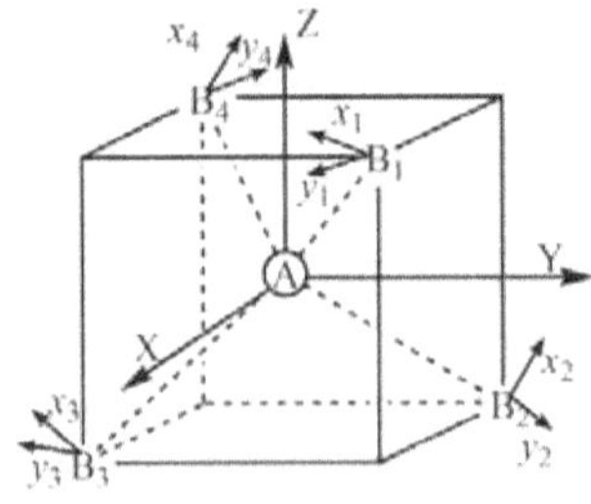

Figure 22. The π-basis set for ligand orbitals in tetrahedral molecules.

The symmetry adapted linear combinations of these fall into two triply and one doubly degenerate irreducible representations labeled as e, t_1 and t_2. The symmetry designations of different ligand orbitals taking part in sidewise overlap in tetrahedral molecules can be given as:

Table 2. Reducible representation based on perpendicular vectors in a tetrahedral geometry.

T_d	E	$8C_3$	$3C_2$	$6S_4$	$6\sigma_d$	Irreducible components
Γ_π	8	-1	0	0	0	$e + t_1 + t_2$

Two of these aforementioned sets are of e and t_2 symmetry. The $d_{x^2-y^2}$ and d_{z^2} orbitals set on the metal also have e-symmetry, and therefore the π-overlap between a central atom and four ligands is possible as far as the generation of molecular orbitals with e-symmetry is concerned. Moreover, p_x, p_y, p_z and d_{xy}, d_{yz}, d_{zx} set are of t_2 symmetry, and therefore, can take part in sidewise overlap. However, the p-subshell largely engaged in σ-bonding and therefore has little or no contribution. While the transition metals show a tendency to use their d_{xy}, d_{yz} and d_{zx} orbitals for $d\pi$–$p\pi$ interactions in octahedral complexes, the main group elements primarily use $d_{x^2-y^2}$ and d_{z^2} as they generally form tetrahedral complexes. The primary reason for this selective behavior is that these two orbitals yield $\sqrt{3}$ times higher $d\pi$–$p\pi$ overlap than that of d_{xy}, d_{yz} and d_{zx} The general scheme for $d\pi$–$p\pi$ overlap for main group compounds with tetrahedral geometry is shown in 'Figure 23'.

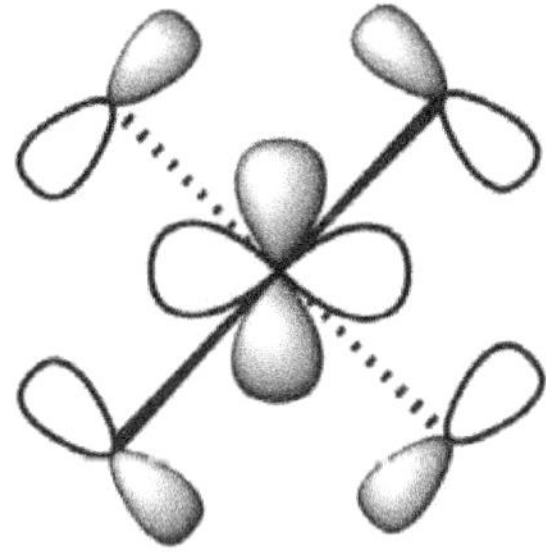 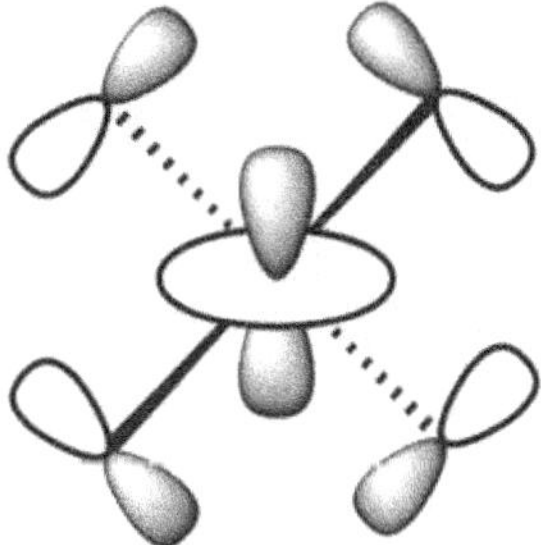

$d_{x^2-y^2}$ overlap with p-Orbitals d_{z^2} overlap with p-Orbitals

Figure 23. The overlap mechanism for the $d\pi$–$p\pi$ interactions when the central atom has valence d-orbitals.

Apart from the tetrahedral molecules, some less symmetrical main group compounds are also found to have $d\pi$–$p\pi$ interactions. Although the exact nature of these interactions is quite difficult to analyze as the lowering of symmetry makes the d-subshell of central atom susceptible to the surrounding groups to a different extent yet the inverse variation of bond length with bond order may be used to approximate the extent of $d\pi$–$p\pi$ overlap. Quantum mechanical calculations have also shown that significant $d\pi$–$p\pi$ interaction is present in molecules like SO_2F_2, PF_3O, ClO_3F, ClO_2^-, ClO_3^-.

> ➤ *Molecules with Central Atom Having p-Valence Shell for Sidewise Overlap*

Sometimes, the existence of $d\pi$–$p\pi$ bonding can be viewed in terms of the molecular geometry. For example, Si_3N and Ge_3N skeleton is planar in $(H_3Si)_3N$, $(H_3Ge)_3N$ and Si/Ge–N bond length is somewhat shorter that what is expected for a single bond. This can be explained by assuming that the electron density from $N(2p_z)$ is overlapping with the $3d$ orbitals of surrounding Si and Ge.

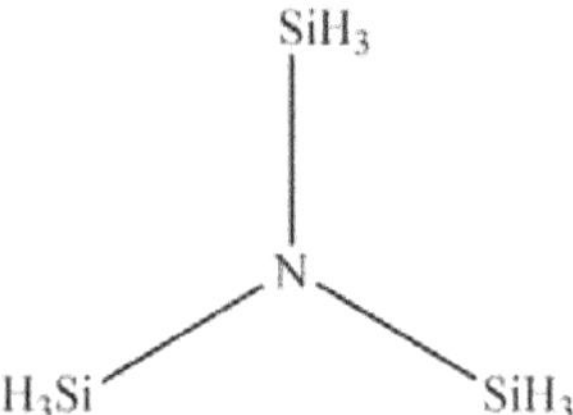

Figure 24. The $d\pi$–$p\pi$ bonding in $N(SiH_3)_3$ molecule.

The character table for the D_{3h} point group is given below.

D_{3h}	E	$2C_3$	$3C_2$	σ_h	$2S_3$	$3\sigma_v$		
A_1'	1	1	1	1	1	1		x^2+y^2, z^2
A_2'	1	1	−1	1	1	−1	R_z	
E'	2	−1	0	2	−1	0	(x, y)	(x^2-y^2, xy)
A_1''	1	1	1	−1	−1	−1		
A_2''	1	1	−1	−1	−1	1	z	
E''	2	−1	0	−2	1	0	(R_x, R_y)	(xz, yz)

The reducible representation based on the s, p and d orbitals of the central atom in trigonal planar geometry is:

Table 3. Reducible representation based on s, p and d orbitals.

D_{3h}	E	$2C_3$	$3C_2$	σ_h	$2S_3$	$3\sigma_v$
Γ_π	9	0	1	3	0	3

Resolving the reducible into irreducible components, the symmetry designations of different orbitals of the central atom taking part in trigonal planar overlap are:

s	–	a_1'
p_z	–	a_2''

p_x, p_y	–	e'
d_z^2	–	a_1'
$d_{xy}, d_{x^2-y^2}$	–	e'
d_{xz}, d_{yz}	–	e''

The symmetry adapted linear combinations of atomic orbitals (SALCs) of surrounding groups for sidewise overlap can be obtained just by resolving the reducible representation based on the displacement vectors perpendicular to the axis of σ-overlap.

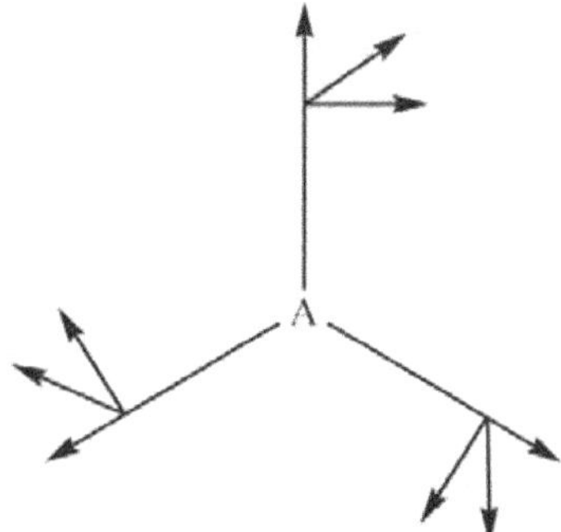

Figure 25. The π-basis set for ligand orbitals in trigonal planar molecules.

The symmetry adapted linear combinations of these fall into two singly and two doubly degenerate irreducible representations labeled as a_2', e', a_2'' and e''. The symmetry designations of different ligand orbitals taking part in sidewise overlap in trigonal planar molecules are:

Table 4. Reducible representation based on perpendicular vectors in trigonal planar geometry.

D_{3h}	E	$2C_3$	$3C_2$	σ_h	$2S_3$	$3\sigma_v$	Irreducible components
Γ_π	6	0	−2	0	0	0	$a_2' + e' + a_2'' + e''$

Three of these aforementioned sets are of a_2'', e' and e'' symmetry. The $d_{x^2-y^2}$ and d_{xy} orbitals set on the central atom also have e'-symmetry, while the d_{xz} and d_{yz} has e''-symmetry; therefore, the π-overlap between a central atom and the ligands is possible as far as the generation of molecular orbitals with e'- and e''-symmetry is concerned. Moreover, p_z orbital of the central atom is of a_2''-symmetry, and therefore, can also take part in sidewise overlap. Now though the symmetry allows the central atom to use $d_{x^2-y^2}, d_{xy}, d_{xz}, d_{yz}$ and p_z; the overlap extent and energy criteria permits largely the p_z to do sidewise overlap.

However, the presence of partially filled p_z orbital of the central atom is prone to overlap with the empty d-subshell of surrounding groups yet it does not assure the sufficient $d\pi$–$p\pi$ bonding leading to a planar structure in all cases. The Si–A–Si bond angles in $P(SiH_3)_3$ and $As(SiH_3)_3$ are 96.5° and 93.8° respectively and both of these compounds exist as pyramidal geometry like $P(GeH_3)_3$ does. This is due to the fact that $3p_z$ orbitals of P and As do not overlap with d-orbitals as efficiently as in the case of $2p_z$ orbital of N atom. Furthermore, $S(SiH_3)_2$ is bent with a Si–A–Si bond angle of 98° resembling its tri-coordinated pyramidal analogs in terms of $d\pi$–$p\pi$ overlap. However, $(H_3Si)_2O$ is also bent in geometry with a bond angle of 144° but shows a small extant of $d\pi$–$p\pi$ overlap which is also confirmed by the shortening of the Si–O bond length. The extant of $d\pi$–$p\pi$ overlap is much larger in flat and linear geometries than that of bent ones.

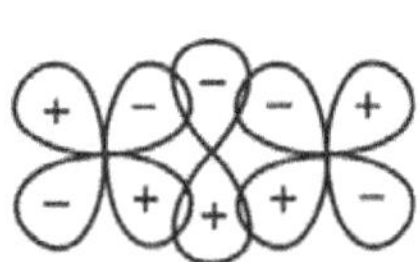

p_z overlap with d-Orbitals in linear molecules

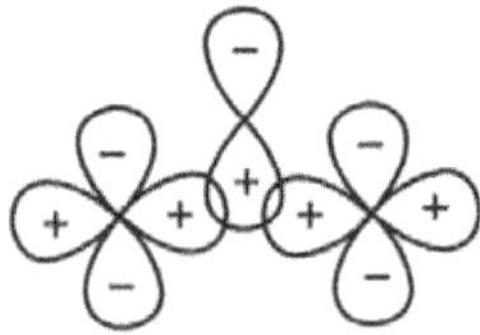

p_z overlap with d-Orbitals in bent molecules

Figure 26. The overlap mechanism for the $d\pi$–$p\pi$ interactions when the central atom has valence p-orbitals

❖ Bent's Rule and Energetic of Hybridization

The success of the valence-shell-electron-pair-repulsion (VSEPR) theory in rationalizing various molecular geometries is quite remarkable as far as simplicity is concerned. However, the failure of VSEPR theory in the case of transition metal complexes or some other serious limitations like inert-pair effect in some main group compounds are also undeniable facts that need to be addressed in a more sophisticated way. The limited applicability of VSEPR theory is actually rooted in the fact that it does not consider any wave-mechanical picture of atomic structure; in other words, as the VSEPR theory does not talk in terms of atomic orbitals, it is quite bound in terms of flexibility required to model the chemical bonds. However, as we know that the Lewis ideas of electron sharing inspired not the Sidgwick and Powell only, but some other physicists like Walter Heitler and Fritz London also. Whilst on the one side the Sidgwick and Powell were working on the VSEPR theory, which was completely immune to the concept of orbitals; Heitler and London were treating the chemical bond in a hundred percent wave-mechanical framework. In 1926, Walter Heitler found out how to use Schrödinger's wave equation to show that the single electron wave-functions of two hydrogen atoms interact with each other to form a covalent bond. Soon after that, he and his associate Fritz London worked on some more theoretical details of the theory and labeled it as valence bond theory (VBT). Thought the VBT was quite effective in the calculation of bonding properties of the H_2 molecule, it was still unable to rationalize complex molecules. Later, an American chemist, Linus Pauling modified the Heitler-London theory by incorporating two key concepts, resonance (1928) and orbital hybridization (1930).

The initial development of the concept of orbital hybridization was to explain the structure of simple molecular geometries like CH_4, BF_3 or $BeCl_2$. The methane molecule had actually been the test case for all bonding theories at that time. Before 1874, the CH_4 molecule was thought to be of square planar geometry with a carbon atom in the center.

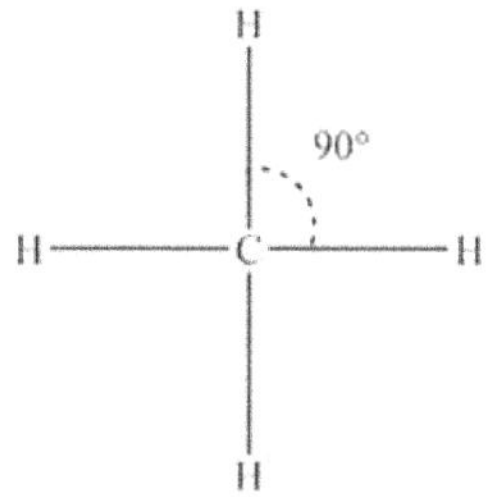

Figure 27. The visualization of the structure of methane before 1874.

Now today it might seem very funny but at that time this structure was quite reasonable because it could explain the observed magnetic moment which is zero for methane. The square planar structure has four C–H bonds of equal length and all adjacent bonds at $90°$; the cancellation of two opposite dipole moments from two trans bonds would also give a zero dipole moment. However, if we attach four different groups to the carbon, the resulting molecule is actually optically active i.e. it would rotate the plane of polarized light in left or right; and therefore, it must exist as two isomers (enantiomeric pair). It is possible only if the arrangement of groups around central carbon is tetrahedral rather square planar; because the later one would exist as three isomers, all optically inactive.

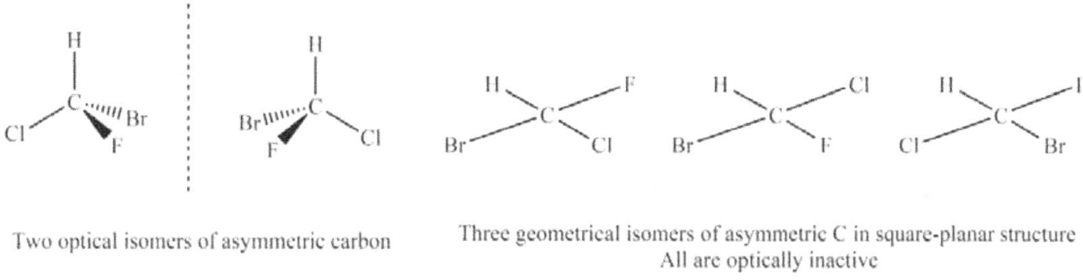

Two optical isomers of asymmetric carbon

Three geometrical isomers of asymmetric C in square-planar structure
All are optically inactive

Figure 28. Isomers for asymmetric carbon in tetrahedral and square-planar geometry.

A Dutch physical chemist, Jacobus Henricus van't Hoff, was the first to formulate the three-dimensional carbon. In 1874, he showed that a tetrahedral arrangement of four different groups around a carbon atom is the only way to give rise to two optical isomers. Using the same theory, he also explained the three isomers of tartaric acid, one enantiomeric pair, and one meso form. The tetrahedral coordination around carbon was also proved in 1913 when X-ray diffraction studies of diamond showed a bond and angle of $109°28'$.

Now although the VSEPR theory, a successor of Lewis idea (1913), explained the tetrahedral coordination in CH_4 molecule by minimum repulsion of four bond pairs; it did not help us to understand the nature of the orbitals involved in the bonding. Linus Pauling in 1931 Pauling pointed out that, in the frame of simple "valence bond theory", a carbon atom can form four bonds by using one s and three p orbitals by promoting its one $2s$ electron to empty $2p_z$ orbital and thus creating four half-filled orbitals. However, it might would mean that a C atom creates three bonds at right angles (from use of p subshell) and a fourth bond (weaker one) using the s wave-function in some arbitrary direction (this fourth C–H bond would be maximum away from the other three bonds if it is put at 135°) as shown below.

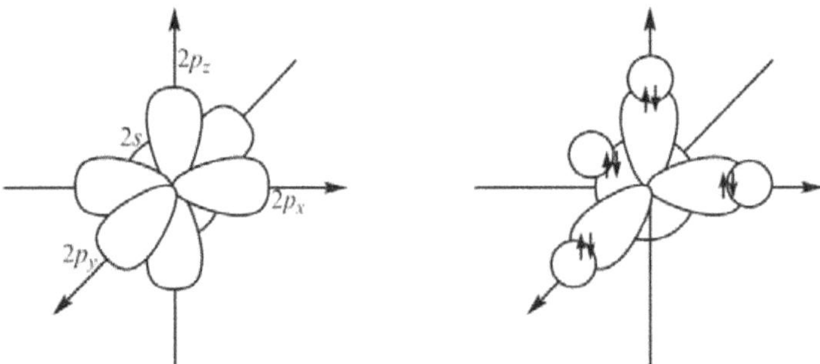

Figure 29. The possible bonding mode in CH_4 molecule from valence bond theory without considering orbital hybridization (one lobe of p-orbitals in the right image are omitted for clarity.

The experimental results like four equal bonds separated by a bond angle of 109.5° or the zero dipole moment cannot be explained by the structure given above. Hence, the aforementioned geometry was a wrong proposal for sure because, besides the bond angle anomaly, it would give rise to small dipole due to the difference in electronegativity between C (2.5) and H (2.2). Moreover, according to simple valence bond theory, as the bonding of hydrogen is happening with one s and three p orbitals of carbon, we should expect one s–s bond to be shorter than three s–p bonds.

To explain this situation, Pauling thought that what would happen if instead of using pure one s and three p wave functions, we use a linear combination of these four. Now though there can be an infinite number of linear combinations of these four single-electron-wave-functions, one should not forget that every single electron-electron-wave-functions must be normalized and orthogonal to others. Linus Pauling found that out of infinite linear combinations, there are only four expressions that are normalized as well as orthogonal to each other. When he plotted these four linearly combined functions in space round carbon nucleus, they were of the same shape, same energy and were oriented at 109°28′. Linus Pauling created the tetrahedral frame of half-filled orbitals, ready for overlap with approaching hydrogens. This was a revolutionary step, he called these functions as hybrid orbitals.

Hybridization is the intermixing of atomic orbitals of different energies and shapes so as to produce new orbitals with the same energy and equivalent shape.

> ➤ *Characteristic Features of Hybridization*

The orbital hybridization in the main group or the transition metal compounds has the following characteristic features:

1. The number of hybrid orbitals is the same as the number of atomic orbitals intermixed.

2. Hybrid orbitals are equivalent in shape and have the same energies.

3. Hybrid orbitals form stronger bonds than their pure atomic counterparts.

4. Hybrid orbitals are directed in specific directions and therefore control the geometry of the molecule.

5. Only valence shell orbitals take part in hybridization.

6. There should be only a small energy difference in the orbitals undergoing hybridization.

7. Electron promotion is not a necessary condition to be followed.

8. In addition to the half-filled orbitals, fully filled and empty orbitals can also undergo hybridization.

> ➤ *Types of Hybridisation*

Hybrid orbitals are assumed to be the mixtures of atomic orbitals, superimposed on each other in various proportions. For example, in the CH_4 molecule, the hybrid orbital of C, which forms a carbon-hydrogen bond consists of 25% s-character and 75% p-character and is thus labeled as sp^3 hybridized. The main types of hybridization are as follows.

1. sp-hybridization: One s and one p-orbitals intermix to form two hybrid orbitals oriented at $180°$ from each other and form the basis for the linear geometry of $BeCl_2$, BeH_2, BeF_2 like molecules.

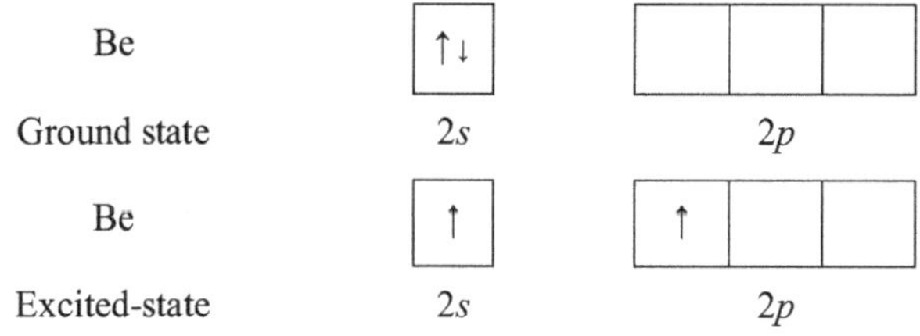

Which gives

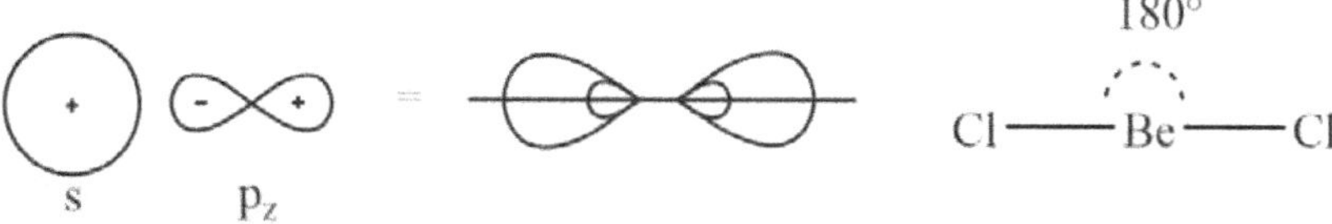

Figure 30. Pictorial representation of sp-hybrid orbitals e.g. $BeCl_2$ molecule

2. sp^2-hybridization: One s and two p-orbitals intermix to form three hybrid orbitals oriented at 120° from each other and form the basis for the trigonal geometry of BF_3 type molecules.

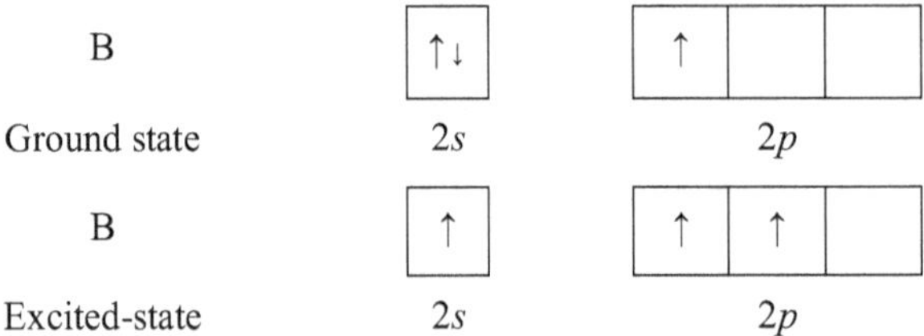

Which gives

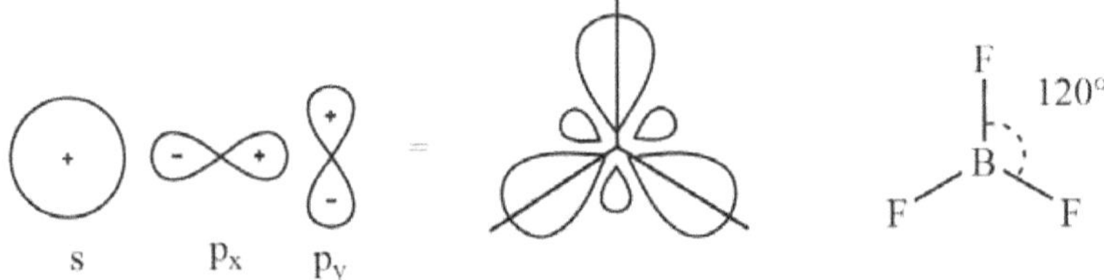

Figure 31. Pictorial representation of sp² hybrid orbitals e.g. BF₃ molecule

3. sp^3-hybridization: One s and three p-orbitals intermix to form four hybrid orbitals oriented at 109°28' from each other and form the basis for the tetrahedral geometry of BF_3 type molecules.

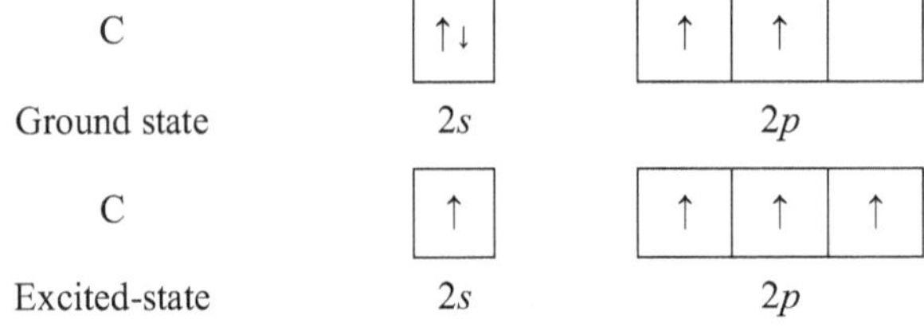

Which gives

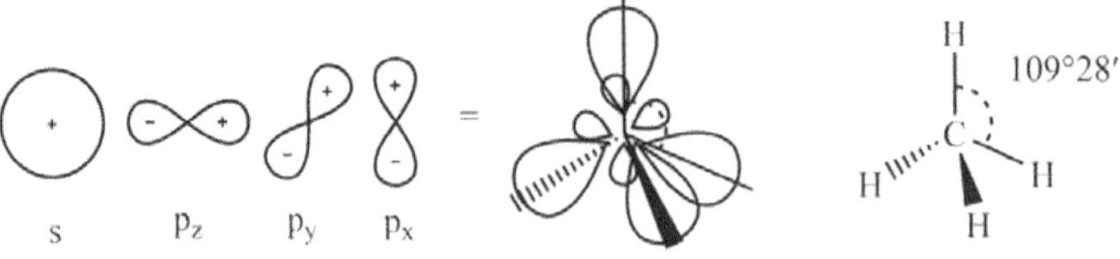

Figure 32. Pictorial representation of sp³ hybrid orbitals e.g. CH₄ molecule

4. sp^3d-hybridization: One s, three p and one d-orbitals intermix to form five hybrid orbitals where three hybrid orbitals are oriented at 120° and at 90° from two axial bonds to form the basis for the trigonal bipyramidal geometry of PF_5 type molecules.

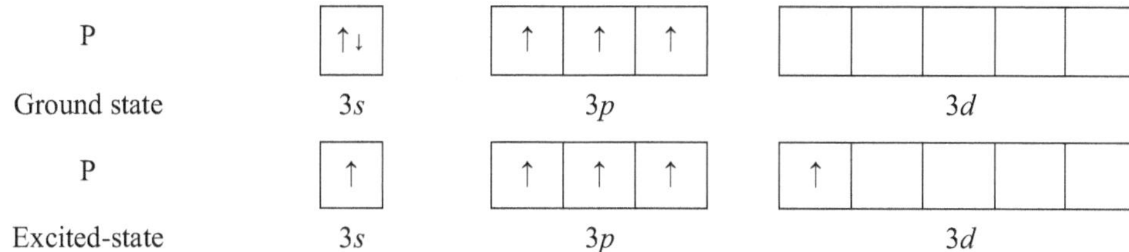

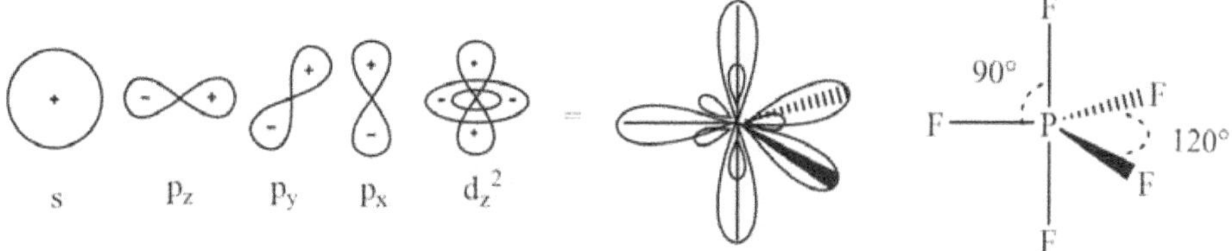

Which gives

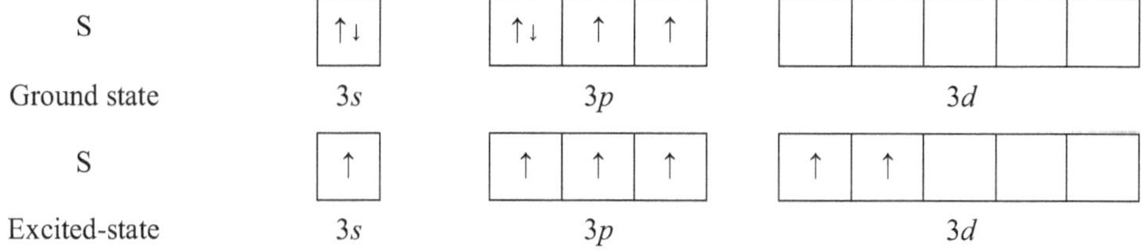

Figure 33. Pictorial representation of sp^3d hybrid orbitals e.g. PF_5 molecule.

5. sp^3d^2-hybridization: One s, three p and two d-orbitals intermix to form six hybrid orbitals where each orbital is at 90° from 4 and at 180° from one to form the basis for the octahedral geometry of SF_6.

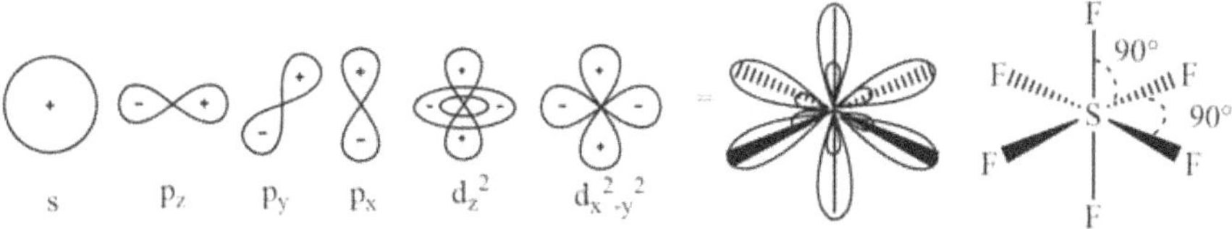

Which gives

Figure 34. Pictorial representation of sp^3d^2 hybrid orbitals e.g. SF_6 molecule.

6. sp^3d^3-hybridization: One s, three p and three d-orbitals intermix to form seven hybrid orbitals where five equatorial hybrid orbitals are oriented at 72° from each other and at 90° from two axial hybrid orbitals to form the basis for the pentagonal bipyramidal geometry of IF_7 type molecules.

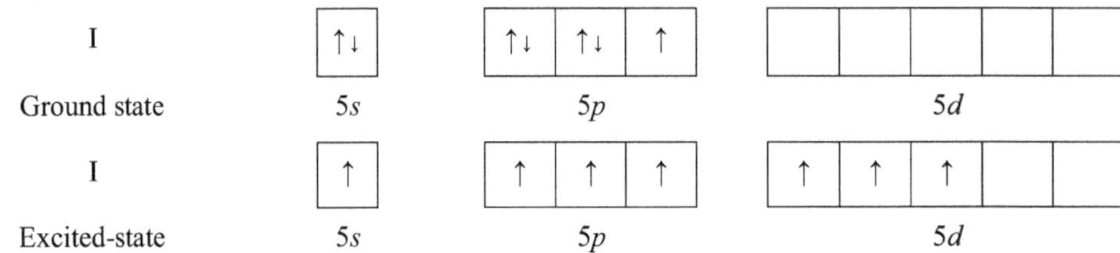

Which gives

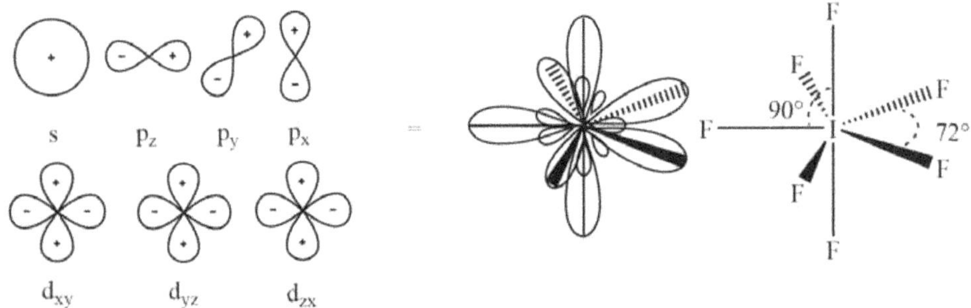

Figure 35. Pictorial representation of sp^3d^3 hybrid orbitals e.g. IF_7 molecule

> ### Some Other Less Common Hybridization Schemes

Besides the simple and hyper-valent molecules like IF_7, there are also some less common hybridization schemes as given below.

1. dsp^2-hybridization: One d, two p and one s-orbitals intermix to form four hybrid orbitals which are oriented at 90° to each other and form the basis for the square planar geometry of tetra-coordinated molecules.

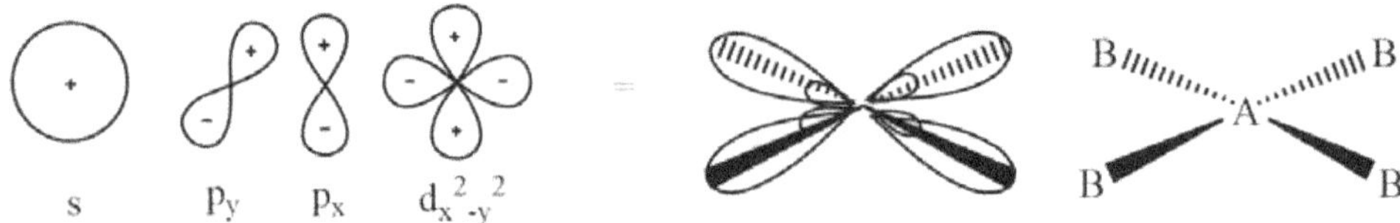

Figure 36. The pictorial representation of dsp^2 orbitals hybridization scheme forming the basis for square planar geometry.

2. sp^3d-hybridization: One s, three p and one d-orbitals intermix to form five hybrid orbitals which are oriented at 90° to each other and form the basis for square pyramidal geometry of penta-coordinated molecules.

Figure 37. The pictorial representation of sp^3d hybrid orbitals forming the basis for square pyramidal geometry.

3. sp^3d^2-hybridization: One s, three p and two d-orbitals intermix to form six hybrid orbitals which are oriented towards the corners of a trigonal prismatic geometry in hexa-coordinated molecules.

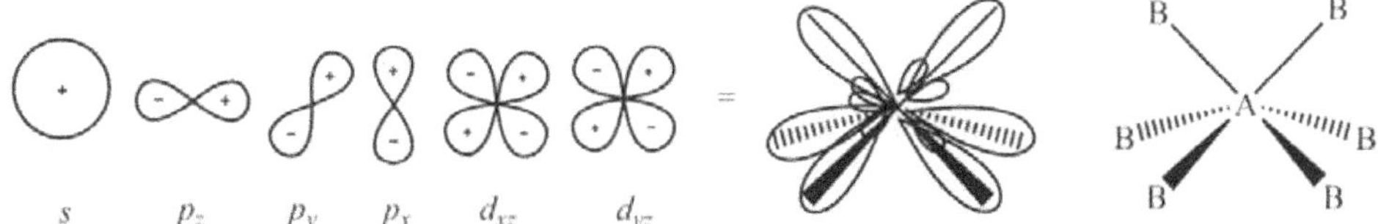

Figure 38. The pictorial representation of sp^3d^2 orbitals hybridization scheme forming the basis for trigonal prismatic geometry.

4. sp^3d^3-hybridization: One s, three p and three d-orbitals intermix to form seven hybrid orbitals which are oriented towards the corners of a mono-capped trigonal-prismatic geometry in hepta-coordinated molecules.

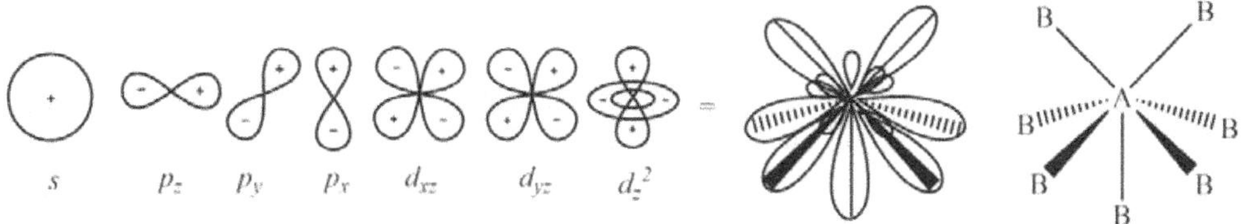

Figure 39. The pictorial representation of sp^3d^3 orbital hybridization scheme for the formation of mono-capped trigonal-prismatic geometry.

However, main group Compounds rarely show these hybridizations like dsp^2-square planar or sp^3d-square pyramidal geometry due to stability issues which makes them more predictable structure than transition metal complexes.

> *Bent's Rule*

The introduction of the concept of hybridization by Linus Pauling was a huge success in the rationalization of the chemical bond, especially in the organic compounds. However, in the 1940s it became apparent that a lot of molecular geometries, including even the simple ones, show slight deviations from the predictions of orbital hybridization concept of valence bond theory. For instance, consider the case of H_2O molecule; where the H−O−H bond angle is 104.5°, which far less than the normal tetrahedral angle of 109.5°. Besides the isovalent species like H_2O or NH_3, the conventional hybridization theory was also unable to explain why the Cl−C−Cl and Me−C−Me bond angles in sp^3-hybridized CCl_2Me_2 are 108.5° and 113°, respectively.

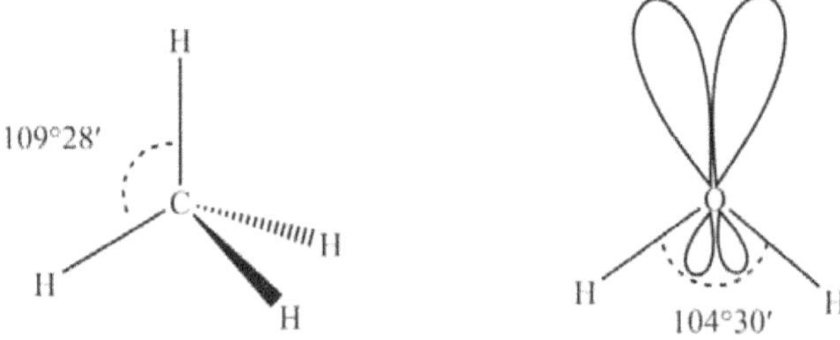

Figure 40. The comparison of the bond angle of methane and water though both have of sp^3 hybridization.

In 1947, a British chemist, A. D. Walsh tried to explain such inconsistencies by describing a correlation between the electronegativity of ligand attached to carbon and the hybridization scheme involved. it was suggested that hybridization may yield orbitals with different magnitude of s and p character. In 1961, Henry A. Bent published a more sophisticated treatment of the problem that correlated the molecular structure, hybridization scheme of the central atom, and the electronegativity strengths of the attached groups. The work is generally labeled as Bent's rule and is based upon the perturbation theory which suggests that isovalent hybridization should transfer more s-character towards more electropositive substituents to maximize the bonding energy. This can be explained in terms of electron density available for bonding. The more electropositive groups would be able to withdraw more electron density from p-orbital than s-orbital.

In the compounds containing mixed substituents and hybridized orbitals, a trend has been observed according to which atomic s-character tends to concentrate in the orbitals that are directed towards electropositive groups and atomic p-character tends to concentrate in the orbitals that are directed towards electronegative groups.

Valence orbitals for main group elements consist of s and p orbitals with s-orbital having lower energy, which means that the bonding orbital is lower in energy and shape more like s-orbital rather than p-orbital. However, in the case where the central atom or ion has a lone pair, it will be having large s-character because s-orbital are closer to the nucleus, allowing greater stabilization for the lone pair. The deviation of molecular geometries from their perfectly tetrahedral, octahedral or other symmetrical counterparts can be explained and predicted by examining the substituent groups and lone pair of electrons.

> ➤ *Applications of Bent's Rule*

The Bent's rule can be used to rationalize many aspects of structure, bonding, spectra and the physical and chemical properties of a wide range of compounds. Some of the important applications of Bent's rule which are quite popular in the field of chemical science are given below.

1. Bond angles: *i)* In $A(Cl)_2(Me)_2$ molecules, where A is C, Si, Ge, Sn or Pb; the Cl–A–Cl bond angle is smaller than that of Me–A–Me bond angle. With high electronegativity halogen substituent Cl, more *p*-character is concentrated in A–Cl bond than A–Me bonds. Subsequently, bonds with greater *p*-character have smaller bond angles than those with *s*-character because *s*-orbitals are closer to the nucleus and hence exert larger inter-electronic repulsion producing higher bond angles.

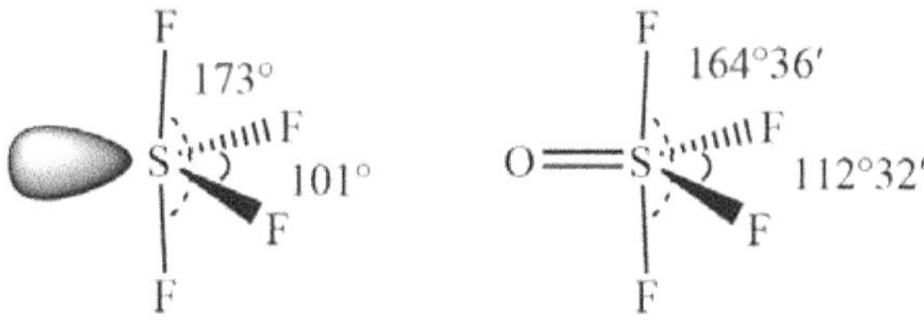

Figure 41. Increasing p-character in A–X bond, decreasing halogen–A–halogen bond angle.

ii) In SF_4, the bond angle of axial F–S–F bond is 173° (ideally 180°) and the equatorial F–S–F bond angle is 101° (ideally 120°). The axial, as well as equatorial bond angles, are decreased slightly due to increased s-character of the lone pair and more *p*-character in bond pairs. Furthermore, thionyl tetrafluoride shows the same trend but the slightly higher bond angle between equatorial fluorines than that of sulfur tetrafluoride. This anomaly can be explained in terms of comparatively larger *s*-character in equatorial fluorines as the lone pair has been replaced by a more electronegative oxygen atom.

Figure 42. SF_4 molecule and SOF_4 molecules.

iii) As one moves left to right in the figure given, the substituents become more electronegative and the bond angle between them decreases. According to Bent's rule, as the electronegativity strength of substituents increase, orbitals rich in *p*-nature will be oriented towards those substituents. From the abovementioned argument, it will reduce the bond angle; which is in good agreement with the experimental data. If we compare this justification with VSEPR theory, the VSEPR model cannot explain why the angle in dimethyl ether is larger than the normal tetrahedral angle of 109.5°. In the prediction of the bond angle in H_2O molecule, Bent's rule proposes that the hybrid orbitals which are rich in s-character should be oriented towards more electropositive lone pairs, whilst the orbitals of more p-nature should point towards the hydrogens. This enhanced p-character in orbitals decreases the corresponding bond angle to less than the normal tetrahedral of 109.5°. The same reason can be applied to NH_3 molecule, the other canonical cases of this kind.

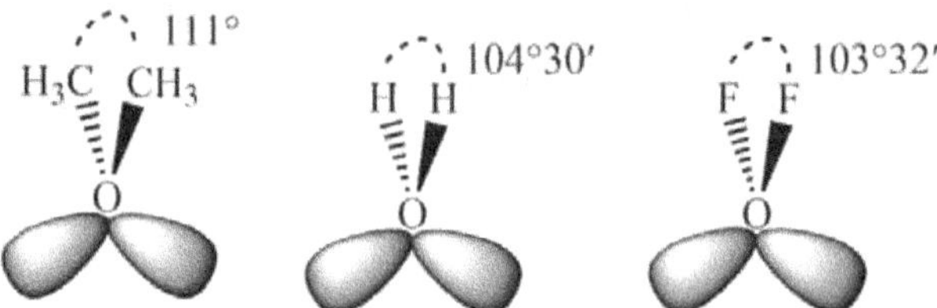

Figure 43. The decreasing trend of the B–A–B bond angle between substituents with increasing electronegativity.

2. Bond lengths: *i)* The bond lengths can also be changed by changing the hybridization or by adding electronegative groups to the central atoms. If a molecule geometry has X–A–Y component, replacement of the group X by a more electronegative substituent would manipulate the hybridization of the central atom and reduces the neighboring A–Y bond length. Now owing to the higher electronegativity of fluorine than hydrogen, the carbon of fluoromethane will direct its hybrid orbitals with the higher s-character towards the three H groups rather than the fluorine. In the case of difluoromethane, only two hydrogens are present, so the s-character in total is pointed towards them is less and more is pointed towards the two fluorine groups. This reduces the carbon–fluorine bond lengths relative to the fluoromethane molecule.

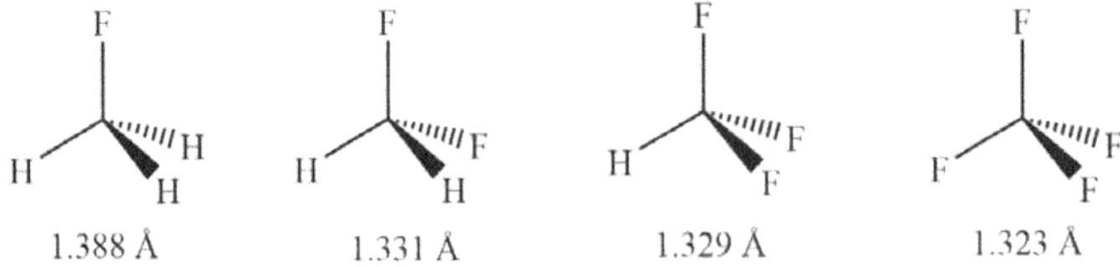

Figure 44. The decreasing trend of average carbon–fluorine bond lengths as we move from CH_3F to CF_4 molecular geometries.

This pattern continues to the CF_4 whose carbon–fluorine bonds have the largest s-character (25%) and the shortest bonds in the whole series.

ii) The same pattern also continues for the chlorinated equivalents of CH_4, though the effect is less efficient because the chlorine is of lower electronegativity strength than the fluorine.

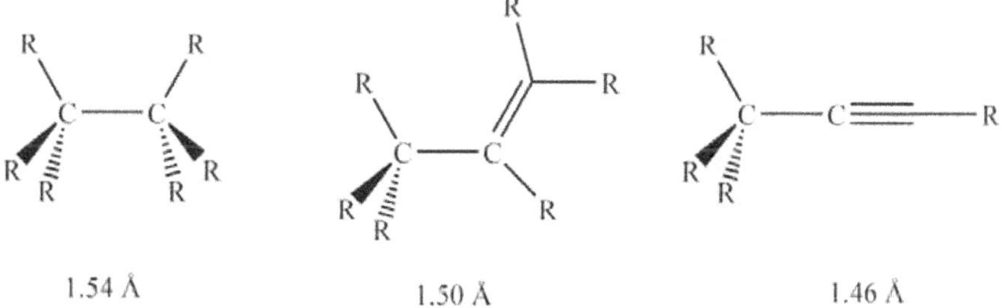

Figure 45. Average carbon–chlorine bond length.

The abovementioned cases seem to prove that the atomic radius of chlorine is less significant than its electronegativity strength because the prediction based on steric factors alone would yield the opposite pattern due to greater setting-apart tendency of larger chlorine groups. As the steric clarification is in contradiction with the experimental results, Bent's rule plays a significant role in structure elucidation.

iii) A similar trend is observed for single, double and triple carbon–carbon bonds, as the change of hybridization scheme from sp^3 to sp-type would lead to more s-character and therefore shorter bond.

Figure 46. Average carbon–carbon bond length.

3. J_{CH} coupling constants: Bent's rule is very much effective in explaining the variation of carbon-hydrogen coupling constant with increasing electronegativity strength of the attached substituents. The ^1H-^{13}C coupling constants (J_{CH}) measured from nuclear magnetic resonance (NMR) spectra is one of the most direct scales that can be used to estimate the magnitude of s character in a bonding overlap between carbon and hydrogen.

Now as we know from the quantum theory that the J_{CH} values will be much greater in bonds with more s character than the one with the less. Therefore, as the electronegative character of the attached group increases, the magnitude of p-character oriented towards the corresponding group increases as well. This induces more s character in the bonds directed towards the methyl protons, which in turn, increases the J_{CH} coupling constants.

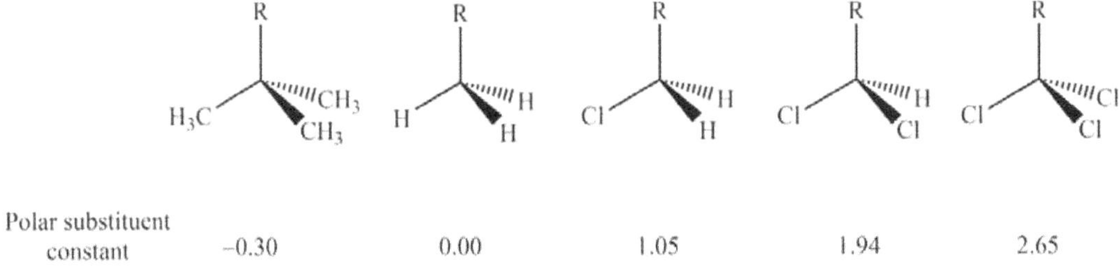

Figure 47. The variation of carbon-hydrogen coupling constants (J_{CH}) with increasing electronegative strength of the attached group.

4. Inductive effect: One of the most important phenomena in the field of organic chemistry, the inductive effect, can also be rationalized with the help of Bent's rule. The phenomenon of inductive effect is actually the passing of charge via the covalent bonds and Bent's rule offers a reasonable mechanism for such observations by exploiting the differences in the hybridization scheme involved. It has been observed that with the increase in the electronegativity strength of the groups attached to the central carbon, the electron-withdrawing power of the central carbon also increases and can be measured by the polar-substituent-constant values. According to the Bent's rule, the increase in the electronegativity of the groups attached would attract more p character towards them; consequently, leaving higher s character for the C−R bond. Now because the s orbitals have higher electron-density near the nuclei than their p counterparts, a shift of electron density towards the carbon in the C−R bond is expected; which in turn would make the central carbon atom more electron-withdrawing to the ligand attached i.e. R.

Polar substituent constant   −0.30   0.00   1.05   1.94   2.65

Figure 48. The variation polar-substituent-constants with increasing electronegative strength of the attached group.

In words, we can say that the electron-withdrawing capability of the substituents-attached is transferred to the neighboring carbon atoms in the carbon chain under consideration. This is exactly the same conclusion what the inductive effect states.

> ### *Exceptions of Bent's Rule*

The general statement of Bent's rule, atomic s-character concentrates in orbitals directed towards electropositive substituents, is true only for main group compounds. However, transition metal complexes show quite strange behavior. It has been observed that the group four transition metals Ti–Hf do not rigorously follow Bent's rule. With these complexes, the more electronegative substituents have larger bond angles indicating greater s-character. This can be explained by the fact that the d-orbitals in transition metals are generally lower in energy than s-orbital. Thus, more electronegative substituents will be attracted to the high lying s-orbitals. Transition metal orbitals are sd^3 hybridized with very little contribution from p-orbitals.

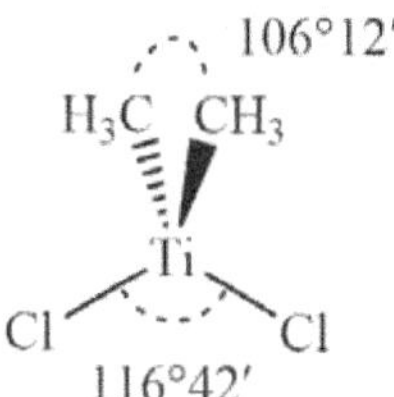

Figure 49. Distorted structure of TiMe$_2$Cl$_2$.

The broad form of Bent's rule can be stated as follows, "The energetically low-lying valence orbital concentrates in bonds directed towards electropositive substituents". This satisfies both the main group and transition metal complexes.

❖ Problems

Q 1. Write down the basic postulates of the valence-shell-electron-pair-repulsion (VSEPR) theory.

Q 2. How would you use the VSEPR theory to explain the molecular geometry of seven electron-pair-domains?

Q 3. Draw and discuss the structure of ClF_3 molecule on the basis of the VSEPR model.

Q 4. The axial bond lengths than the equatorial one in trigonal bipyramidal geometries, while shorter in pentagonal bipyramidal structure. Why?

Q 5. Write down the five main limitations of the VSEPR theory.

Q 6. How would you explain the nature of the orbitals used by main-group elements for $d\pi$–$p\pi$ bonding in tetrahedral molecules?

Q 7. What are the characteristic features of orbital hybridization?

Q 8. Draw and explain the orbital hybridization in hypervalent molecules using suitable examples.

Q 9. Discuss the orbital hybridization of NH_3 and H_2O molecules in a comparative framework.

Q 10. Discuss the stereochemistry of the following: (i) $SnCl_2$ (ii) ClF_3.

Q 11. Define the Bent's rule and explain Why the bond angle of PH_3 is less than that of PF_3?

❖ Bibliography

[1] R. J. Gillespie, *Electron Densities and the VSEPR Model of Molecular Geometry*, E. Can. J. Chem., 70, 1992, 742.

[2] I. Hargittai and B. Chamberland, *The VSEPR Model of Molecular Geometry*, Comp. & Maths. with Appls., 12B, 1986, 1021.

[3] R. J. Gillespie and R. S. Nyholm, *Inorganic Stereochemistry*, Q. Rev. Chem. Soc., 11, 1957, 339.

[4] J. D. Lee, *Concise Inorganic Chemistry*, Chapman & Hall, New York, USA, 1994.

[5] B. R. Puri, L. R. Sharma, K. C. Kalia, *Principals of Inorganic Chemistry*, Milestone Publishers, Delhi, India, 2012.

[6] J. E. Huheey, E. A. Keiter, R. L. Keiter, *Inorganic Chemistry: Principals of Structure and Reactivity*, HarperCollins College Publishers, New York, USA, 1993.

[7] A. F. Wells, *Structural Inorganic Chemistry*, Oxford University Press, London, UK, 1975.

[8] F. A. Cotton, G. Wilkinson, C. A. Murillo, M. Bochmann, *Advanced Inorganic Chemistry*, John Wiley & Sons, New Jersey, USA, 1999.

[9] W. U. Malik, G. D. Tuli, R. D. Madan, *Selected Topics in Inorganic Chemistry*, S. Chand Publishers, New Delhi, India, 2014.

CHAPTER 2

Metal-Ligand Equilibria in Solution:

❖ Stepwise and Overall Formation Constants and Their Interactions

The formation of a complex between a metal ion and a bunch of ligands is in fact usually a substitution reaction. However, ignoring the aquo ions, the formation of the complex can be written as:

$$M + nL \overset{\beta}{\rightleftharpoons} ML_n \tag{1}$$

Where M represents the metal center, L is the ligand type involved, n represents the number of ligands, and β is the equilibrium constant for the whole process. The expression for β (or β_n) for the above equilibria can simply be written as:

$$\beta_n = \frac{[ML_n]}{[M][L]^n} \tag{2}$$

Now because the magnitude of β_n is proportional to the molar concentration of complex formed, the equilibrium constant β_n is also called formation constant of the metal complex.

The formation constant or stability constant may be defined as the equilibrium constant for the formation of a complex in solution.

The magnitude of β_n is actually a measure of the strength of the interaction between the ligands, which come in contact to form the complex, and the metal center. However, it has also been observed that the complex formation in the solution phase occurs via a step-to-step addition of the ligands to the metal center used. For instance, the chemical equation (1), which shows the formation of a complex ML_n, can also be written as a combination of many other equations representing a corresponding series of individual steps. In other words, the overall formation process of ML_n complex can be resolved into the following steps:

$$M + L \overset{K_1}{\rightleftharpoons} ML \quad K_1 = \frac{[ML]}{[M][L]} \tag{3}$$

$$ML + L \overset{K_2}{\rightleftharpoons} ML_2 \quad K_2 = \frac{[ML_2]}{[ML][L]} \tag{4}$$

$$ML_2 + L \overset{K_3}{\rightleftharpoons} ML_3 \quad K_3 = \frac{[ML_3]}{[ML_2][L]} \tag{5}$$

The equations (3–5) and corresponding equilibrium constants can further be extended for the attack of n number of ligands as given below.

$$ML_{n-1} + L \overset{K_n}{\rightleftharpoons} ML_n \quad K_n = \frac{[ML_n]}{[ML_{n-1}][L]} \tag{6}$$

Where K_1, K_2, K_3......K_n are the equilibrium constants for different steps, which in turn also imparted their conventional label of stepwise stability or the stepwise formation constants. The magnitude of these individual equilibrium constants indicates the extent of the formation of different species in a particular step.

Nevertheless, the stepwise stability constant of any particular step does not include the information about the previous ones. Therefore, to include the extent of formation of a complex up to a particular step, say 3rd, the overall formation constant β_3 should be used as it indicates the extent of formation of ML_3 as a whole. Moreover, it can also be shown that the overall formation constant up to the 3rd step (β_3) can be represented as the product of K_1, K_2, K_3.

$$\beta_3 = K_1 \times K_2 \times K_3 \tag{7}$$

$$\beta_3 = \frac{[ML]}{[M][L]} \times \frac{[ML_2]}{[ML][L]} \times \frac{[ML_3]}{[ML_2][L]} \tag{8}$$

$$\beta_3 = \frac{[ML_3]}{[M][L]^3} \tag{9}$$

$$\beta_n = K_1 \times K_2 \times K_3 \times K_4 \times K_5 \times K_6 \times ... K_n \tag{10}$$

The overall stability constant is generally reported in logarithmic scale as $\log \beta$ as given below

$$\log \beta_n = \log K_1 + \log K_2 + \log K_3 + \log K_4 + \log K_5 + \log K_6 + \cdots \log K_n \tag{11}$$

Or

$$\log \beta_n = \sum_{i=1}^{i=n} \log K_i \tag{12}$$

The whole process of calculating the overall formation constant can be exemplified by taking the case of $[Cu(NH_3)_4]^{2+}$ complex.

$$Cu^{2+} + NH_3 \overset{K_1}{\rightleftharpoons} [Cu(NH_3)]^{2+} \quad K_1 = \frac{[[Cu(NH_3)]^{2+}]}{[Cu^{2+}][NH_3]} \tag{13}$$

$$[Cu(NH_3)]^{2+} + NH_3 \overset{K_2}{\rightleftharpoons} [Cu(NH_3)_2]^{2+} \quad K_2 = \frac{[[Cu(NH_3)_2]^{2+}]}{[[Cu(NH_3)]^{2+}][NH_3]} \tag{14}$$

$$[Cu(NH_3)_2]^{2+} + NH_3 \overset{K_3}{\rightleftharpoons} [Cu(NH_3)_3]^{2+} \quad K_3 = \frac{[[Cu(NH_3)_3]^{2+}]}{[[Cu(NH_3)_2]^{2+}][NH_3]} \tag{15}$$

$$[Cu(NH_3)_3]^{2+} + NH_3 \overset{K_4}{\rightleftharpoons} [Cu(NH_3)_4]^{2+} \quad K_4 = \frac{[[Cu(NH_3)_4]^{2+}]}{[[Cu(NH_3)_3]^{2+}][NH_3]} \tag{16}$$

The overall reaction with overall formation constant can be given by the equation (17) as:

$$Cu^{2+} + 4NH_3 \overset{\beta_4}{\rightleftharpoons} [Cu(NH_3)_4]^{2+} \quad \beta_4 = \frac{[[Cu(NH_3)_4]^{2+}]}{[Cu^{2+}][NH_3]^4} \tag{17}$$

Now putting the experimental values of log K_1 = 4.0, log K_2 = 3.2, log K_3 = 2.7 and log K_4 = 2.0 in equation (12); the value of log β_4 can be calculated as follows:

$$\log \beta_4 = 4.0 + 3.2 + 2.7 + 2.0 \tag{18}$$

$$\log \beta_4 = 11.9 \tag{19}$$

Finally, it should also be noted that the thermodynamic stability of metal complexes is calculated by the overall formation constant. If the value of log β is more than 8, the complex is considered as thermodynamically stable; suggesting pretty much high stability for $[Cu(NH_3)_4]^{2+}$ complex. Moreover, the term dissociation or instability constant of a metal complex may also be defined here as the reciprocal of the stability constant.

❖ Trends in Stepwise Constants

The values of stepwise equilibrium constants for the formation of a particular metal-complex decrease successively in most of the cases i.e. $K_1 > K_2 > K_3 > K_4 > K_5 >....> K_n$. This regular decrease in the values of stepwise formation constants may be attributed to the decrease in the number of coordinated H_2O ligands that are available for the replacement by the attacking ligands. Besides, the continuous decline in the values of successive stepwise stability constant values may also be attributed to the decreasing ability of metal ions with a progressive intake of ligands, Coulombic factors and steric hindrance. Consider the following ligand displacement reaction:

$$[Cu(H_2O)_6]^{2+} + 4NH_3 \rightleftharpoons [Cu(H_2O)_2(NH_3)_4]^{2+} + 4H_2O \tag{20}$$

$$\beta_4 = \frac{[[Cu(H_2O)_2(NH_3)_4]^{2+}][H_2O]^4}{[[Cu(H_2O)_6]^{2+}][NH_3]^4} \tag{21}$$

The overall process can be supposed to take place through the following steps:

$$[Cu(H_2O)_6]^{2+} + NH_3 \overset{K_1}{\rightleftharpoons} [Cu(H_2O)_5(NH_3)]^{2+} + H_2O \tag{22}$$

$$K_1 = \frac{[[Cu(H_2O)_5(NH_3)]^{2+}][H_2O]}{[[Cu(H_2O)_6]^{2+}][NH_3]} \tag{23}$$

$$[Cu(H_2O)_5(NH_3)]^{2+} + NH_3 \overset{K_2}{\rightleftharpoons} [Cu(H_2O)_4(NH_3)_2]^{2+} + H_2O \tag{24}$$

$$K_2 = \frac{[[Cu(H_2O)_4(NH_3)_2]^{2+}][H_2O]}{[[Cu(H_2O)_5(NH_3)]^{2+}][NH_3]} \tag{25}$$

$$[Cu(H_2O)_4(NH_3)_2]^{2+} + NH_3 \overset{K_3}{\rightleftharpoons} [Cu(H_2O)_3(NH_3)_3]^{2+} + H_2O \tag{26}$$

$$K_3 = \frac{[[Cu(H_2O)_3(NH_3)_3]^{2+}][H_2O]}{[[Cu(H_2O)_4(NH_3)_2]^{2+}][NH_3]} \tag{27}$$

$$[Cu(H_2O)_3(NH_3)_3]^{2+} + NH_3 \overset{K_4}{\rightleftharpoons} [Cu(H_2O)_2(NH_3)_4]^{2+} + H_2O \tag{28}$$

$$K_4 = \frac{[[Cu(H_2O)_2(NH_3)_4]^{2+}][H_2O]}{[[Cu(H_2O)_3(NH_3)_3]^{2+}][NH_3]} \tag{29}$$

It has been observed that log K values for K_1, K_2, K_3 and K_4 are 4.3, 3.6, 3.0 and 2.3, respectively. This regular decrease in stepwise stability constants can be attributed to the decreasing site availability for the attack of the incoming ligand.

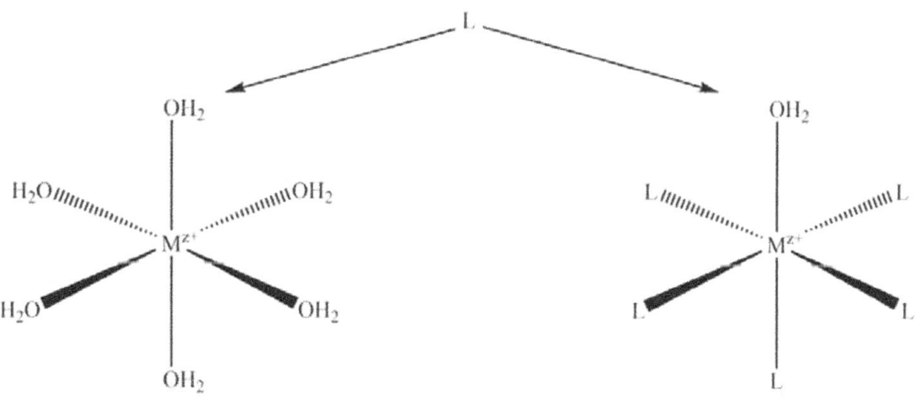

Figure 1. Decreasing availability in of coordinated water molecules to get replaced in first to the last step in octahedral complexes.

In the first step, NH_3 can attack all the six-coordination sites while in the last step the number of coordination sites available for the attack is reduced to three. Hence, stepwise constant, as well as the ease of formation of the complexes, also decreases as we move from first to the last step of ligand displacement.

Now although the decreasing trend in stepwise constants is pretty much common in most of the complex formation processes, still some exceptions do exist in which it is found that $K_{n+1} > K_n$. This weird behavior in stepwise equilibrium constants may be explained in terms of some unusual structural deviations and variations in the electronic structure of the metal center. The deviations in electronic configurations cause the change in crystal field stabilization energy (CFSE), and therefore, also affect the overall stability of the metal complex i.e. complex with a large magnitude of CFSE will be more stable and, consequently, will have higher value of the stepwise formation constants. Consider the following ligand displacement reaction:

$$[Cd(H_2O)_6]^{2+} + 4Br^- \rightleftharpoons [CdBr_4]^{2-} + 6H_2O \tag{30}$$

$$\beta_4 = \frac{[[CdBr_4]^{2-}][H_2O]^6}{[[Cd(H_2O)_6]^{2+}][Br^-]^4} \tag{31}$$

The overall process can be supposed to take place through the following steps:

$$[Cd(H_2O)_6]^{2+} + Br^- \overset{K_1}{\rightleftharpoons} [Cd(H_2O)_5Br]^{1+} + H_2O \tag{32}$$

$$[Cd(H_2O)_5Br]^{1+} + Br^- \overset{K_2}{\rightleftharpoons} [Cd(H_2O)_4Br_2] + H_2O \tag{33}$$

$$[Cd(H_2O)_4Br_2] + Br^- \overset{K_3}{\rightleftharpoons} [Cd(H_2O)_3Br_3]^- + H_2O \tag{34}$$

$$[Cd(H_2O)_3Br_3]^- + Br^- \overset{K_4}{\rightleftharpoons} [CdBr_4]^{2-} + 3H_2O \tag{35}$$

It has been observed that log K values follows the order $K_1 > K_2 > K_3 < K_4$, instead of $K_1 > K_2 > K_3 > K_4$. This unusually high value of K_4 is because the last step is actually pretty much favored by the release of three aquo ligands and some simultaneous structural and electronic changes.

Similarly, consider the formation of $[Fe(bpy)_3]^{2+}$ complex:

$$[Fe(H_2O)_6]^{2+} + bpy \overset{K_1}{\rightleftharpoons} [Fe(H_2O)_4(bpy)]^{2+} + 2H_2O \tag{36}$$

$$[Fe(H_2O)_4(bpy)]^{2+} + bpy \overset{K_2}{\rightleftharpoons} [Fe(H_2O)_2(bpy)_2]^{2+} + 2H_2O \tag{37}$$

$$[Fe(H_2O)_2(bpy)_2]^{2+} + bpy \overset{K_3}{\rightleftharpoons} [Fe(bpy)_3]^{2+} + 2H_2O \tag{38}$$

It has been observed that log K values follow the order $K_1 > K_2 < K_3$, instead of $K_1 > K_2 > K_3$. This unusually high value of K_3 is because the complexes formed during first two steps are high spin due to weak H_2O ligands with a CFSE of $-0.4\Delta_O$ ($t_{2g}^4 e_g^2$), while the last complex $[Fe(bpy)_3]^{2+}$ is low spin with a CFSE value of $-2.4\Delta_O$ ($t_{2g}^6 e_g^0$). Hence, large crystal field stabilization in the last step makes K_3 even greater than K_2.

❖ Factors Affecting Stability of Metal Complexes with Reference to the Nature of Metal Ion and Ligand

The stability of transition metal complexes depends upon a number of factors but is largely governed by the nature and the coordinative environment of the ligands-attached and the nature of the central metal ion or atom itself. A detailed discussion on both is given below.

> ### *Nature of the Metal Ion*

The following properties of the central metal atom or ion affect the stability of the transition metal complexes to a significant extent.

1. Size of the central metal ion: The stability of metal complexes decreases with the increase in the size of central metal ion provided the valency and ligands the same. Thus, the stability of isovalent complexes decreases down the group and increases along the period as the size varies in the reverse order. Stability order of hydroxide complexes of alkali metal ions and alkaline earth metal ions is:

M^{1+}	Li^{1+}	>	Na^{1+}	>	K^{1+}	>	Rb^{1+}	>	Cs^{1+}
r (Å)	0.60		0.95		1.33		1.48		0.95

Similarly,

M^{2+}	Be^{2+}	>	Mg^{2+}	>	Ca^{2+}	>	Sr^{2+}	>	Ba^{2+}	>	Ra^{2+}
r (Å)	0.31		0.65		0.99		1.13		1.35		1.40

Similarly,

M^{3+}	Sc^{3+}	>	Y^{3+}	>	La^{3+}
r (Å)	0.81		0.93		1.15

Besides the stability order of hydroxide complexes of 3rd group metal ions, there is a very popular stability order of metal complexes formed by bivalent metal ions of the first transition series, which is known as Irving-William series are given below.

M^{2+}	Mn^{2+}	<	Fe^{2+}	<	Co^{2+}	<	Ni^{2+}	<	Cu^{2+}	>	Zn^{2+}
r (Å)	0.91		0.83		0.82		0.78		0.69		0.74

The trend given by Irving-William for high spin octahedral metal complexes is actually independent of the ligand used. For instance, the variation of over first stepwise stability constant for the formation of ethylenediamine-complex on the logarithmic scale is given below. It can be clearly seen that the sequence actually starts from bivalent manganese rather Sc^{2+}, which is obviously due to the lack of data for the first two bivalent members because M(II) oxidation states are pretty much unstable.

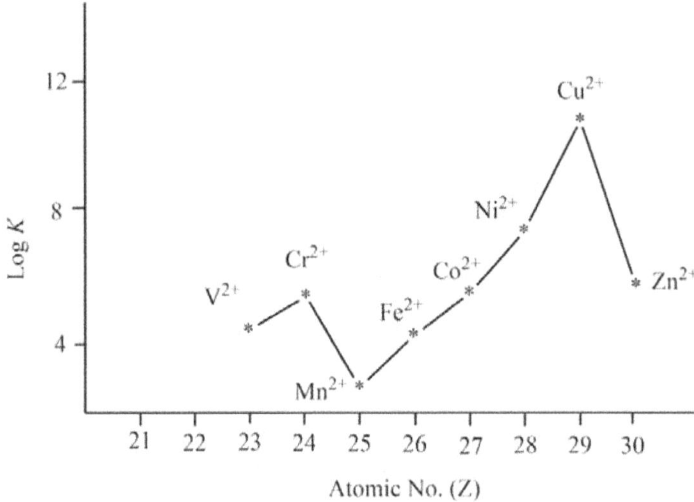

Figure 2. The first stepwise stability constants for the formation of ethylenediamine-complex.

The exceptionally special position of bivalent copper may be attributed to the formation distorted octahedral complexes Jahn-Teller effect and is discussed later in the book.

2. Charge on the central metal ion: The stability of transition metal complexes, with the same ligands and similar coordinative environment, increases with the increase of the charge on the central metal atom or ion.

M^{n+}	La^{3+}	>	Sr^{2+}	>	K^{1+}
r (Å)	1.15		1.13		1.33

Similarly,

M^{n+}	Th^{3+}	>	Y^{3+}	>	Ca^{2+}	>	Na^{1+}
r (Å)	0.95		0.93		1.14		1.16

Therefore, the greater is the charge on the central ion, the higher will be the stability of the metal complex.

3. Charge to size ratio of the metal ion: Now although we have already studied the effect of size and charge of the metal center on the overall stability of the complex, the more precise parameter, however, to do so is the ratio of the two. The charge to size ratio can be used to rationalize all the orders discussed previously on the charge or the size factor only.

M^{n+}	Co^{3+}	>	Co^{2+}
Charge/size	4.76		2.70

Hence, the greater is the charge and smaller the size i.e. large charge/size ratio increases the stability of these complexes quite significantly. A more illustrative presentation using different metal ions from different groups or periods is given below.

Table 1. The variation of charge-to-size ratio for different types of the metal ion.

Metal ion	Charge	Ionic radius	Charge/size
Li^+	+1	0.6	1.6
Ca^{2+}	+2	0.99	2.0
Ni^{2+}	+2	0.72	2.97
Y^{3+}	+3	0.93	3.22
Th^{4+}	+4	0.95	4.20
Al^{2+}	+2	0.50	6.0
Ba^{2+}	+2	0.31	6.45

The data listed above infers that the charge-to-size ratio is increasing 1.6 to 6.45 as we move from Li^+ to Ba^{2+}; and remarkably the stability of complexes also follows the same order, confirming our initial statement.

Furthermore, the charge-to-size ratio can also be used to rationalize the effect of electronegativity of the metal ion. As the bonding between metal ion and ligands is considered in the electron donation ability of the ligand, the electronegativity of the central metal ion also plays a key role in the stability of metal complexes. Conclusively, the greater is the positive charge density on the central metal ion, greater will be the electronegativity and consequently greater stability of the complexes.

4. Class of the metal ion: It has been observed that class *a* metals e.g. alkali metal ion, alkaline earth metal ions and metals from first transition series are shown to form stable complexes with ligands having N, O or F as the donor site. On the other hand, class *b* metals e.g. metals from second and third transition series are shown to form stable complexes ligands having P, S or Cl like donor atoms. However, borderline metals do not show unique behavior towards complex formation as far as stability is concerned.

This trend is explainable by the hard-soft acid-base principle which states that hard acid prefers hard base while soft acid prefers soft base for binding to yield stable systems. Metals such as Li^{1+}, Ba^{2+}, Mg^{2+} and Al^{3+}, which have large negative reduction potential have a lesser tendency to attract electrons, and hence, form stable complexes with highly electronegative groups like N, F or O so that they become unable to draw the unwanted electron density due to polarization. However, Metals like Pd^{2+} or Pt^{2+} which have large positive reduction potential have a greater tendency to accept electron and hence form stable complexes with less electronegative groups like P so that they can easily grab the electron density by polarizing the surrounding.

Table 2. The class a, class b and borderline metals.

Class a metals	Class b metals	Borderline metals
H^+, Li^+, Na^+, K^+	Cu^+, Ag^+, Au^+, Tl^+	
Be^{2+}, Mg^{2+}, Ca^{2+}, Sr^{2+}	Hg^{2+}, Pd^{2+}, Pt^{2+}	Mn^{2+}, Fe^{2+}, Co^{2+}, Sr^{2+} Ni^{2+}, Cu^{2+}, Zn^{2+}
Al^{3+}, Ga^{3+}, In^{3+} Cr^{3+}, Mn^{3+}, Fe^{3+}, Co^{3+}, La^{3+}, Ce^{3+}, Gd^{3+}	Tl^{3+}	
In^{4+}, Zr^{4+}, Hf^{4+}, Th^{4+}, U^{4+}, Pu^{4+}	Pt^{4+}	

For class a metals, the stability of different metal complexes with different ligands follows the order:

F^-	$>$	Cl^-	$>$	Br^-	$>$	I^-
O	$\gg$	S	$>$	Se	$>$	Te
N	$\gg$	P	$>$	As	$>$	Sb

For class b metals, the stability of different metal complexes with different ligands follows the order:

F^-	$<$	Cl^-	$<$	Br^-	$<$	I^-
O	$\ll$	S	$<$	Se	$<$	Te
N	$\ll$	P	$<$	As	$<$	Sb

For borderline metals, the stability of different metal complexes is pretty much independent of the nature of the ligand and follows the order given by the Irving-William series.

> ➤ *Nature of the Ligand*

The following properties of ligands-attached affect the stability of the transition metal complexes to a significant extent.

1. Charge and size of the ligand: Just like the metal, the charge and size of the ligand also play a significant role in deciding the stability of the transition metal complexes. Smaller size ligands are expected to form more stable complexes as they can approach the metal ion more closely and ligands with higher charges are expected with the same trend as they would form a strong bond with the central metal ion. However, this is true only for class a metal ions and this order gets a reverse sweep for class b metal ion. This can be illustrated as follows:

The stability order of halide complexes with class a metal ion is:

F^-	$>$	Cl^-	$>$	Br^-	$>$	I^-

The stability order of halide complexes with class *b* metal ion is:

$$F^- \quad < \quad Cl^- \quad < \quad Br^- \quad < \quad I^-$$

2. Basicity of the ligand: Stability of the metal complexes increases with the increase in the basic nature of the ligands as the donation of electron pair becomes more favorable. Thus, NH_3 should be a better ligand than H_2O which in turn should form more stable complexes than HF. This trend is quite robust for alkali metals, alkaline earth metals, $3d$ transition series, lanthanides and actinides. For instance, the stability order for metal complexes of a bivalent transition metal is:

$$F^- \quad < \quad H_2O \quad < \quad NH_3$$

3. Back-bonding capacity of ligand: The ligands such as CO, CN^-, PR_3, NO, alkenes and alkynes have special ability to form π-bonding with the central metal ion usually form more stable complexes than the others.

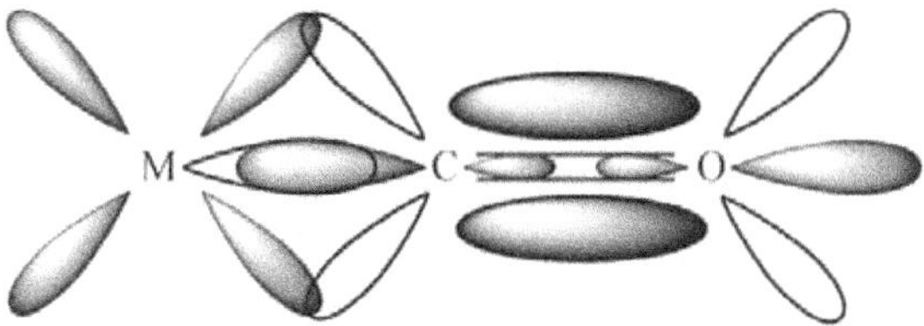

Figure 3. The back bonging mode in metal carbonyl complexes enhances stability.

4. Steric effects of the ligand: Metal complexes with bulky groups are usually less stable due to the weakening of the metal–ligand bond arising from the steric hindrances. For instance, consider the case of bivalent nickel complexes with 8-hydroxy quinoline and 2-methyle-8-hydroxy quinoline; the former is more stable in comparison to the other due to less steric hindrance.

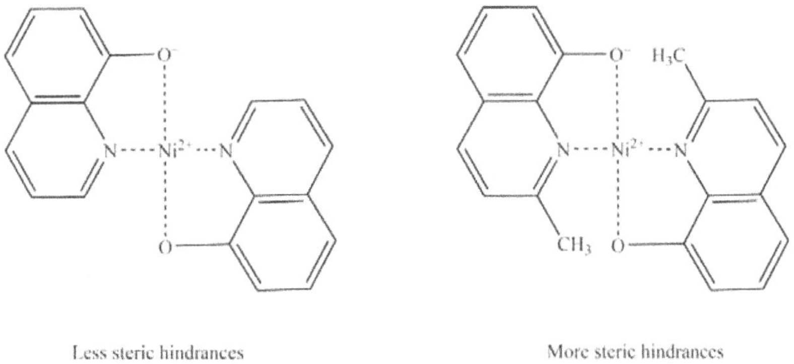

Figure 4. Bivalent nickel complexes with 8-hydroxy quinoline and 2-methyle-8-hydroxy quinoline.

5. Dipole moment of the ligand: Neutral ligands are shown to produce more stable complexes as their permanent dipole moment increases. For example, consider the stability order for amine complexes:

Ammonia　　　>　　　Ethylamine　　　>　　　Diethylamine　　　>　　　Triethylamine

6. Special configuration of the ligand: The ligands like porphyrin and their derivatives usually form very stable organometallic complex as they are stabilized by the aromatic character which extends over its entire structure due to exclusive planarity.

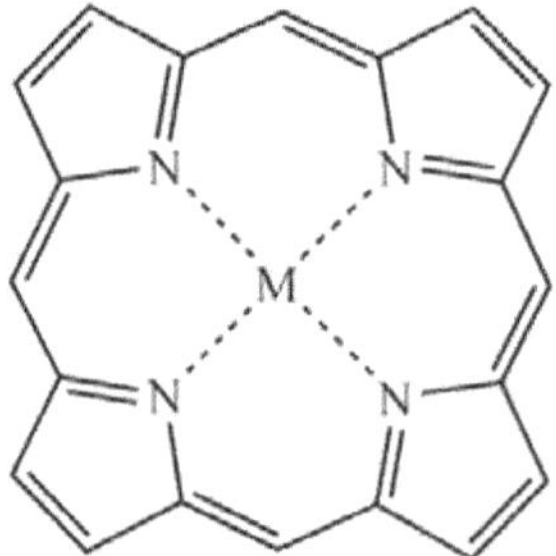

Figure 5. Structure the simplest porphyrin complex.

7. Chelating effect of ligand: The ligands which can form five or six-membered ring structure with the metal center usually form more stable complexes than the others. This effect is called as the chelation and the ligands are called as chelating ligands.

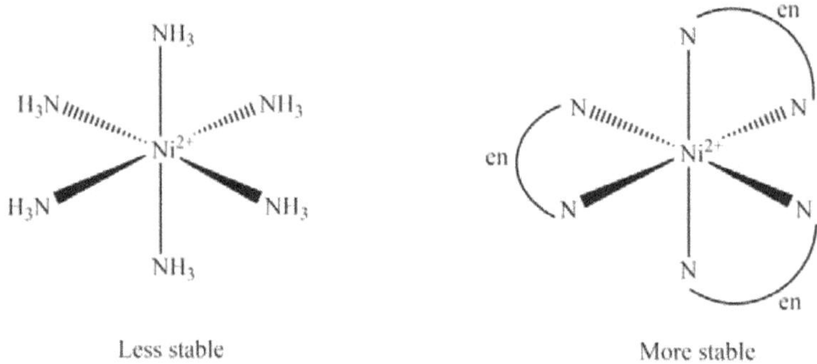

Figure 6. Complexes of bivalent nickel with monodentate amine and bidentate ethylene diamine ligands to form the chelate in the latter case.

The log β values for $[Ni(NH_3)_6]^{2+}$ and $[Ni(en)_3]^{2+}$ complexes are 8.6 and 18.6, respectively; which in turn confirms the higher stability of the chelate one.

8. Macrocyclic effect of ligand: It has been observed that nine or more membered cyclic ring systems with three or more donor atoms form extensively more stable metal complexes than their acyclic analogues. This effect is called as macrocyclic effect and these types of ligands are called as macrocyclic ligands. The very strong affinity of macrocyclic groups can be considered as a combination of the entropic effect like in the chelation, joined with an extra energetic contribution which comes from the pre-organized nature of the ligating groups that is, no additional strains are introduced to the ligand on coordination.

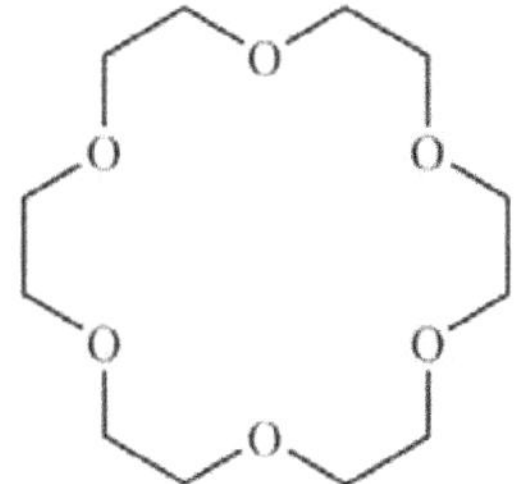

Figure 7. Structure of 18-Crown-6

9. Concentration of ligand: It has been observed that sometimes complexation occurs only at a high concentration of the ligands otherwise solvent molecules tend to bind the metal ion more preferably. For example, $[Co(SCN)_4]^{2-}$ exists only at a high concentration of thiocyanate ions and as we dilute the aqueous solution, the blue color complex starts disappearing and pink colored $[Co(H_2O)_6]^{2+}$ become dominant.

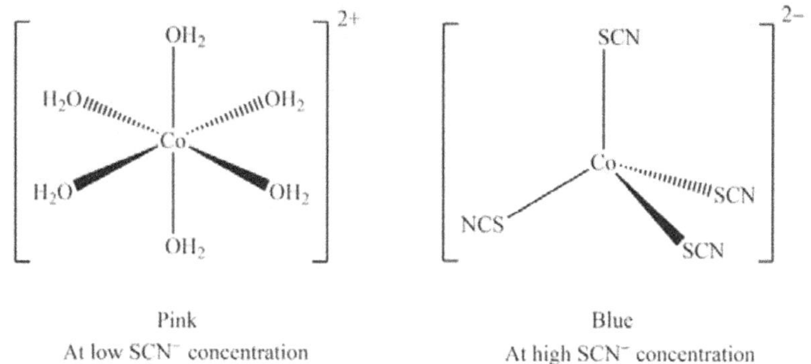

Figure 8. Structure of $[Co(H_2O)_6]^{2+}$ and $[Co(SCN)_4]^{2-}$.

As the octahedral $[Co(H_2O)_6]^{2+}$ complex transform into T_d-symmetry $[Co(SCN)_4]^{2-}$ at high ligand concentration, the huge shift in the intensity and the color from pale pink to very deep blue is obviously due to the removal center of symmetry, making *d-d* transitions as Laporte selection allowed in nature.

❖ Chelate Effect and Its Thermodynamic Origin

The chelate effect or chelation may simply be defined as an equilibrium reaction between the complexing agent and a metal ion, characterized by the formation of two or more bonds between metal and the complexing agent, resulting in the formation of a ring structure including the metal ion.

The term 'chelates' was first proposed by Morgan and Drew in 1920 for the cyclic structure arising from the assembly of metals with organic ligands. The chelate effect results in the enhanced affinity of chelating ligands for a metal ion compared to the affinity of a collection of similar non-chelating monodentate ligands for the same metal. This, in turn, results in the larger values of overall formation constants for the chelate complexes than their monodentate counterparts.

Figure 9. Popular chelating ligands; (a) ethylenediamine (symmetrical) and (b) glycinato (unsymmetrical).

➢ Characteristic Features of Chelates

The Chelate complexes generally have low melting points and are soluble in the organic solvents. Now though the phenomenon of chelation follows pretty much all rules of metal-ligand bonding as followed by simple complexes; two characteristic features of the chelate complexes are unique and can be given as:

1. Ligand-type: One of the most obvious features of the chelates is that only polydentate ligands are capable of forming chelate complexes. The ligand can be bidentate, tridentate or even hexa-dentate; but it must have more than one donor sites to form the chelate.

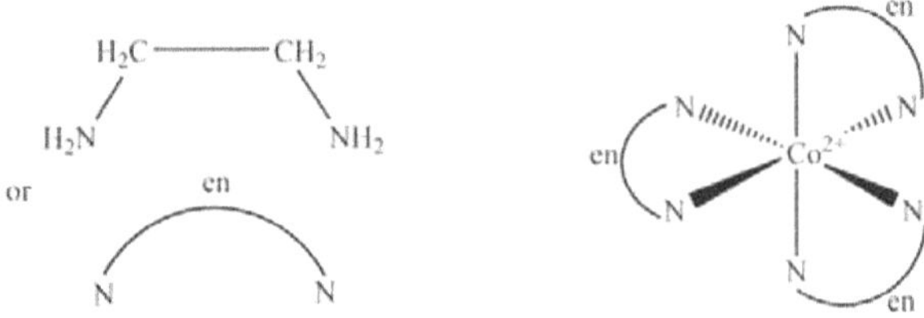

Figure 10. Structure of chelate complex of Co^{2+} with the bidentate ligand.

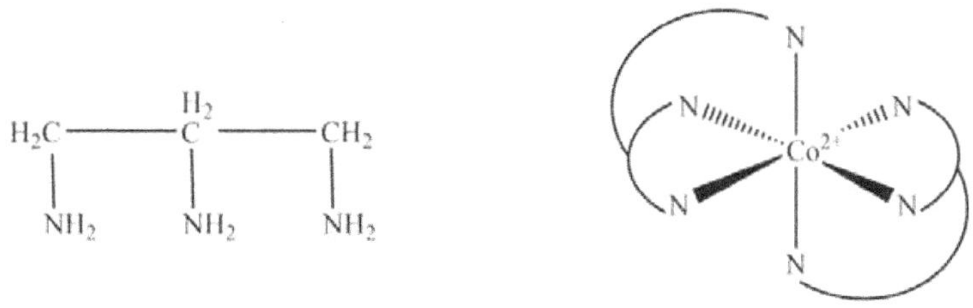

Figure 11. Structure of chelate complex of Co^{2+} with a tridentate ligand.

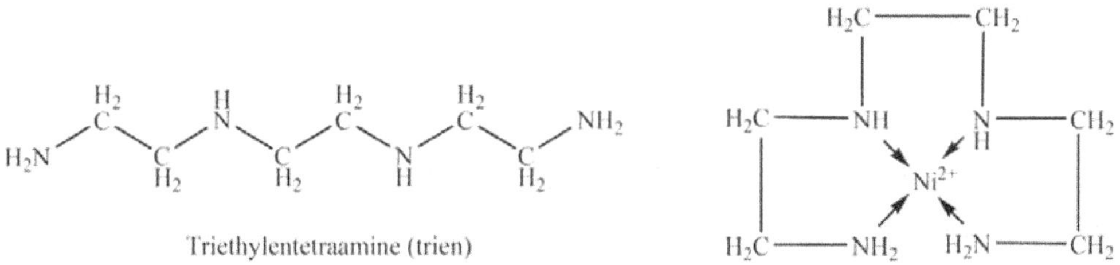

Triethylentetraamine (trien)

Figure 12. Structure of chelate-complex of Ni^{2+} with a tetradentate ligand.

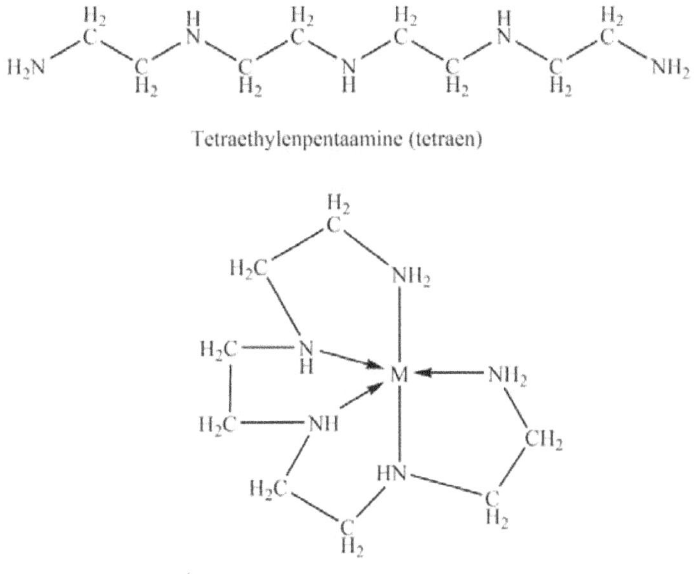

Tetraethylenpentaamine (tetraen)

Figure 13. Structure of metal chelate-complex with a pentadentate ligand.

Ethylenediaminetetraacetate (EDTA^{4-})

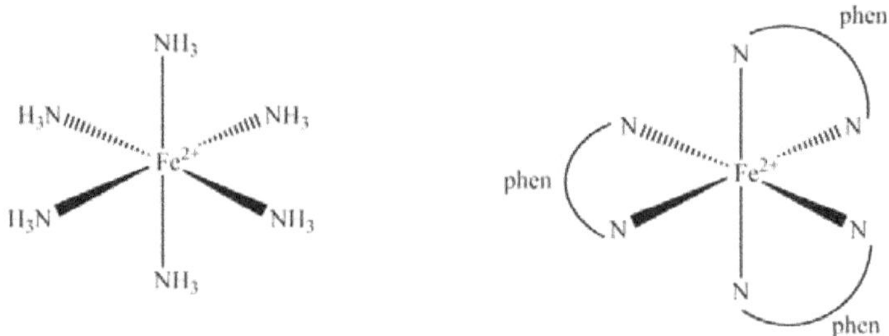

Figure 14. Structure of chelate complex of Fe^{3+} with hexadentate ligand.

2. Stability: The chelating agent binds with the metal-center of suitable size firmly and the bond can be covalent or coordinate. This makes the chelate complexes as more stable than their non-chelating counterparts.

$[Fe(NH_3)_6]^{2+}$ Non-chelate complex (Less stable) $[Fe(phen)_3]^{2+}$ Chelate complex (More stable)

Figure 15. Structure of chelate complex of Fe^{2+} with a bidentate ligand.

Also, it has also been observed that five and six-membered ring chelate complexes are more stable than the other due less steric strain. The chelating ligands like ethylenediamine, which do not have multiple bonds, prefer to form five-membered rings complexes; while the chelating ligands like acetylacetonate, which do have double bonds, prefer form six-membered rings complexes.

| Log K | Very small | 5.7 | 4.7 | 3.6 |

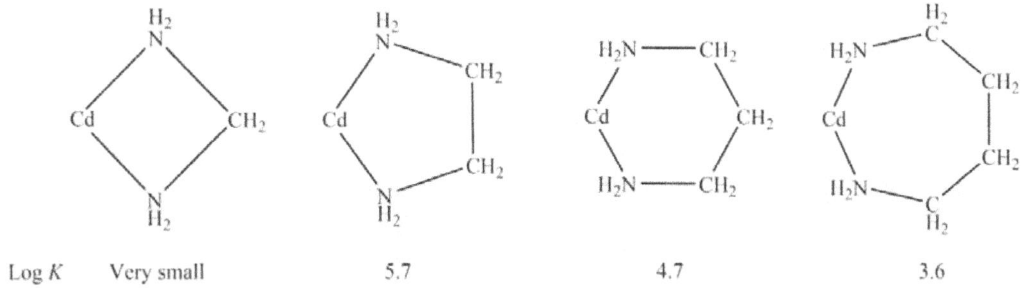

Figure 16. Stability order for the bivalent cadmium complexes of bidentate amines.

Furthermore, the chelating ligands with bulky groups usually form less stable complexes due to the weakening of the metal–ligand bond arising from the steric hindrances. For instance, consider the case of metal complexes involving 2,2-bipyridyl and its derivatives as the chelating ligand. The structures of bivalent iron complexes with normal 2,2-bipyridyl and 4,4'-Dimethyl-2,2'-bipyridyl as the chelating ligand clearly shows that the former has less steric hindrance, and therefore, is more stable.

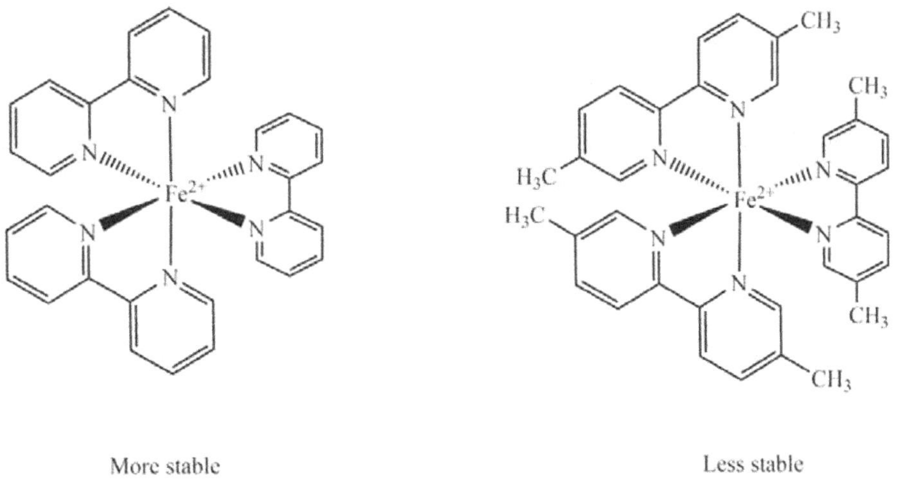

More stable   Less stable

Figure 17. Structures of bivalent iron complexes with normal 2,2-bipyridyl and 4,4'-Dimethyl-2,2'-bipyridyl as the chelating ligand.

> *Thermodynamic Explanation of Chelation*

The principal driving force for the chelation phenomena is the stability gain arising from the increasing entropy during the complexation. The free energy change for a general reaction is given by:

$$\Delta G° = \Delta H° - T\Delta S° \quad or \quad \Delta G° = -RT \ln K \tag{39}$$

Now, as the enthalpy change between non-chelate and chelates is almost negligible because of the similar metal–ligand donor type, the free energy change is primarily governed by the value of $\Delta S°$. Furthermore, it follows from the equation (39) that $\Delta G°$ will become more negative with a more positive value of $\Delta S°$, which in turn would increase the value of stability constant K. Hence, the stability of metal complexes increases with chelation as it occurs through an increment in the entropy of the system.

$$[Cd(H_2O)_4] + 4CH_3NH_3 \rightleftharpoons [Cd(CH_3NH_3)_4]^{2+} + 4H_2O \tag{40}$$

$$1 + 4 = 5 \qquad\qquad\qquad 1 + 4 = 5$$

$$[Cd(H_2O)_4] + 2en \rightleftharpoons [Cd(en)_2]^{2+} + 4H_2O \tag{41}$$

$$1 + 2 = 3 \qquad\qquad\qquad 1 + 4 = 5$$

Thus, chelation is a thermodynamically favorable process as the number of species in the product are higher than that of what it was in the starting materials leading to enhanced randomness.

$$[Cd(H_2O)_4] + trien \rightleftharpoons [Cd(trien)]^{2+} + 4H_2O \tag{42}$$

$$1 + 1 = 2 \qquad\qquad\qquad 1 + 4 = 5$$

It is obvious that entropy increase in case of trien is higher than ethylenediamine ligand making the trien chelate complex as more stable. The stability of chelate complexes increases with the denticity of the chelating agent.

> *Factors Affecting the Stability of Chelate Complexes*

1. The size of the chelate ring: Five or six membered chelate rings are more stable than the four, seven, eight or higher membered chelate rings due to less steric strain.

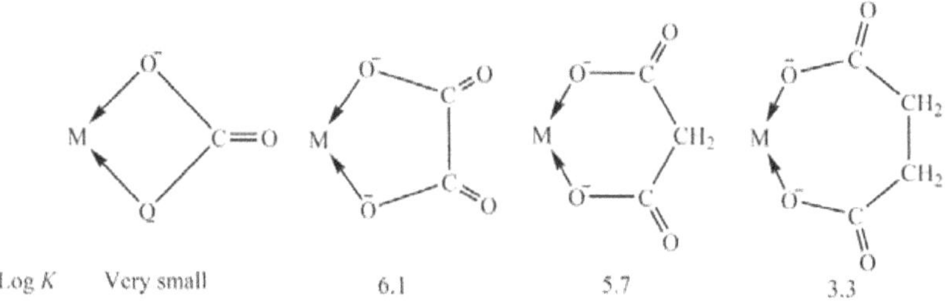

Figure 18. Stability order for the Cu^{2+} complexes with chelating ligands.

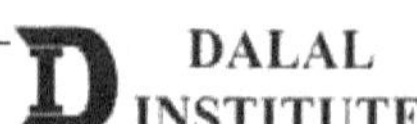

2. Number of chelate rings and nature of the metal ion: it has been observed that as the number of donor sites increase, the stability of metal chelates also increases. Furthermore, the stability of chelate complexes was also found in accordance with the Irving-William series.

Table 3. The values of overall formation constants (log β) for different types of metal ions as a function of the number of chelate rings.

Metal complex	No. of chelate rings	Values of overall formation constants (log β)						
		Mn^{2+}	Fe^{2+}	Co^{2+}	Ni^{2+}	Cu^{2+}	Zn^{2+}	Cd^{2+}
$M(NH_3)_4$	0	–	3.7	5.31	7.79	12.59	9.06	6.92
$M(en)_2$	2	4.9	7.7	10.9	14.5	20.2	11.2	10.3
$M(trien)$	3	4.9	7.8	11.0	14.1	20.2	12.1	10.0
$M(tren)$	3	5.8	8.8	12.8	14.0	18.8	14.6	12.3
$M(dien)_2$	4	7.0	10.4	14.1	18.9	21.3	14.4	13.8
$M(penten)$	5	9.4	11.2	15.8	19.3	22.4	16.2	16.2

In addition to stability prediction by Irving Williams series, transition metal ions usually form more stable chelate complexes and follow the order given by Maley and Mellor.

$$Pd > Cu > Ni > Co > Zn > Cd > Fe > Mn > Mg$$

3. Resonance effect: The resonance stabilization of the chelate ring also enhances complex stability. For example, the exceptionally high log β for [Fe(acac)$_3$] is due to resonance stabilization in the chelating ligand.

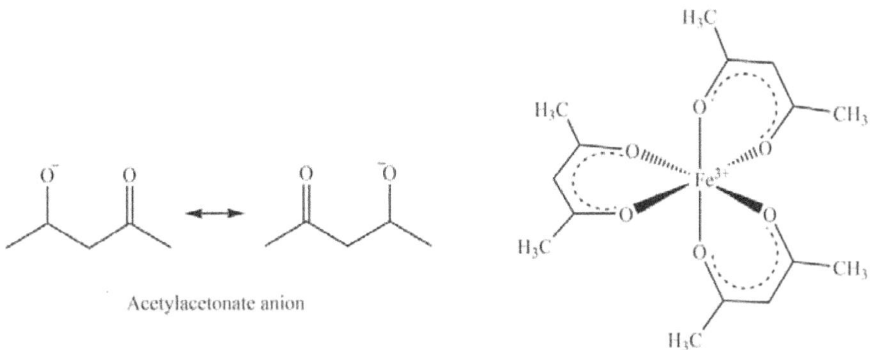

Figure 19. The resonance stabilization in tris(acetylacetonato) iron(III) or [Fe(acac)$_3$].

4. Lewis base strength and charge of ligand: Calvin and Wilson studied the Cu^{2+} chelate complexes and proved that the stability of the chelate complexes increases with the increase of the basic character of the ligand. Moreover, it has also been observed that the uncharged ligands like ethylenediamine or triethylenetetraamine form more stable complexes than the anionic ligands like oxalato or glycinato.

Figure 20. Stability order for the Ni^{2+} complexes with neutral and anionic chelating ligands.

5. Steric hindrance and strain due to large ligands: The chelating ligands with large groups usually form less stable complexes due to the weakening of the metal–ligand bond arising from the steric hindrances. For example, ethylenediamine form more stable complexes with bivalent metal ions than N−tetramethylethylenediamine, which obviously due to greater steric hindrance in the later one. Moreover, the overall stability can also be affected by the strain in metal-chelates arising from the particular geometry of the chelating ligand. For instance, trien-complexes of bivalent copper are more stable than their tren-complexes due to less steric strain as trien-ligand can adapt square planar coordination much easily.

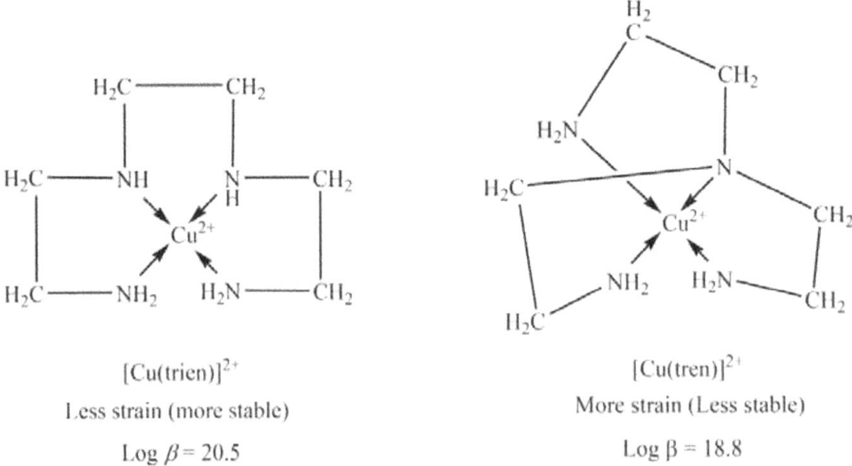

Figure 21. Stability order for the Cu^{2+} complexes with chelating ligands.

> ### *Main Applications of Chelates*

1. The colored chelate complex formation with Al^{3+}, Ni^{2+}, Mg^{2+} can be used to quantify these metals from their solutions.

2. The hardness of water can be removed by synthesizing the water-soluble chelate complexes of Mg^{2+} and Ca^{2+} ion which otherwise undergoes precipitations reaction.

3. Some neutral chelate complexes are sparingly soluble in water but have high solubility in an organic solvent that can be used to purify metal in the organic phase.

4. The chelating agents like EDTA can be used to eliminate harmful radioactive metal from the body forming their water-soluble complexes.

5. Chelating agents can also act food preservatives by binding or sequestering metal ions like iron or copper in certain foods in order to prevent the metals from oxidizing and speeding up spoilage.

❖ Determination of Binary Formation Constants by pH-metry and Spectrophotometry

The stability constants for metal complexes can be calculated using various techniques but the two of most popular experimental methods are pH-metry and spectrophotometric analysis owing to their simple setup and good accuracy.

> ### *pH-metric Method*

The alteration of hydronium ion concentration during the complex formation can be used to calculate the stability constant. Actually, H^+ ions are in direct competition with metal ions for the association with ligand species present in the solution.

Consider the ligand displacement reaction in which a metal ion along with some weak acid is added to ligand solution.

$$M^+ + L \xrightleftharpoons{K_f} ML^+ \quad K_f = \frac{[ML^+]}{[L][M^+]} \tag{43}$$

$$H^+ + L \xrightleftharpoons{K_a} HL^+ \quad K_a = \frac{[HL^+]}{[L][H^+]} \tag{44}$$

Where K_f and K_a are formation constants for the metal complex and acid association constant, respectively. Now let C_M, C_L and C_H as the molar concentration for metal ion, ligand and acid, we have

$$C_H = [H^+] + [HL^+] \tag{45}$$

$$C_L = [L] + [ML^+] + [HL^+] \tag{46}$$

$$C_M = [M^+] + [ML^+] \tag{47}$$

The above three equations can be solved as follows, From equation (47):

$$[M^+] = C_M - [ML^+] \tag{48}$$

The total ligand concentration can be calculated by using equation (44) and putting the value of $[HL^+]$ from equation (45) as:

$$[L] = \frac{C_H - [H^+]}{K_a[H^+]} \tag{49}$$

Now subtracting equation (45) from equation (46) and putting the value of [L] from equation (49), we get:

$$[ML^+] = C_L - C_H + [H^+] - \frac{C_H - [H^+]}{K_a[H^+]} \tag{50}$$

Thus, using the values of [M], [L] and $[ML^+]$ from equation (48), (49) and (50) in equation (43); we would be to calculate the formation constant by knowing C_M, C_L, C_H, K_a and the concentration of $[H^+]$ ion which is generally given the by the pH-meter. It has also been observed that the accuracy of the formation constant is high if the value to K_f is within the range of 10^5 times than that of K_a.

> ➤ *Spectrophotometric Method*

1. Conventional Spectrophotometric Method: The change in the absorbance during the complex formation can be used to find out the stability constant very easily. According to Beer-Lambert law:

$$A = \varepsilon.c.l \tag{51}$$

where A is the absorbance, l is the optical path length, ε is a molar extinction coefficient and c is a concentration of the solution. Thus, by measuring the absorbance at a particular wavelength for a metal complex solution of known path length and known concentration; the value of the molar extinction coefficient can be calculated. Consider the general metal-ligand equilibria:

$$M + L \underset{}{\overset{K_f}{\rightleftharpoons}} ML \quad K_f = \frac{[ML]}{[M][L]} \tag{52}$$

Now let C_M and C_L as the molar concentration for metal ion and ligand, respectively. We have

$$C_M = [M] + [ML] \tag{53}$$

$$C_L = [L] + [ML] \tag{54}$$

Now if we measure the absorbance of ML complex at a known path length and molar absorption coefficient; its concentration can be given by using equation (51) as:

$$[ML] = c = \frac{A}{\varepsilon l} \tag{55}$$

Putting the value of equation (55) in equation (53) and (54), we get:

$$[M] = M_L - [ML] = M_L - \frac{A}{\varepsilon l} \tag{56}$$

$$[L] = C_L - [ML] = C_L - \frac{A}{\varepsilon l} \tag{57}$$

Here and now, we can calculate the value of formation constant (K_f) by put the values of [ML], [M] and [L] from equation (55), (56) and (57) in equation (52).

For example, the above method can successfully be used to calculate the stability constant for the following reaction:

$$Fe^{3+} + NCS^- \overset{K_f}{\rightleftharpoons} [Fe(NCS)]^{2+}; \quad K_f = \frac{[[Fe(NCS)]^{2+}]}{[Fe^{3+}][NCS^-]} \tag{58}$$

Ferric ion and thiocynate ions are colorless in aqueous solution but their metal-ligand equilibria generate instance blood-red color and its λ_{max} is at 450 nm. In order to find the value of ε, we will have to measure the absorbance of $[Fe(H_2O)_5(NCS)]^{2+}$ complex prepared by dissolving Fe^{3+} ions in excess of NCS^- ligand so that all the ferric ions convert into the complex. Once the value of ε for $[Fe(H_2O)_5(NCS)]^{2+}$ is known, its equilibrium concentrations can easily be obtained by recording the absorption spectra. Then, using these equilibrium concentrations in equation (58) will give you the value of formation constant.

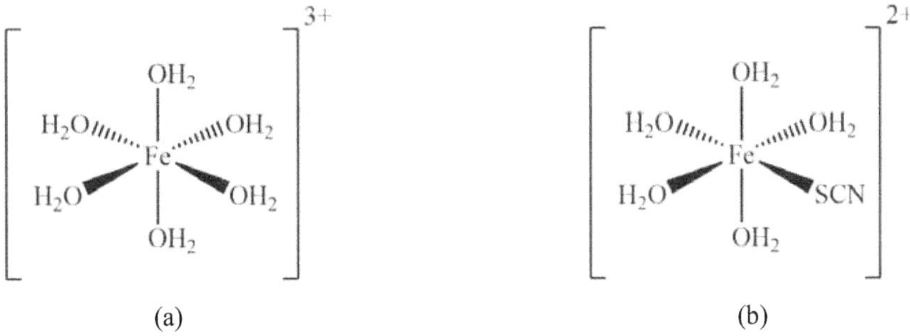

(a)   (b)

Figure 22. (a) Fe^{3+} ion in water (b) $[Fe(SCN)]^{2+}$ ion in water.

2. Method of continuous variation or Job's method: This method is the variation of the spectroscopic method and is very useful in determining the composition of the metal complexes. This method is used when only one complex is formed under the given experimental conditions and the volume of the solution remains constant. Consider the metal-ligand equilibria:

$$M + nL \underset{}{\overset{K_f}{\rightleftharpoons}} ML_n \quad K_f = \frac{[ML_n]}{[M][L]^n} \tag{59}$$

Prepare ten metal-ligand solutions with varying concentrations as given below.

Metal ion (*ml*)	0	1	2	3	4	5	6	7	8	9
Ligand (*ml*)	10	9	8	7	6	5	3	2	1	0

Now, let C_M and C_L as the molar concentration for metal ion and ligand, respectively. we have:

$$C_M + C_L = C \tag{60}$$

$$x = \frac{C_L}{C} \tag{61}$$

$$1 - x = \frac{C_M}{C} \tag{62}$$

Where C is a constant; while x and $(1-x)$ are the mole fractions for ligand and metal ion, respectively. Now by assuming that the complex absorbs much stronger than that of metal ion or ligand in the visible region, spectrophotometry can be used to determine the relative quantities of all three-species present at equilibrium. Thus, the maximum amount of complex and consequently maximum absorbance will be observed when the metal ion and the ligands are present in the stoichiometric ratio yielding the value of n for ML_n.

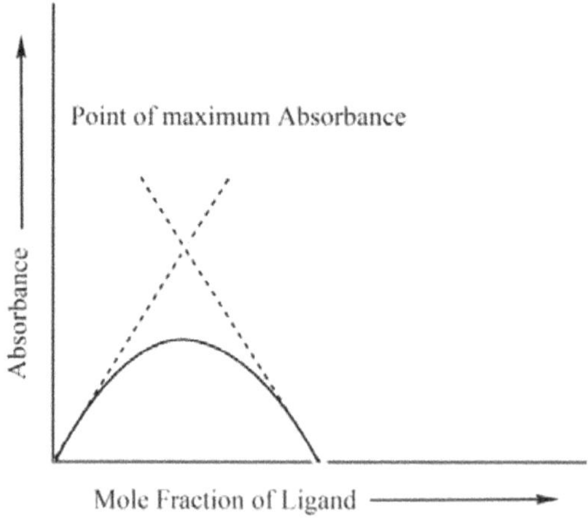

Figure 23. Continuous method of variation (Job's Plot)

$$\text{For complex } ML_n \qquad n = \frac{C_L}{C_M} \tag{63}$$

Dividing equation (61) by (62) we get:

$$n = \frac{C_L}{C_M} = \frac{C_L}{C} \times \frac{C}{C_M} = \frac{x}{1-x} \tag{64}$$

Thus, the composition of the complex is obtained by knowing the value of n. The absorbance calculated at the crossover point is the theoretical maximum amount of the complex formed that would occur if the equilibrium constant is very large. Here and now, as the absorbance is proportional to concentration, the extrapolated absorbance gives the maximum concentration of the complex while observed absorbance at the same mole fraction gives the concentration of the complex actually formed. Hence, we have:

$$\frac{[ML_n]_{actual}}{[ML_n]_{maximum}} = \frac{A_{observed}}{A_{maximum}} \tag{65}$$

Now the concentrations of all three species at equilibrium can be found and the equilibrium constant can be calculated using the relationship for K_f given in equation (59).

❖ Problems

Q 1. Discuss the thermodynamic stability of a complex and correlate stepwise constant with overall stability constant.

Q 2. How chelation and nature of donor atom affect the stability of metal complexes.

Q 3. Discuss the determination of formation constant by pH-metry.

Q 4. What is log β? How is it related to the stability of complexes?

Q 5. Illustrate with suitable example the trends in stepwise formation constant.

Q 6. Discuss a spectrophotometric method for the determination of stability constants of complexes.

Q 7. How does the nature of the central metal ion affect the stability of the complexes?

❖ Bibliography

[1] J. D. Lee, *Concise Inorganic Chemistry*, Chapman & Hall, New York, USA, 1994.

[2] B. R. Puri, L. R. Sharma, K. C. Kalia, *Principals of Inorganic Chemistry*, Milestone Publishers, Delhi, India, 2012.

[3] J. E. Huheey, E. A. Keiter, R. L. Keiter, *Inorganic Chemistry: Principals of Structure and Reactivity*, HarperCollins College Publishers, New York, USA, 1993.

[4] A. F. Wells, *Structural Inorganic Chemistry*, Oxford University Press, London, UK, 1975.

[5] F. A. Cotton, G. Wilkinson, C. A. Murillo, M. Bochmann, *Advanced Inorganic Chemistry*, John Wiley & Sons, New Jersey, USA, 1999.

[6] S. F. A. Kettle, *Physical Inorganic Chemistry: A Coordination Chemistry Approach*, Springer-Verlag, Berlin, Germany, 1996.

[7] M. Gerloch, E. G. Constable, *Transition Metal Chemistry*, VCH Verlagsgesellschaft mbH, Weinheim, Germany, 1994.

[8] W. U. Malik, G. D. Tuli, R. D. Madan, *Selected Topics in Inorganic Chemistry*, S. Chand Publishers, New Delhi, India, 2014.

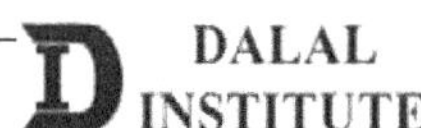

CHAPTER 3

Reaction Mechanism of Transition Metal Complexes – I:

❖ Inert and Labile Complexes

The metal complexes in which the rate of ligand displacement reactions is very fast and hence show high reactivity are called as labile Complexes and this property is termed as lability. On the other hand, the metal complexes in which the rate of ligand displacement reactions is very slow and hence show less reactivity are called as inert complexes and this property is termed as inertness.

Thermodynamic stability is the measure of the extent to which the complex will form or will be transformed into another complex when the system has reached equilibrium while kinetic stability refers to the speeds at which these transformations take place. The thermodynamic stability depends upon the energy difference of the reactant and product if the product has less energy than that of the reactant, it will be more stable than the reactant. The thermodynamic stability of metal complexes is calculated by the overall formation constant. If the value of log β is more than 8, the complex is considered as thermodynamically stable.

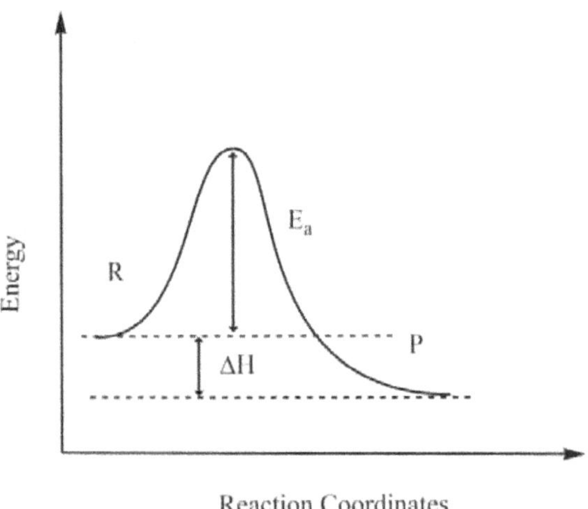

Figure 1. The general reaction coordinates diagram.

The kinetic stability of the complex depends upon the activation energy of the reaction. If the activation energy barrier is low, the reaction will take place at a higher speed. These types of complexes are also called as kinetically unstable. If the activation energy barrier is high, the substance will react slowly and will be called as kinetically stabilized or inert. There is no correlation between thermodynamic and kinetic stability. Thermodynamically stable products may labile or inert and the vice versa is also true.

> *Labile and Inert Complexes on the Basis of Valence Bond Theory*

According to the valence bond theory of chemical bonding, octahedral metal-complexes can be divided into two types.

1. Outer orbital complexes: These complexes have sp^3d^2 hybridization and are generally labile in nature. Valence bond theory proposed that the bonds in sp^3d^2 hybridization are generally weaker than that of $(n\text{-}1)d^2sp^3$ orbitals and therefore they show labile character. For example, octahedral complexes of Mn^{2+}, Fe^{2+}, Cr^{2+} complexes show fast ligand displacement.

2. Inner orbital complexes: Since d^2sp^3 hybrid orbitals are filled with six electron pairs donated by the ligands, d^n electron of metal will occupy d_{xy}, d_{yz} and d_{xz} orbitals. These d^2sp^3 hybrid orbitals can form both inert or labile complexes. In order to show lability, one orbital out of d_{xy}, d_{yz}, d_{xz} must be empty so that it can accept another electron pair and can form seven coordinated intermediate which is a necessary step for the associative pathway of ligand displacement. On the other hand, if all the d_{xy}, d_{yz}, d_{xz} orbitals contain at least one electron, it will not be able to accept electron pair from the incoming ligand and hence is expected to show inert character.

> *Labile and Inert Complexes on the Basis of Crystal Field Theory*

Octahedral complexes react either by SN_1 or SN_2 mechanism in which the intermediates are five and seven-coordinated species, respectively. In both cases, the symmetry of the complex is lowered down and due to this change in crystal field symmetry, the crystal field stabilization (CFSE) value also changes. The cases for lability and inertness are:

1. Labile complexes: If the CFSE value for the five or seven-membered intermediate complex is greater than that of the reactant, the complex will be of labile nature as there is zero activation energy barrier.

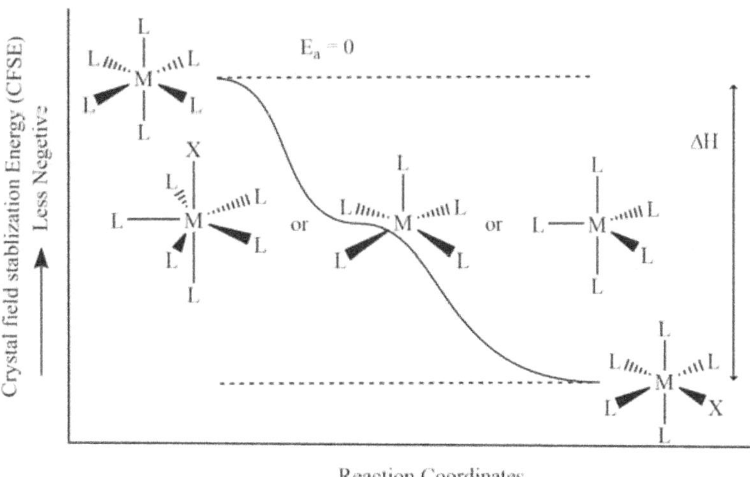

Figure 2. The reaction coordinates diagram for ligand displacement reactions in labile metal complexes.

DALAL INSTITUTE

2. Inert complexes: If the CFSE value for the five or seven-membered intermediate complex is less than that of the reactant, the metal complex will be of inert nature as loss of CFSE will become the activation energy barrier.

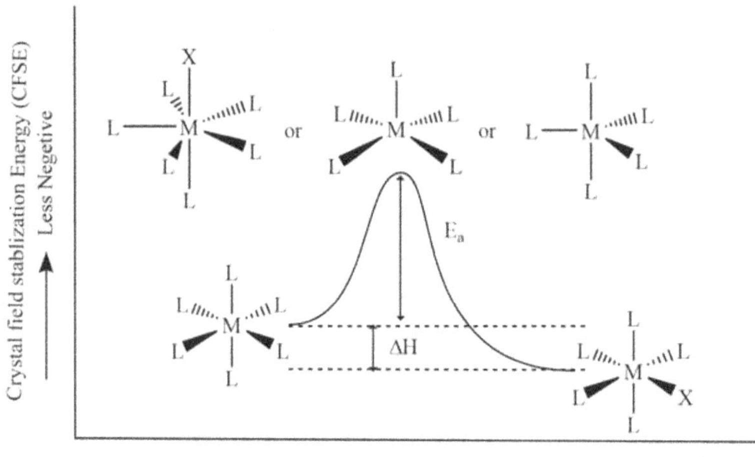

Reaction Coordinates

Figure 3. The general reaction coordinates diagram for ligand displacement reactions in inert metal complexes.

Hence, the gain of crystal field stabilization energy will make complex labile while the loss of CFSE will make complex inert. The calculation of CFSE relies upon the following assumptions:

1. All six-coordinated complexes should be treated as perfect octahedral even if the mixed ligands are present.

2. The inter-electronic repulsive forces arising from d-subshell can simply be neglected.

3. The Dq-magnitude for reacting as well as the intermediate complexes are assumed to be the same though they might have considerably different values.

4. The Jahn-Teller distortion should also be neglected in all calculations.

Evidence for the lability and inertness: The ligand displacement can be dissociative or associative depending upon the nature of the reaction.

(i) For SN$_1$ or dissociative pathway, the 5-coordinate intermediate can be trigonal-bipyramidal or square-pyramidal. However, it has been observed that the dissociative mechanism occurs through a square-pyramidal intermediate. Hence, the gain or loss of the crystal field stabilization energy can be calculated as:

CFSE gain or losss = CFSE of Squarepyramidal Intermediate – CFSE of Octahedral Reactant

If the CFSE gain-or-loss is negative, it means that activation energy is zero because it cannot be negative.

Table 1. CFSE values of high-spin (HS) and low-spin (LS) octahedral complexes undergoing ligand displacement reactions through SN_1 mechanism

Configuration	CFSE for octahedral reactant (Coordination No = 6)	CFSE for square-pyramidal intermediate (Coordination No = 5)	Gain or loss of CFSE Negative = gain Positive = loss	Kinetic stability
d^0	0	0	0	Labile
d^1	–4	–4.57	–0.57	Labile
d^2	–8	–9.14	–1.14	Labile
d^3	–12	–10	+2.00	Inert
d^4 (HS)	–6	–9.14	–3.14	Labile
d^4 (LS)	–16	–14.57	+1.43	Inert
d^5 (HS)	0	0	0	Labile
d^5 (LS)	–20	–19.4	+0.86	Inert
d^6 (HS)	–4	–4.57	–0.57	Labile
d^6 (LS)	–24	–20	+4.00	Inert
d^7 (HS)	–8	–9.14	–1.14	Labile
d^7 (LS)	–18	–19.14	–1.14	Labile
d^8	12	–10	+2.00	Inert
d^9	–6	–9.14	–3.14	Labile
d^{10}	0	0	0	Labile

From Table 1, we can say that:

Metal complexes with d^0, d^1, d^2, d^{10} are labile in nature and undergo fast ligand displacement through the dissociative pathway. High spin metal complexes with d^4, d^5, d^6, d^7 are also labile in nature and react quickly through the dissociative pathway. Low spin complexes of d^7 metal ions are also found to be labile due to CFSE gain.

On the other side, d^3 and d^8 metal complexes are inert in nature and undergo slow ligand displacement through the dissociative pathway. Moreover, low spin complexes with d^4, d^5 and d^6 metal complexes are also

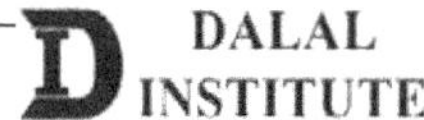

inert due to loss of CFSE during the SN_1 mechanism. Using the order of CFSE loss, the reactivity can be represented as $d^5 > d^4 > d^8 > d^3 > d^6$.

(i) For SN_2 or associative pathway, the 7-coordinate intermediate can be a pentagonal-bipyramidal or octahedral wedge. However, it has been observed that the associative mechanism preferably occurs through an octahedral-wedge intermediate. Hence, the gain or loss of the CFSE can be calculated as:

CFSE gain or loss = CFSE of Octahedral-Wedge Intermediate − CFSE of Octahedral Reactant

Table 2. CFSE values of high-spin (HS) and low-spin (LS) octahedral complexes undergoing ligand displacement reactions through SN_2 mechanism

Configuration	CFSE for octahedral reactant (Coordination No = 6)	CFSE for octahedral-wedge intermediate (Coordination No = 7)	Gain or loss of CFSE Negative = gain Positive = loss	Kinetic stability
d^0	0	0	0	Labile
d^1	−4	−6.08	−2.08	Labile
d^2	−8	−8.68	−0.68	Labile
d^3	−12	−10.20	+1.80	Inert
d^4 (HS)	−6	−8.79	−2.79	Labile
d^4 (LS)	−16	−16.26	−0.26	Labile
d^5 (HS)	0	0	0	Labile
d^5 (LS)	−20	−18.86	+1.14	Inert
d^6 (HS)	−4	−6.08	−2.08	Labile
d^6 (LS)	−24	−20.37	+3.63	Inert
d^7 (HS)	−8	−8.68	−0.68	Labile
d^7 (LS)	−18	−18.98	−0.98	Labile
d^8	−12	−10.20	+1.80	Inert
d^9	−6	−8.79	−2.79	Labile
d^{10}	0	0	0	Labile

The following conclusions can be drawn from the data listed in Table 2.

Metal complexes with d^0, d^1, d^2, d^{10} are labile in nature and undergo fast ligand displacement through the associative pathway. High spin metal complexes with d^4, d^5, d^6, d^7 are also labile in nature and react quickly through the associative pathway. Low spin complexes of d^7 metal ions are also found to be labile due to CFSE gain. It can be seen that d^4 low spin are also labile in nature.

On the other side, d^3 and d^8 metal complexes are inert in nature and undergo slow ligand displacement through the associative pathway. Moreover, low spin complexes with d^5 and d^6 metal complexes are also inert due to the loss of CFSE during the SN$_1$ mechanism. Using the order of CFSE loss, the reactivity can be represented as $d^5 > d^8 > d^3 > d^6$.

> ### Factors Affecting the Kinetic Stability or Lability of Non-Transition Metal Complexes

The kinetic stability of non-transition metal complexes can be rationalized from the valence bond theory (VBT) as well as from the perspectives of crystal field theory (CFT). According to the valence bond model, all of the non-transition metal complexes are outer-orbital in nature; and therefore, are expected to show labile behavior. Similarly, the predictions of kinetic stability of octahedral complexes of non-transition metals are also labile because whatever the path is followed, associative or dissociative, the loss of CSFE will always zero. Nevertheless, the overall trend kinetic stability of transition metal complexes depends upon a number of factors discussed below.

1. Charge on the central metal ion: The Lability of a complex decreases with the increasing charge on the central metal ion. For example, lability order of the following complexes:

$$[AlF_6]^{3-} \quad > \quad [SiF_6]^{2-} \quad > \quad [PF_6]^- \quad > \quad [SF_6]^0$$

$$+3 \qquad\qquad\qquad +4 \qquad\qquad\qquad +5 \qquad\qquad\qquad +6$$

Similarly, the rate of water exchange increases with the decrease of cationic charge as:

$$[Al(H_2O_n)]^{3+} \quad < \quad [Mg(H_2O_n)]^{2+} \quad < \quad [Na(H_2O)_n]^{+1}$$

$$+3 \qquad\qquad\qquad +2 \qquad\qquad\qquad +1$$

Hence, we can say that the complexes with the highest oxidation state would be most stable kinetically; while complexes with the lowest oxidation state of the metal center would be least stale kinetically.

2. Radii of central metal ion: It has been also observed that the kinetic stability greatly depends upon the radius of the metal center in a complex. As the radius of metal ion decreases, the lability of its complex decreases. This can be attributed to smaller metal–ligand bond, which in turn, results in stronger attraction between the metal and ligands involved. In other words, the lability of the complex is proportional to the radius of the cation. For example:

$$[Sr(H_2O_6)]^{2+} \quad > \quad [Ca(H_2O_6)]^{2+} \quad > \quad [Mg(H_2O)_6]^{2+}$$

$$1.13 \qquad\qquad\qquad 0.99 \qquad\qquad\qquad 0.65$$

3. Charge to ionic size ratio: It has been observed that for a series of octahedral complexes having the same ligands, the lability of the complexes decreases with the increase of charge to ionic size ratio. For instance, consider the replacement of H_2O^{16} by H_2O^{18} or simply H_2O^*.

$$[M(H_2O)_6]^{n+} + H_2O^* \rightarrow [M(H_2O)_5(H_2O^*)]^{n+} + H_2O \tag{1}$$

The order of lability is:

$$[Na(H_2O)_6]^{+1} \qquad > \qquad [Mg(H_2O_6]^{2+} \qquad > \qquad [Al(H_2O_6]^{3+}$$

This trend can be rationalized in terms of the highest charge to ionic-size ratio for hexaaquo complex of trivalent aluminum (6.0), lowest ionic-size ratio for hexaaquo complex of trivalent sodium (1.05). Similarly,

$$[Sr(H_2O_6]^{2+} \qquad > \qquad [Ca(H_2O_6]^{2+} \qquad > \qquad [Mg(H_2O)_6]^{2+}$$

4. Geometry of the complex: Four-coordinated complexes, tetrahedral as well as square planar, reacts more rapidly than that of six-coordinated complexes. This can be explained in terms of lesser steric repulsion and the availability of more sites to the incoming ligand.

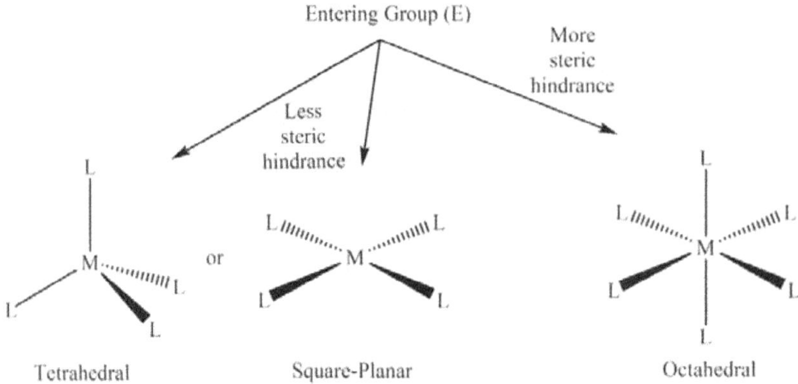

Figure 4. The general diagram comparing the steric and site availability for ligand attack in four-coordinated and six coordinated complexes of non-transition metals.

Now although the lability of transition metal complexes mainly depends upon the gain or loss of CFSE during the formation intermediate (what we have already discussed in this section previously), the geometry also plays some role in the same. In other words, in addition to the lability of non-transition metal complexes, 'Figure 4' may also be used to explain some lability profiles in transition metal complexes. For instance, the rate of exchange of CN^- by $^{14}CN^-$ in $[Ni(CN)_6]^{2-}$ is greater than what is observed for $[Mn(CN)_6]^{3-}$ and $[Co(CN)_6]^{3-}$ complexes. This can be attributed to the easy formation of an activated complex with the incoming ligand, which in turn, facilitates the removal of the previously attached ligand.

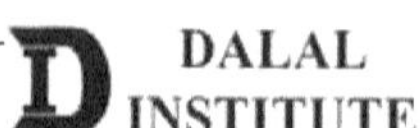

❖ Mechanisms for Ligand Replacement Reactions

The ligand displacement in metal complexes is said to have been taken place if one of the previously attached ligands got replaced by another ligand from its coordination sphere. The scheme can be shown as:

$$MA_nL \quad + \quad E \quad \longrightarrow \quad MA_nE + L \tag{2}$$

Where ligand L is the leaving group present in the complex, E is the entering ligand which is nucleophilic in nature. The coordination number of the complex remains the same.

➢ *Ligand Displacement Mechanism in Octahedral Complexes*

In octahedral complexes, the replacement of the ligand can occur through dissociative, associative or by interchange mechanism. It has also been observed that most of the ligand displacement takes place through the interchange rout rather than purely associative or dissociative.

1. Dissociative or SN_1 Mechanism (D): In this mechanism, first of all, a metal-ligand bond breaks and the coordination number of the complex reduces from six to five forming a penta-coordinated intermediate complex. After that, the entering group attacks this intermediate and the coordination number again gets restored to six giving octahedral geometry. The whole process can be shown as

$$MA_5L \quad \underset{-L}{\overset{slow}{\longrightarrow}} \quad MA_5 \quad \underset{fast}{\overset{+E}{\longrightarrow}} \quad MA_5E \tag{3}$$

The first step is the slow step and hence it is also the rate-determining step for the process. The overall rate is:

$$Rate = k[MA_5L] \tag{4}$$

The reaction is of the first order and is independent of the concentration of the entering ligand. These types of reactions are also called as the unimolecular nucleophilic substitution or SN_1 reactions.

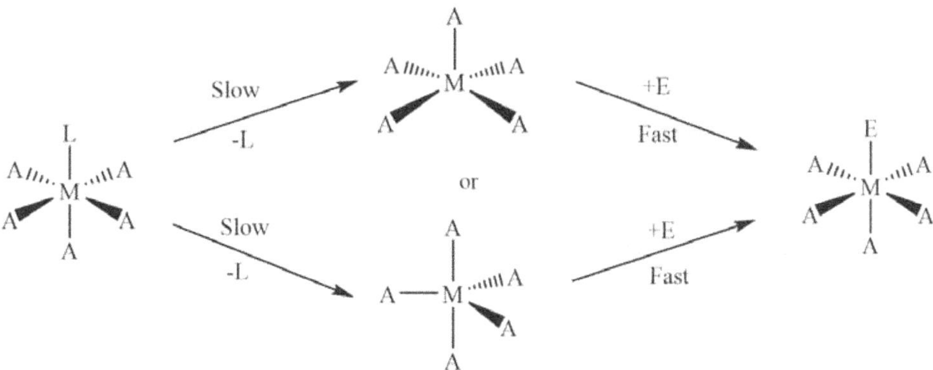

Figure 5. Pathway for ligand displacement reactions in octahedral metal complexes through SN_1.

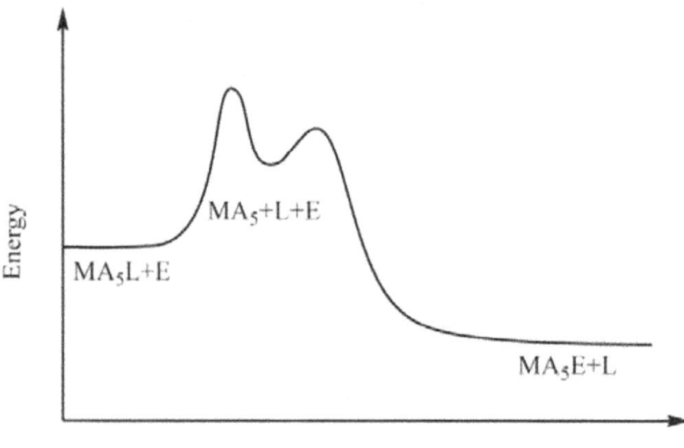

Figure 6. The typical reaction coordinate diagram for ligand displacement reactions in octahedral metal complexes through SN$_1$ mechanism.

Most of the ligand substitution reactions in octahedral complexes occur through dissociative or by interchange dissociative mechanism which in turn can be evidenced by the following rules.

i) The rate of the ligand substitution is almost independent of the concentration of the entering ligand.

ii) The rate of the ligand substitution increases as the steric bulk around the metal center increase.

iii) The entropy of activation, $\Delta S^†$, is positive as there are more species in activated complex than in reactant.

iv) The volume of activation, $\Delta V^†$, for the reaction is also found to be positive.

2. Associative or SN$_2$ Mechanism (A): In this mechanism, firstly the bond making with the entering group takes place, and therefore, the coordination number of the metal complex increases from six to seven forming a hepta-coordinated intermediate complex. After that, the leaving group dissociates itself from the intermediate complex completely and the coordination number of the complex again gets restored to six giving octahedral geometry. The whole process can be shown as

$$\underset{k}{\overset{\text{slow}}{MA_5L + E \longrightarrow}} \underset{-L}{\overset{\text{fast}}{MA_5LE \longrightarrow}} MA_5E \tag{5}$$

The first step is the slow step and hence it is also the rate-determining step for the process. The overall rate is

$$\text{Rate} = k[MA_5L][E] \tag{6}$$

The reaction is of second order and it depends of the concentration of the reactant complex as well as the concentration of the entering ligand. These types of reactions are also called as bimolecular nucleophilic substitution or SN$_2$ reactions.

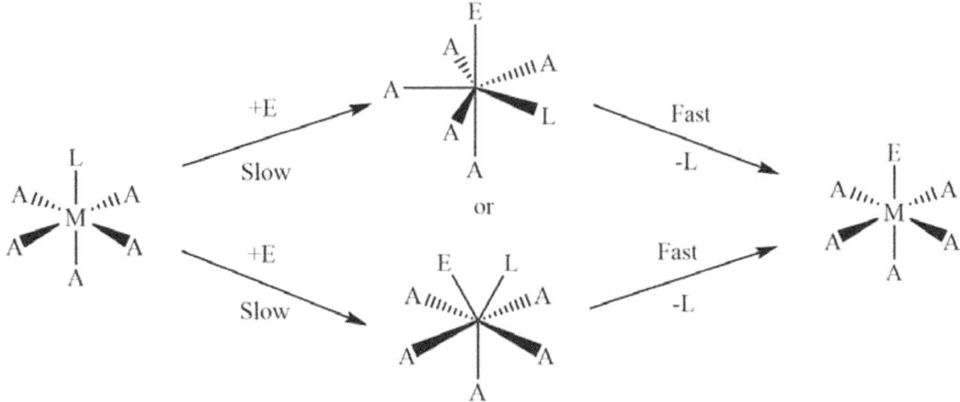

Figure 7. The general reaction mechanism for ligand displacement reactions in octahedral metal complexes through SN$_2$ pathway.

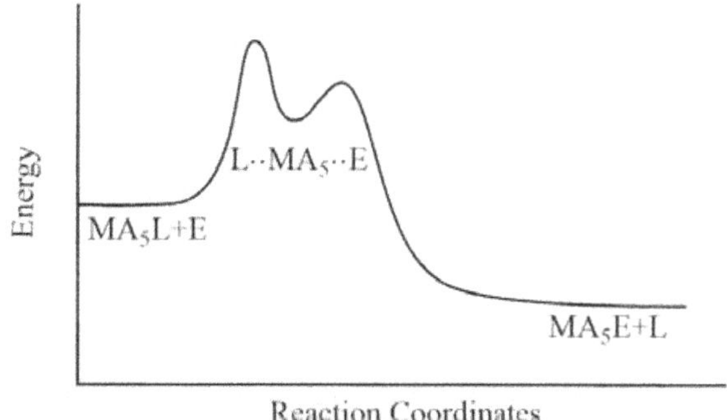

Figure 8. The typical reaction coordinate diagram for ligand displacement reactions in octahedral metal complexes through SN$_2$ mechanism.

i) The rate of the ligand substitution is increased with the concentration of the entering ligand.

ii) The rate of the ligand substitution decreases as the steric bulk around the metal center increase.

iii) The entropy of activation, $\Delta S^\dagger$, is negative as there are lesser number species in activated complex than in reactant.

iv) The volume of activation, $\Delta V^\dagger$, for the reaction is also found to be negative.

3. Interchange mechanism (I): It has been observed that most of the ligand displacement reactions are neither purely associative or dissociative but follow an intermediate mechanism in which bond breaking and bond making takes place simultaneously and no penta-coordinated or hepta-coordinated intermediates have actually been isolated. These types of reactions proceed via a transition state just like in organic SN_2 reactions.

$$MA_5L + E \longrightarrow MA_5LE \longrightarrow MA_5E \tag{7}$$

However, if the rate of the reaction is strongly dependent on the concentration of the entering ligand which indicates that bond making is more important in determining the rate of the reaction than the displacement is said to have been taken place via interchange associative or I_a mechanism. On the other hand, if the rate of the reaction is almost independent of the concentration of the entering ligand which clearly indicates that bond breaking is more important in determining the rate of the reaction than the displacement is said to have been taken place via interchange dissociative or I_d mechanism.

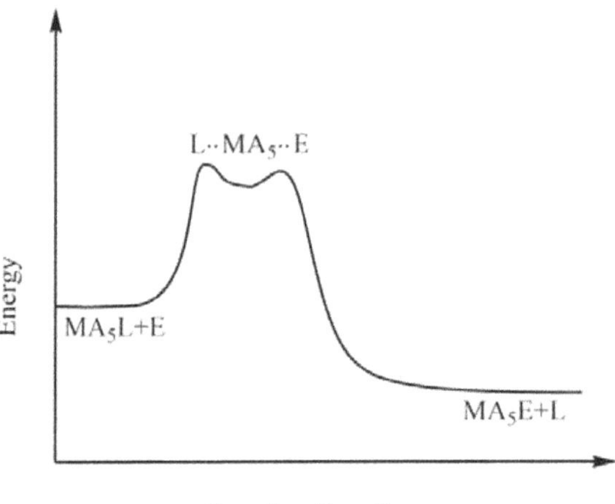

Figure 9. The typical reaction coordinate diagram for ligand displacement reactions in octahedral metal complexes through the interchange mechanism.

> ### Ligand Displacement Mechanism in Square-planar Complexes

In square-planar complexes, the ligand displacement is much more favorable through the associative route than that of dissociative which can be understood in terms of low steric crowding due to lesser coordination number. The general ligand displacement can be written as:

$$MA_3L + E \underset{k}{\overset{slow}{\longrightarrow}} MA_3LE \underset{-L}{\overset{fast}{\longrightarrow}} MA_3E \tag{8}$$

The intermediate state is trigonal-bipyramidal and undergoes rapid Berry-pseudo-rotation followed by the elimination of the leaving group.

Figure 10. The general reaction mechanism for ligand displacement reactions in square-planar complexes.

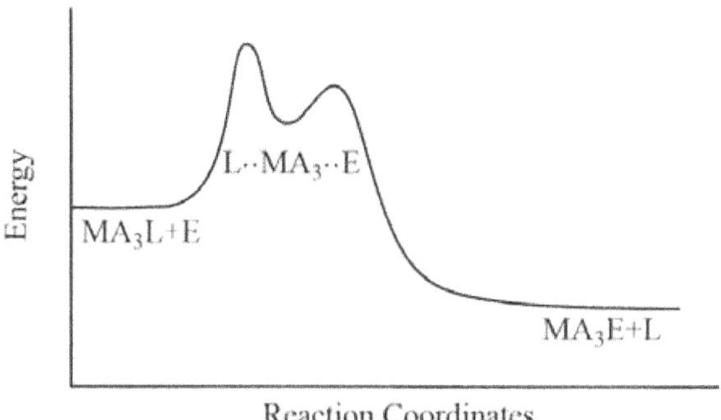

Figure 11. The typical reaction coordinate diagram for ligand displacement reactions in square-planar complexes through the associative mechanism.

A more in-depth visualization of ligand displacement reactions in square-planar complexes is given below in which the involvement of Berry-pseudorotation is depicted more precisely.

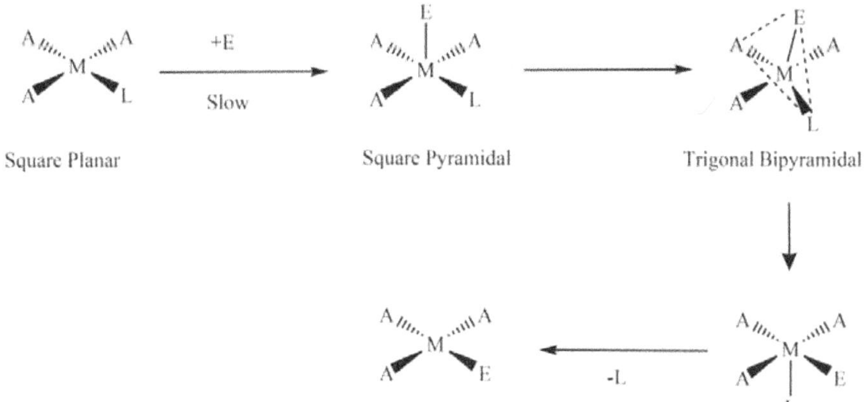

Figure 12. Involvement of Berry-pseudorotation in the ligand displacement in square-planar complexes.

❖ Formation of Complexes from Aquo Ions

The complex formation from the aquo ions yields the assembly containing metal ions with only water as ligands. These complexes are the major components in aqueous solutions of many metal salts, like metal sulphates, perchlorates and nitrates. The formula for metal-aquo complexes $[M(H_2O)_n]^{z+}$ where the value of z generally varies from +2 to +4. The metal-aquo complexes play some very important roles in biological, industrial and environmental aspects of chemistry. Now though the homoleptic aquo complexes (only with H_2O as the ligands attached) are very common, there are many other complexes that are known to have a mix of aquo and other ligand types.

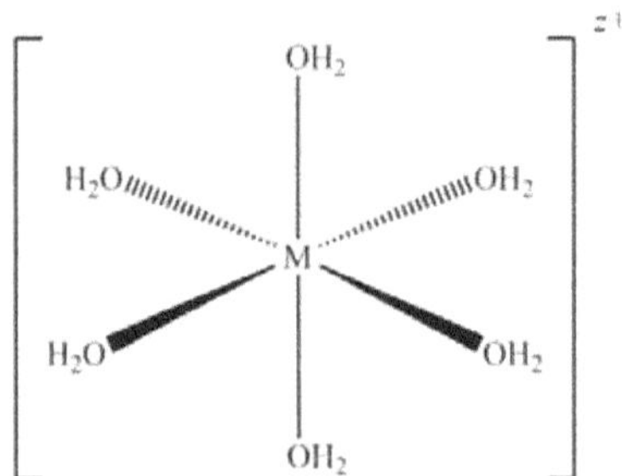

Figure 13. Structure of an octahedral metal aquo complex.

The most common stereochemistry for metal-aquo complexes is octahedral with the formula $[M(H_2O)_6]^{2+}$ and $[M(H_2O)_6]^{3+}$; nevertheless, some square-planar and tetrahedral complexes with the formula $[M(H_2O)_4]^{2+}$ are also known. A general discussion on the different types and other properties is given below.

> ### ➢ *Six-Coordinated Metal-Aquo Complexes*

Most of the transition metal elements from the first transition series and some alkaline earth metals form hexa-coordinated complexes when their corresponding salts are dissolved in water. Some of the most studied hexa-coordinated complexes are given below.

Figure 14. Structure of six-coordinated metal-aquo complexes.

> ### *Four-Coordinated Metal-Aquo Complexes*

The metal-aquo complexes that exist with coordination numbers lower than six are very uncommon but not absent from the domain. For instance, Pd^{2+} and Pt^{2+} form $[M(H_2O)_4]^{2+}$ complexes with the square-planar stoichiometry; and a rare tetrahedral aquo complex $[Ag(H_2O)_4]^+$ is also known. Some of the most studied hexa-coordinated complexes are given below.

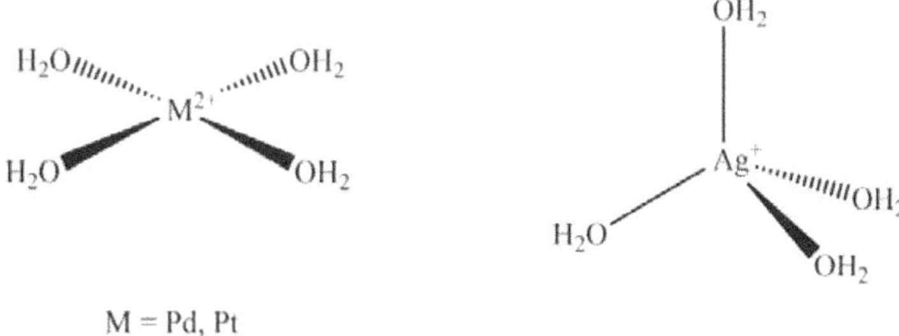

Figure 15. Structure of four-coordinated metal-aquo complexes.

> ### *Eight and Nine Coordinated Metal-Aquo Complexes*

The metal-aquo complexes of the trivalent lanthanides are eight- and nine-coordinated, which is obviously due to the large size of the metal ions. In past, the coordination number of Ln^{3+} ions in their aquo complexes was somewhat more or less controversial; however, nowadays, advanced characterization techniques like O^{17} NMR or density functional studies number of coordinated water molecules decreases from nine to eight with the decrease of the ionic radius i.e La^{3+} to Lu^{3+}. Some of the most studied eight and nine-coordinated complexes are given below.

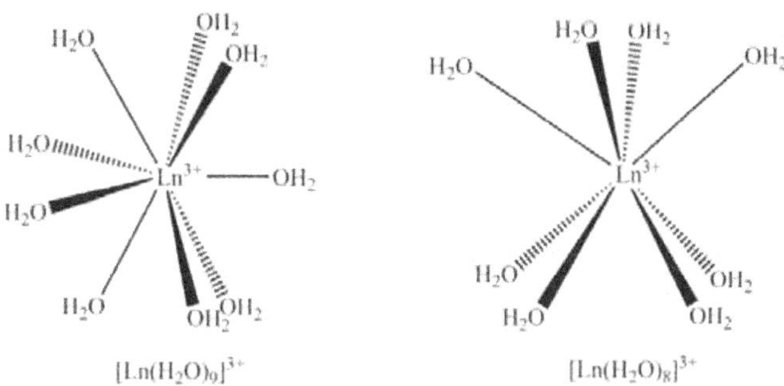

Figure 16. Structure of eight and nine-coordinated metal-aquo complexes.

The $[Ln(H_2O)_9]^{3+}$ ions have a trigonal triprismatic geometry with a slightly distorted D_3 symmetry, while the $[Ln(H_2O)_8]^{3+}$ ions possess a square antiprismatic geometry with a slightly distorted S_8 symmetry.

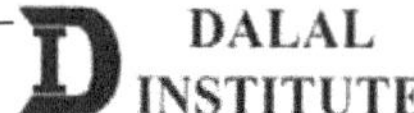

> *Metal-Aquo Complexes with Metal-Metal Bond*

There are some metal-aquo complexes which do possess metal-metal bonds. Two of the most studied examples are $[Mo_2(H_2O)_8]^{4+}$ and $[Rh_2(H_2O)_{10}]^{4+}$ which have eclipsed and staggered conformations, respectively. It should also be noted that metal-metal in $[Mo_2(H_2O)_8]^{4+}$ is of quadruple nature (four bond order).

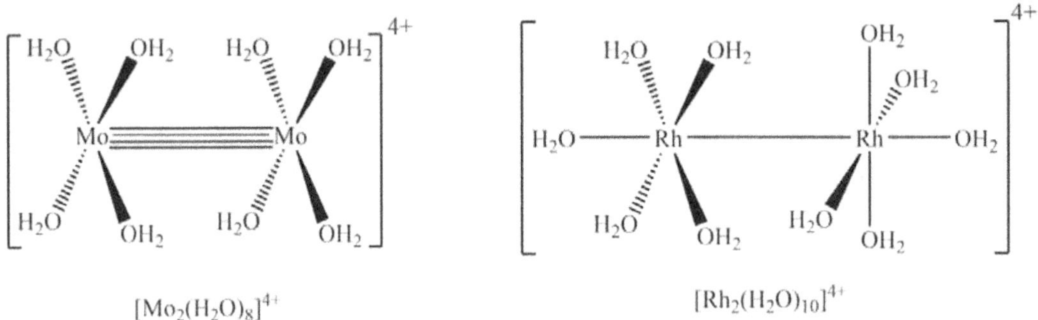

$$[Mo_2(H_2O)_8]^{4+} \qquad\qquad [Rh_2(H_2O)_{10}]^{4+}$$

Figure 17. Structure of some representative metal aquo complex with metal–metal bond.

> *Important Reactions of Metal-Aquo Complexes*

The main reactions shown by metal-aquo complexes are the electron-transfer, ligand exchange, and acid-base reactions of the O-H bonds. A general discussion on all three is given below.

1. Ligand exchange: The process of ligand-exchange means the replacement of a previously coordinated H_2O molecule with one from the solution. However, the detection of such phenomena is possible only by labeling the normal H_2O with H_2O^*. The ligand-exchange generally occurs via a dissociative route, which infers that rate constants must be in accordance to the first-order kinetics. The general chemical equation for the ligand-exchange in metal-aquo complexes can be given as follows:

$$[M(H_2O)_n]^{z+} + H_2O^* \rightarrow [M(H_2O)_{n-1}(H_2O^*)]^{z+} + H_2O \qquad (9)$$

It is also worthy to mention that the above reaction would be of zero free energy change if isotopic labeling is avoided. The rate of ligand-exchange reactions is mainly affected by the charge of the species under consideration. Water-exchange rates are found to be lower for highly charged metal-aquo complexes, while the singly charged species show a much faster rate. For instance, the rates of water exchange in $[Na(H_2O)_6]^+$ is 10^9 to what is observed in case of $[Al(H_2O)_6]^{3+}$. However, the rates of ligand-exchange in $[Al(H_2O)_6]^{3+}$ and $[Ir(H_2O)_6]^{3+}$ also differs by a factor of 10^9 indicating that electronic configuration is also a major element in deciding the reaction kinetics.

2. Electron exchange: The second major types of reaction shown by the metal-aquo complexes are the electron-exchange case of electron transfer reactions. The common electron-exchange or self-exchange involves the interconversion of trivalent and bivalent metal ions via the exchange of one electron only. To track the metal ions, isotopic labeling in the self-exchange process is used and can be written as:

$$[M(H_2O)_6]^{2+} + [M^*(H_2O)_6]^{3+} \rightarrow [M(H_2O)_6]^{3+} + [M^*(H_2O)_6]^{2+} \tag{10}$$

Moreover, the standard redox potentials for different M^{2+}/M^{3+} redox couples are:

Metal ion	V	Cr	Mn	Fe	Co
Standard redox potential (V)	−0.26	−0.41	+1.51	+0.77	+1.82

It can be clearly seen from the redox potential data that the stability of the lower oxidation state increases with the increase of the atomic number. The exceptionally large magnitude of standard reduction potential for Mn^{2+}/Mn^{3+} pair is because octahedral complexes of bivalent manganese have zero crystal field stabilization energy (CFSE) but manganese(III) has −6Dq CFSE. The electron exchange rates depend mainly upon the reorganization energies. In other words, the process of electron-exchange occurs via an outer sphere electron transfer and would be slow if there is a large structural difference between the bivalent and trivalent metal centers. A detailed discussion on this topic is given in the next chapter of this book.

3. Acid-base reactions: The ionizable nature of the protons of coordinated ligand makes these metal-aquo complexes as acidic. Consider the case of aquo-complex of trivalent chromium:

$$[Cr(H_2O)_6]^{3+} \rightleftharpoons [Cr(H_2O)_5(OH)]^{2+} + H^+ \tag{11}$$

Therefore, in comparison to acetic acid (pKa of about 4.8) the Cr(III) aquo complex is a weaker acid (4.3). Moreover, The acidity of these aquo-complexes is also influenced by the electronic configuration; as $[Ru(H_2O)_6]^{3+}$ (pKa = 2.7) is more acidic than $[Rh(H_2O)_6]^{3+}$ (pKa =4), regardless of the fact that trivalent rhodium is expected to be of higher electronegativity. This can be explained in terms of the stabilization of the π-donor hydroxide ligand by the $Ru^{3+}(t_{2g})$. It has also been observed that aquo-complex of bivalent metal ions arc less acidic than those of trivalent cations. The metal-hydroxo complexes so formed usually undergo olation, a condensation phenomenon that results in formation of polymeric species concentrated solutions. The properties of hydrolyzed species (formed by metal-aquo complexes) are pretty much different from the starting hexa-aquo complexes.; for instance, The rate of ligand-exchange in $[Al(H_2O)_6]^{3+}$ is twenty thousand times faster than $[Al(H_2O)_5OH]^{2+}$.

Finally, about one-third of all transition metals like Zr, Hf, Nb, Ta, W, Tc, Re, Os and Au; the metal-aquo complexes are either unknown or rarely explained in detail. Besides, the metal-aquo complexes of tetravalent metal ions (M^{4+}) are expected to be extraordinarily acidic; which in turn, makes their existence highly unfavorable. For instance, the aquo complex $[Ti(H_2O)_6]^{4+}$ is unknown, but $[Ti(H_2O)_6]^{3+}$ is well reported in many papers. The stoichiometry of aquo-complexes Zr^{4+} also controls the acidification as in $[Zr_4(OH)_{12}(H_2O)_{16}]^{8+}$. Similarly, $[VO(H_2O)_5]^{2+}$ is highly stable and well-characterized but $[V(H_2O)_6]^{5+}$ is still unknown. Monovalent metal ions such as Rh^+ or Cu^+ rarely form isolable complexes with H_2O as the ligand.

❖ Ligand Displacement Reactions in Octahedral Complexes- Acid Hydrolysis, Base Hydrolysis

The general scheme for the ligand displacement reactions in octahedral complexes can be shown as:

$$MA_5L \; + \; E \quad \longrightarrow \quad MA_5E \; + \; L \tag{12}$$

Where ligand L is the leaving group present in the complex, E is the entering ligand which is nucleophilic in nature. The coordination number of the complex remains the same. Moreover, if the entering group E is H_2O or OH^- in aqueous solution, the study of ligand displacement become more important due to extremely wide application domain. Some of the most prominent reactions in ligand substitution in six-coordinated complexes are discussed in detail.

➤ Acid Hydrolysis

Acid hydrolysis or aquation reactions may be defined as the reactions in which an aquo complex is formed due to the replacement of a ligand by water molecule.

It has been observed that NH_3, ammines like ethylene diamine or its derivatives coordinated to Co^{3+} are displaced at a very small rate. Hence, displacement of the ligand other than ammonia takes place during the course of acid hydrolysis. Consider the following reaction

$$[Co(NH_3)_5L]^{2+} + H_2O \; \longrightarrow \; [Co(NH_3)_5H_2O]^{3+} + L^- \tag{13}$$

As reaction media is the water itself, H_2O concentration (55.5 M) remains almost constant and the change in water concentration cannot be detected at all. Hence, rate law cannot be used to predict whether the reaction takes place via the associative or dissociative pathway. SN_1 reactions follow first-order kinetics while SN_2 reactions follow second-order kinetics. However, if the complexing agent is in the excess, SN_2 reactions also become pseudo first-order reactions. Hence, it is difficult to tell whether the reaction occurs through the SN_1 or SN_2 mechanism.

Now, it has also been found experimentally that divalent monochloro complexes of Co(III) react at much slower than monovalent dichloro complexes.

$$[Co(NH_3)_4Cl_2]^+ \underset{-Cl}{\overset{slow}{\longrightarrow}} [Co(NH_3)_4Cl]^{2+} \underset{fast}{\overset{+H_2O}{\longrightarrow}} [Co(NH_3)_4(H_2O)(Cl)]^+ \tag{14}$$

and

$$[Co(NH_3)_5Cl]^{2+} \underset{-Cl}{\overset{slow}{\longrightarrow}} [Co(NH_3)_5]^{3+} \underset{fast}{\overset{+H_2O}{\longrightarrow}} [Co(NH_3)_5(H_2O)]^{2+} \tag{15}$$

The reaction (14) is 1000 times faster than reaction (15) suggesting that both of the reactions occur through dissociative or SN_1 pathway. This is because the separation of a negatively charged Cl^- is much more difficult

---

from a complex of high charge density. There are also some other ground evidences which support the dissociative mechanism.

1. Solvation energy of the intermediate: The rate of acid hydrolysis in cis-$[Co(en)_2(NH_3)Cl]^{2+}$ is five times less than in $[Co(NH_3)_5Cl]^{2+}$ which can be explained in term of the lesser solvation energy of the intermediate. Owing to the larger chelate ring in cis-$[Co(en)_2(NH_3)Cl]^{2+}$ the intermediate $[Co(en)_2(NH_3)]^{3+}$ also possesses the larger size and hence less solvation energy and thus by making its formation unfavorable slows down the rate of acid hydrolysis. On the other hand, the smaller sized $[Co(NH_3)_5]^{3+}$ complex has a smaller size and high solvation energy making its formation more favorable. Thus, by comparing the rate of acid hydrolysis in cis-$[Co(en)_2(NH_3)Cl]^{2+}$ and $[Co(NH_3)_5Cl]^{2+}$, we can conclude that chelation stability is somewhat less important than that of the extent of solvation the intermediate undergoes in this case.

2. Steric Hindrance: The rate of acid hydrolysis in cis-$[Co(en)_2(NH_3)Cl]^{2+}$ is smaller in cis-$[Co(pn)_2(NH_3)Cl]^{2+}$. The concept of solvation energy of the intermediate would give just the opposite order as the $[Co(pn)_2(NH_3)]^{3+}$ is larger and has less solvation energy. However, after the dissociation of Cl^-, the gain of steric relief is much greater in cis-$[Co(pn)_2(NH_3)Cl]^{2+}$ due to bulky groups. It has been observed that the rate becomes almost double as the ethylenediamine (en) is replaced by propylene diamine group in cis-$[Co(en)_2(NH_3)Cl]^{2+}$.

3. Effect of the leaving group: The rate of acid hydrolysis is directly proportional to ease of the breaking of the bond between the metal ion and the leaving group. Batter the leaving group faster is the acid hydrolysis rate. Consider the reaction:

$$[Co(NH_3)_5L]^{2+} + H_2O \longrightarrow [Co(NH_3)_5(H_2O)]^{2+} + L^- \tag{16}$$

The rate of aquation is different for different L. For example, consider the acid hydrolysis of the following

$$[Co(NH_3)_5I]^{2+} > [Co(NH_3)_5Cl]^{2+} > [Co(NH_3)_5SCN]^{2+} > [Co(NH_3)_5(NO_2)]^{2+} \tag{17}$$

The nature of the leaving group has a pronounced effect on the aquation rate as the bond breaking is the rate-determining step. The reactivity of the leaving group decreases in the order:

$$HCO_3^- > NO^{3-} > I^- > Br^- > Cl^- > SO_4^{2-} > F^- > CH_3COO^- > SCN^- > NO_2^-$$

>   ***Base Hydrolysis***

    Base hydrolysis reactions may be defined as the reactions in which a hydroxo complex is formed due to the replacement of a ligand by hydroxyl ion.

    Base hydrolysis reactions occur in solutions having pH greater than ten. Consider the following reaction

$$[Co(NH_3)_5Cl]^{2+} + OH^- \longrightarrow [Co(NH_3)_5(OH)]^{2+} + X^- \tag{18}$$

In order to predict whether the reaction takes place via the associative or dissociative pathway, the value of the rate constant and reaction order must be examined very carefully. The possibility of a simple SN_1 mechanism can be ruled out on the basis of an exceptionally fast rate of the reaction at higher pH. If the base hydrolysis

had taken place via simple SN_1 pathway, mono-chloro complexes of Co(III) would not have shown such fast rates as the rate-determining step involves the dissociation on Cl^- from $[Co(NH_3)_5Cl]^{2+}$ is quite slow due to higher charge on the complex. Moreover, at low or moderate concentrations, the rate of the reaction also depends upon the concentration OH^- ions which doesn't go well according to SN_1 pathway. The proposed SN_2 mechanism for base hydrolysis can be given as:

$$[Co(NH_3)_5Cl]^{2+} \quad \underset{slow}{\overset{+OH^-}{\longrightarrow}} \quad [Co(NH_3)_5(OH)(Cl)]^+ \quad \underset{-Cl^-}{\overset{fast}{\longrightarrow}} \quad [Co(NH_3)_5(OH)]^{2+} \tag{19}$$

Hence, the rate law for the reaction should be given by:

$$Rate = k[Complex][OH^-] \tag{20}$$

However, the rate of reaction becomes independent of OH^- at high concentration and the reaction starts to follow first-order kinetics. Furthermore, it has also been observed that the ligands like NCS^-, N^{3-}, NO^{2-} are as strong nucleophile as OH^- and hence are expected to show almost the same rate of hydrolysis of Co(III) amine complexes but these ligands show very slow displacement rate and is independent of the concentration of these ligands. SN_2 pathway could not explain why the rate of hydrolysis of Co(III) amine complexes depends only upon OH^- but not on ligands like NCS^-, N^{3-}, NO^{2-}. Hence, the exact mechanism must be sought elsewhere.

The whole process can successfully be explained via SN_1CB or substitution nucleophilic unimolecular conjugate base mechanism as

$$[Co(NH_3)_5Cl]^{2+} + OH^- \quad \underset{fast}{\overset{K}{\rightleftharpoons}} \quad [Co(NH_3)_4(NH_2)(Cl)]^+ + H_2O \tag{21}$$

$$\quad\quad Acid \quad\quad\quad\quad Base \quad\quad\quad\quad Conjugate\ Base \quad\quad Conjugate\ Acid$$

The equilibrium constant is given by

$$K = \frac{[CB][H_2O]}{[Co(NH_3)_5Cl]^{2+}[OH^-]} \quad or \quad [CB] = \frac{K[Co(NH_3)_5Cl]^{2+}[OH^-]}{H_2O}$$

The conjugate base as obtained is more labile than the original complex $[Co(NH_3)_5Cl]^{2+}$ and hence undergoes SN_1 dissociative pathway by losing Cl^- and gives penta-coordinated intermediate.

$$[Co(NH_3)_4(NH_2)(Cl)]^+ \quad \underset{-Cl}{\overset{slow}{\longrightarrow}} \quad [Co(NH_3)_4(NH_2)]^{2+} \tag{22}$$

The above step is the rate-determining step.

$$[Co(NH_3)_4(NH_2)]^{2+} + H_2O \longrightarrow [Co(NH_3)_5(OH)]^{2+} \tag{23}$$

There are some ground evidences which support dissociative mechanism through the formation conjugate base.

1. Hydrolysis of the complex-ions without acidic protons: The complexes like $[Co(CN)_5Br]^{3-}$, which do not have N–H hydrogens undergo hydrolysis much slowly in basic solutions at a rate which is almost independent the OH^- concentration over a wide range. Hence, the acidic proton is a must.

2. Hydrolysis of $[Co(NH_3)_5Cl]^{2+}$ by nucleophiles with same strength as OH^-: Although the anions like NCS^-, N_3^-, NO_2^- are as strong nucleophile as OH^- yet they do not show fast hydrolysis of $[Co(NH_3)_5Cl]^{2+}$ because these anions are much weaker base than that of OH^- and hence unable to extract the acidic proton from the complex to form conjugate base. Therefore, the hydrolysis of $[Co(NH_3)_5Cl]^{2+}$ by NCS^-, N_3^-, NO_2^- cannot take place via SN_1CB mechanism and takes place possible either via simple SN_1 or SN_2 pathway.

3. Hydrolysis of anionic complexes: Consider the base hydrolysis of anionic complexes like $[Fe(CN)_5(NH_3)]^{3-}$ which have acidic proton in NH_3 but also have a high negative charge. In these types of complexes, the rate of hydrolysis is considerably slow and independent of the OH^- ion concentration which may be attributed to the highly unfavorable formation on the conjugate base.

❖ Racemization of Tris Chelate Complexes

The tris chelate metal complexes exist in two enantiomeric forms, called as Λ and Δ configurations.

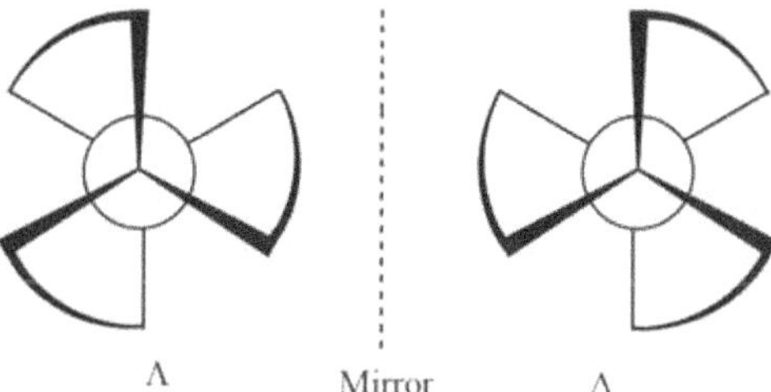

Figure 18. Enantiomeric forms of octahedral tris chelate complexes.

The point group symmetry for the above system D_3. There are also geometrical isomers when the bidentate ligands are unsymmetrical in nature like glycinato. The total number isomer, in that case, is four as:

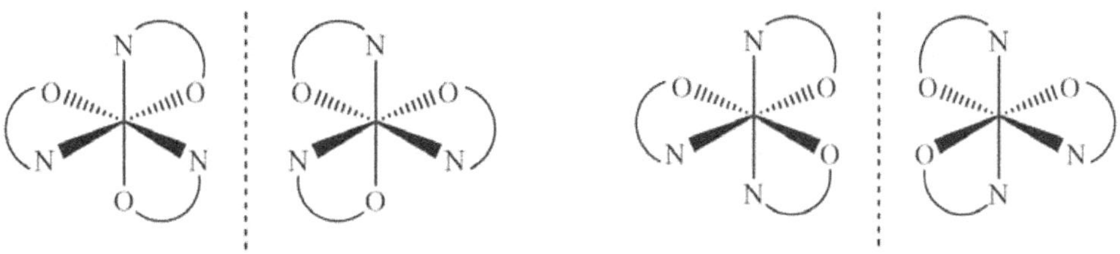

Figure 19. Enantiomeric forms of octahedral tris chelate complexes with the unsymmetrical ligand.

D DALAL
INSTITUTE

The interconversion or the racemization of tris chelate complexes in case symmetrical ligands can take place via with or without the rupturing of the metal-ligand bond.

1. Racemization without the breakage of metal-ligand bond: The two most common processes suggested for the interconversion of the two enantiomeric forms are the trigonal or Bailer twist and the rhombic or Ray-Dutt twist. Both of the processes can be visualized from the following mechanism.

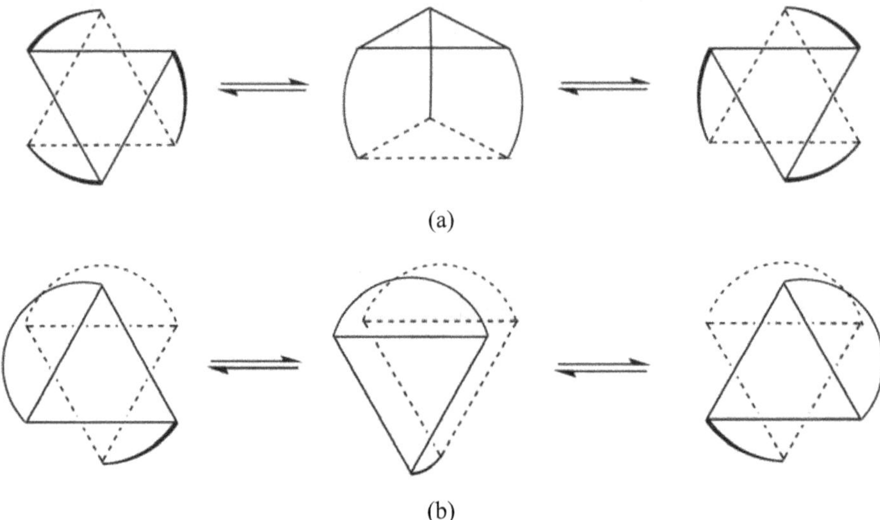

(a)

(b)

Figure 20. Racemization of tris chelate complexes through (a) trigonal or Bailer twist (b) rhombic or Ray-Dutt twist.

The Bailar twist mechanism proposes that the racemization of octahedral metal complexes with three bidentate rings normally occurs via the formation of an intermediate of trigonal prismatic symmetry (D_{3h} point group). In honor of John C. Bailar, Jr., the inventor of the process, the pathway is called as Bailar twist. The second route is called the Ray-Dutt twist, a mechanism proposed also for the racemization of octahedral metal complexes with three bidentate chelate rings. These complexes usually adopt an octahedral geometry in their ground states and are therefore optically active. The Rây-Dutt pathway includes the formation of an intermediate species with C_{2v} point group symmetry. The name Rây-Dutt twist is in the honor of P. C. Râ018y and N. K. Dutt, the inorganic chemists who suggested this mechanism.

2. Racemization with the breakage of metal-ligand bond: There are four different possible pathways suggested for the interconversion of the two enantiomeric forms of tris chelate complexes through the dissociative mechanism. After the detachment of one of the atoms of the bidentate ligand, a five-coordinated intermediate with trigonal-bipyramidal or square-pyramidal geometry is formed which converts into another enantiomer afterward. The whole process can be visualized as follows However, it is also observed that the square-pyramidal intermediate may have a weakly bonded solvent molecule at the sixth site in some cases.

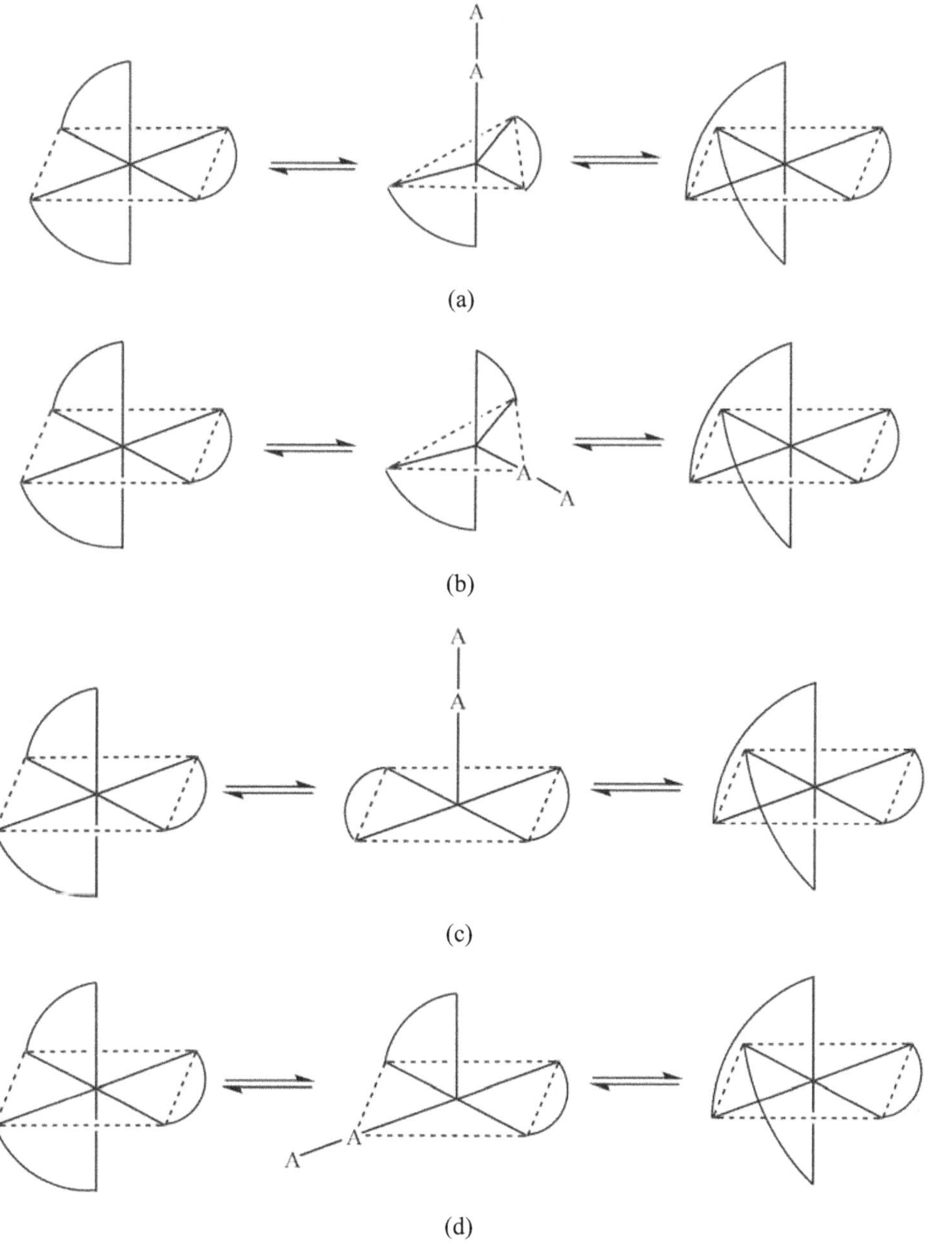

Figure 21. Racemization of tris chelate complexes through ring-opening mechanism via (a,b) trigonal-bipyramidal and (c,d) square-pyramidal intermediate.

❖ Electrophilic Attack on Ligands

There is a special class of ligand substitution reactions in which ligand displacement occurs without the breaking of metal-ligand bonds. Experimental studies suggested that these reactions take place via an electrophilic attack on the ligand already attached to the metal center. Consider the preparation of $[Co(NH_3)_5(H_2O)]^{3+}$ complex from $[(NH_3)_5\text{–}Co\text{–}CO_3]^+$ cationic species:

$$[(NH_3)_5Co - O - CO_2]^+ + (H_3O^{18})^+ \; \rightarrow \; [Co(NH_3)_5(H_2O)]^{3+} + H_2O^{18} + CO_2 \qquad (24)$$

$$\text{Carbonate Complex} \qquad\qquad\qquad \text{Aquo Complex}$$

It has been observed that the no isotopically labeled oxygen is found in the aquo complex which indicates that no metal-ligand bond-breaking has actually been taken place during the course of the whole substitution process.

Mechanism: The whole process can be summed up into the following steps

(1) Electrophilic or the proton attack on the oxygen atom of the ligand bonded to the trivalent cobalt.

(2) Expulsion of CO_2 and H_2O to form hydroxo complex.

(3) Protonation of the hydroxo complex.

It is also worthy to mention that the reaction described above is a decarboxylation rather than the acid hydrolysis.

The similar behavior can be observed in the reaction of NO_2^- group with pentaammineaquocobalt(III) ion. When the studies involving isotopic labelling were carried out, we came to know that show that the oxygen

of bound water is actually the same to what is used by NO_2^- to bind with metal centre. This unusual result can be explained by the following mechanism.

$$2[(NH_3)_5Co(H_2O^*)]^{3+} + 2NO_2^- \rightleftharpoons 2[(NH_3)_5Co-O^*H]^{2+} + N_2O_3 + H_2O$$

$$[(NH_3)_5Co-O^*H]^{2+} + N_2O_3 \longrightarrow \left[\begin{array}{c} (NH_3)_5Co-O^*\cdots H \\ \vdots \quad\quad \vdots \\ ON\cdots ONO \end{array} \right]^{2+}$$

Transition State

$\Big\downarrow$ Fast

$$[(NH_3)_5Co-(NOO^*)]^{2+} \xleftarrow{\text{Slow}} [(NH_3)_5Co-O^*NO]^{2+} + HNO_2$$

Hence, in both of the reactions given above, the displacement of previously attached ligand octahedral complexes occurs without metal-ligand bond breaking.

❖ Problems

Q 1. How does the ligand field affect the choice between SN1 and SN2 paths in the substitution reactions of octahedral complexes? Also, explain how is this choice influenced by the basicity and π bonding capacity of a non-reacting ligand?

Q 2. Discuss the mechanism of acid hydrolysis taking the example of the octahedral complex of Co(III).

Q.3 What do you understand by SN_1CB? Explain with example.

Q 4. Explain the stereochemistry of SN_2 substitution reactions of octahedral complexes.

Q 5. What is base hydrolysis? Discuss the possible mechanisms.

Q 6. How does the racemization of tris chelate complexes take place?

Q 7. Explain the mechanism of nucleophilic substitution reactions in octahedral complexes.

Q 8. What are metal-aquo complexes? Also, draw and discuss the structure of eight and nine coordinated aquo complexes of trivalent lanthanide ions.

Q 9. Give a brief discussion on important reactions of metal-aquo complexes.

❖ Bibliography

[1] J. D. Lee, *Concise Inorganic Chemistry*, Chapman & Hall, New York, USA, 1994.

[2] B. R. Puri, L. R. Sharma, K. C. Kalia, *Principals of Inorganic Chemistry*, Milestone Publishers, Delhi, India, 2012.

[3] J. E. Huheey, E. A. Keiter, R. L. Keiter, *Inorganic Chemistry: Principals of Structure and Reactivity*, HarperCollins College Publishers, New York, USA, 1993.

[4] A. F. Wells, *Structural Inorganic Chemistry*, Oxford University Press, London, UK, 1975.

[5] F. A. Cotton, G. Wilkinson, C. A. Murillo, M. Bochmann, *Advanced Inorganic Chemistry*, John Wiley & Sons, New Jersey, USA, 1999.

[6] S. F. A. Kettle, *Physical Inorganic Chemistry: A Coordination Chemistry Approach*, Springer-Verlag, Berlin, Germany, 1996.

[7] K. Djanashvili, C. P. Iglesiasb, J. A. Peters, *The Structure of the Lanthanide Aquo Ions in Solution as Studied by ^{17}O NMR Spectroscopy and DFT Calculations*, Dalton Trans., 2008, 602.

[8] J. A. McCleverty, T.J. Meyer, Comprehensive Coordination Chemistry II: From Biology to Nanotechnology, Newnes, Oxford, UK, 2003.

[9] A. A. Holder, *Molybdenum: Its Biological and Coordination Chemistry and Industrial Applications*, Nova Science Publishers, New York, USA, 2013.

CHAPTER 4

Reaction Mechanism of Transition Metal Complexes – II:

❖ Mechanism of Ligand Displacement Reactions in Square Planar Complexes

The ligand displacement in square-planar complexes is much more favorable via the associative pathway than that of the dissociative one which can be explained in terms of low steric crowding due to the lesser coordination number. In order to understand this claim, consider the general ligand displacement in square planar complexes:

$$MA_3L + E \;\longrightarrow\; MA_3E + L \tag{1}$$

Now, if the concentration of the entering ligand E is very large, the observed pseudo first-order rate law for the above reaction can be given by:

$$Rate = k_1[MA_3L] + k_2[MA_3L][E] \tag{2}$$

On rearrangement

$$Rate = (k_1 + k_2[E])\,[MA_3L] \tag{3}$$

or

$$Rate = k_o[MA_3L] \tag{4}$$

Where k_o is the observed rate constant and is equal to

$$k_o = k_1 + k_2[E] \tag{5}$$

The value of k_1 and k_2 can be calculated from the intercept and slope of the plot of the observed rate constant vs concentration of the entering group E. It has been observed that, for all type of entering ligands, the values of both k_1 and k_2 are nonzero which suggests a possibility of dissociative pathway too.

> ### ➤ *Solvent Assisted SN₂ Pathway*

The non-zero value of k_1 can also be interpreted in some other form of the associative pathway in which the solvent molecules also act as the nucleophile and compete with E to form MA_3S. The following process is responsible for the first term in equation (2).

$$MA_3L + S \;\underset{-L}{\overset{k_1}{\longrightarrow}}\; MA_3S \;\underset{-S}{\overset{+E}{\longrightarrow}}\; MA_3E \tag{6}$$

Now, as the concentration of the solvent is practically constant, the rate of the reaction depends upon the concentration of MA_3L only and becomes of first-order kinetics.

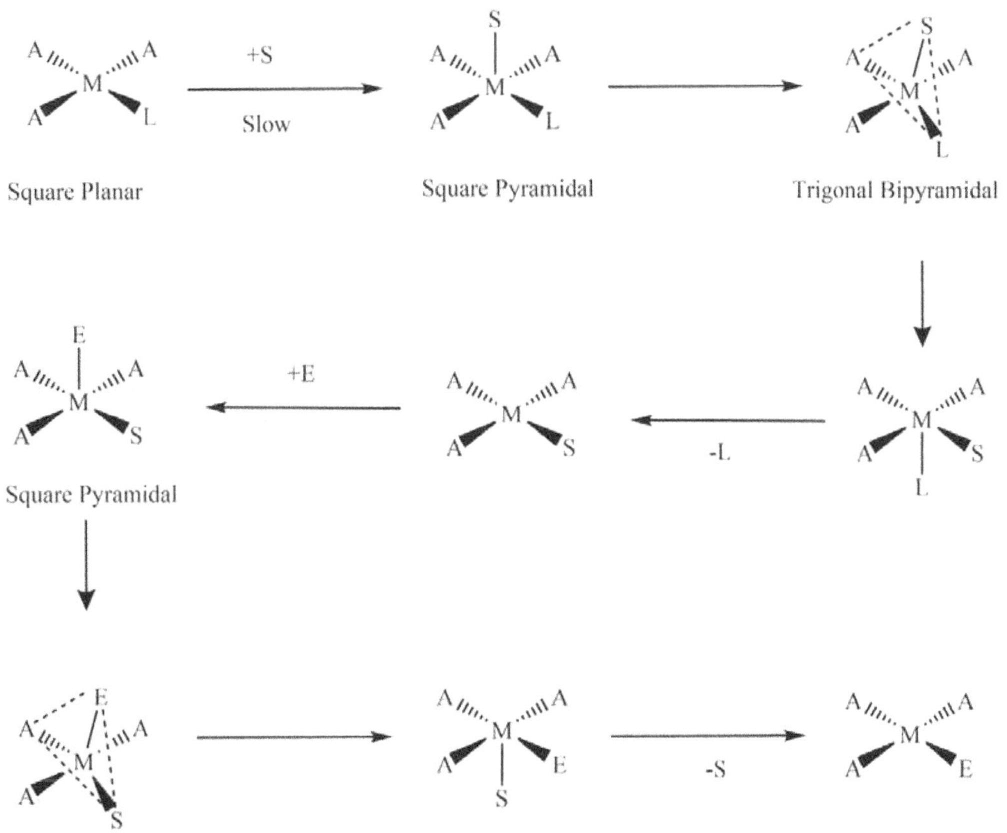

Figure 1. The systematic mechanism of solvent-assisted ligand displacement reactions in square planar complexes.

> ### *Normal SN₂ Pathway*

The non-zero value of k_2 can be explained in terms of the normal associative route in which the entering ligand replaces the leaving group via a five-membered transition state. The reaction scheme can be shown as:

$$MA_3L + E \xrightarrow[-L]{k_2} MA_3E \tag{7}$$

It has also been observed that the rate of the direct associative route is generally higher than the rate via solvent assistance. Moreover, the rate has also been found dependent on the nucleophilicity of the solvent suggesting a key role of the solvent attack in ligand displacement.

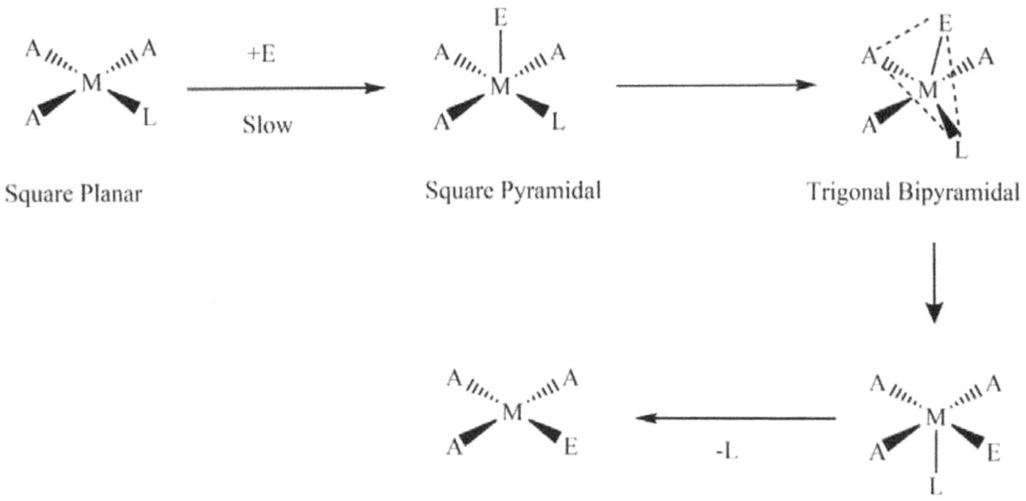

Figure 2. Mechanism of ligand displacement in square planar complexes via normal associative pathway.

❖ The Trans Effect

The tendency of an already attached group to direct the incoming ligand to its trans position in ligand displacement reactions in square-planar complexes is called as the trans effect and such groups are labeled as the trans-directing ligands.

It is worthy to mention that various ligands have different trans directing effects and when these ligands are arranged in increasing order of their trans effect, the order is termed as the trans-effect series which is given below.

$$CN^- > C_2H_4 > CO > NO > SCN^- > I^- > Br^- > Cl^- > Py > NH_3 > OH^- > H_2O$$

In other words, the trans-effect may also be defined as the labilization (the easier displacement) of the ligands trans to other trans-directing ligands. Moreover, the trans-directing ligands are also called as the spectator ligands as they are neither the entering nor the leaving group yet affect the rate of the ligand substitution considerably. Most of the people are actually quite confused about the nature of the trans-effect, weather is it kinetic or thermodynamic. The trans-effect may be classified into two types.

➤ *Kinetic Trans Effect or Trans Effect*

The kinetic trans-effect or trans-effect proper in square-planar complexes is the phenomenon in which certain ligands increase the rate of displacement of the ligands positioned trans to them. Now as this effect deals with the rate only, it must not be confused with trans-influence which involves a weakening of the trans bond and hence is thermodynamic in nature. If E is the entering ligand, T_1 and T_2 as the trans-directing groups, k_1 and k_2 as the rate constants and L is the leaving group; the process can be depicted as:

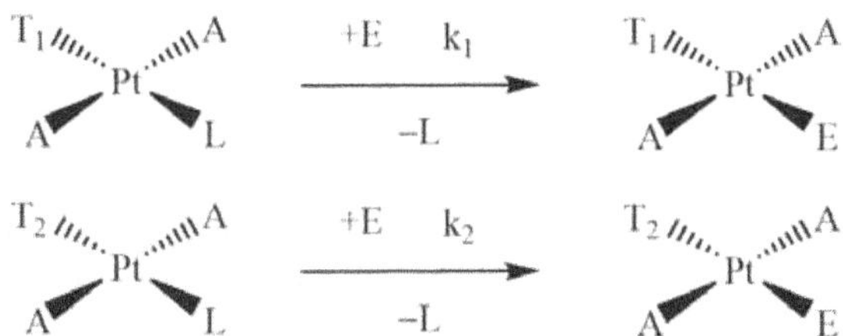

Figure 3. Kinetic trans-effect in action with $k_2 \gg k_1$ showing that T_2 ligand is having greater trans-effect strength than T_1.

> ### Thermodynamic Trans Effect or Trans Influence

The thermodynamic trans-effect or trans influence may be defined as the impact of a ligand on the length of the bond trans to it in the ground state of a square-planar complex. Like trans-effect, trans-influence is also independent of metal ion but depends primarily upon the geometry of the metal center.

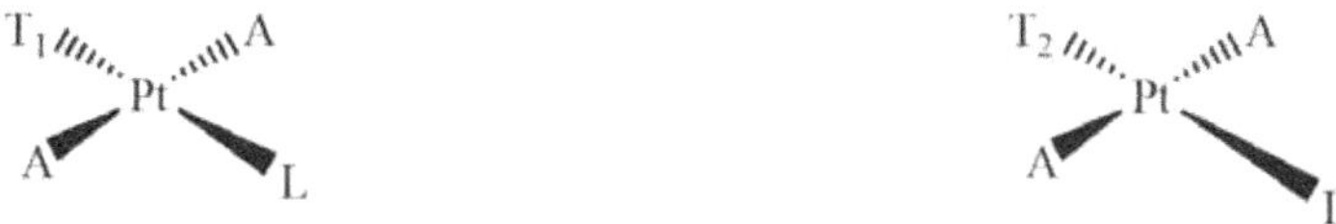

Figure 4. Thermodynamic trans-effect in action showing that T_2 ligand is having greater trans-influence strength than T_1.

It can be seen very clearly that elongation of trans-bond-length is done more profoundly by T_2 than T_1; which infer that the thermodynamic trans-effect of T_2 ligand is higher than T_1.

> ### Applications of the trans effect

1. Synthesis of the cis and trans platin: $[PtCl_2(NH_3)_2]$ exists in two isomeric forms as given below.

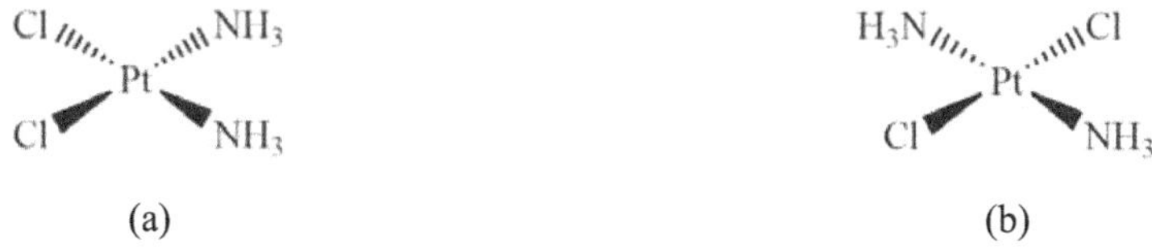

(a)   (b)

Figure 5. (a) cis-$[PtCl_2(NH_3)_2]$ (b) trans-$[PtCl_2(NH_3)_2]$

These two isomers can be synthesized via a number of number chemical routes by exploiting the trans effect. However, the most common are given below.

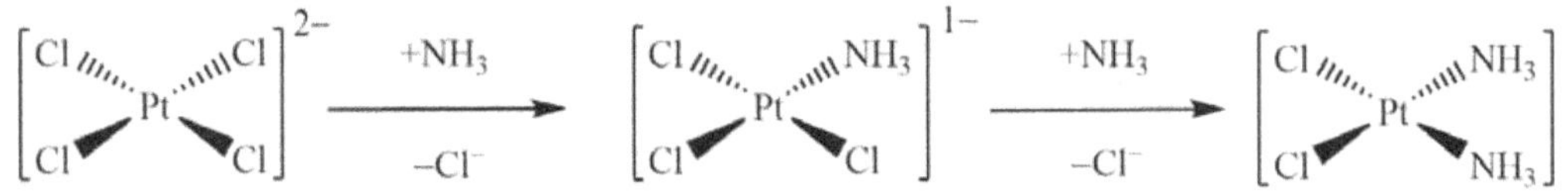

(a)

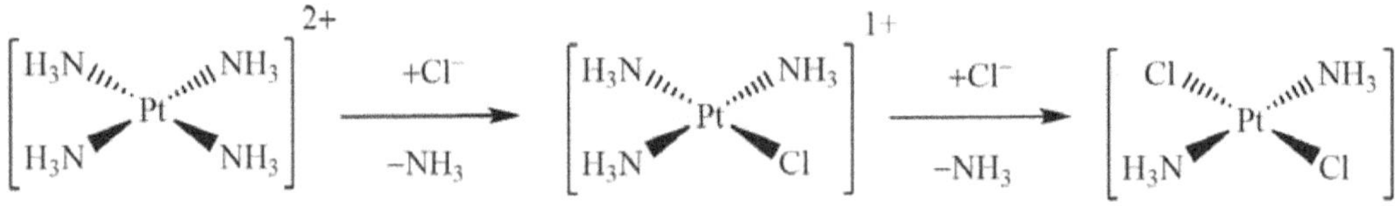

(b)

Figure 6. The general synthesis route for the preparation of cis-[PtCl$_2$(NH$_3$)$_2$] or cis-platin and Trans-[PtCl$_2$(NH$_3$)$_2$] or trans-platin complexes.

The initial raw materials for the synthesis of isomeric forms are different as the Cl$^-$ has a higher trans effect than NH$_3$. Hence, the trans effect plays a major role in product formation. It is worth noting that the cis-platin is having very important therapeutic uses in the treatment of cancer.

2. Synthesis of the isomers of [Pt(Cl)(NH$_3$)(Br)(Py)]: The [Pt(Cl)(NH$_3$)(Br)(Py)] complex can exist in three different isomeric forms which can successfully be synthesized by using the trans-effect order of the participating ligands.

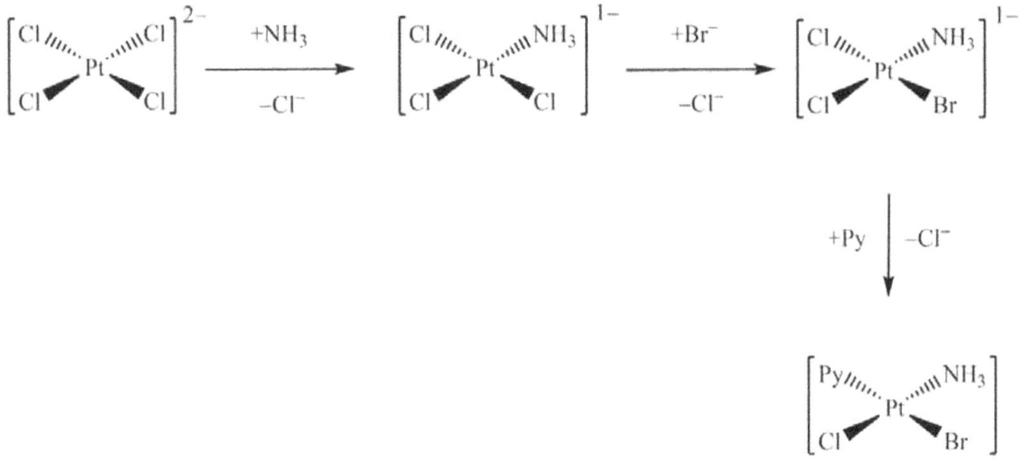

Figure 7. Continued on the next page...

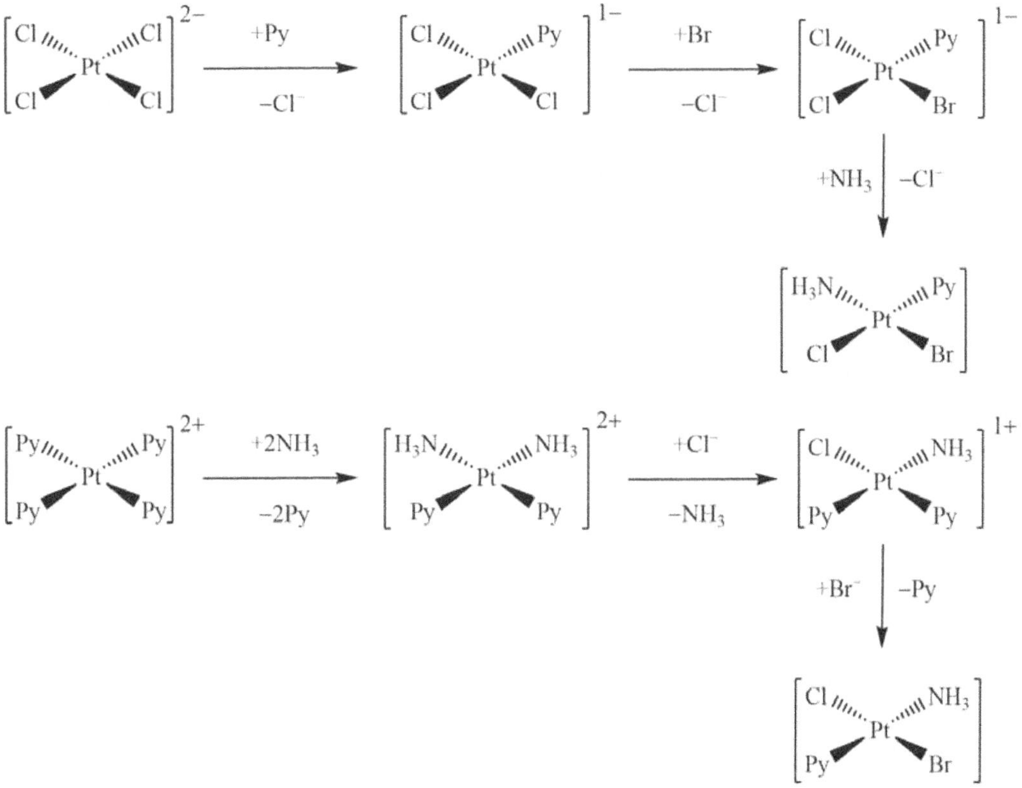

Figure 7. Synthesis route for [Pt(Cl)(NH₃)(Br)(Py)] complexes.

3. Differentiating between cis and trans isomers of [PtCl₂(NH₃)₂]: A Russian scientist Kurnakov has used the trans effect to distinguish between the cis and trans platin. The experimental route he invented is named after him as the Kurnakov test in which thiourea is used as a primary component. The trans effect of thiourea is greater than chloride and amine ligands. The addition of thiourea to the trans platin results in the replacement of trans chloride ions and the reaction stops. However, when the thiourea is added to the cis isomer, all the four ligands are displaced and [Pt(tu)₄]²⁺ is formed.

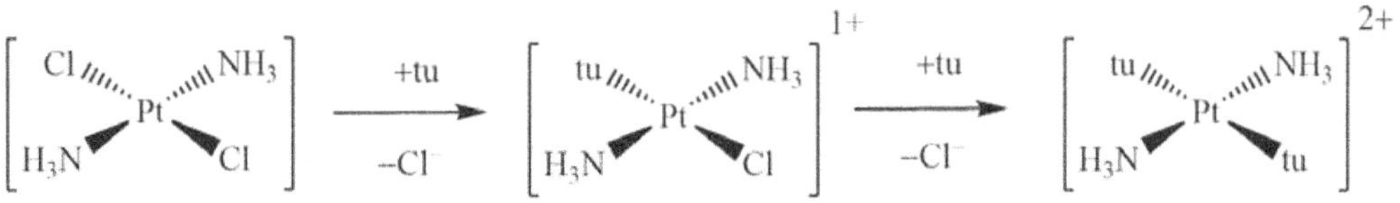

Figure 8. Continued on the next page…

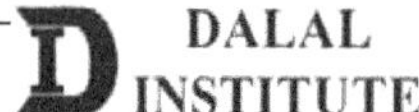

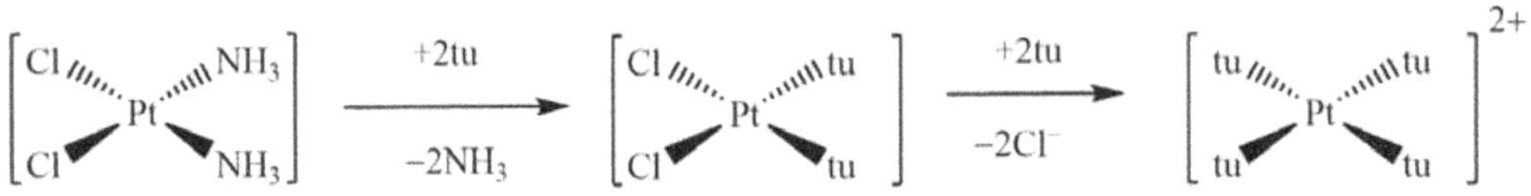

Figure 8. Differentiating route for cis and trans-platin complexes.

4. Synthesis of the isomers of $[Pt(NH_3)(NO_2)Cl_2]^{1-}$: The trans-effect order of the three groups in $[Pt(NH_3)(NO_2)Cl_2]^{1-}$ complex is $NO_2^- > Cl^- > NH_3$.

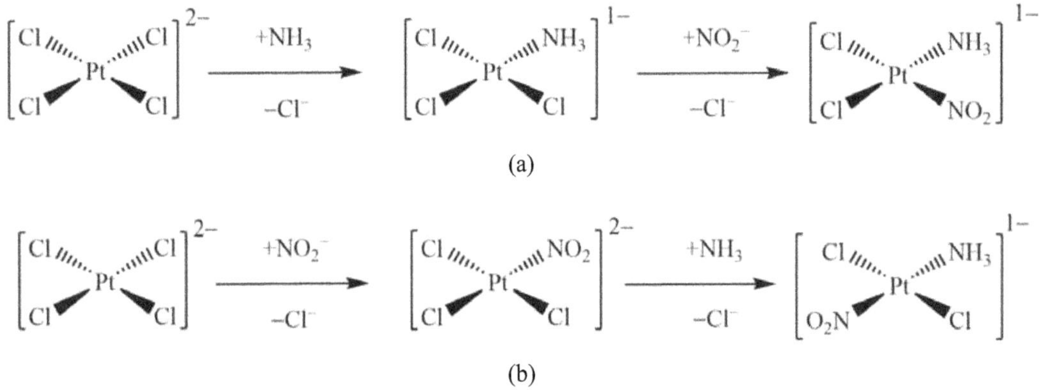

(a)

(b)

Figure 9. Synthesis route for (a) cis-$[Pt(NH_3)(NO_2)Cl_2]^{1-}$ and trans-$[Pt(NH_3)(NO_2)Cl_2]^{1-}$ complexes.

5. Synthesis of the isomers of $[Pt(C_2H_4)(NH_3)Cl_2]$: The trans-effect order of the three groups in $[Pt(C_2H_4)(NH_3)Cl_2]^{1-}$ complex is $C_2H_4 > Cl^- > NH_3$.

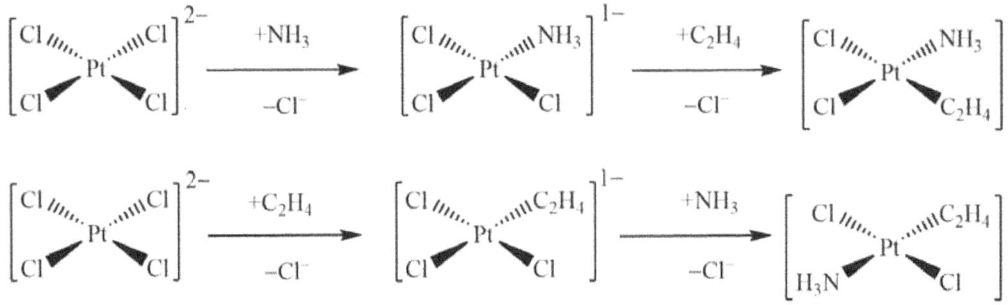

Figure 10. Synthesis route for (a) cis-$[Pt(C_2H_4)(NH_3)Cl_2]^{1-}$ and trans-$[Pt(C_2H_4)(NH_3)Cl_2]^{1-}$ complexes.

6. Synthesis of the isomers of [Pt(PR$_3$)$_2$Cl$_2$]: The trans-effect order of the two groups in [Pt(PR$_3$)$_2$Cl$_2$] complex is PR$_3$ > Cl$^-$.

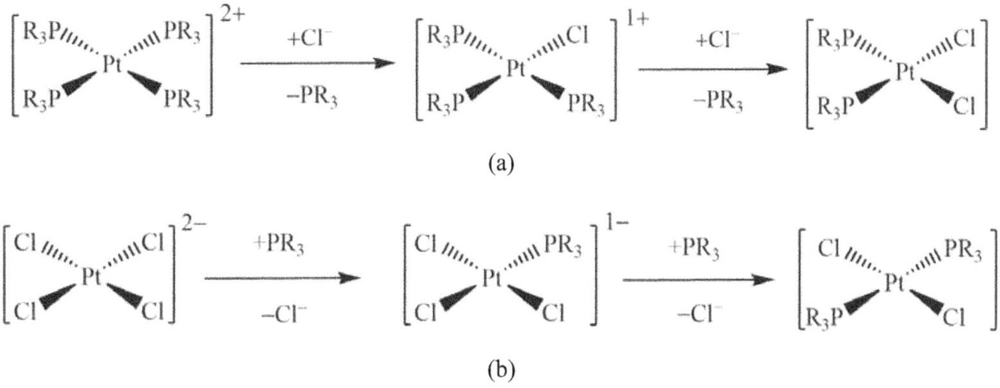

(a)

(b)

Figure 11. Synthesis route for (a) cis-[Pt(PR$_3$)$_2$Cl$_2$] and trans-[Pt(PR$_3$)$_2$Cl$_2$] complexes.

❖ Theories of Trans Effect

Owing to the kinetic nature of the trans-effect, it is quite sensible to think of activation energy in terms of the ground state (the complex before substitution) as well as activated complex. Hence, all the factors which can affect the energy of the ground state or activated complex are expected to have a profound effect on the activation energy of the ligand displacement reaction and thus expected to govern the trans effect of the attached groups. The two theories which are sought for the explanation of trans-effect are given below.

➢ *The Polarization Theory*

This theory mainly deals with the ground state of the complex and proposes that the metal center has a tendency to induce a dipole moment in the surrounding ligands by polarizing them according to Fajans' rule. In the case of MA$_4$ type complexes, metal ion induces an equal dipole moment in all the four surrounding ligands which are, in turn, cancel out each other due to square-planar geometry. However, in the case of MA$_3$B type complexes, the situation is quite different as the polarizability of all the four ligands is not anymore, the same. If the polarizability of B type ligand is higher than that of A-type then the primary charge of the metal ion will polarize the electronic cloud of A more effectively and thus will induce a strong dipole moment in A-type ligand. Furthermore, this dipole moment is also bound to induce an alternate dipole in the metal center also. The orientation of this dipole is such that it repels the negative charge on the ligand, A-type, situated trans to B. This results in the weakening and consequently lengthening of the metal-ligand bond trans to the B-type group. Therefore, according to this concept, the trans effect is directly proportional to the polarizability of the ligand. It is also worth noting that the trans-effect is more prominent with a large and more polarizable metal center. The general order can be given as Pt(II) > Pd(II) > Ni(II).

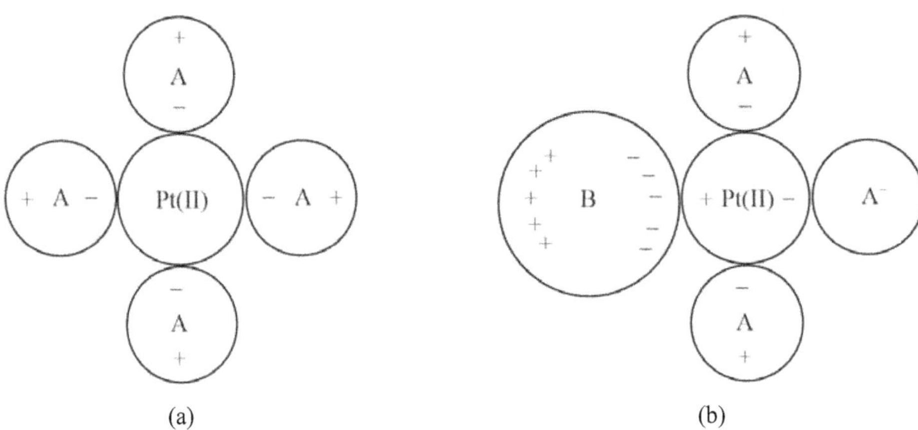

Figure 12. The effect of polarization in (a) MA₄ and (b) MA₃B complexes.

> ### The π-Bonding Theory

The nature of the trans-effect is electronic instead of steric which clearly means that the electronic profile of the ligand is primarily governing trans-effect strength of various groups. Considering the trans-effect order of halide ions, $I^- > Br^- > Cl^- > F^-$, it seems that the increasing electronegativity makes them poor σ-donor or σ-base which in turn also decreases their trans-effect strength. However, the exceptionally high trans-effect of the ligands like CO, C_2H_4 or PR_3 cannot be explained by σ-donation ability as they are not very good σ-donor but π-acceptor in nature. Therefore, we can conclude that a stronger trans-effect is the combination of both, either it should be a strong σ-base or it should show a good π-acid character.

Now, in order to understand the whole process by which the trans-directing groups speed-up the ligand displacement in square-planar complexes, we will have to recall the mechanism of ligand substitution. Let T be the trans-directing group, L as the leaving group and E as the entering group. The entering ligand binds to 16-electron Pt(II) complex to form 18-electron complex which in turn again converted into a new 16-electron complex as:

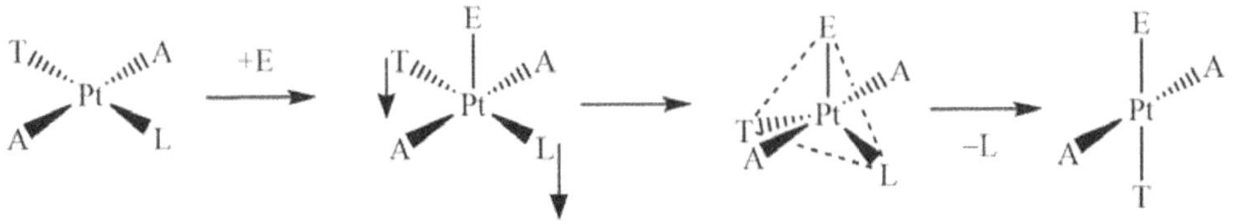

Figure 13. The mechanism of associative ligand substitution of Pt(II) complexes.

The characteristic features of this mechanism are:

i) The entering ligand always sits at the equatorial position of the activated complex which is trigonal-bipyramidal in nature.

ii) The trans-directing group and the leaving group are pushed down to create the equatorial plane of the activated complex.

iii) As the entering group E attacks at the equatorial site, the leaving must also be from the equatorial plane; which is followed directly from the principle of microscopic reversibility.

The rate-determining step is when the entering ligand "E" pushes down the trans-director T and the leaving group L. Now as the equatorial sites of the trigonal-bipyramidal intermediate are richer in electron density than the axial ones, ligands with greater π-acidity like to be pushed down to get this privilege of stronger back-bonding. This forces the leaving group "L" to no other choice but to detach from the activated complex.

The transition state is stabilized by the overlap of empty π^* orbital of "T" and the filled d_{xz} orbital of the metal center. This results in a decrease in the electron density in metal-leaving group bond which makes the displacement of L by E much easier.

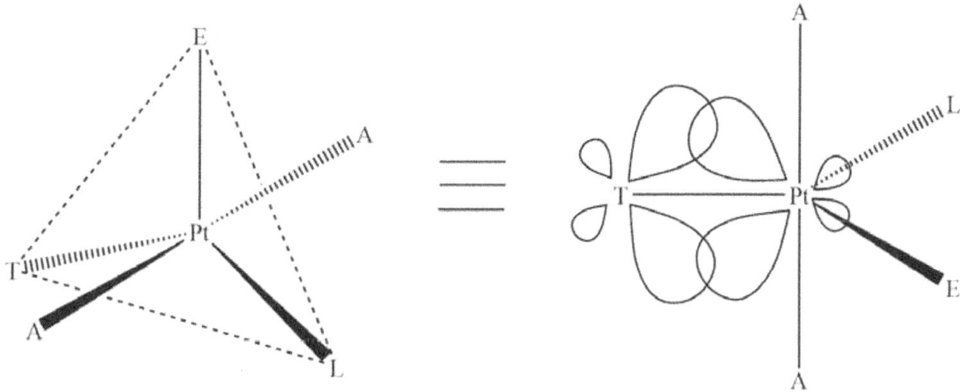

Figure 14. The transition state of Pt(II) complexes is stabilized by the overlap between empty π^* orbital of the trans-directing group and filled d_{xz} orbital of Pt(II) metal center.

Thus, the stronger trans-effect of π-acid ligands is explained in terms of the stabilization of the activated complex (which in turn decreases the activation energy). However, it has also been observed that the bond length between metal ion the leaving group is generally larger than the expected values when ligands with strong trans-effects are there. This suggests that these π-acid ligands affect the ground state as well and a combination on both contributes to the over trans-effect strength.

In addition to the polarization theory, the trans-effect strength of σ-donor ligands can also be explained in the activated complex framework. It can be assumed that trans-directing group T and leaving group L are in direct competition for the σ-donation to the same metallic d-orbital. Thus, the greater σ-donation ability of the trans-director weakens the opposite bond. Though the effect is thermodynamic in nature (trans-influence), the destabilization of the ground state decreases the value of activation required for the ligand substitution making it a kinetic phenomenon too. In other words, the strong σ-donors will decrease the height of the activation barrier and consequently increase the rate of ligand displacement reactions in transition metal complexes with square-planar geometry.

❖ Mechanism of Electron Transfer Reactions – Types; Outer Sphere Electron Transfer Mechanism and Inner Sphere Electron Transfer Mechanism

The process of electron transfer from one species to another species leads to the oxidation of the donor and the reduction of the acceptor. The electron donor acts as the reducing agent and called as reductant while the electron acceptor acts as the oxidizing agent and called as the oxidant. We have already studied the redox reaction of simple species like:

$$Na + Cl \longrightarrow Na^+ + Cl^- \tag{8}$$

However, the ligand displacement reactions, we have studied so far, do not involve any change in the oxidation state of the metal center but the substitution reactions involving the electron transfer between complex species do exist. The oxidation-reduction reaction may or may not occur through the net chemical change. A simple example of electron transfer between complex species can be given as:

$$[Fe(H_2O)_6]^{2+} + [Fe^*(H_2O)_6]^{3+} \longrightarrow [Fe(H_2O)_6]^{3+} + [Fe^*(H_2O)_6]^{2+} \tag{9}$$

The mechanism by which the electron transfer occurs in transition metal complexes can be classified into two types as given below.

➤ *Outer Sphere Electron Transfer Mechanism*

In this mechanism, there is a direct transfer of electrons from the reductant to oxidant and the coordination sphere remains intact. The ligands in both the reactants remain as such and the bond making or bond-breaking does not take place. In other words, the complexes do not undergo ligand substitution and no new bonds are formed or broken. Consider the following example.

$$[Cr(H_2O)_6]^{2+} + [Co(H_2O)_6]^{3+} \longrightarrow [Cr(H_2O)_6]^{3+} + [Co(H_2O)_6]^{2+} \tag{10}$$

If R is the reductant and O is the oxidant, the outer sphere electron transfer can be shown as:

$$\left[R+O\right] \longrightarrow \left[R\text{----}O\right] \longrightarrow \left[R\text{----}O\right]^* \longrightarrow \left[R^+\text{----}O^-\right] \longrightarrow \left[R^+ +O^-\right]$$

I	II	III	IV	V
Reactants	Precursor Complex	Transition State	Successor Complex	Products

It has been observed that the formation of precursor complex and the dissociation of the successor complex is very fast but the electron transfer is quite slow. The course of the complete reaction is shown below.

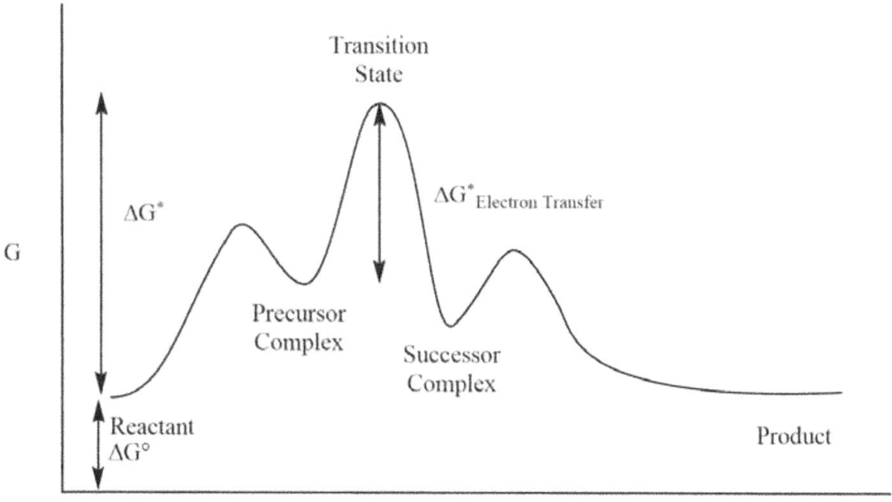

Figure 15. The general reaction coordinate diagram for the mechanism involved outer-sphere electron transfer reactions.

In the transition state of the complex, the electrons have equal probability to be present on either of the metal center. In order to proceed further, we need to recall the concept of inert and labile complexes first.

Inert Complexes	Labile Complexes
d^3, d^4 LS, d^5 LS, d^6 LS, d^8	d^0, d^1, d^2, d^4 HS, d^5 HS, d^7, d^6 HS, d^9, d^{10}

It is worth remembering that low-spin complexes are generally formed by the strong field ligands like CO, PR_3, CN^- or by the transition metal ions from 2nd and 3rd transition series. Co(III) complexes are also generally low spin in nature. On the other hand, high-spin complexes are generally formed by the weak field ligands like H_2O, NH_3, Cl^- or by the transition metal ions of 1st transition series. Co(II) complexes are also generally high-spin in nature.

The overall activation energy and hence the rate dependence of outer-sphere electron transfer reactions can be given as:

$$\Delta G^* = \Delta G_t^* + \Delta G_O^* + \Delta G_i^* \tag{11}$$

Where

i) ΔG_t^* is the energy required to bring the reactant molecules closer to each other against the Coulombic repulsion.

ii) ΔG_o^* is the energy required to reorganize the solvent molecules. As the solvent interactions like H-bonding with the complex molecules increase, the rate of the reaction decreases. On the other hand, solvents with less or no interaction will increase the reaction rate.

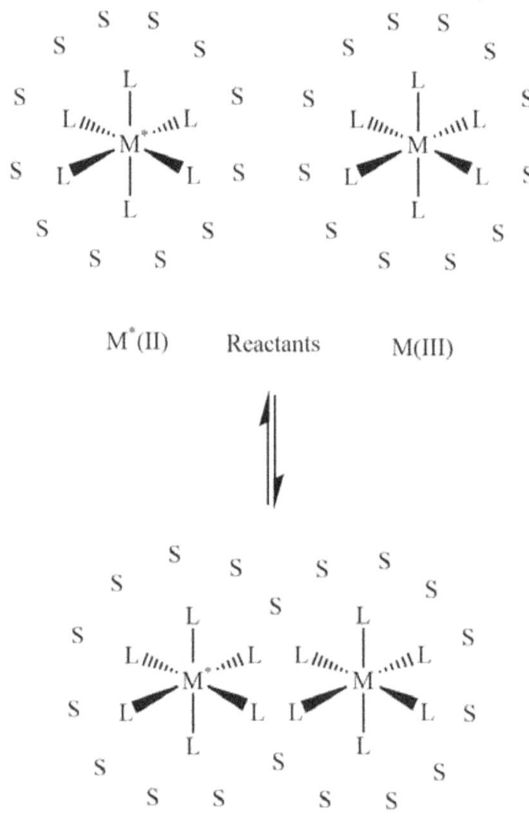

Figure 16. The solvent reorganization during the formation of the precursor complex.

iii) ΔG_i^* is energy required for the reorganization of various bond lengths to make the interacting orbitals of the approximately same energy. This can be explained in terms of the Frank-Condon principle which states that the electronic transitions occur at a much faster rate than the nuclei can respond. Now as we know that the metal-ligand bond lengths are highly dependent on the oxidation state of the metal center, it is pretty obvious that complexes must adjust their bond lengths for the electron transfer to occur.

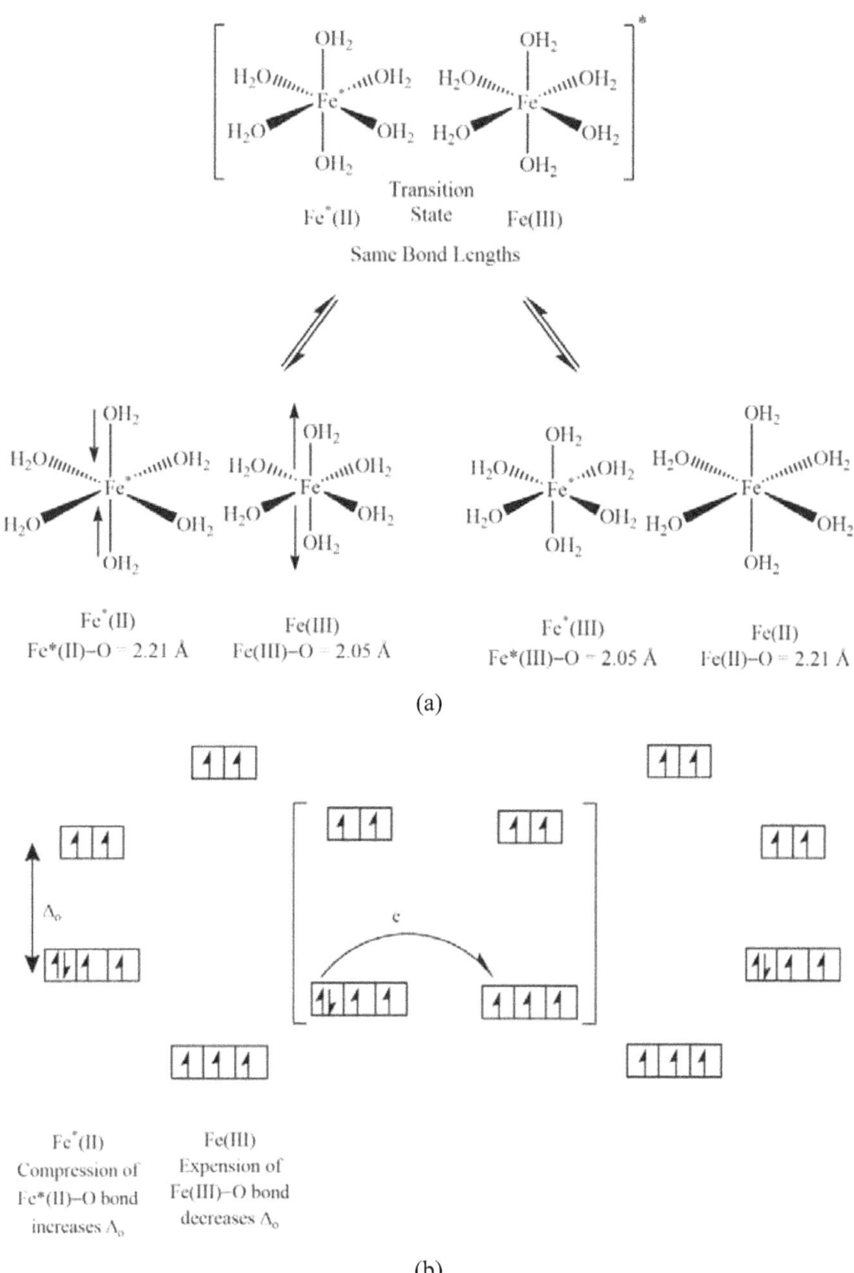

(a)

(b)

Figure 17. Formation of (a) internal reorganization and (b) corresponding ligand field splitting.

The main factors affecting the outer sphere electron transfer are:

1. Orbital Symmetry: Electron transfer outer-sphere mechanism requires orbital overlap and occurs between orbitals of the same symmetry.

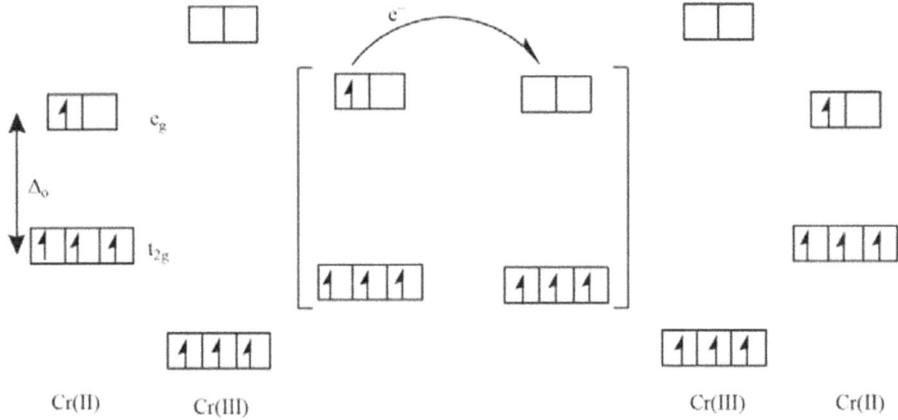

Figure 18. The electron transfer in $e_g \longrightarrow e_g$ ($\sigma^* \longrightarrow \sigma^*$) transition.

In the octahedral fields, $e_g \longrightarrow e_g$ ($\sigma^* \longrightarrow \sigma^*$) transition requires a large change in bond lengths and orbital overlap is also very small due to the ligand steric. This results in an electron transfer at a very slow rate. On the other side, $t_{2g} \longrightarrow t_{2g}$ ($\pi/\pi^* \longrightarrow \pi/\pi^*$) transition involves a small change in bond lengths and the orbital overlap is also very good which depends upon the nature of the ligands. This results in an electron transfer at a very fast rate.

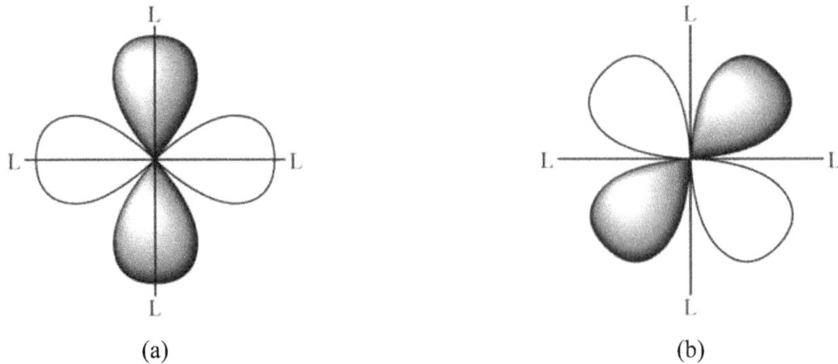

Figure 19. The electron transfer in (a) $e_g \longrightarrow e_g$ ($\sigma^* \longrightarrow \sigma^*$) transitions is having poor overlap; while (b) the $t_{2g} \longrightarrow t_{2g}$ ($\pi/\pi^* \longrightarrow \pi/\pi^*$) transitions involve batter overlap.

2. Orbital Overlap: It has been observed that outer-sphere electron transfer is generally faster for 2nd and 3rd transition series metal ion than that 1st which can be explained in terms of batter overlap of $4d$ and $5d$ orbitals as compared to the $3d$ one. Moreover, the stronger crystal field also results in less bond length distortion. Ligands with extended π-system also speed up the transfer rate.

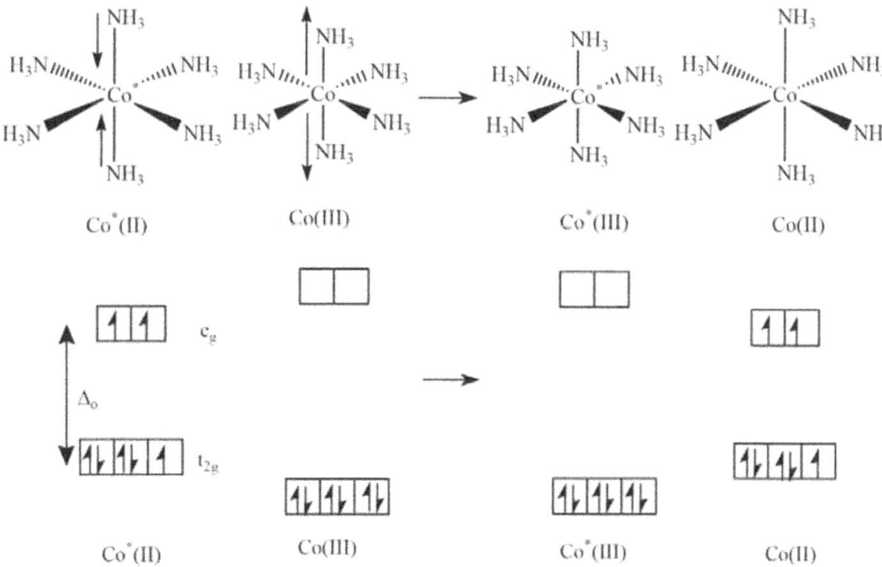

(a)   (b)

Figure 20. Ligands with extended π-conjugation supporting orbital overlap (a) 1,10-phenanthroline (b) 2,2'-bipyridine.

3. Electronic Configuration: If the electron transfer occurs through a change in electronic configuration (high-spin to low-spin or low-spin to high-spin), the activation energy barrier will be very high and a slow electron transfer is expected. In another view, electron reorganization induces a large change in bond lengths and hence slows down the rate of electron transfer.

Figure 21. The electron transfer phenomena in transitions involving a reorganisation of electronic configuration.

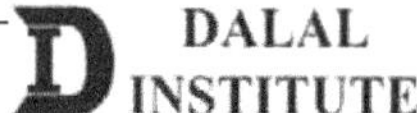

DALAL INSTITUTE

> ### *Inner Sphere Electron Transfer Mechanism*

In this mechanism, the electron is transferred from the reductant to oxidant through a bridge activated complex. The bridging ligand shared in the precursor, successor and activated complex may or may not be transferred during the course of the reaction. The octahedral complexes require the dissociation of one ligand so that the bridging ligand can play its role in creating an intimate contact between the reductant and oxidant so that electron transfer can occur. Inner sphere reactions generally have one labile and one inert reactant. Consider the following example.

$$[Cr(H_2O)_6]^{2+} + [Co(NH_3)_6]^{3+} \longrightarrow [Cr(H_2O)_6]^{3+} + [Co(H_2O)_6]^{2+} + 6NH_3 \qquad (12)$$

The electron transfer reaction shown by equation (12) is extremely slow with a rate constant of 10^{-3} M^{-1}s^{-1} and can easily be explained in terms of outer-sphere mechanism. However, the electron transfer results in the formation of $[Co(NH_3)_6]^{2+}$ which is labile in nature (d^7 high-spin) and immediately hydrolyzed to $[Co(NH_3)_6]^{2+}$ yielding six NH$_3$.

$$[Cr(H_2O)_6]^{2+} + [Co(NH_3)_5Cl]^{2+} \longrightarrow [Cr(H_2O)_5Cl]^{2+} + [Co(H_2O)_6]^{2+} + 5NH_3 \qquad (13)$$

On the other side, the value of the rate constant for electron transfer shown by equation (13) is 6×10^5 M^{-1}s^{-1} which is too high to be explained via outer-sphere mechanism. Moreover, if we use $[Co(NH_3)_5Cl]^{2+}$ at the start and add Cl^{*-} afterward, Cl^{*-} is not found in the final product. It suggests that the electron transfer has occurred via a bridge activated complex.

The course of the complete reaction is shown below.

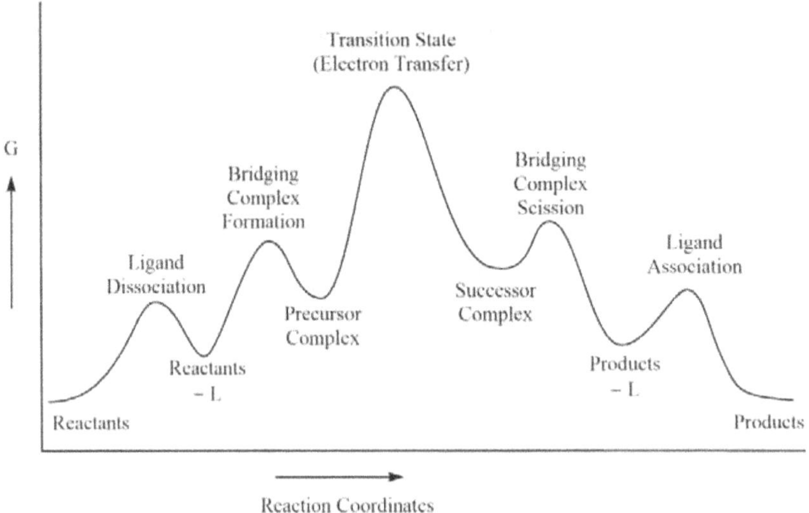

Figure 22. The reaction coordinate diagram for inner sphere electron transfer mechanism.

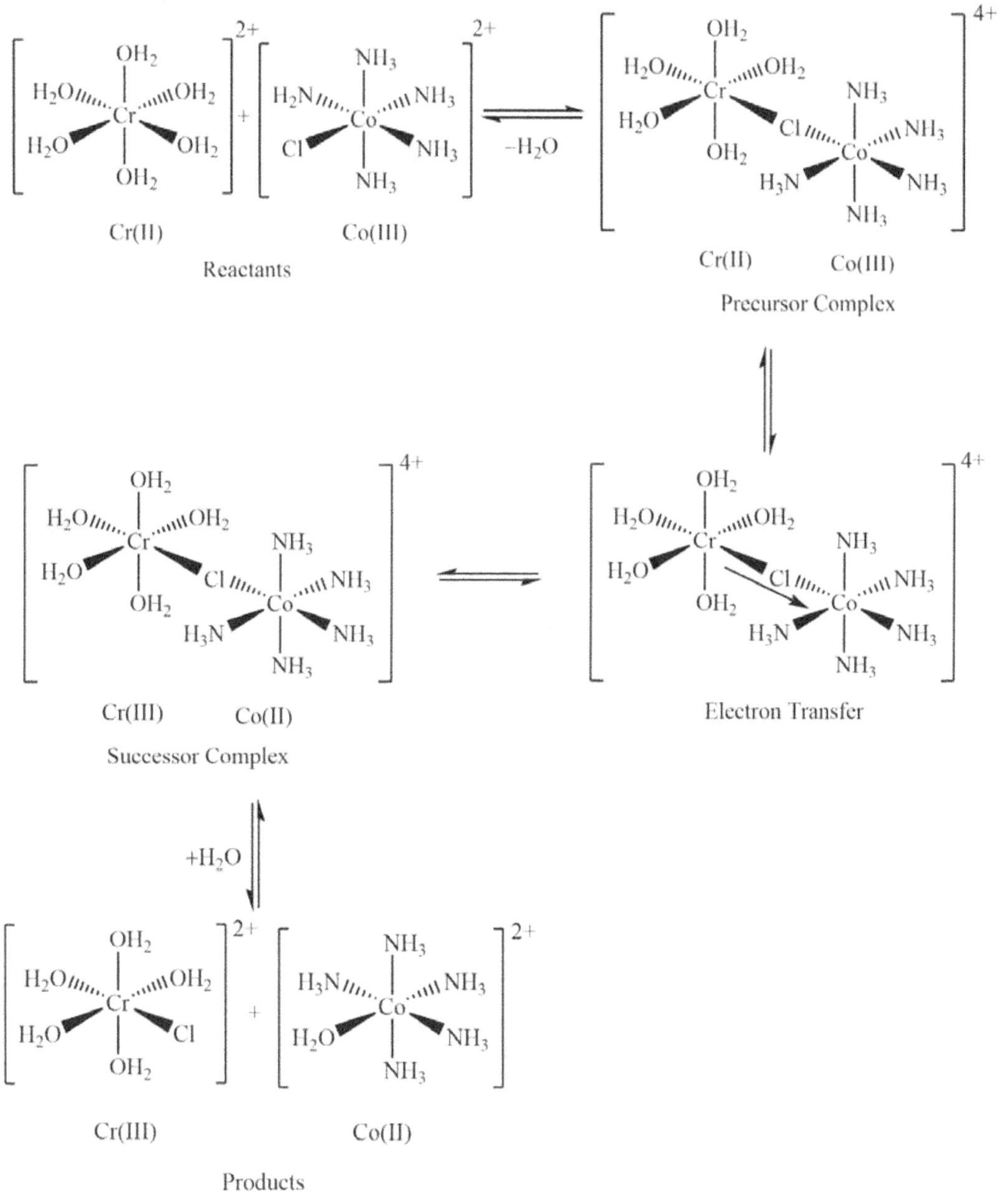

Figure 23. The formation of precursor complex, successor complex and the state of electron transfer in inner sphere electron transfer mechanism.

The electron transfer results in the formation of $[Co(NH_3)_5(H_2O)]^{2+}$ which is labile in nature (d^7 high-spin) and immediately hydrolyzed to $[Co(H_2O)_6]^{2+}$ releasing five NH_3. The main factors affecting the inner sphere electron transfer are:

1. Bridging Complex: It has been observed that instead of the electron transfer step, the formation of the bridging complex can also be the rate-determining step, sometimes. This will primarily be dependent on how labile or inert the complexes are. Furthermore, it is also possible that the scission of the bridging complex is actually playing as the rate-limiting step.

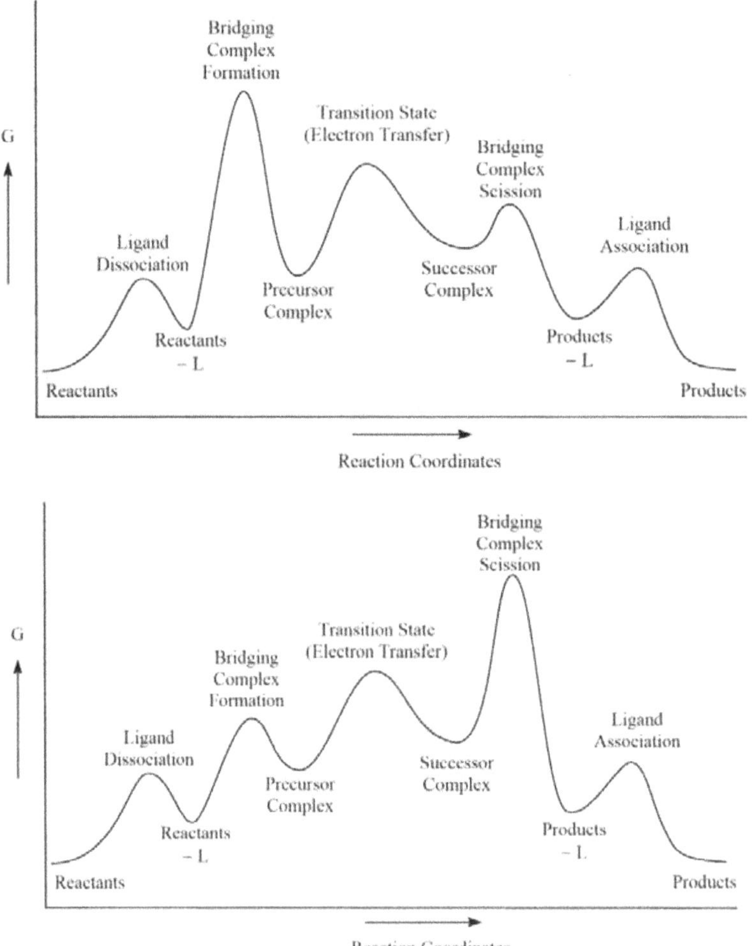

Figure 24. The reaction coordinate diagram for inner-sphere electron transfer mechanism with bridging complex formation and scission as the rate-determining step.

2. Electronic Configuration: The interaction between e_g orbitals and the bridging ligand is greater than their t_{2g} counterparts which can be attributed to the orientation of the lobes along the bonding axis. Hence, an acceleration in rates is observed as we go from outer to inner sphere mechanism due to the facilitation of electron transfer via bridging ligands and orbital symmetries.

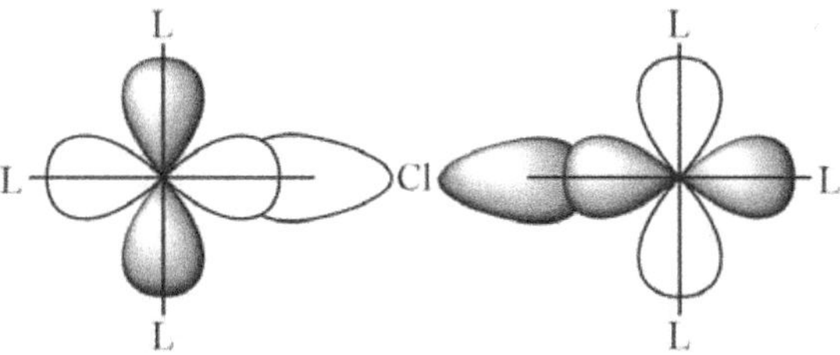

Figure 25. The interaction between e_g orbitals through the bridging ligand in the inner-sphere electron transfer mechanism.

Table 1. The correlation between the frontier molecular orbitals (FMOs) and the acceleration in going from outer-sphere to inner-sphere mechanism.

Reaction	HOMO	LUMO	Acceleration in going from outer-sphere to inner-sphere
$Cr^{2+} \longrightarrow Co^{3+}$	σ^*	σ^*	$\sim 10^{10}$
$Cr^{2+} \longrightarrow Ru^{3+}$	σ^*	π^*/π	$\sim 10^2$
$V^{2+} \longrightarrow Co^{3+}$	π^*	σ^*	$\sim 10^4$
$V^{2+} \longrightarrow Ru^{3+}$	π	π	All outer-sphere

Large reorganization for Co^{3+} outer-sphere. Inner-sphere is much faster.

Reorganisation of Ru^{3+} less than Co^{3+}

Figure 26. Continued on the next page…

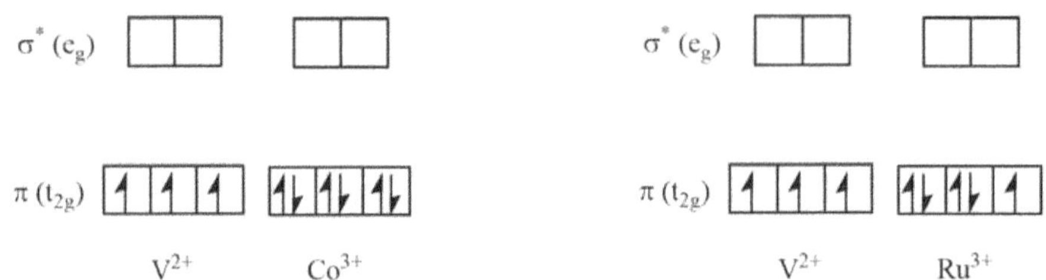

Large reorganization for Co^{3+} outer-sphere. Inner-sphere is much faster.

No reorganization is involved. Allowed in outer-sphere.

Figure 26. The solvent reorganization various redox pairs.

3. Nature of the bridging ligand: The rate of inner-sphere electron transfer is mechanism is highly sensitive to the nature of the bridging ligand. The rate of the reaction generally increases with the increase in the nucleophilic character of the bridging ligand. Furthermore, the substitution on the bridging ligand can also affect the rate significantly. Consider the following example.

1st step

$$[Co(NH_3)_5(COOR)]^{2+} \;+\; [Cr(H_2O)_6]^{2+} \longrightarrow \left[\begin{array}{c} (NH_3)_5Co \underset{O}{\diagup} R \\ \vdots \\ Cr^{3+}(aq) \end{array} \right]^{4+}$$

Oxidant　　　　　　Reductant

2nd step

$$\left[\begin{array}{c} (NH_3)_5Co \underset{O}{\diagup} R \\ \vdots \\ Cr^{3+}(aq) \end{array} \right]^{4+} \xrightarrow{\text{Hydrolysis}} \begin{array}{c} [Co(NH_3)_5(COOR)]^{1+} \\ + \\ [Cr(H_2O)_6]^{3+} \end{array}$$

When

R =	H	Me	Ph	^{t}Bu
k (M^{-1}s^{-1})	7.2	0.35	0.15	9.6×10^{-3}

Hence, as the bulkiness of the R group increases, the binding capacity of the bridging decreases which in turn slows down the overall rate of the electron transfer.

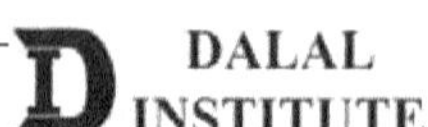

❖ Electron Exchange

On the basis of the mechanism involved, the redox reactions of transition metal complexes have been divided into two types; outer-sphere and inner-sphere. We have already studied the same in the previous section. However, these electron-transfer reactions can also be classified on the basis of the metal centers involved as given below.

i) The electron transfer between two different oxidation states of the same metal ion. These reactions are generally called as self-exchange reactions.

ii) The electron transfer between two completely different metal complexes. These reactions are called as the cross-reactions.

The *electron exchange or self-exchange reactions may simply be defined as the redox reactions involving the transfer of electrons between two complexes having the same ligands and metal ions but with different oxidation states.*

In other words, the reactants in electron-exchange reactions are indistinguishable from the products. These reactions result in no net chemical change and can be studied only by special experimental techniques.

$$[Fe(H_2O)_6]^{2+} + [Fe^*(H_2O)_6]^{3+} \longrightarrow [Fe(H_2O)_6]^{3+} + [Fe^*(H_2O)_6]^{2+} \tag{14}$$

$$[Co(NH_3)_6]^{2+} + [Co^*(NH_3)_6]^{3+} \longrightarrow [Co(NH_3)_6]^{3+} + [Co^*(NH_3)_6]^{2+} \tag{15}$$

$$[Mn(CN)_6]^{4-} + [Mn^*(CN)_6]^{3-} \longrightarrow [Mn(CN)_6]^{3-} + [Mn^*(CN)_6]^{4-} \tag{16}$$

All of the above are electron-exchange reactions proceeding via the outer-sphere electron transfer mechanism. The rate measurement of these reactions is quite difficult to record as the reactants are chemically equivalent to the products. Out of various techniques such as isotopic tracer, optical activity and EPR-NMR spectroscopy; isotopic labeling is the most convenient and widely accepted method to study the process of self-exchange. After mixing the reactants, the rate at which the radioactivity disappears in M^*(III) complex is the rate of electron exchange in the two complexes. Similarly, the rate at which the radioactivity appears in the M^*(II) complexes can also be used to estimate the rate of electron exchange.

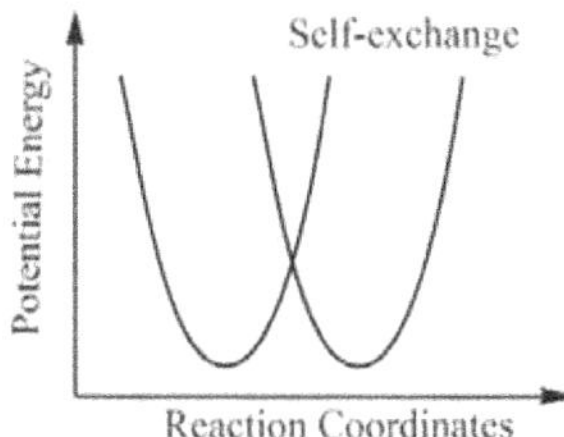

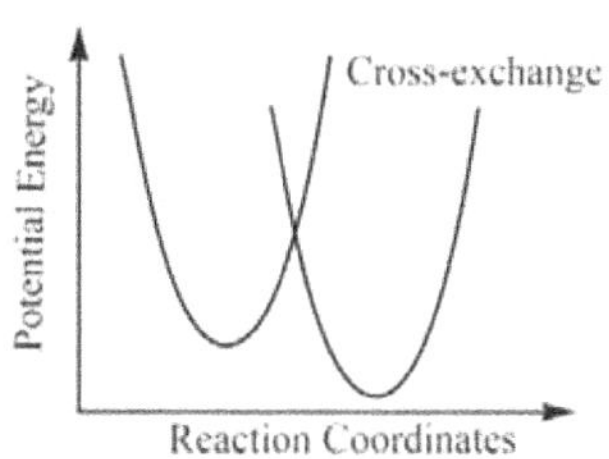

Figure 27. The comparison of potential energy curves of self-exchange and cross-exchange.

Table 2. The values of rate constants for some electron-exchange processes of type $[ML_6]^{2+} + [M^*L_6]^{3+} \rightarrow [ML_6]^{3+} + [M^*L_6]^{2+}$.

Complex	Electronic configuration $M(II) \rightarrow M^*(III)$		k (M^{-1}s^{-1})	T(°C)
Cr(H$_2$O)$_6$	$t_{2g}^3\,e_g^1$	$t_{2g}^3\,e_g^0$	2×10^{-5}	25
Fe(H$_2$O)$_6$	$t_{2g}^4\,e_g^2$	$t_{2g}^3\,e_g^2$	4.2	25
Co(H$_2$O)$_6$	$t_{2g}^5\,e_g^2$	$t_{2g}^4\,e_g^2$	5	25
Co(en)$_3$	$t_{2g}^5\,e_g^2$	$t_{2g}^6\,e_g^0$	7.7×10^{-5}	25
Fe(phen)$_3$	$t_{2g}^6\,e_g^0$	$t_{2g}^5\,e_g^0$	1.3×10^7	3
Co(phen)$_3$	$t_{2g}^6\,e_g^1$	$t_{2g}^6\,e_g^0$	12	25

Now, as the electron-exchange reactions are the simplest type of outer-sphere mechanism; factors like transition-type, orbital overlap, and reorganization of electronic configuration can be used to explain the large differences in the electron-exchange rates listed in Table 2.

In order to get a wider view of the mechanism involved in electron-exchange, we need to recall the Frank-Condon principle which says the electronic rearrangements are so rapid that the nuclei can be considered as stationary. Moreover, the electron transfer from one complex to another is expected only when it has the same energy in both of the sites. Hence, the activation energy to achieve this goal is mainly governed by nuclear rearmaments. Consider the following reaction:

$$[Fe(H_2O)_6]^{2+} + [Fe^*(H_2O)_6]^{3+} \rightarrow [Fe(H_2O)_6]^{3+} + [Fe^*(H_2O)_6]^{2+} \tag{17}$$

The potential energy curves for the reactants and products should be parabolic because the metal–ligand deformations are almost harmonic in nature. The activated complex is present at the intersection point of the two curves.

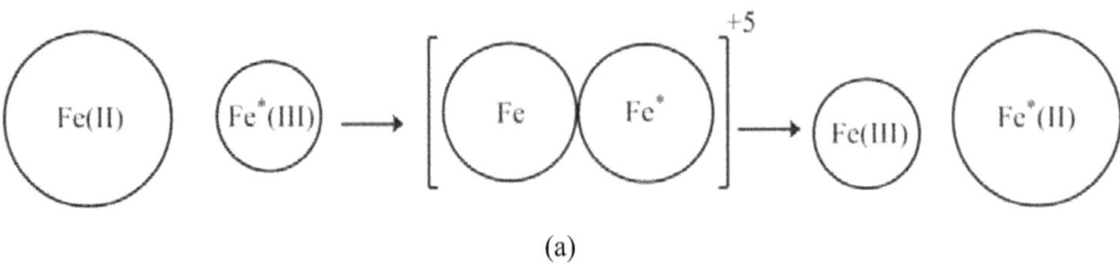

Figure 28. Continued on the next page…

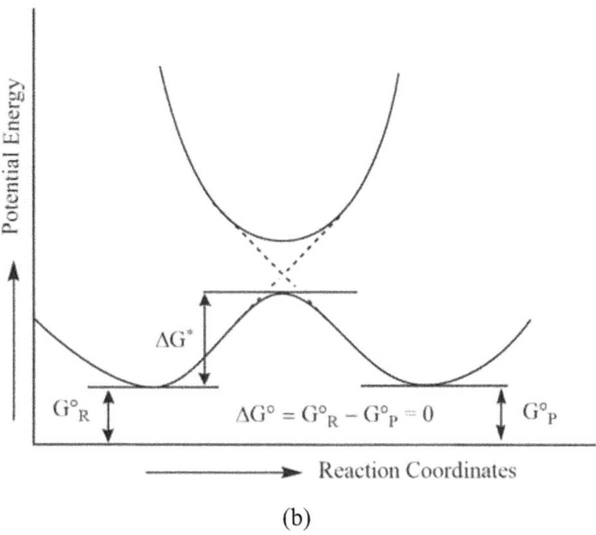

Figure 28. The (a) reaction course and (b) potential energy curve for electron-exchange reactions.

During the formation of activated complex, the Fe(II) bonds shorten while the Fe*(III) bonds lengthen which facilitates the transfer of electron. It is worth noting that the Gibbs free energy change for self-exchange reactions is zero but free energy of activation is still a non-zero term.

Rudolph A. Marcus derived an equation using the first principle to predict the overall rate constant for outer-sphere electron transfer reactions from the self-exchange rate constants of each of the redox couples involved and overall equilibrium constant. The equation can be given as:

$$k = \lfloor k_1 k_2 Kf \rfloor^{1/2} \tag{18}$$

Where k_1 and k_2 are rate contents for two electron-exchange reactions and K is the equilibrium constant for the overall reaction. The factor f *is* a complex parameter composed of the rate constants and has been incorporated to correct the free energy differences of the two reactants. Consider the example,

$$[Fe(CN)_6]^{4-} + [Fe^*(CN)_6]^{3-} \longrightarrow [Fe(CN)_6]^{3-} + [Fe^*(CN)_6]^{4-} \tag{19}$$

$k_1 = 7.4 \times 10^2 \ M^{-1}s^{-1}$

$$[Mo(CN)_6]^{4-} + [Mo^*(CN)_6]^{3-} \longrightarrow [Mo(CN)_6]^{3-} + [Mo^*(CN)_6]^{4-} \tag{20}$$

$k_2 = 3 \times 10^4 \ M^{-1}s^{-1}$

$$[Fc(CN)_6]^{4-} + [Mo^*(CN)_6]^{3-} \longrightarrow [Fe(CN)_6]^{3-} + [Mo^*(CN)_6]^{4-} \tag{21}$$

$K = 1.0 \times 10^2$

Now, putting the values of k_1, k_2 and K from equation (19), (20) and (21) in equation (18); the value of rate constant for the overall reaction is found to be 4×10^4 $M^{-1}s^{-1}$; and it is obviously pretty much comparable to the experimental value of 3×10^4 $M^{-1}s^{-1}$.

In 1992, R. A. Marcus received the Nobel Prize in the field of Chemistry for his work on the theory of electron transfer reactions occurring in chemical systems. His theory is also named after him, the Marcus theory, and it provides a kinetic and thermodynamic framework for describing outer-sphere electron transfer reactions involving one electron.

❖ Problems

Q 1. Explain the mechanism of ligand displacement reactions in square-planar complexes.

Q 2. Write a short note on trans-effect and also explain the synthesis route of trans-platin.

Q 3. How would you explain the exceptionally high trans-effect of strong π-acceptor ligands like CN^{-1}, CO or ethylenediamine?

Q 4. Discuss the π-bonding theory of trans effect in detail.

Q 5. What do you understand from the outer-sphere electron transfer reactions? Also, discuss the factors affecting in detail.

Q 6. Discuss the nature of the inner-sphere electron transfer mechanism with special reference of electronic configuration.

Q 7. Define electron-exchange reactions. How can the Marcus equation be used to study cross-reaction?

❖ Bibliography

[1] J. D. Lee, *Concise Inorganic Chemistry*, Chapman & Hall, New York, USA, 1994.

[2] B. R. Puri, L. R. Sharma, K. C. Kalia, *Principals of Inorganic Chemistry*, Milestone Publishers, Delhi, India, 2012.

[3] J. E. Huheey, E. A. Keiter, R. L. Keiter, *Inorganic Chemistry: Principals of Structure and Reactivity*, HarperCollins College Publishers, New York, USA, 1993.

[4] A. F. Wells, *Structural Inorganic Chemistry*, Oxford University Press, London, UK, 1975.

[5] F. A. Cotton, G. Wilkinson, C. A. Murillo, M. Bochmann, *Advanced Inorganic Chemistry*, John Wiley & Sons, New Jersey, USA, 1999.

[6] S. F. A. Kettle, *Physical Inorganic Chemistry: A Coordination Chemistry Approach*, Springer-Verlag, Berlin, Germany, 1996.

[7] B. W. Pfennig, *Principles of Inorganic Chemistry*, John Wiley & Sons, New Jersey, USA, 2015.

CHAPTER 5

Isopoly and Heteropoly Acids and Salts:

❖ Isopoly and Heteropoly Acids and Salts of Mo and W: Structures of Isopoly and Heteropoly Anions

The term polymetalate acid or simply poly acid may be defined as the condensed or polymerized form of the weak acids of amphoteric metals like vanadium, niobium, tantalum (VB group metals) or chromium, molybdenum, and tungsten (VIB group metals) in the +5 and +6 oxidation states. The anions of these poly acids contain several molecules of the acid anhydride and the corresponding salts are called as polysalts.

Furthermore, if these polymerized acids contain only one type of acid anhydride, they are called as isopoly acids. However, these anhydrides can also condense with some other acids like phosphoric or silicic acid to form heteropoly acids. In other words, isopoly acids contain only one metal along with hydrogen and oxygen while heteropoly acids contain two elements other than hydrogen and oxygen. The corresponding salts of isopoly and heteropoly acids are called as isopoly and heteropoly salts, respectively.

Consider the polymerization of chromate ion to form different isopoly chromates anions. CrO_3 dissolves in an alkali to give yellow colored CrO_4^{2-} ions solution. At very high pH, above 8, the chromate ions, CrO_4^{2-}, exist as the discrete entities but as the pH is lowered down, the protonation and dimerization take place. For instance:

$$2H_2CrO_4 \xrightarrow[-H_2O]{} H_2Cr_2O_7$$

and

$$3H_2CrO_4 \xrightarrow[-2H_2O]{} H_2Cr_3O_{10}$$

and

$$4H_2CrO_4 \xrightarrow[-3H_2O]{} H_2Cr_4O_{13}$$

The polymeric anions $Cr_2O_7^{2-}$, $Cr_3O_{10}^{2-}$ and $Cr_4O_{13}^{2-}$ produced by the polyacids $H_2Cr_2O_7$, $H_2Cr_3O_{10}$ and $H_2Cr_4O_{13}$, can successfully be isolated from their aqueous as sodium or potassium polysalts like $K_2Cr_2O_7$, $K_2Cr_3O_{10}$ and $K_2Cr_4O_{13}$, respectively. Among the isopoly-anions of V^{5+}, Nb^{5+}, Ta^{5+}, Cr^{6+}, Mo^{6+} and W^{6+}, only Cr^{6+} is found to have tetrahedral CrO_4^{2-} units joined through the corners. The other metal ions form isopolyanions by the sharing of edges of octahedral MO_6 units. This may be attributable to the small size of Cr^{6+} which can afford only four oxide ions around itself.

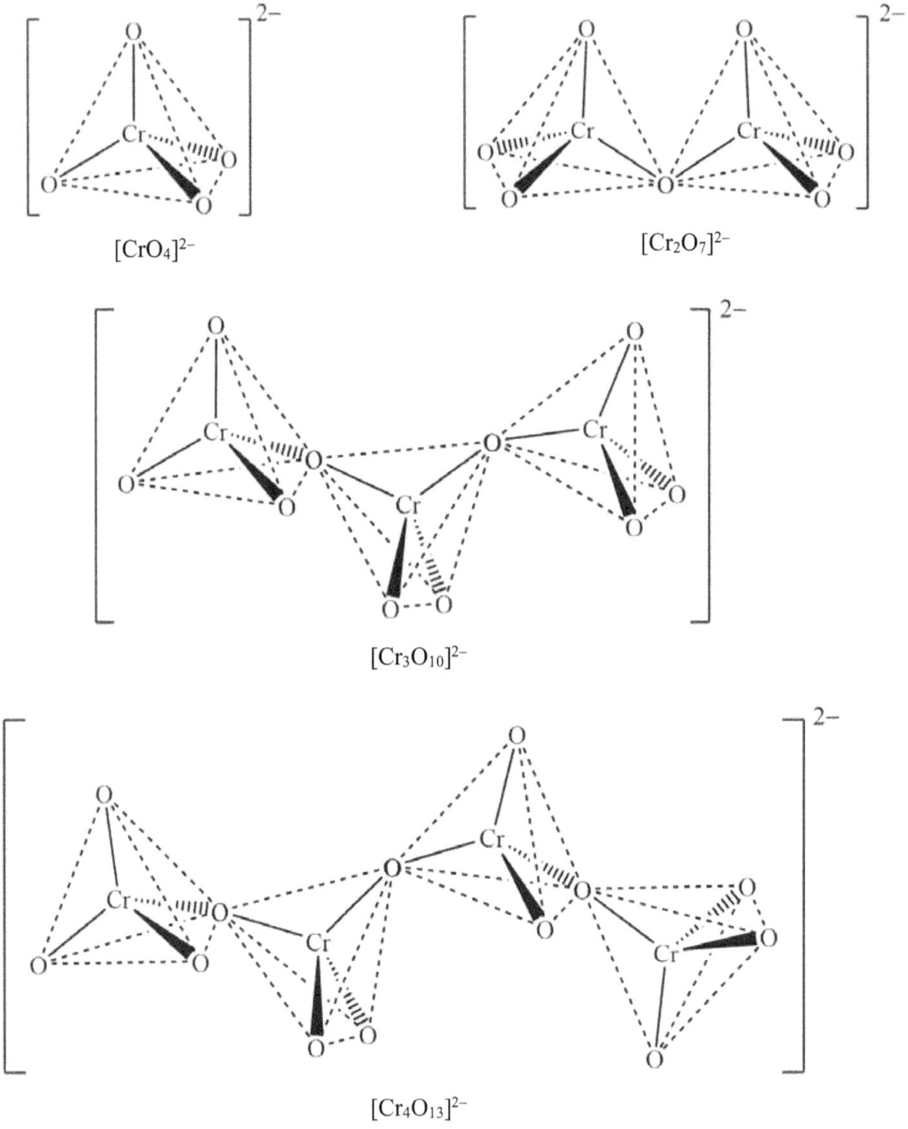

$$[CrO_4]^{2-}$$

$$[Cr_2O_7]^{2-}$$

$$[Cr_3O_{10}]^{2-}$$

$$[Cr_4O_{13}]^{2-}$$

Figure 1. The structure of chromate and isopoly anions of chromium.

The tri-chromate and tetra-chromate anions can be crystallized as their alkali metal salts only from strongly acidic solution and no polymerization beyond the tetrameric entity is observed. The Cr–O–Cr bond angle of all polychromates is approximately 120°.

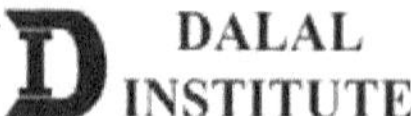

> ### *Isopoly Acids and Salts of Mo and W*

When molybdenum trioxide is dissolved in highly basic aqueous solutions of sodium hydroxide or potassium hydroxide (alkali solutions), molybdate ions with tetrahedral geometry are formed as:

$$MoO_3 + 2NaOH \longrightarrow Na_2MoO_4 + H_2O$$

These normal molybdates, Na_2MoO_4, containing discrete MoO_4^{2-} units, can easily be crystallized out of them.

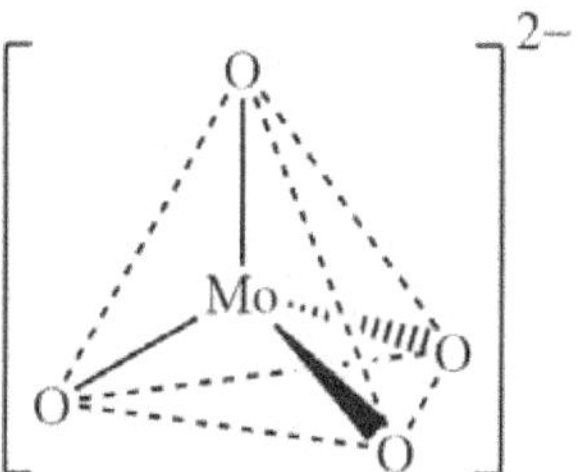

Figure 2. The discrete MoO_4^{2-}.

When the pH of the solution is lowered down, the protonation of molybdate ions starts followed by the condensation yielding the first major polyanion i.e. paramolybdate. The whole process of condensation can be depicted as follows:

$$MoO_4^{2-} \; + \; H_3O^+ \; \rightleftharpoons \; [HO{-}MoO_3]^- \; + \; H_2O$$

$$[HO{-}MoO_3]^- \; + \; H_3O^+ \; + \; H_2O \; \rightleftharpoons \; Mo(OH)_6$$

Although the entropy (ΔS) of the second reaction is negative yet it is as fast as the first reaction which may be attributed to the large negative enthalpy for the second reaction, compensating the entropy loss. The $Mo(OH)_6$, thus formed during the course of the second reaction, reacts with $[MoO_3(OH)]^-$ ions present in the acidic media as:

$$Mo(OH)_6 \;\underset{-H_2O}{\overset{HMoO_4^-}{\rightleftharpoons}}\; [(HO)_5{-}Mo{-}(OMoO_3)]^- \;\underset{-H_2O}{\overset{HMoO_4^-}{\rightleftharpoons}}\; [(HO)_4{-}Mo{-}(OMoO_3)_2]^{2-}$$

$$\underset{-H_2O}{\overset{HMoO_4^-}{\updownarrow}}$$

$$[(HO){-}Mo{-}(OMoO_3)_5]^{5-} \xrightleftharpoons[-H_2O]{HMoO_4^-} [(HO)_2{-}Mo{-}(OMoO_3)_4]^{4-} \xrightleftharpoons[-H_2O]{HMoO_4^-} [(HO)_3{-}Mo{-}(OMoO_3)_3]^{3-}$$

$$-H_2O \updownarrow HMoO_4^-$$

$$\text{Isomerisation}$$

$$[Mo{-}(OMoO_3)_6]^{6-} \rightleftharpoons [Mo_7O_{24}]^{6-}$$

The complete reactions giving different isopoly molybdates can be written as:

$$7MoO_4^{2-} + 8H^+ \longrightarrow [Mo_7O_{24}]^{6-} + 4H_2O$$

$$8MoO_4^{2-} + 12H^+ \longrightarrow [Mo_8O_{26}]^{4-} + 6H_2O$$

It is worth noting that the hepta-molybdate or para-molybdate is the first stable form of isopoly-molybdates and when the acidification of the solution further continues, $Mo_8O_{26}^{4-}$ is obtained. If the pH of the solution is dropped below 1, yellow colored, molybdic acid $(MoO_3.2H_2O)$ is formed which can be converted into monohydrate form just by warming it up.

$$[MoO_4]^{2-} \xrightarrow[6]{pH} [Mo_7O_{24}]^{6-} \xrightarrow[1.5-3.0]{pH} [Mo_8O_{26}]^{6-} \xrightarrow[>1]{pH} MoO_3.2H_2O$$

It is worth noting that during the condensation process, from MoO_4^{2-} to $Mo_7O_{24}^{4-}$, the coordination number of molybdenum ion changes from four to six and the building block unit of the polyhedral entity becomes MoO_6 octahedron.

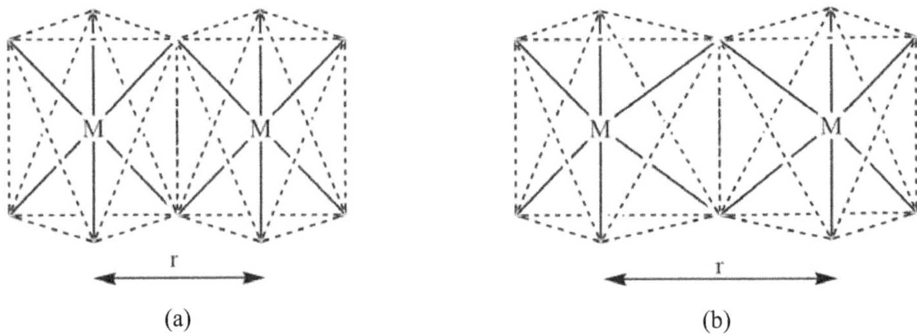

Figure 3. The structure of (a) ideal edge sharing octahedrons and (b) distorted edge sharing octahedrons.

These octahedral units bind to each other by the sharing of corners or edges. Moreover, the terminal Mo–O bond lengths are less than that of Mo–O bond lengths in Mo–O–Mo bridge and hence the two joining

octahedra are somewhat more or less distorted. This increment in the distance between two metal centers can be explained in terms of reduced inter-metal ion repulsion.

Keeping in mind that electrostatic repulsion between is metal centers is a governing factor in justifying the structures of isopoly and heteropoly anions and acids, corner-sharing among octahedrons is expected more frequently than edge-sharing as it separates the metal centers more profoundly. But in actual practice, edge-sharing is preferred over the corner-sharing because the edge-sharing reduces the number of oxide ions required to pack the polyhedral units which in turn decreases the unwanted anionic charge from O^{2-} ions. Therefore, the edge-sharing stabilizes the polyanions to a greater extent than corner-sharing. The order of removal of O^{2-} anionic charge by different types of sharing, face > edge > corner, suggests a face sharing of the octahedron units as the most favorable choice for the condensation but it is actually the least favored due to the incorporation of highest inter-metal ion repulsion.

The structure and properties of different types of isopoly-molybdate are discussed below.

1. Dimolybdate: The general formula for dimolybdates system is $M_2O.2MoO_3.xH_2O$ (M = Na, K or NH_4) and are observed in the pH range of 5-6.

$$Na_2O.\,2MoO_3.\,xH_2O \rightleftharpoons Na_2Mo_2O_7$$

$Na_2Mo_2O_7$ is not stable and hence dimolybdate systems do not contain $Mo_2O_7^{2-}$ ions. The compounds like $M_2Mo_2O_7$ have been isolated from anhydrous melts or aqueous solutions but discrete ions like $Mo_2O_7^{2-}$ are found to be absent. These systems are the mixture of simple (MoO_4^{2-}) and paramolybdate ($Mo_7O_{24}^{6-}$) ions in general. However, molybdate ions as a salt of tetrabutylammonium, $[(C_4H_9)_4N]_2Mo_2O_7$, is one of the very few compounds which is known to have discrete $Mo_2O_7^{2-}$ ions.

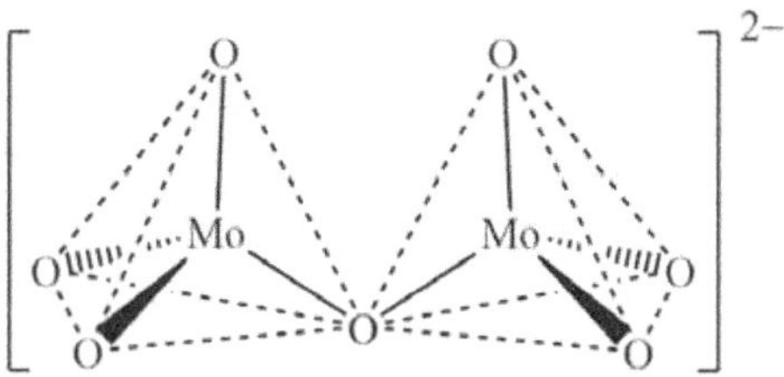

Figure 4. $Mo_2O_7^{2-}$ ion in tetrabutylammonium dimolybdate.

2. Trimolybdate: The general formula for trimolybdate system is $M_2O.3MoO_3.xH_2O$ where M represents Na or K.

$$Na_2O.\,3MoO_3.\,xH_2O \rightleftharpoons Na_2Mo_3O_{10}$$

$Na_2Mo_3O_{10}$ is not stable and hence trimolybdate systems do not contain $Mo_3O_{10}^{2-}$ ions. These systems are prepared by crystallising paramolybdate in the presence acetic acid or by saturating molybdenum oxide in the presence of alkali.

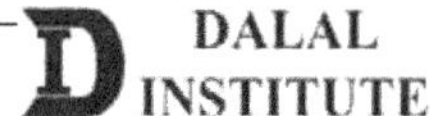

3. Tetramolybdate or metamolybdate: The general formula for tetramolybdate system is $M_2O.4MoO_3.xH_2O$ where M represents Na or K.

$$Na_2O.\,4MoO_3.\,xH_2O \rightleftharpoons Na_2Mo_4O_{13}$$

$Na_2Mo_4O_{13}$ is not stable and hence tetramolybdate systems do not contain $Mo_4O_{13}^{2-}$ ions. These systems are the mixture of simple (MoO_4^{2-}) and octamolybdate ($Mo_8O_{26}^{4-}$) ions in general.

4. Heptamolybdate or paramolybdate: The general formula for paramolybdates is $M_6Mo_7O_{24}$ where M represents Na, K or NH_4. $M_6Mo_7O_{24}$ is quite stable and contains discrete $Mo_7O_{24}^{6-}$ ions.

The edge connections of seven ideal MoO_6 octahedrons in paramolybdate ion can be visualized in terms of one front unit (composed of four MoO_6) and one rear unit (composed of three MoO_6). As the name suggests, the front unit is placed above the rear unit in such a way that each octahedron shares three edges with its neighbors.

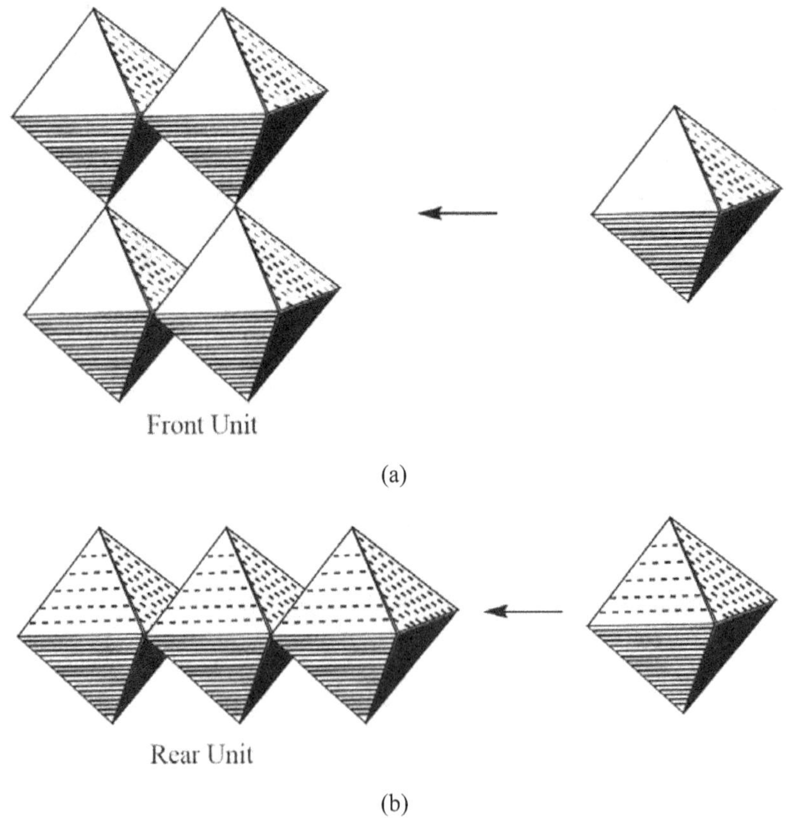

Figure 5. Continued on the next page…

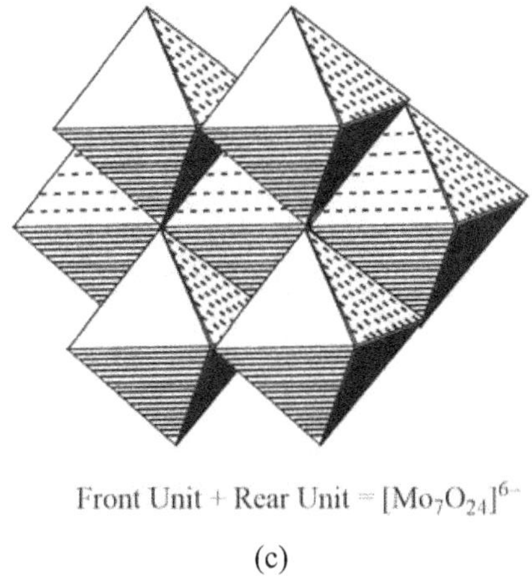

Front Unit + Rear Unit = $[Mo_7O_{24}]^{6-}$

(c)

Figure 5. The (a) front unit and (b) rear unit are combined to produce the (c) overall structure of paramolybdate ion.

The structure shown (Figure 5) is an idealized one with perfect octahedrons; but in actual practice, the distortions play its role giving a range of Mo–O bond lengths in MoO_6 units.

5. Octamolybdate: The general formula for octamolybdates is $M_4Mo_8O_{24}$ where M represents Na, K or NH_4. $M_4Mo_8O_{26}$ is quite stable and contains discrete $Mo_8O_{26}^{4-}$ ions. These molybdates can be synthesized by adding stoichiometric amount of MoO_3 in paramolybdate solution or by treating alkali molybdate solutions with 1.75 molar hydrochloric acid. These molybdates exist in three forms.

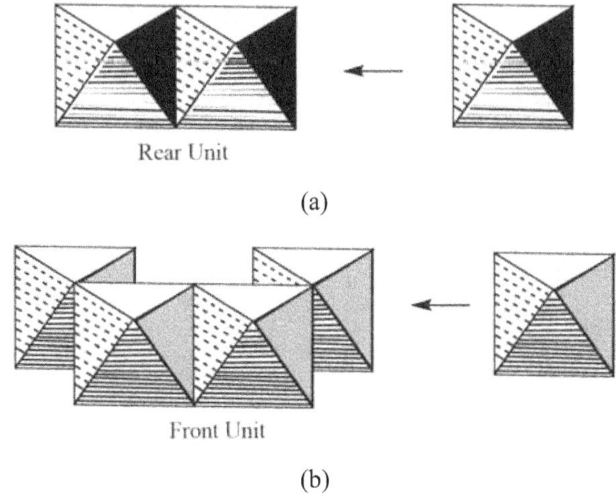

(a)

(b)

Figure 6. Continued on the next page…

Front Unit + Rear Unit + Two Tetrahedral Units = $\alpha-[Mo_8O_{26}]^{4-}$

(c)

Figure 6. The (a) front unit and (b) rear unit and two tetrahedral units as capping are combined to produce the (c) overall structure of α–octamolybdate ion.

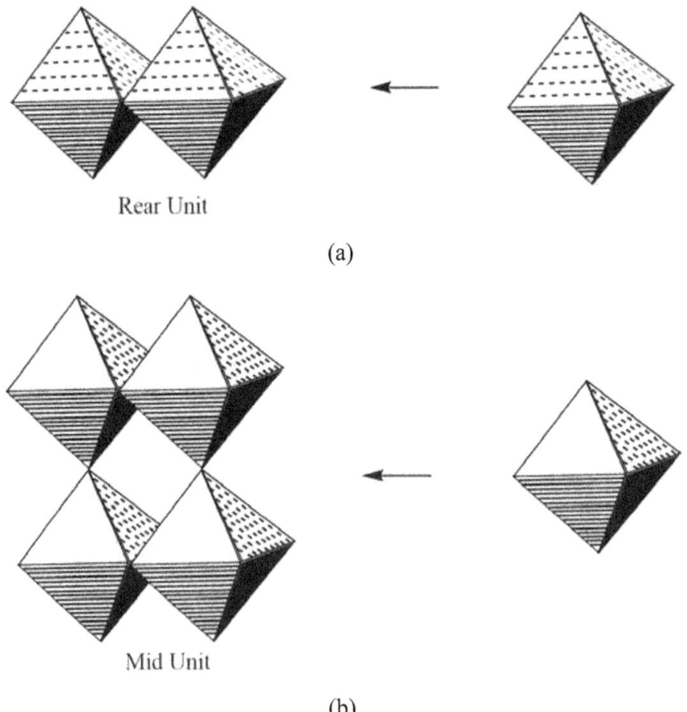

Figure 7. Continued on the next page…

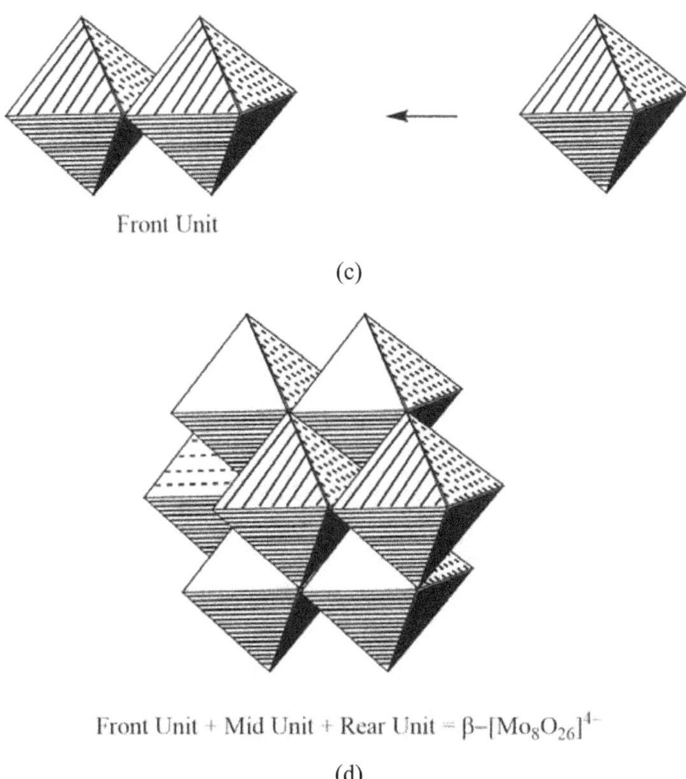

(c)

(d)

Figure 7. The (a) rear unit and (b) mid-unit and (c) front unit are combined to produce the (d) overall structure of the β–octamolybdate ion.

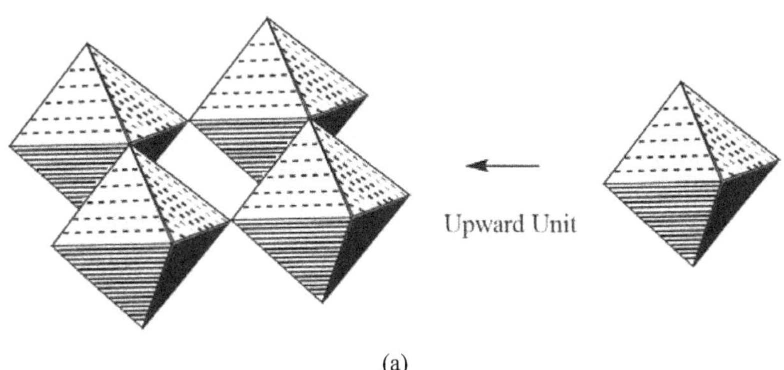

(a)

Figure 8. Continued on the next page…

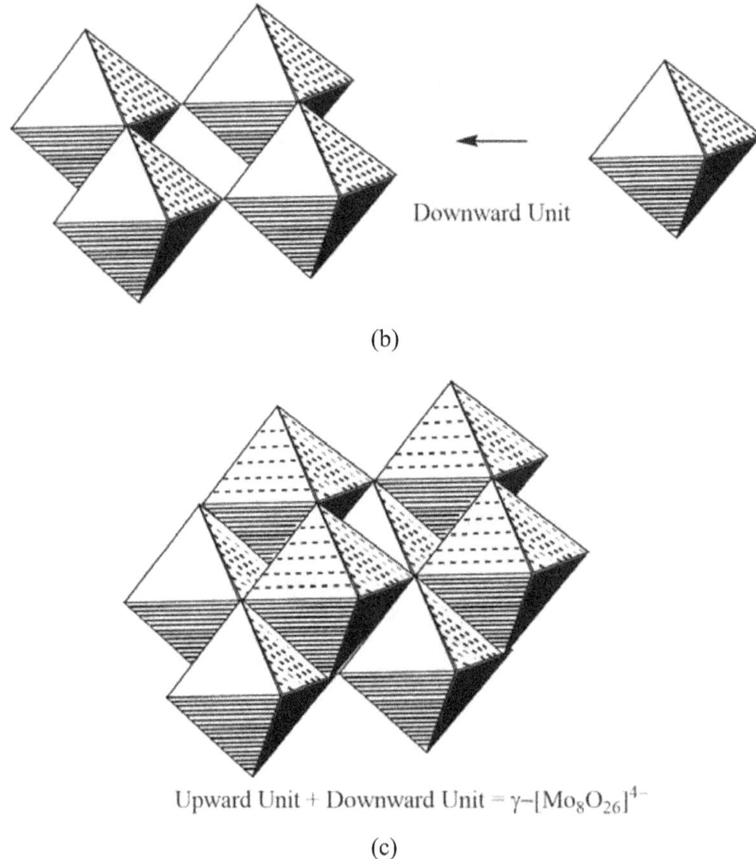

Figure 8. The (a) upward unit and (b) downward units are combined to produce the (c) overall structure of the γ–octamolybdate ion.

The first two isomeric forms of octamolybdate anion, α–[Mo$_8$O$_{26}$]$^{4-}$ and β–[Mo$_8$O$_{26}$]$^{4-}$, are in equilibrium with each other in solution phase. α– and β– isomers have been obtained from non-aqueous and aqueous solutions respectively. A γ– form has also been suggested as an intermediate between α– and β– isomers.

The edge connections of eight ideal MoO$_6$ octahedrons in β–octamolybdate ion can be visualized in terms of one front unit (composed of two MoO$_6$), one mid-unit (composed of four MoO$_6$) and one front unit (composed of two MoO$_6$). It must be noted down that one octahedra of the rear unit is completely hidden in the overall idealized structure of β–octamolybdate.

6. Hexamolybdate: The hexamolybdates are derived from discrete Mo$_6$O$_{19}^{2-}$ ions and are not as conventional as hepta or octamolybdates. Crystallographic studies have shown that these isopoly anions are formed by

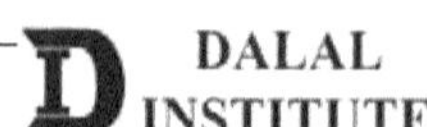

molybdenum ions in octahedral pattern and each ion is further surrounded by six oxide ions. Each MoO_6 unit has only one terminal O^{2-} ion.

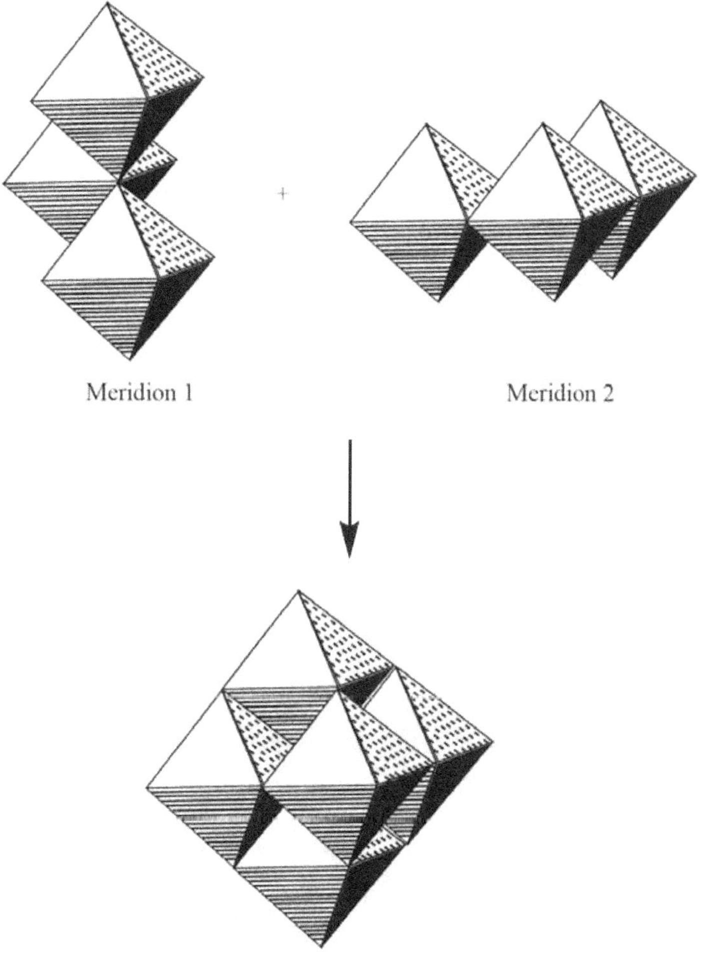

Figure 9. The hexamolybdate anion.

7. Decamolybdate: The decamolybdates are obtained as $(NH_4)_8Mo_{10}O_{34}$ and the corresponding molybdate ion $Mo_{10}O_{34}^{8-}$ are found to have one Mo_8O_{28} unit and two MoO_4 tetrahedral units attached at corners. The structure of decamolybdate can be visualized as an extension of $\gamma-[Mo_8O_{26}]^{8-}$ in which two additional tetrahedral units are attached at corners of two opposite octahedrons.

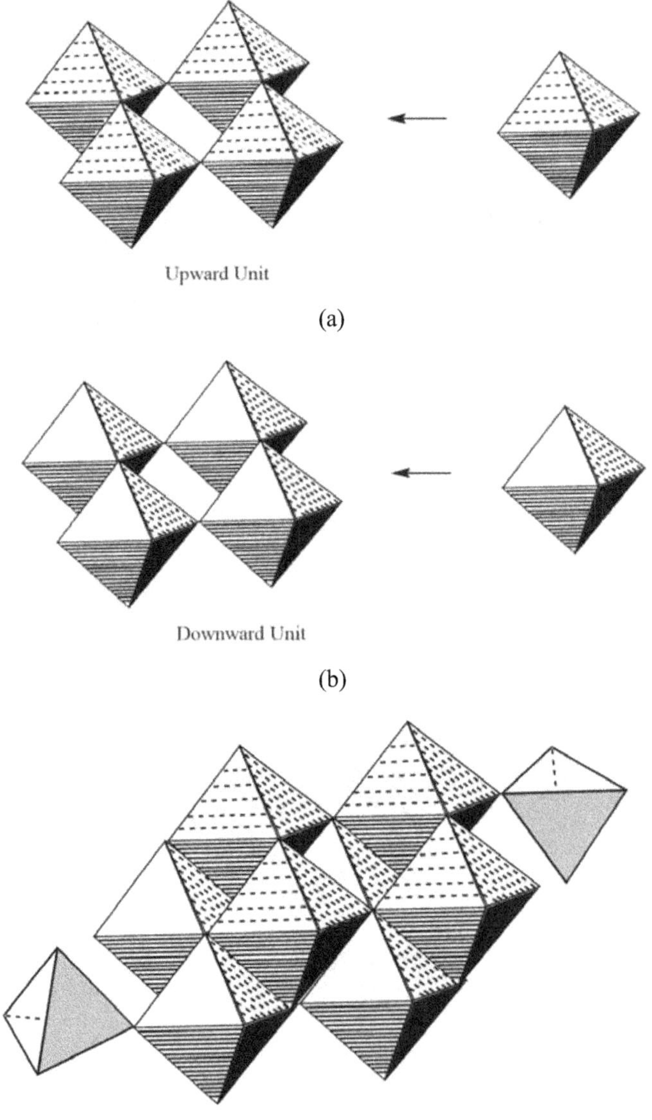

Upward Unit + Downward Unit + Two tetrahedral Unit = $[Mo_{10}O_{34}]^{8-}$

(c)

Figure 10. The (a) upward unit and (b) downward unit and two tetrahedral units are combined to produce the (c) overall structure of dacamolybdate ion.

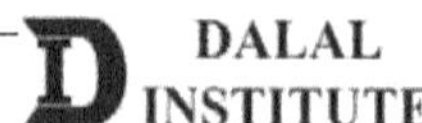 **DALAL INSTITUTE**

8. Mo₃₆-polymolybdate: The largest polymolybdate under non-reducing conditions known till recent times is $[Mo_{36}O_{112}(H_2O)_{16}]^{8-}$ and can be synthesized at a pH of 1.8 as:

$$36MoO_4^{2-} + 64H^+ \longrightarrow [Mo_{36}O_{112}]^{8-} + 32H_2O$$

The structure of Mo₃₆-polymolybdate contains the [(Mo)Mo₅]-type unit comprising a central MoO_7 pentagonal bipyramid sharing edges with five MoO_6 octahedra.

When tungsten trioxide is dissolved in highly basic aqueous solutions of sodium hydroxide or potassium hydroxide (alkali solutions), tungstate ions with tetrahedral geometry are formed as:

$$WO_3 + 2NaOH \longrightarrow Na_2WO_4 + H_2O$$

These normal tungstates, Na_2WO_4, containing discrete WO_4^{2-} units, can easily be crystallized out of aqueous solution. The normal tungstate of other metals can be synthesized from sodium tungstate via double decomposition.

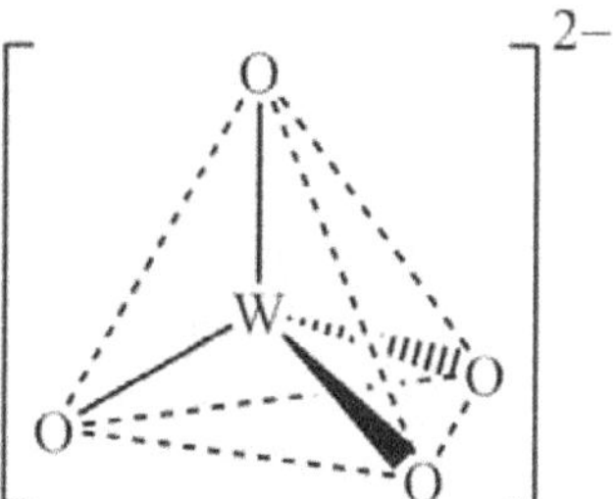

Figure 11. The discrete WO_4^{2-}.

When the pH of the solution is lowered down in the range of 9-5, the protonation of tungstate ions starts followed by the condensation yielding the $W_7O_{24}^{6-}$ i.e. paratungstate A. The whole process of condensation can be depicted as follows:

$$WO_4^{2-} + H_3O^+ \rightleftharpoons [HO-WO_3]^- + H_2O$$

$$[HO-WO_3]^- + H_3O^+ + H_2O \rightleftharpoons W(OH)_6$$

Although the entropy (ΔS) of the second reaction is negative yet it is as fast as the first reaction which may be attributed to the large negative enthalpy for the second reaction, compensating the entropy loss. The $W(OH)_6$, thus formed during the course of the second reaction, reacts with $[WO_3(OH)]^-$ ions present in the acidic media as:

$$W(OH)_6 \underset{-H_2O}{\overset{HWO_4^-}{\rightleftharpoons}} [(HO)_5{-}W{-}(OMoO_3)]^- \underset{-H_2O}{\overset{HWO_4^-}{\rightleftharpoons}} [(HO)_4{-}W{-}(OMoO_3)_2]^{2-}$$

$$\Big\downarrow\; {-H_2O} \;\Big|\; HWO_4^-$$

$$[(HO){-}W{-}(OMoO_3)_5]^{5-} \underset{-H_2O}{\overset{HWO_4^-}{\rightleftharpoons}} [(HO)_2{-}W{-}(OWO_3)_4]^{4-} \underset{-H_2O}{\overset{HWO_4^-}{\rightleftharpoons}} [(HO)_3{-}W{-}(OWO_3)_3]^{3-}$$

$$\Big\uparrow\; {-H_2O} \;\Big|\; HWO_4^-$$

$$[W{-}(OWO_3)_6]^{6-} \xrightarrow{\text{Isomerisation}} [W_7O_{24}]^{6-}$$

In the early times, it was believed that the paratungstate A is formed as a hexamer in solution but later studies showed that it is actually a heptamer. If the acidification of the solution further continues, metatungstate, $[H_2W_{12}O_{40}]^{6-}$, is obtained. Metatungstate is more soluble than paratungstate and can be crystallized out by prolonged heating of the solution or by standing for a long time.

If the pH of the solution is maintained in the range of 2–1, a yellow-colored tungstate $[W_{10}O_{32}]^{4-}$, tungstate Y, is obtained. Moreover, if pH is lowered down below 1, yellow-colored, tungstic acid ($WO_3.2H_2O$) is formed which can be converted into monohydrate form just by warming it up.

The complete reactions giving different isopoly molybdates can be written as:

$$7WO_4^{2-} + 8H^+ \longrightarrow [W_7O_{24}]^{6-} + 4H_2O$$

$$12WO_4^{2-} + 14H^+ \longrightarrow [H_2W_{12}O_{42}]^{10-} + 6H_2O$$

$$12WO_4^{2-} + 18H^+ \longrightarrow [H_2W_{12}O_{40}]^{6-} + 8H_2O$$

The has been proven that $W_2O_7^{2-}$ never exists in the solution phase and no discrete $W_2O_7^{2-}$ ions in the crystal phase are obtained. However, $Na_2W_2O_7$ can be prepared by fusing tungsten oxide with sodium oxide in the required stoichiometry. Recent studies have shown that no W_6 is obtained in solution phase but discrete salts with $[W_6O_{19}]^{6-}$ ions can be crystallized out.

The reaction scheme for the condensation of tungstate ions in aqueous solutions (along with the pH range required) is given below:

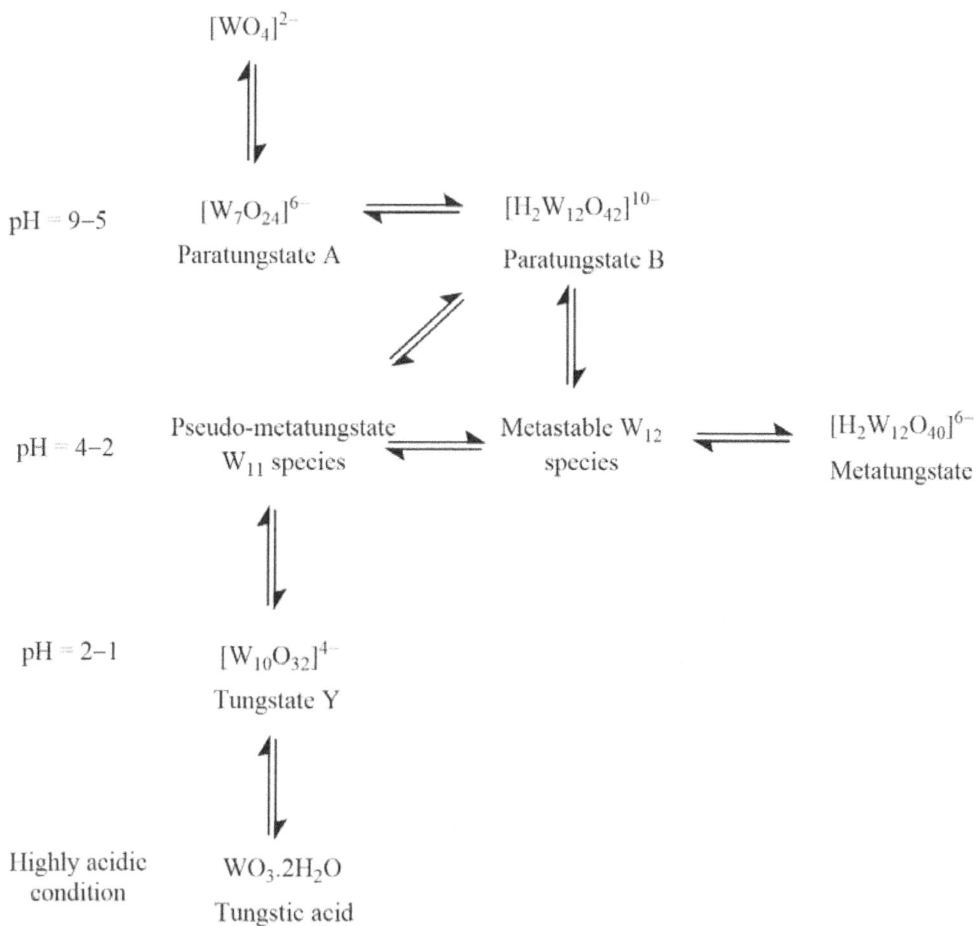

Just like in isopoly-molybdates, the condensation process changes the coordination number of tungsten ion from four to six and the building block unit of polyhedral entity becomes WO_6 octahedron. These WO_6 octahedrons are joined together by the sharing of edges and corners rather than the faces which we have already discussed in isopoly anions of molybdenum.

The structure and properties of different types of isopoly-molybdate are discussed below.

1. Paratungstate A: The general formula for paratungstate A is $M_6W_7O_{24}$ where M is normally Na, K or NH_4 and contains discrete $W_7O_{24}^{6-}$ ions. The edge connections of seven ideal WO_6 octahedrons in paratungstate-A ion can be visualized in terms of one front unit (composed of four WO_6) and one rear unit (composed of three WO_6). As the name suggests, the front unit is placed above the rear unit in such a way that each octahedron shares three edges with its neighbors. It should be noted that $W_6O_{24}^{6-}$ is isostructural with $Mo_6O_{24}^{6-}$.

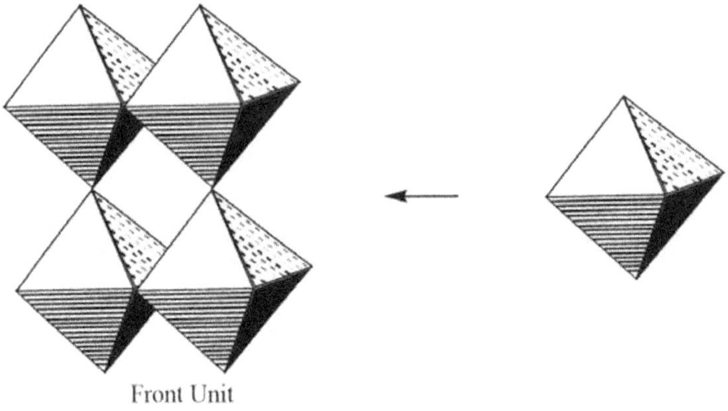

Front Unit

(a)

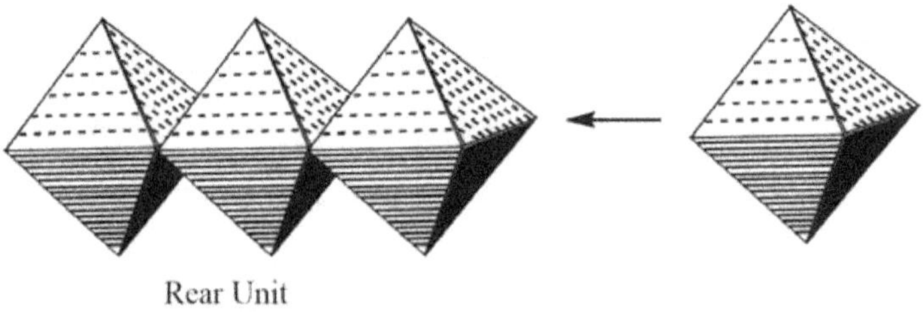

Rear Unit

(b)

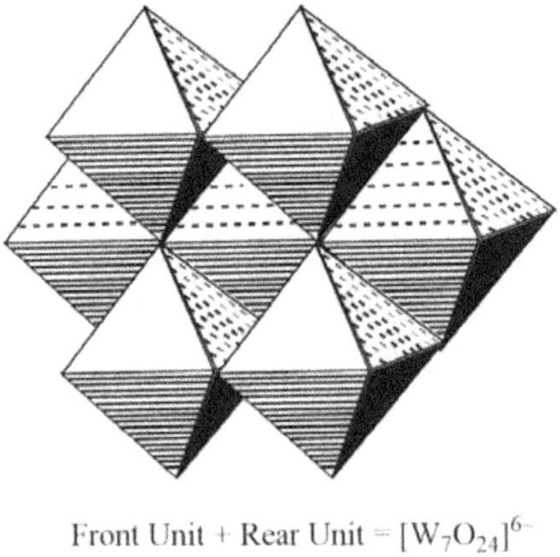

Front Unit + Rear Unit = $[W_7O_{24}]^{6-}$

(c)

Figure 12. The (a) front unit and (b) rear unit are combined to produce the (c) overall structure of the paratungstate A ion.

2. Paratungstate B: The paratungstate B is derived as the alkali metal salts of dodecameric anions, $[H_2W_{12}O_{42}]^{10-}$, from the acidification of normal tungstate aqueous solutions. The protons reside in the polyhedral cavity and are involved in rapid exchange solvent water.

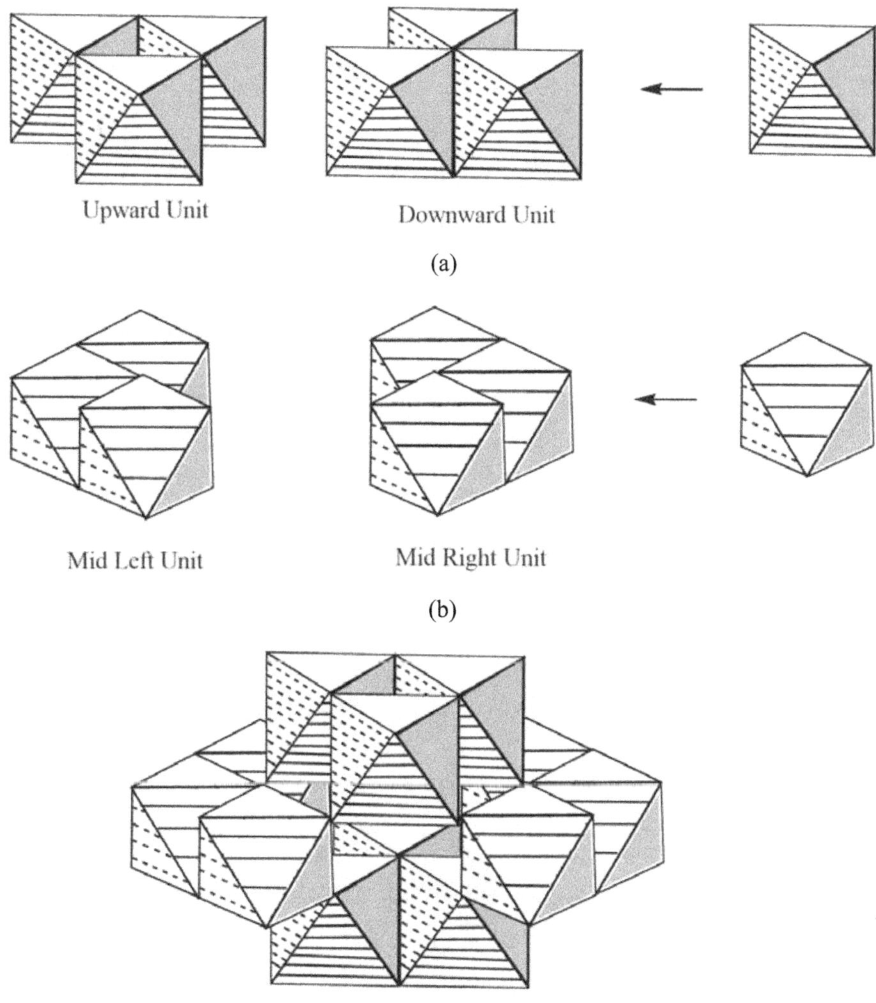

Figure 12. The (a) upward - downward units and (b) mid left - mid units are combined to produce the (c) overall structure of the metatungstate anion.

3. Metatungstate: The metatungstates are derived as the alkali metal salts of dodecameric anions, $[H_2W_{12}O_{40}]^{6-}$, from the acidification of normal tungstate aqueous solutions. The protons reside in the polyhedral cavity and are not involved in rapid exchange with solvent water as in peratungstate B.

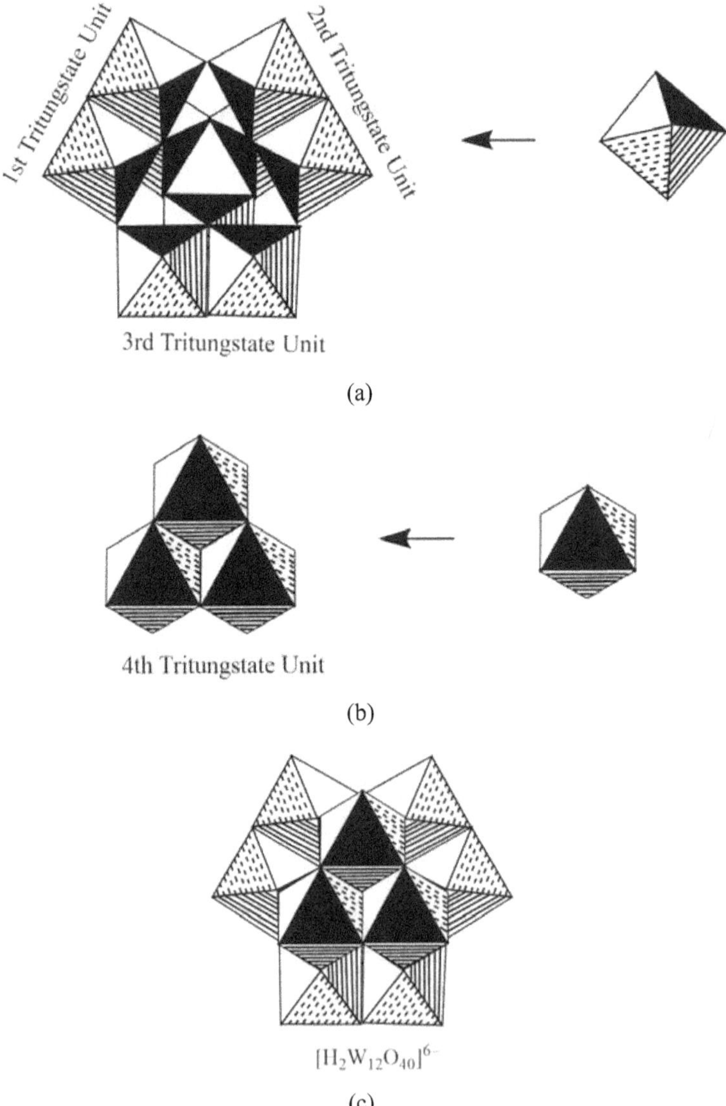

Figure 13. The structure of metatungstate anion is obtained by joining four equivalent tritungstate units along tetrahedral faces.

4. Tungstate Y: These isopoly tungstates are derived as the alkali metal salts of $[W_{10}O_{32}]^{4-}$ anions and are yellow in color. The discrete $[W_{10}O_{32}]^{4-}$ anions are found to be present in both solution and crystallized form.

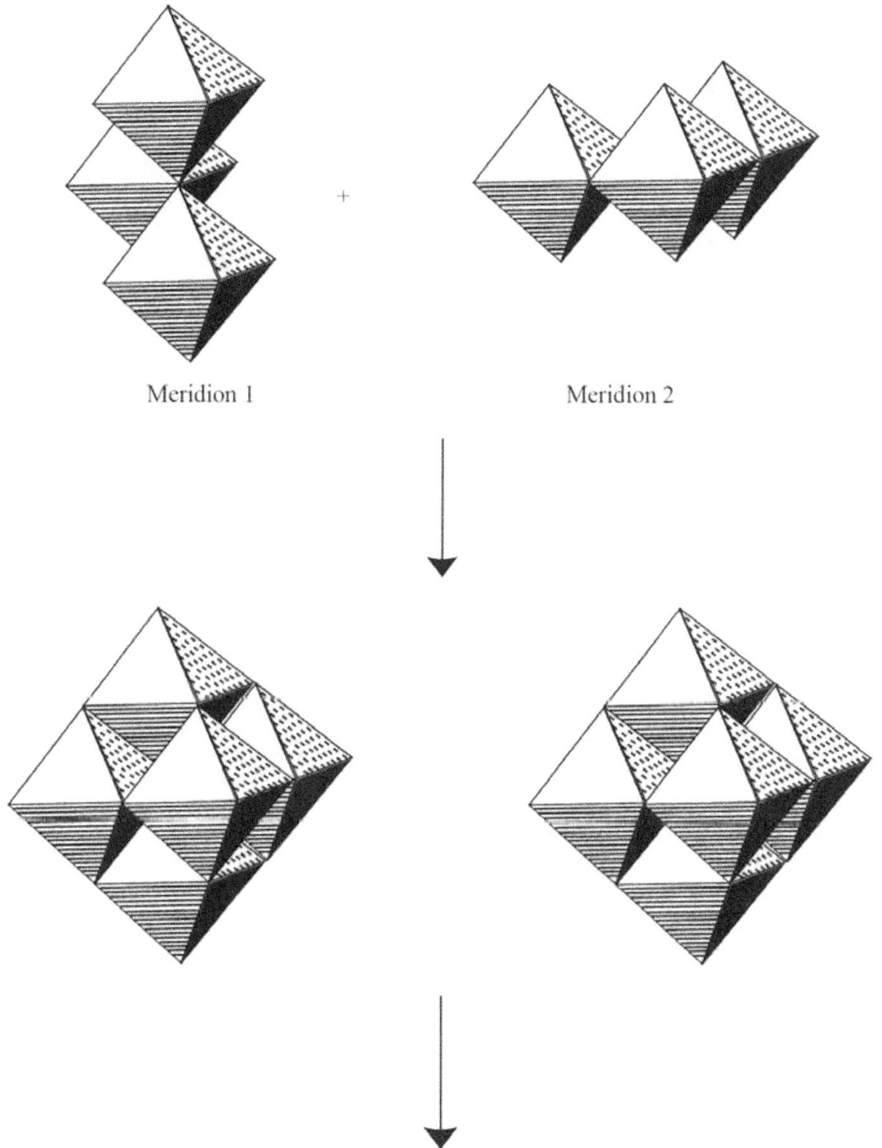

Figure 14. Continued on the next page…

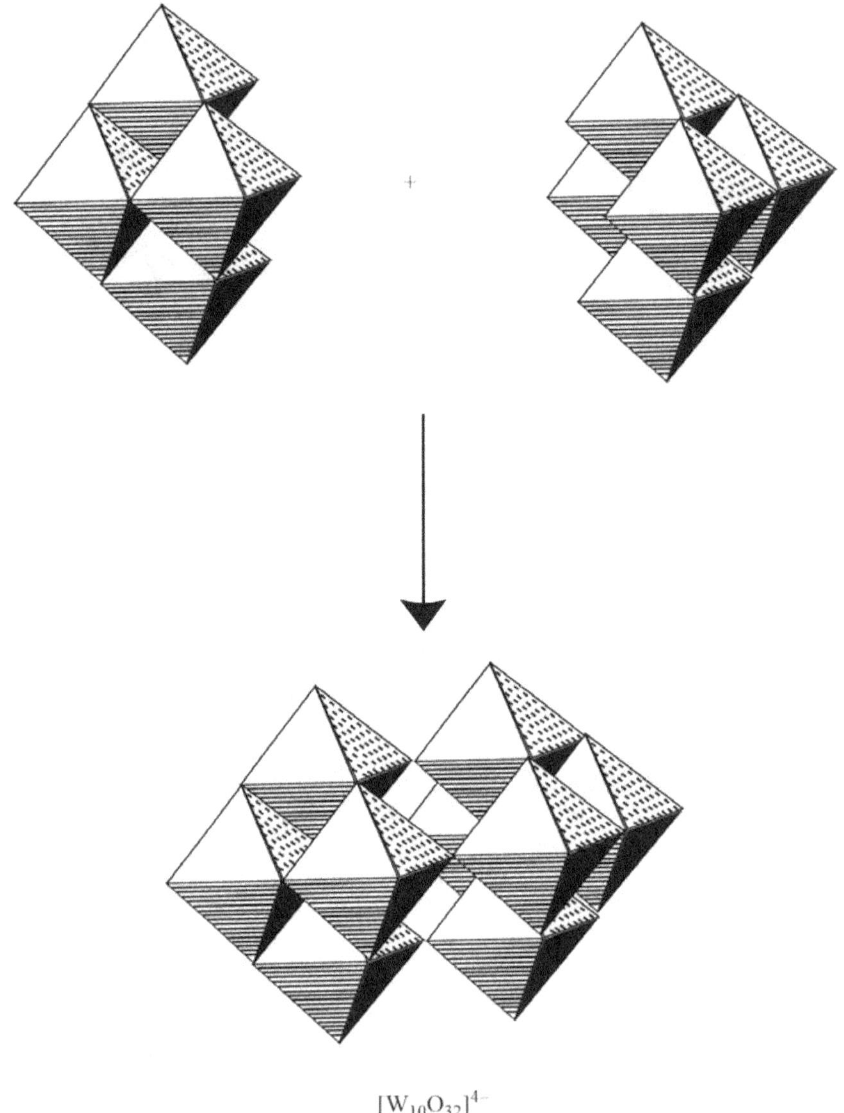

$$[W_{10}O_{32}]^{4-}$$

Figure 14. The structure of tungstate Y.

The structure of tungstate Y can be visualized from hexa-tungstate systems in which two meridians, composed of WO_6 octahedral units are joined to form W6–polyhedra. When one octahedron from each $[W_6O_{10}]^{2-}$ unit is removed, the resulting structural units can be combined through the corners to form the overall geometry of decameric isopoly anion.

5. Hexatungstate: These isopoly hexatungstates are obtained from methanolic solutions and contain discrete $[W_6O_{19}]^{2-}$ ions both in solution and in crystallized form.

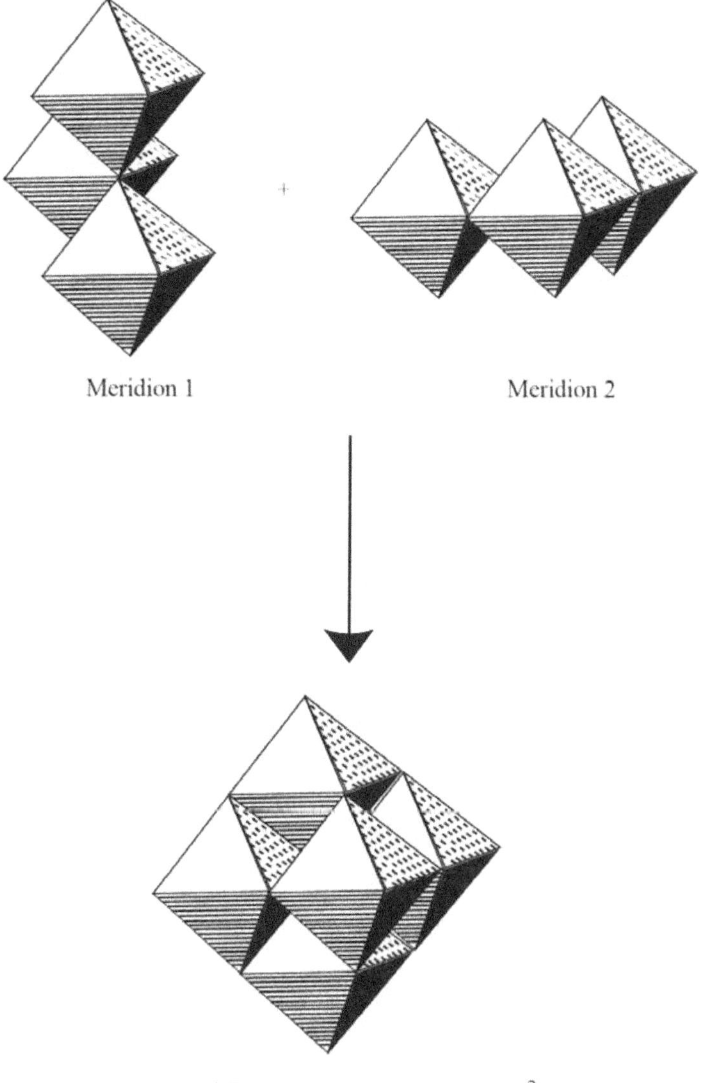

Figure 15. The hexatungstate anion.

It is worth noting that hexatungstate is isostructural with hexamolybdate anion in which the coordination number of the oxide ion situated at the center of the $[W_6O_{19}]^{2-}$ anion is six.

6. Tetratungstate: The crystal form of $Li_{14}(WO_4)_3(W_4O_{16}).4H_2O$, obtained from aqueous solution, is found to have discrete $W_4O_{16}{}^{8-}$. However, the presence of this isopoly anion in the solution phase is still unknown.

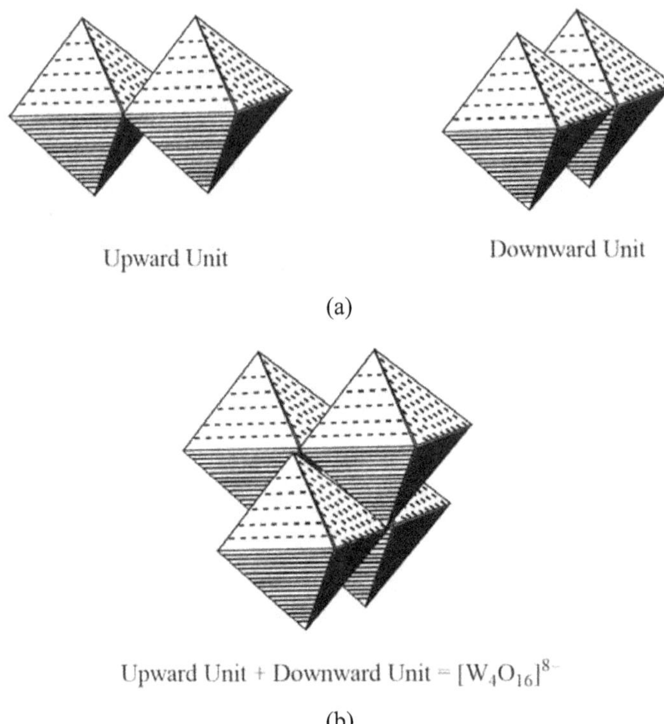

(a)

(b)

Figure 16. The (a) upward and downward units are combined to produce the (b) overall structure of $[W_4O_{16}]^{8-}$ anion.

> ### *Heteropoly Acids and Salts of Mo and W*

One of the founders of modern chemistry, a Swedish chemist, Jacob Berzelius found that the acidification of the solution containing phosphate and molybdate results in a yellow-colored precipitate. This crystalline substance was actually the first example of a heteropoly anion, $[PMo_{12}O_{40}]^{3-}$, which has been frequently used in the quantitative estimation phosphates. A large number of heteropoly anions with a wide range of hetero-atoms, metal as well as non-metal, have been prepared in the following years. The thermal stability of heteropoly salts is found to be greater than their isopoly analogs. One of the major application of these heteropoly anions is in the petrochemical industry where they are used as a catalyst, as flame retardants and as precipitants for dyes.

The hetero-atoms of these polymetallates generally reside in the baskets or cavities formed by the parent MO_6 octahedron units. The hetero-atoms are bonded to the neighboring oxygens of surrounding octahedrons. The shape of the cavity, generally decided by the ratio of hetero to parent atoms, greatly influences the stereochemistry of heteroatom. The acidification of simple monomeric anions can yield condensed heteropoly systems as:

$$HPO_4^{-2} + 12MoO_4^{2-} + 23H^+ \longrightarrow [PMo_{12}O_{40}]^{3-} + 12H_2O$$

$$HPO_4^{-2} + 12WO_4^{2-} + 23H^+ \longrightarrow [PW_{12}O_{40}]^{3-} + 12H_2O$$

The structure and properties of different types of heteropoly-metallates are discussed below.

1. 1:12 (Tetrahedral heteroatom): The general formula for these types of heteropolyanions is $[X^{n+}M_{12}O_{40}]^{(8-n)-}$, where M is Mo or W and X represents the heteroatom which is generally P, Si, As, Ge or Ti. The labeling of this category of polyanions is justified on the basis of the number ratio of heteroatom to the parent atom which is one to twelve. The small heteroatom is present in the inner tetrahedral cavity of the polyanion and hence is surrounded by six oxygen atoms of different MO_6 octahedron units.

The structure of this type of heteropolyanions is of either T_d symmetry (Keggin structure) or a C_{3v} one which can be obtained as a derivative of Keggin structure by rotating one of the four sets of trimetallate (composed of three MO_6) units through 60°. The first structure is perfect tetrahedral with four three-fold axes of rotation but the second structure contains only one three-fold axis of rotation. The stability of both the structures is the same. $[XM_{12}O_{40}]^{4-}$, with X as Ge^{4+} or Si^{4+}, is found to be present in both of the isomeric forms.

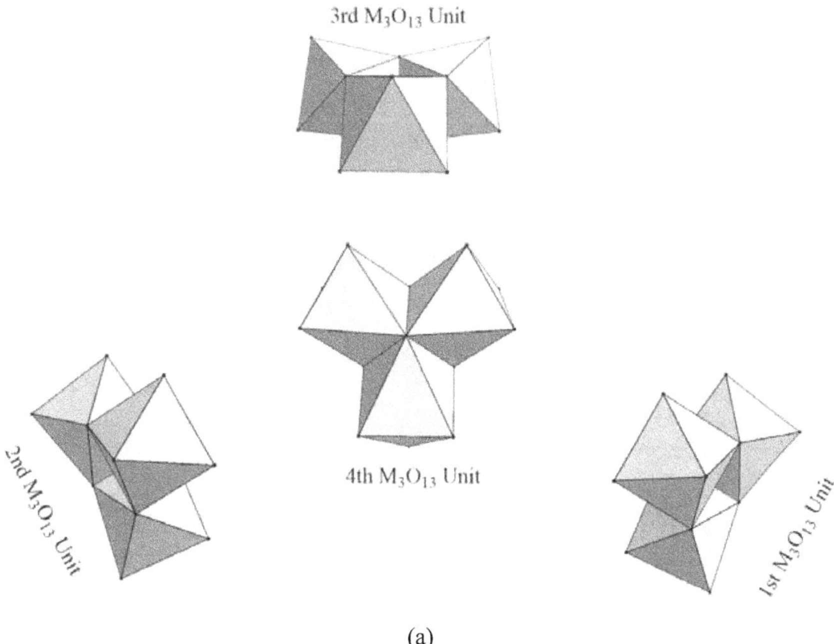

(a)

Figure 17. Continued on the next page…

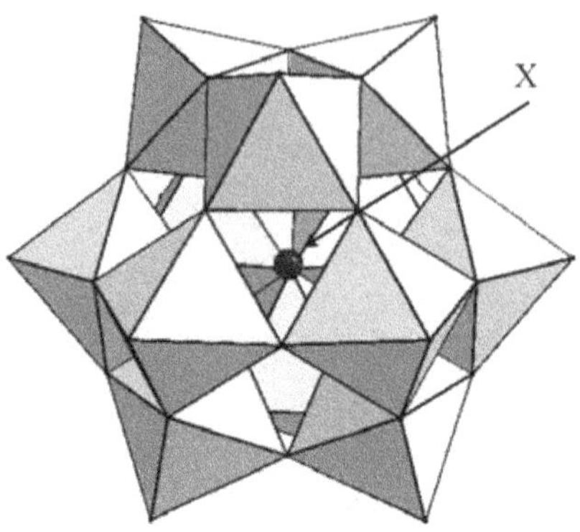

The Keggin Structure of $[X^{n+}M_{12}O_{40}]^{(8-n)-}$

(b)

Figure 17. (a) four M_3O_{13} units are combined to produce a tetrahedral cavity occupied by the heteroatom. (b) The overall structure of $[X^{n+}M_{12}O_{40}]^{(8-n)-}$ anion.

One of the most common examples of these types of heteropoly anions is $[Co^{2+}W_{12}O_{40}]^{6-}$, which can be oxidized to $[Co^{3+}W_{12}O_{40}]^{5-}$. In both of the anions, the heteroatom is present the tetrahedral void and is surrounded by four oxygen atoms of nearby MoO_6 octahedrons.

2. 2:18 (Tetrahedral heteroatom): The general formula for this type of heteropolyanions is $[X_2^{5+}M_{18}O_{62}]^{6-}$, where M is Mo or W and X represents the heteroatom which is generally P or As. The labeling of this category of polyanions is justified on the basis of the number ratio of heteroatom to the parent atom which is two to eighteen. The potassium or ammonium salts of these polyanions can be crystallized out by standing a mixture of salts of 1:12 $[X^{n+}M_{12}O_{40}]^{(8-n)-}$ heteropoly anions.

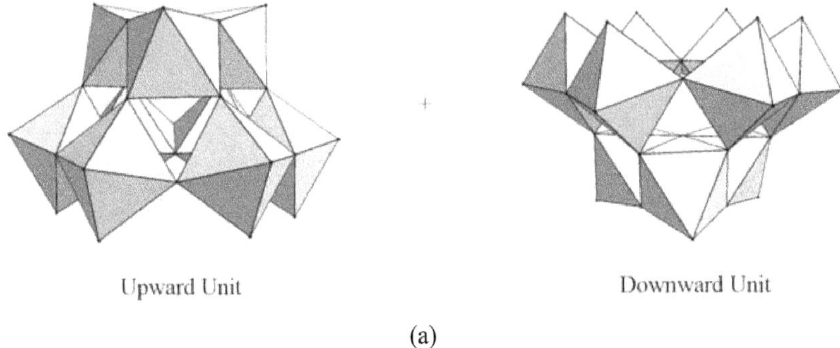

(a)

Figure 18. Continued on the next page…

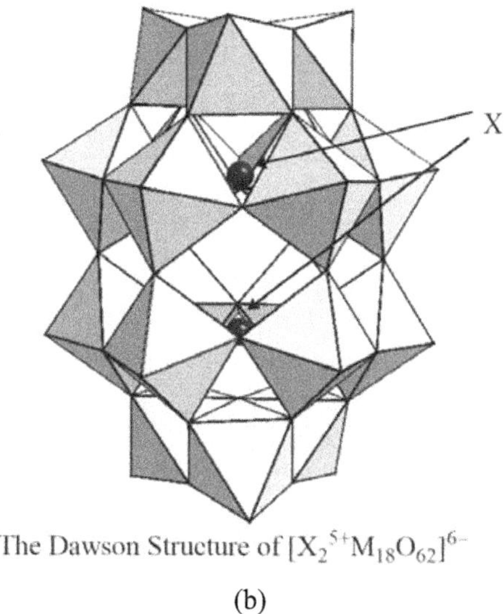

The Dawson Structure of $[X_2^{5+}M_{18}O_{62}]^{6-}$

(b)

Figure 18. The (a) upward and downward units are combined to produces the (b) overall structure of $[X_2^{5+}M_{18}O_{62}]^{6-}$ anion.

The structure is generally referred as the "Dawson structure" and can be visualized from Keggin structure easily. The $[X_2^{5+}M_{18}O_{62}]^{6-}$ can be considered as a fusion of two M_9–units each of which is derived from the Keggin structure by the removal of three corner linked Mo_6 octahedrons. Both of the heteroatoms are present in the inner cavity of the polyanion made by different MO_6 octahedron units.

3. 1:6 (Octahedral heteroatom): The general formula for this type of heteropolyanions is $[X^{n+}M_6O_{24}]^{(12-n)-}$, where M is Mo or W and X represents the heteroatom which is generally Te, Co, I or Al. The labeling of this category of polyanions is justified on the basis of the number ratio of heteroatom to the parent atom which is one to six. The tendency form 1:6 heteropoly anions is found to be greater in the case of Mo than in W. The heteroatoms are generally larger in size and are surrounded by six oxygen atoms of edge-sharing MO_6 octahedron units. It is worth noting down that all the six MO_6 octahedron units are parallel to each other.

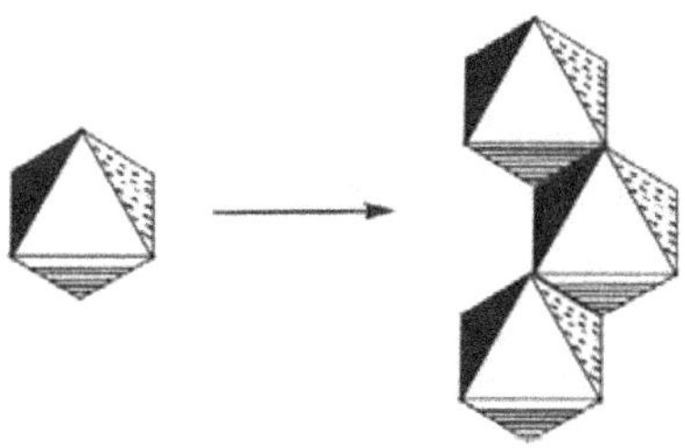

Figure 19. Continued on the next page…

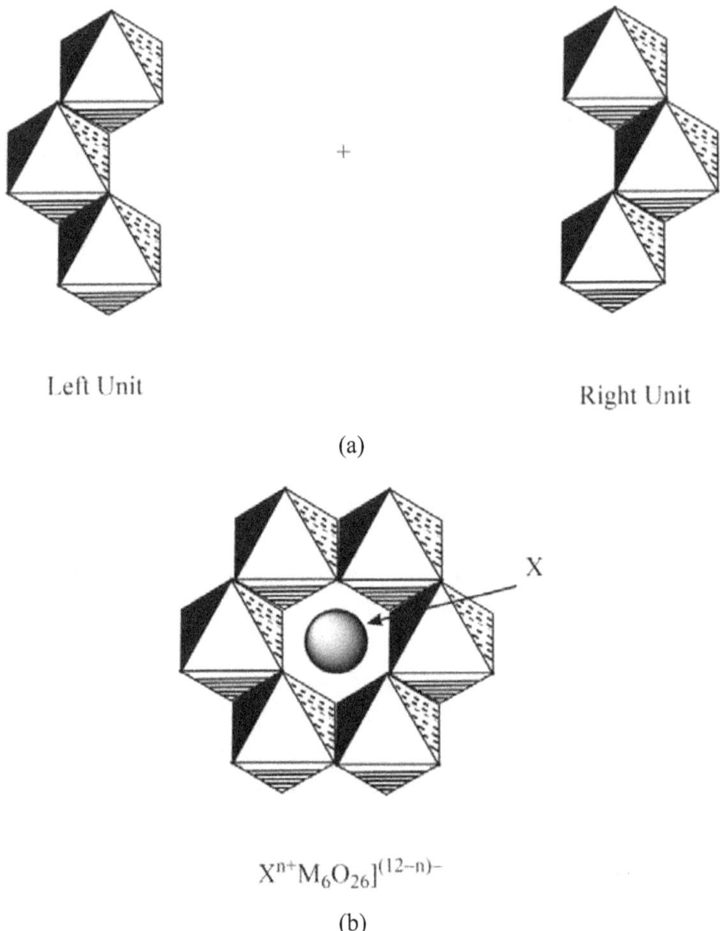

$$X^{n+}M_6O_{26}]^{(12-n)-}$$

Figure 19. The (a) left and right units are combined to produce the (b) overall structure of $[X^{n+}M_6O_{26}]^{(12-n)-}$ heteroploy anion.

4. 1:9 (Octahedral heteroatom): The general formula for this type of heteropolyanions is $[X^{n+}M_9O_{32}]^{(10-n)-}$, where M is Mo or W and X represents the heteroatom which is generally Mn or Ni. The labeling of this category of polyanions is justified on the basis of the number ratio of heteroatom to the parent atom which is one to nine.

The nine MO_6 octahedron units are joined through the edges to form an octahedral cavity for the heteroatom present. Two of the most studied examples of these types of heteropoly anions are $[NiMo_9O_{32}]^{6-}$ and $[MnMo_9O_{32}]^{6-}$. It must also be noted down that Ni and Mn are present in +4 oxidation state which is highly unusual.

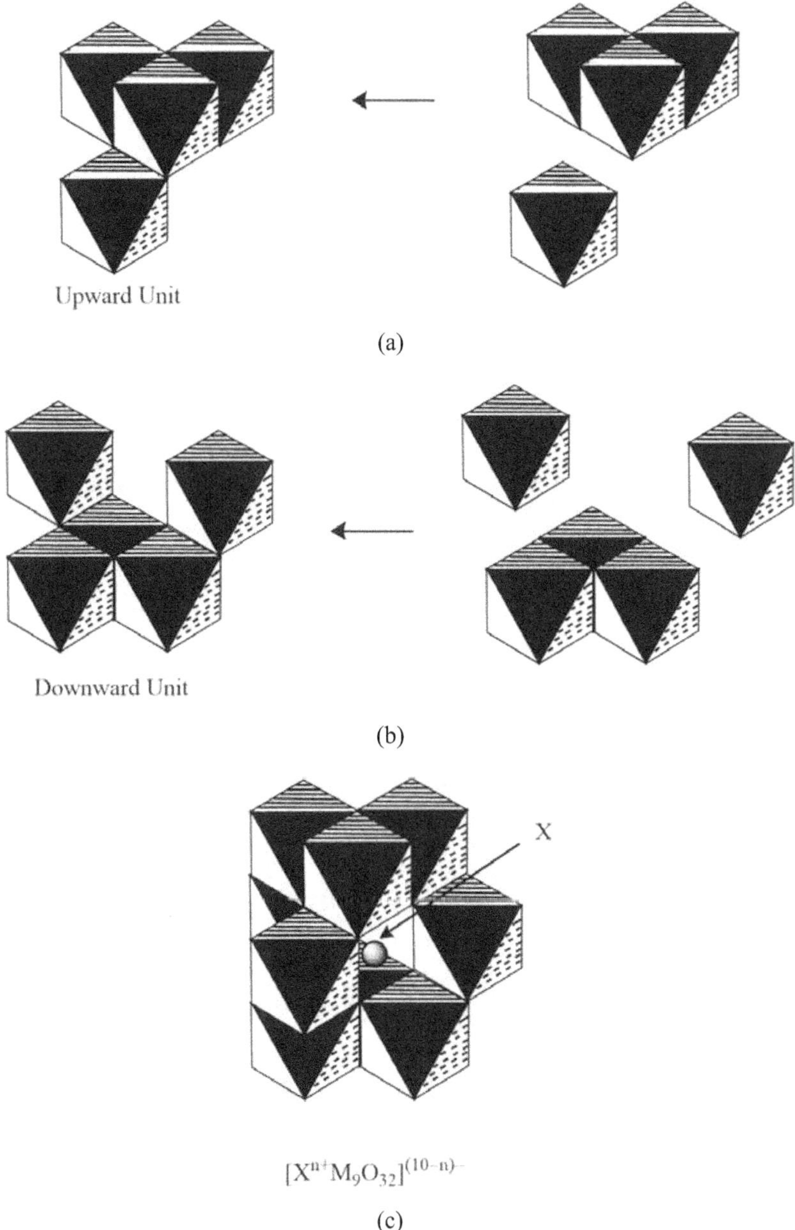

$[X^{n+}M_9O_{32}]^{(10-n)-}$

Figure 20. The (a) upward and (b) downward units are combined to produces the (c) overall structure of $[X^{n+}M_9O_{32}]^{(10-n)-}$ anion.

5. 1:12 (Icosahedral heteroatom): The general formula for this type of heteropolyanions is $[X^{n+}M_{12}O_{42}]^{(12-n)-}$, where M is Mo or W and X represents the heteroatom which is generally Ce, Th or U. The labeling of this category of polyanions is justified on the basis of the number ratio of heteroatom to the parent atom which is one to twelve. The heteroatom is present in an icosahedral void and is surrounded by 12 oxygen atoms of the nearby polyhedral unit of the parent atoms.

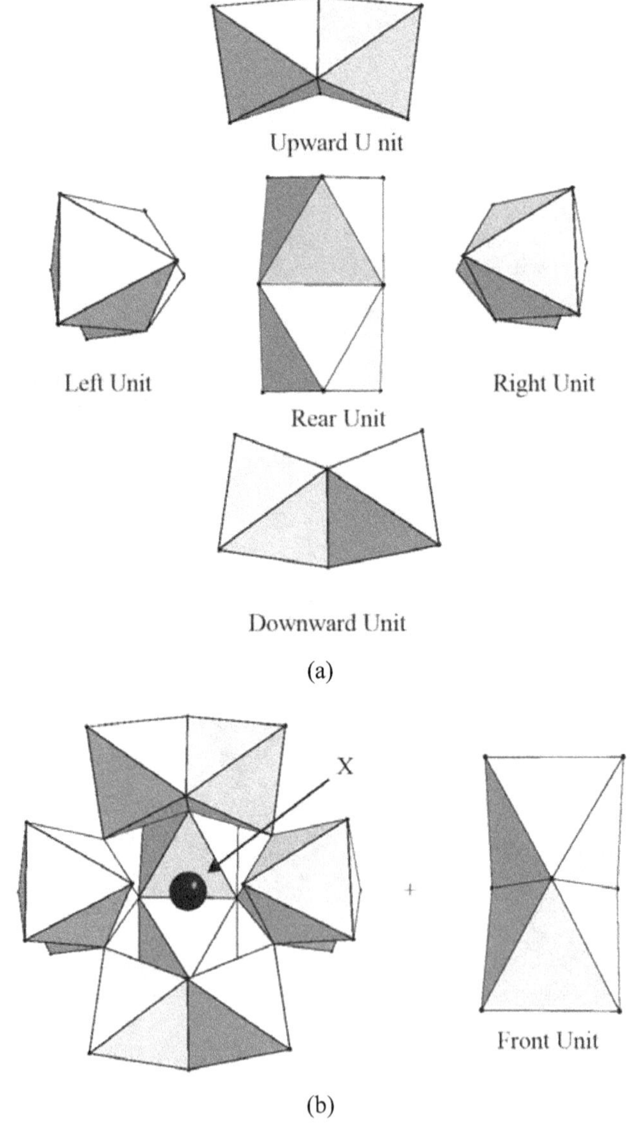

Figure 21. Continued on the next page…

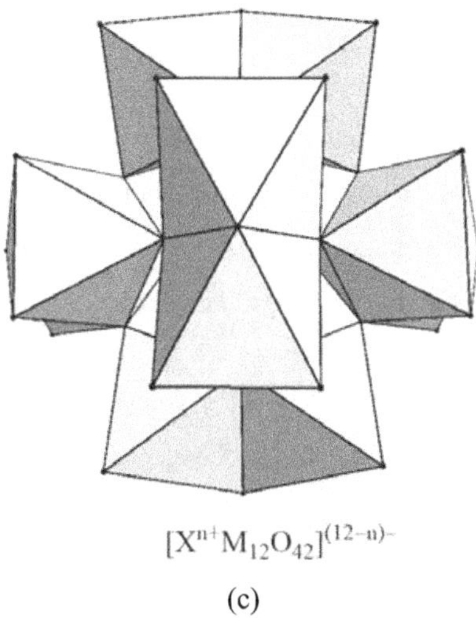

$$[X^{n+}M_{12}O_{42}]^{(12-n)-}$$

(c)

Figure 21. (a) Five M_2O_{10} units are fused to create (b) $M_{10}O_{30}$ fragment which, in turn, covers by the sixth M_2O_{10} unit resulting in the (c) overall structure of $[X^{n+}M_{12}O_{42}]^{(12-n)-}$ anion.

In recent years, bigger heteropoly anions with crown i.e. $[Mo_{36}O_{110}(NO)_4(H_2O)_{14}]$ and double crown i.e. structure $[NaP_5W_{36}O_{110}]^{14-}$, $[H_7P_8W_{48}O_{184}]^{33-}$, have also been prepared.

❖ Problems

Q 1. Define isopoly and heteropoly metallates.

Q 2. Explain the condensation process involved in the formation of paramolybdates.

Q 3. Draw and discuss the structure of paramolybdate anion, $Mo_7O_{24}^{6-}$, in detail.

Q 4. Draw and discuss the structure of octamolybdate anion.

Q 5. Write down the reaction scheme giving the tungstic acid from normal tungstate with special reference to paramolybdate A.

Q 6. Explain the structure of metatungstate, $[H_2W_{12}O_{40}]^{6-}$, in detail.

Q 7. How can a Keggin structure with T_d symmetry be changed into a C_{3v} isomer?

Q 8. Draw and explain the structure of $[NiMo_9O_{32}]^{6-}$ and $[MnMo_9O_{32}]^{6-}$.

Q 9. Why do the isopoly and heteropoly acids of molybdenum and tungsten prefer edge-sharing over the corner connection? Also, explain the highly unusual face sharing.

Q 10. Write down the general formula for 1:12 and 2:18 heteropoly anions.

❖ Bibliography

[1] B. R. Puri, L. R. Sharma, K. C. Kalia, *Principals of Inorganic Chemistry*, Milestone Publishers, Delhi, India, 2012.

[2] J. E. Huheey, E. A. Keiter, R. L. Keiter, *Inorganic Chemistry: Principals of Structure and Reactivity*, HarperCollins College Publishers, New York, USA, 1993.

[3] A. F. Wells, *Structural Inorganic Chemistry*, Oxford University Press, London, UK, 1975.

[4] F. A. Cotton, G. Wilkinson, C. A. Murillo, M. Bochmann, *Advanced Inorganic Chemistry*, John Wiley & Sons, New Jersey, USA, 1999.

[5] J. D. Lee, *Concise Inorganic Chemistry*, Chapman & Hall, New York, USA, 1994.

[6] N. N. Greenwood, A. Earnshaw, *Chemistry of the Elements*, Butterworth-Heinemann, Oxford, Britain, 1998.

[7] M. Pope, *Heteropoly and Isopoly Oxometalates*, Springer Verlag GmbH, Heidelberg, Germany, 2013.

CHAPTER 6

Crystal Structures:

❖ **Structures of Some Binary and Ternary Compounds Such as Fluorite, Antifluorite, Rutile, Antirutile, Cristobalite, Layer Lattices - CdI_2, BiI_3; ReO_3, Mn_2O_3, Corundum, Perovskite, Ilmenite and Calcite.**

The structure of crystalline compounds can be understood only after knowing the close packing of solid spheres as all the constituent particles (whether they are atoms, ions or molecules) may be treated as somewhat more or less spherical in shape.

The problem of spheres packed closely was first studied in a mathematical framework by Thomas Harriot in 1587. The idea was initiated by a question on piling the cannonballs on ships, and the question was posed to T. Harriot by Sir W. Raleigh when they were on their expedition to the land of America. The usual scheme of piling the cannonballs included the placement in a triangular or rectangular wooden frame, which forms a three or four-sided pyramidal shape. Both arrangements produce a face-centered cubic (fcc) lattice with different orientations to the base. However, the hexagonal-close-packing (hcp) scheme would result in a six-sided pyramidal shape with a hexagonal base. These two simple regular lattices achieve the highest average density and are based upon sheets of spheres placed at the vertices trigonal with a tiling pattern. However, they vary in the stacked style of these sheets.

The packing of identical spheres in two dimensions can be visualized in two arrangements as given below.

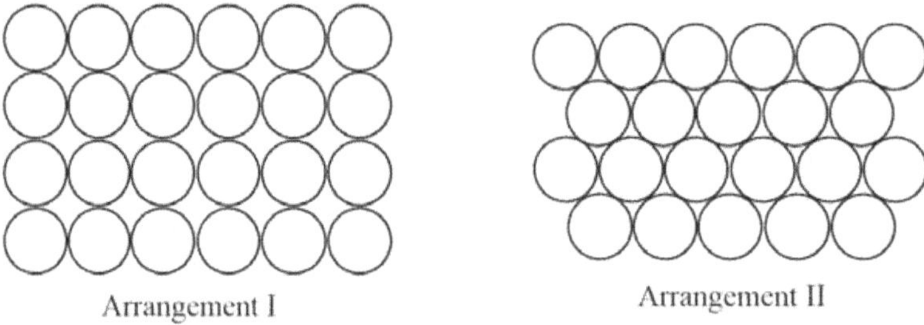

Figure 1. The close packing of identical spheres in two dimensions.

It has been observed that 52.4% of available space is occupied in the arrangement I while 60.4% of available space is occupied in arrangement II. The remaining space is empty and is labeled as void volume. Hence, the arrangement II has greater efficiency than arrangement I. Furthermore, it can be noticed that the coordination

---

number of each sphere in arrangement II is six but four in case of arrangement I and the lines joining the centers of nearest spheres in the arrangement I form a square but a tringle in arrangement II.

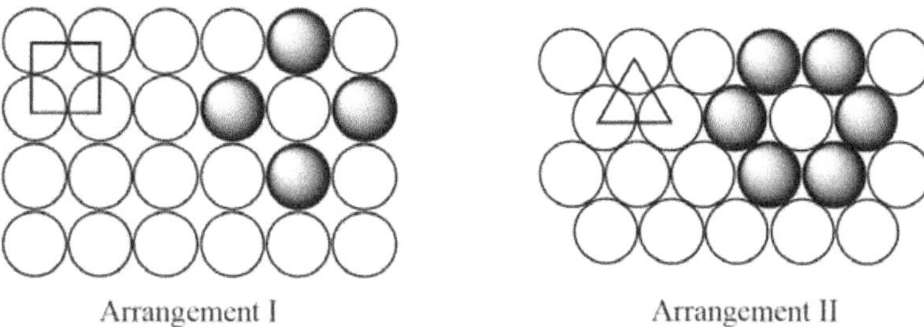

Figure 2. The coordination number and nearest sphere system in close packing of identical spheres in two dimensions.

The two-dimensional packing can successfully be extended to three-dimensional packing just by placing the other layers of packed identical spheres over it. This can be achieved in two ways as depicted below.

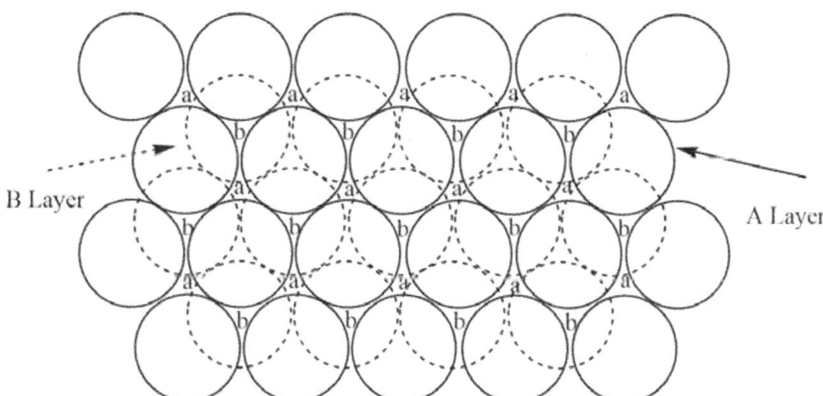

Figure 3. The extension of close packing of identical sphere systems in two dimensions to three dimensions.

The first layer is labeled as A, while the second layer B is placed over the first layer in such a way that its spheres are covering the b-voids. Now, if we create a third layer directly over the layer A, then the hcp lattice is formed and if the third layer is placed over a-voids in the first layer, then the fcc lattice is built. In both the arrangements, the fcc as well as hcp, each sphere has twelve neighboring spheres. For every sphere, there is one void surrounded by six neighboring spheres (octahedral site) and two smaller voids surrounded by four neighboring spheres (tetrahedral site). The distances to the centers of these gaps from the centers of the surrounding spheres are $\sqrt{3}/2$ for the tetrahedral, and $\sqrt{2}$ for the octahedral when the sphere radius is unity.

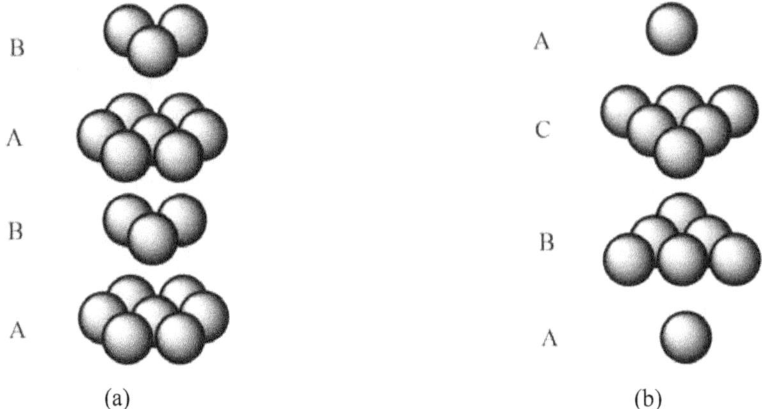

Figure 4. The illustration of the close-packing of equal spheres in both (a) hcp and (b) fcc lattices.

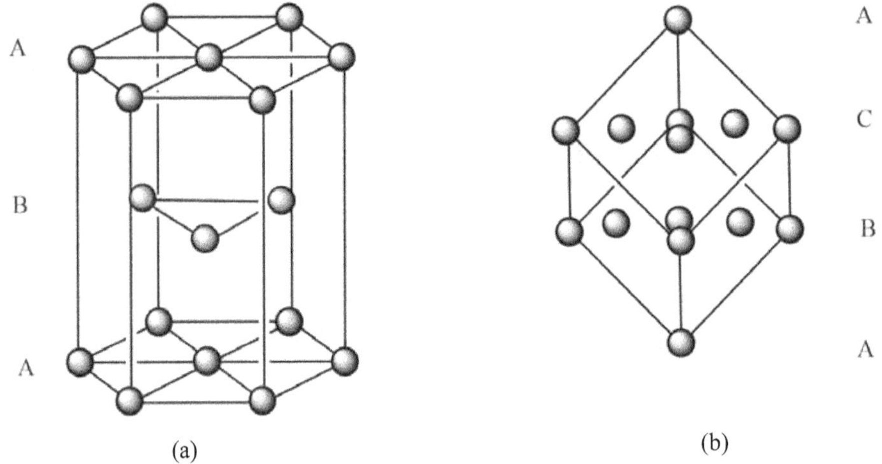

Figure 5. The ball and stick model of the close-packing of equal spheres in both (a) hcp and (b) fcc lattices.

The number of interstitial sites, tetrahedral as well as octahedral, is the same for fcc and hcp lattice systems. However, the tetrahedral sites are double in number than octahedral ones. A number of crystal structures are obtained by the close-packing of one kind of atom, or by a close-packing of bigger ions having smaller ions

occupying the void spaces. The hexagonal and cubic patterns of packing are quite close to each other in energy, and therefore, it might be problematic to predict the form preferred using first-principle calculations.

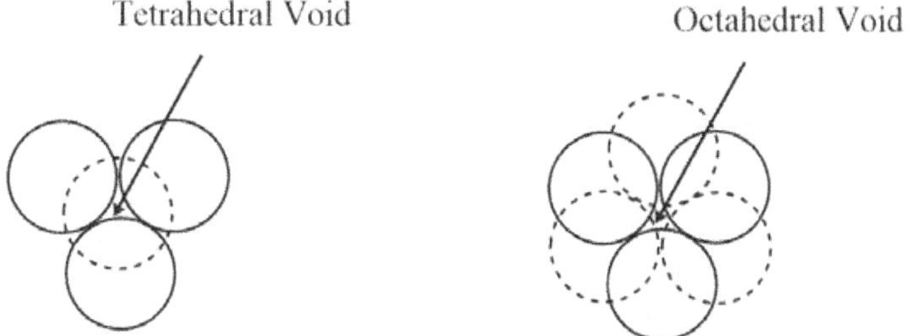

Figure 6. The formation of a tetrahedral and octahedral void in close packing of identical spheres in three dimensions.

The location of interstitial sites in hexagonal close packing and cubic close packing is quite important as it is expected to be occupied by smaller ion which is generally cation. Nonetheless, the anions can also occupy these tetrahedral octahedral voids if they are smaller in size.

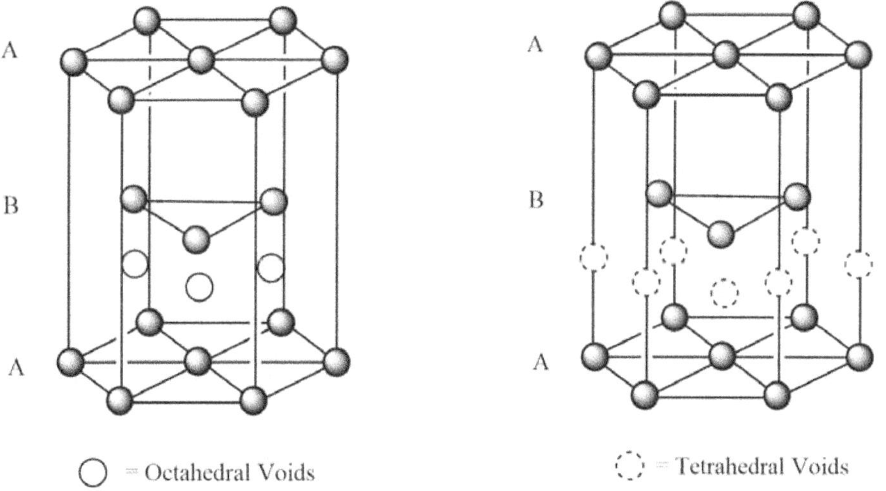

Figure 7. The location of octahedral and tetrahedral sites in hexagonal close packing or hcp lattices formed by the identical spheres.

In a hcp structure, the arrangement of the third layer is the same as the spheres' placement of first layer and covers all the tetrahedral voids. The hcp and fcc type packings of equal spheres are of the highest density with the highest symmetry. Though the denser packings are known; they, however, involve spheres of unequal size. Only a packing of non-spherical shapes, like honeycombs, is capable of occupying a hundred percent space.

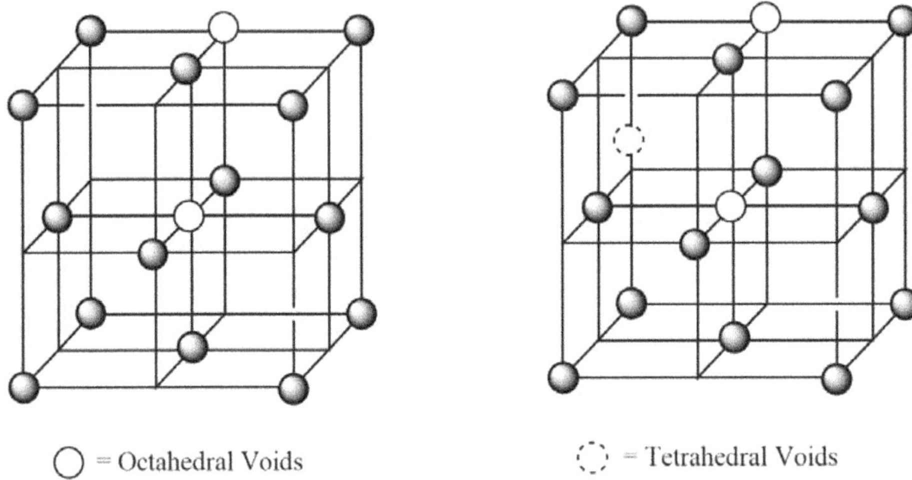

Figure 8. The location of octahedral and tetrahedral sites in cubic close packing or fcc lattices.

In addition to the hcp and fcc arrangements, one more common lattice-type also exists, called as body-centered cubic or simply the bcc lattice. The placement of the second layer opens the spheres of the first layer and the third layer is exactly eclipsed to the first layer. Each sphere is in contact with spheres of the first layer and four spheres of the third layer giving an overall coordination number of eight.

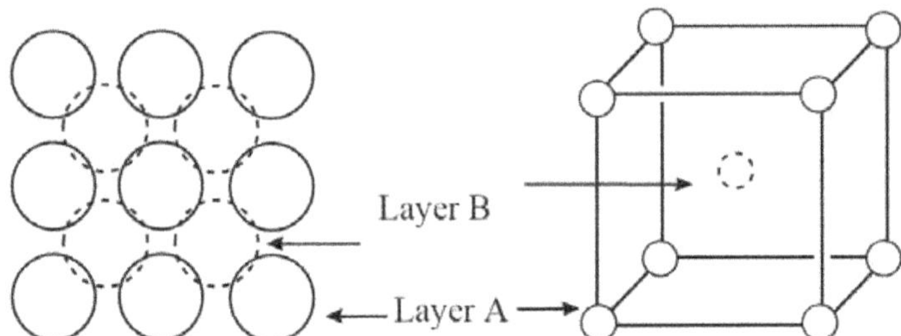

Figure 9. Body-centered close packing of equal spheres.

This arrangement can also be visualized as a derivative of a simple cubic unit cell in which eight spheres are present at the corners of the cube and one extra sphere is placed in the center. The packing efficiency of body-centered cubic packing is lesser than hcp or fcc lattice system and only 68% of the available volume is actually occupied.

Now, as we have studied the packing of equal spheres and the resulting void types, we are ready to discuss the crystal structure of some important binary and ternary compounds.

> *Fluorite (CaF₂)*

Fluorite or fluorspar is the mineral form of calcium fluoride, CaF_2. The unit cell of CaF_2 is shown below.

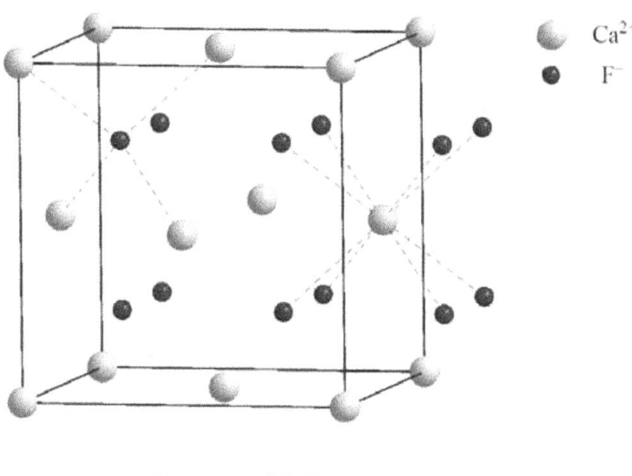

Figure 10. The crystal structure of calcium fluoride or CaF_2.

The main features of this crystal structure are as follows:

1. The Ca^{2+} ions are present in cubic close packing or fcc lattice in which they are present at all the corners and at the center of each face.

2. The F^- ions are occupying all the tetrahedral sites.

3. The number of tetrahedral sites for the ccp packing of N spheres is 2N. This means that two tetrahedral sites are formed for each Ca^{2+} ion. Now, because all the tetrahedral sites are occupied by F^- anions, the stoichiometry of the compound becomes 1:2.

4. The coordination number of Ca^{2+} is eight while each F^- ion is surrounded by four calcium ions. The surrounding geometry of Ca^{2+} ion is tetrahedral and of F^- is cubical in nature. Therefore, the coordination number ratio of Ca^{2+} and F^- is 8:4.

The other common examples of this structural prototype of are CuF_2, BaF_2, PbF_2, HgF_2, SrF_2. It is worthy to note that ideally packed samples are colorless but the mineral form of fluorite is usually deep in color, which is obviously due to the presence of F-centres. The high purity fluorite is a source of fluoride for the manufacturing of hydrofluoric acid, an intermediate source of most chemicals that contain fluorine. The transparent fluorites of high optical purity have low dispersion, and therefore, are used to make lenses with less chromatic aberration. This makes them valuable for the telescopes and microscopes devices.

> ➢ *Antifluorite (Na₂O)*

Antifluorite is the mineral form of sodium oxide, the unit cell of antifluorite lattice is shown below.

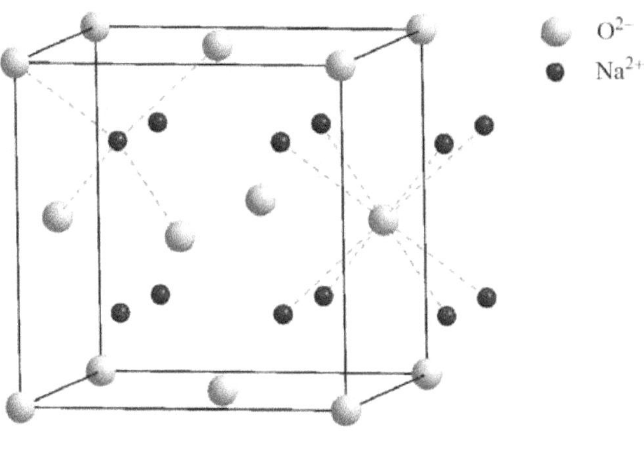

Structure of Na₂O

Figure 8. The crystal structure of calcium fluoride or Na₂O.

The main features of this crystal structure are as follows:

1. The O^{2-} ions are present in cubic close packing or fcc lattice in which they are present at all the corners and at the center of each face.

2. The Na^+ ions are occupying all the tetrahedral sites.

3. The number of tetrahedral sites for the ccp packing of N spheres is 2N. This means that two tetrahedral sites are formed for each O^{2-} ion. Now, because all the tetrahedral sites are occupied by Na^+ anions, the stoichiometry of the compound becomes 2:1.

4. The coordination number of O^{2-} is eight while each Na^+ ion is surrounded by four oxide ions. The surrounding geometry of O^{2-} ion is tetrahedral and of Na^+ is cubical in nature. Therefore, the coordination number ratio of O^{2-} and Na^+ is 4:8.

The other examples of this type of structure are Li_2O, K_2O, Rb_2O. It is worth noting that in this motif the positions of the anions and cations are reversed relative to their positions in CaF_2, with sodium ions tetrahedral coordinated to 4 oxide ions and oxide cubically coordinated to 8 sodium ions. The Na₂O is a primary component of windows and glasses though it is added in the form of "soda" (Na_2CO_3). The Na₂O does not exist in glasses explicitly since glasses are cross-linked polymers of complex profile. The Na_2CO_3 also serves as a flux to lower the temperature for melting the silica. The melting temperature of soda glass is much lower than pure silica but has a slightly higher value of elasticity.

> ### *Rutile (TiO₂)*

Rutile is the mineral form of titanium oxide, TiO_2. The unit cell of titanium oxide is shown below.

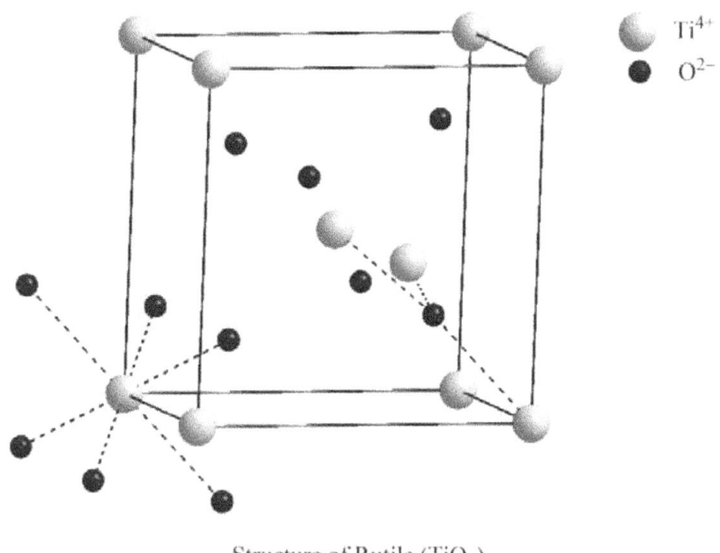

Figure 8. The crystal structure of rutile or TiO_2.

The main features of this crystal structure are as follows:

1. The Ti^{4+} ions are present in distorted body-centered cubic close packing in which they are present at all the corners and at the center of the distorted cube. The unit cell cannot be labeled as cubic because all the three sides are not equal to each other. One of the edges is different from the other two. Hence, the structure is described as tetragonal in nature.

2. The Ti^{4+} ions are surrounded by six O^{2-} ions while each oxide ion is surrounded by three Ti^{4+} ions. occupying all the tetrahedral sites. The coordinating geometry of Ti^{4+} ion is octahedral and of O^{2-} is trigonal in nature. Therefore, the coordination number ratio of Ti^{4+} and O^{2-} is 6:3.

The refractive index of rutile at visible wavelengths is highest than any known crystal, and also shows a particularly high dispersion and large birefringence. These properties make it useful for the manufacturing of some particular optical elements like polarization optics for longer visible and infrared radiations. Now because the rutile is also a large band-gap semiconductor, it has been in the limelight for the research area of finding its applications as a functional oxide for dilute magnetism and photocatalysis.

➢ *Antirutile (Ti₂N)*

   Antirutile is the mineral form of titanium nitride, Ti_2N. The unit cell of Ti_2N lattice is shown below.

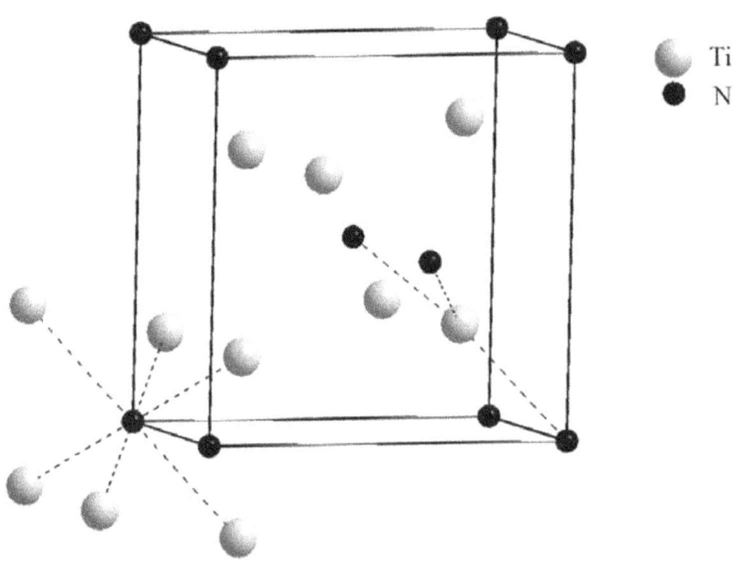

Structure of Antirutile (Ti₂N)

Figure 9. The crystal structure of antirutile or Ti_2N.

The main features of this crystal structure are as follows:

1. The nitride ions are present in distorted body-centered cubic close packing in which they are present at all the corners and at the center of the distorted cube. The unit cell cannot be labeled as cubic because all the three sides are not equal to each other. One of the edges is different from the other two. Hence, the structure is described as tetragonal in nature.

2. The nitride ions are surrounded by six titanium ions while each titanium ion is surrounded by three titanium ions. The coordinating geometry of titanium ion is trigonal and of nitride is octahedral in nature. Therefore, the coordination number ratio of Ti and N is 3:6.

   It is worth noting that in this motif the positions of the anions and cations are reversed relative to their positions in TiO_2 (rutile), with titanium ions trigonally coordinated to three nitride ions and nitrides are cubically coordinated to six titanium ions. Therefore, the name antrutile is assigned due to the reversal of the positions of nitride and titanium ions. Seldom, the antirutile structure is encountered where the metal and non-metals have changed places. The number of formula units presents per unit cell is two, same as in case rutile structure.

> *Cristobalite (SiO₂)*

The mineral cristobalite has the same structural formula to quartz i.e. SiO_2, but a different crystal structure. The cristobalite phase is stable above 1470°C (β–Cristobalite), but can also be crystallized metastably at lower temperatures in α–Cristobalite. The crystal structure of the two forms can be discussed as:

1. β–Cristobalite: The unit cell of this high-temperature form is shown below.

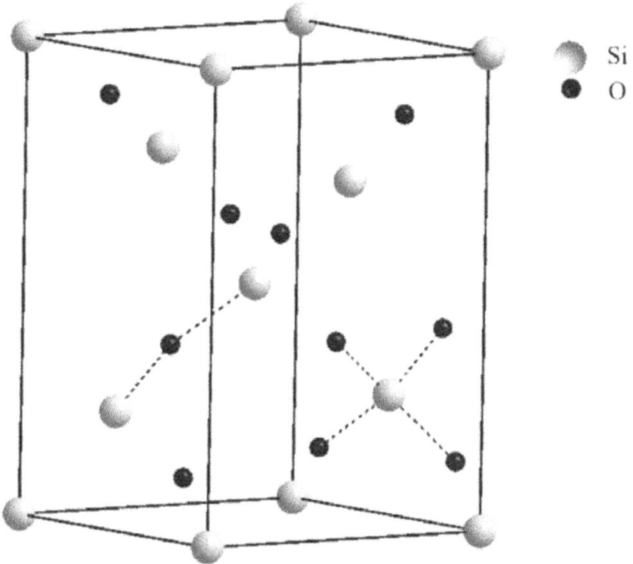

Structure of β–Cristobalite

Figure 10. The crystal structure of β–cristobalite.

The main features of this crystal structure are as follows:

1. β–cristobalite crystallizes in the tetragonal lattice in which silicon atoms are present at all the corners and at four faces unit cell.

2. The oxygen atoms are surrounded by two silicon atoms while each silicon atom is coordinated by four oxygen atoms. The coordinating geometry of silicon ion is tetrahedral and of oxygen is V–shaped in nature. Therefore, the coordination number ratio of Si and O is 4:2.

3. The number of formula units present per unit cell is 4.

Cristobalite is found as white spherulites or octahedra in acidic volcanic rocks and in converted diatomaceous deposits in California.

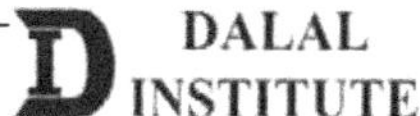

2. α–Cristobalite: The unit cell of this high-temperature form is shown below.

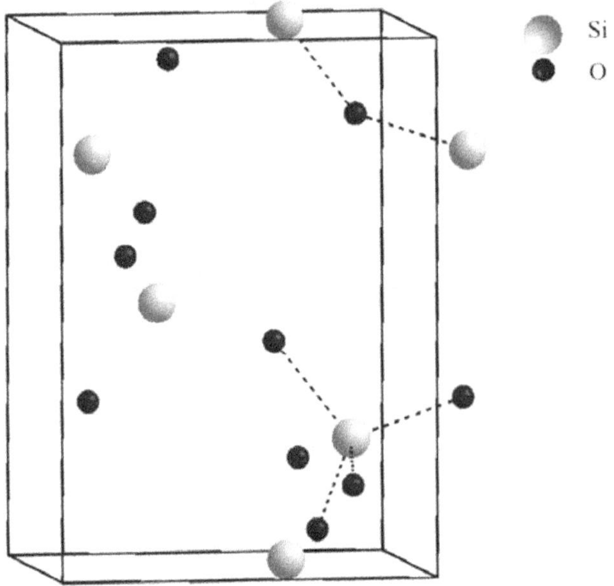

Structure of α–Cristobalite

Figure 10. The crystal structure of α–cristobalite.

The main features of this crystal structure are as follows:

1. α–cristobalite crystallizes in the tetragonal lattice in which three silicon atoms are present inside while two silicon atoms are situated at two opposite faces unit cell.

2. The oxygen atoms are surrounded by two silicon atoms while each silicon atom is coordinated by four oxygen atoms. The coordinating geometry of silicon ion is tetrahedral and of oxygen is V–shaped in nature. Therefore, the coordination number ratio of Si and O is 4:2.

3. Like β–cristobalite, the number of formula units presents per unit cell is also 4.

The high temperatures α–cristobalite form transforms into β–cristobalite if it is cooled below 250°C at ambient pressure. This transition is variously called the low-high or α − β transition. In the transitioning α–β phase, only one of the three degenerate cubic crystallographic axes preserve a 4-fold axis of rotation in the tetragonal phase. Moreover, various twins can form within the same grain due to the arbitrary choice of axis. The coupling of discontinuous nature of the transition with different twin orientations can cause significant mechanical damage to the substance in which the cristobalite phase is present and that pass recurrently through the transition temperature.

> ***CdI₂***

Cadmium iodide is a layered lattice of cadmium and iodine with formula CdI_2. It is pretty much notable for its lattice profile, which is quite common for the MX_2 type compounds with strong polarization effects. The unit cell of CdI_2 is shown below.

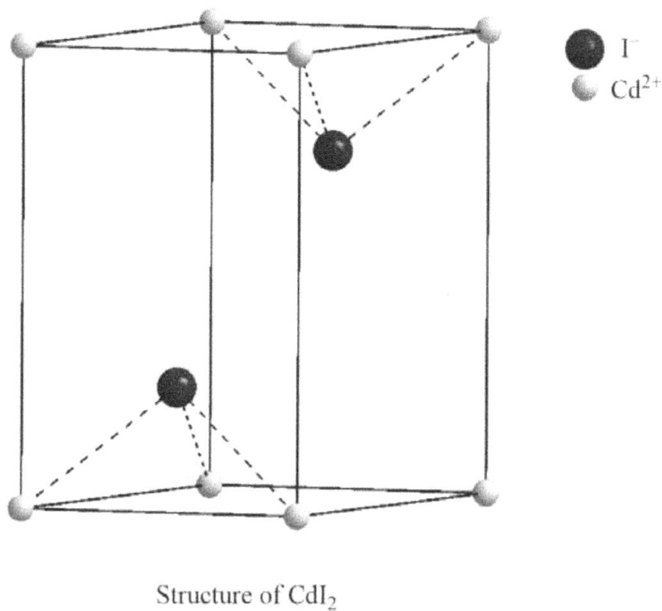

Structure of CdI_2

Figure 11. The unit cell of CdI_2 crystal system.

The main features of this crystal structure are as follows:

1. In cadmium iodide, the I^- anions form a hcp arrangement, while the Cd^{2+} cations fill all of the octahedral voids in the alternate layers.

2. Each Cd^{2+} ion is surrounded by six iodide ions while each I^- is coordinated by three cadmium cations. The coordinating geometry of cadmium ion is octahedral and of iodide ion is a trigonal pyramid in nature. Therefore, the coordination number ratio of Cd^{2+} and I^- is 6:3.

3. The resultant structure comprises of a layered lattice in which the Cd^{2+} ions are occupying all octahedral sites but in between the alternate layers of I^-. It can be visualized as if the cadmium ions layer is sandwiched between the two layers of iodide ions. Moreover, these sandwiched are piled over each other in such a way that the upward iodide ion layer of one sandwich is in contact with the downward iodide ion layer of the other sandwich. The same holds for the downward iodide ion layer. This same basic structure is found in many other salts and minerals. This can be depicted as:

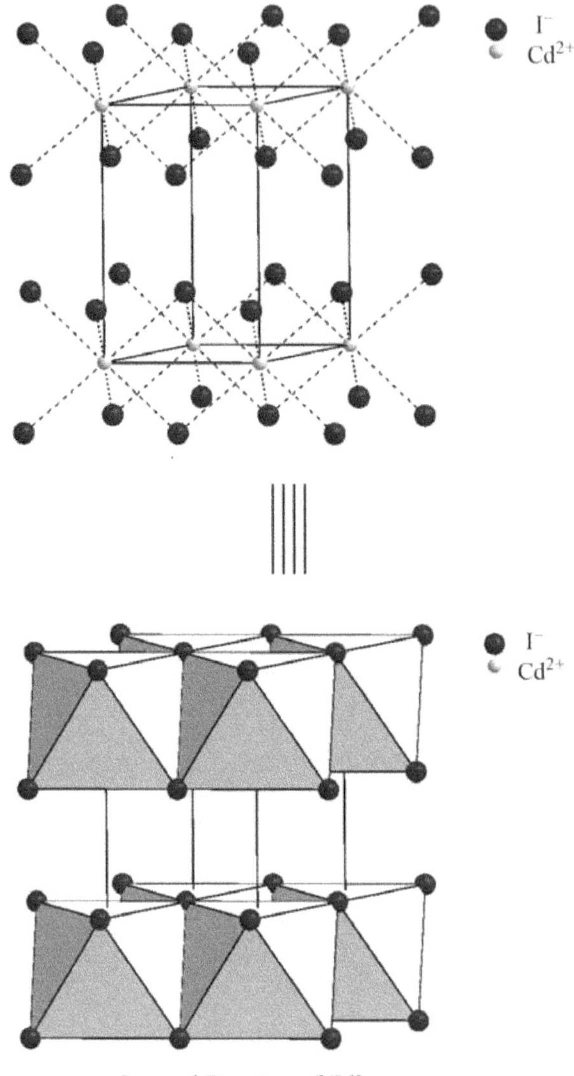

Figure 12. The crystal structure of CdI₂.

4. Cadmium iodide is bonded mostly in ionic fashion but does have partial covalent character too.

The other examples of this type of structure are PbI_2, MgI_2, TiI_2, VI_2, $TiCl_2$, VCl_2, CoI_2, FeI_2, $Mg(OH)_2$, $Ni(OH)_2$ and $Ca(OH)_2$. Cadmium iodide is used in photography, electroplating, lithography and the synthesis of phosphors.

> ➤ *BiI₃*

Bismuth iodide, BiI₃, is a layer lattice of bismuth and iodine. It is notable for its crystal structure, which is typical for MX₃ type compounds. The unit cell of BiI₃ is shown below.

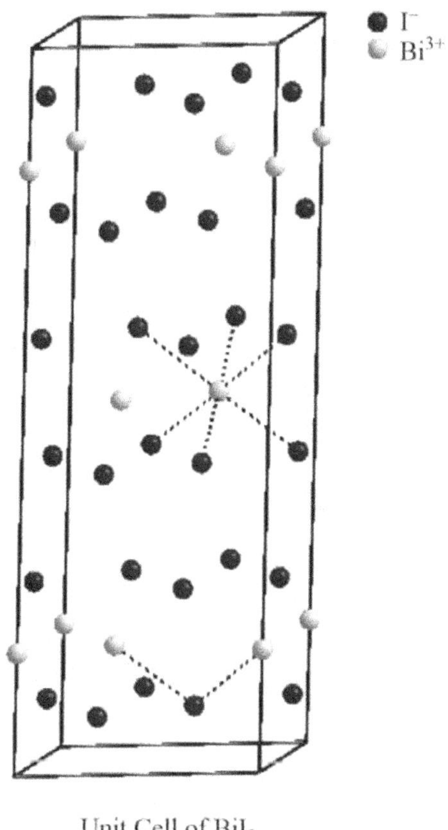

Figure 13. The unit cell of BiI₃ crystal system.

The main features of this crystal structure are as follows:

1. Bismuth iodide adopts a distinctive crystal structure, with I⁻ occupying a hcp lattice while bismuth centers occupy either none or two-thirds of the octahedral sites in the alternating layers. Hence, we can say that one-third of all octahedral sites are actually occupied.

2. Each Bi^{3+} ion is surrounded by six iodide ions while each I⁻ is coordinated by two bismuth cations. The coordinating geometry of bismuth ion is octahedral and of iodide ion is V–shaped in nature. Therefore, the coordination number ratio of Bi^{3+} and I⁻ is 6:2.

3. The resultant structure consists of a layered lattice. The Bi^{3+} ions are occupying one-third of all octahedral sites but in between the alternate layers of I^-. However, only two-third of all octahedral voids in every other layer are actually occupied. It can be visualized as if the bismuth ions layer is sandwiched between the two layers of iodide ions. Moreover, these sandwiched are piled over each other in such a way that the upward iodide ion layer of one sandwich is in contact with the downward iodide ion layer of the other sandwich. The same holds for the downward iodide-ion layer. This can be depicted as:

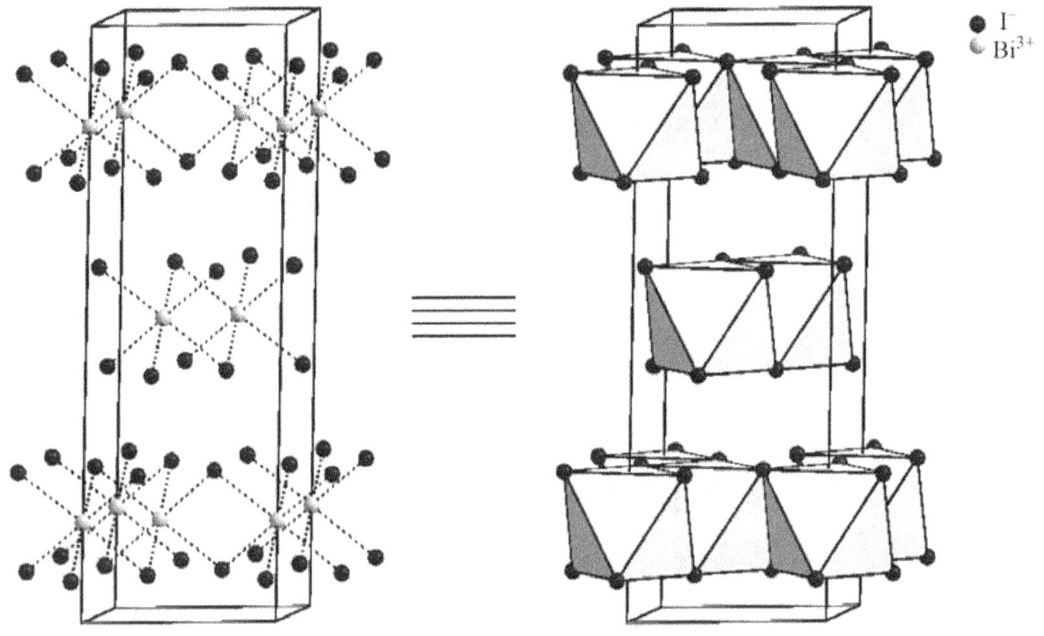

Figure 14. The layered lattice of BiI₃ crystal system.

4. The number of formula units present per unit cell of bismuth iodide is six.

5. The third side of the trigonal (hexagonal) lattice is approximately three times to that of the remaining two sides which are equal to each other.

The BiI₃ crystal belongs to a class of layered lattices of heavy metal semiconductors with fascinating anisotropic optical and electronic properties. The single crystals and thin films and of bismuth iodide have been studied for X-ray imaging due to the relatively wide bandgap, high mass density and high atomic numbers of the constituent elements. It can also be used as hard radiation detectors. Moreover, because BiI₃ can be processed in the solution phase; cost-efficient, large-scale device fabrication could be facilitated.

> ***ReO₃***

The rhenium oxide or ReO_3 is a reddish inorganic solid with a metallic luster, which looks like copper. It is the only trioxide of group seven which is stable. The unit cell of ReO_3 is shown below.

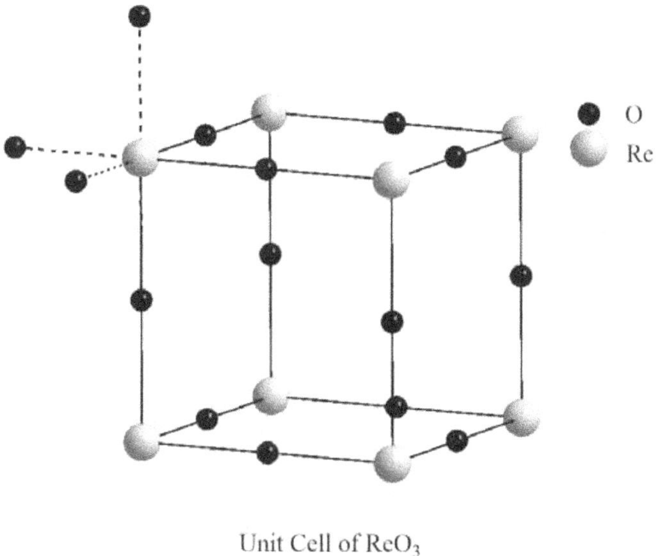

Unit Cell of ReO₃

Figure 15. The unit cell of ReO_3 crystal system.

The main features of this crystal structure are as follows:

1. Rhenium oxide crystallizes with a primitive cubic unit cell Re atoms are present at all the corners while the oxygen atoms are situated center of at all the edges.

2. Each Re atom is surrounded by six oxygen atoms while each O atom is coordinated by two rhenium atoms. The coordinating geometry of rhenium atom is octahedral and of oxygen atom is linear in nature. Therefore, the coordination number ratio of Re and O is 6:2.

3. The number of formula units present per unit cell is one.

4. The octahedron units, ReO_6, share corners to form the 3-dimensional structure.

5. The ReO_3structure is similar to perovskite (ABO_3), just large A cation absents at the unit-cell-center.

Rhenium trioxide finds some use in organic synthesis as a catalyst for amide reduction. Rhenium trioxide (ReO_3) is a thermally stable but highly insoluble substance that is quite suitable for optic, glass and ceramic based other devices.

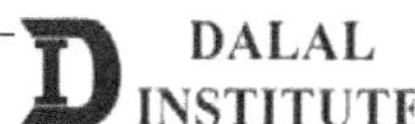 **DALAL INSTITUTE**

➢ *Mn₂O₃*

Manganese oxide is an inorganic solid of manganese and oxygen. Mn_2O_3 generally exists in two forms, α-Mn_2O_3 and γ-Mn_2O_3. The heating of MnO_2 in the air at below 800 °C produces α-Mn_2O_3 (higher temperatures produce Mn_3O_4) while γ-Mn_2O_3 can be produced by oxidation followed by dehydration of $Mn(OH)_2$. The crystal structure of the two forms can be discussed as:

1. α-Mn₂O₃: The unit cell of α-Mn_2O_3 crystal form is shown below.

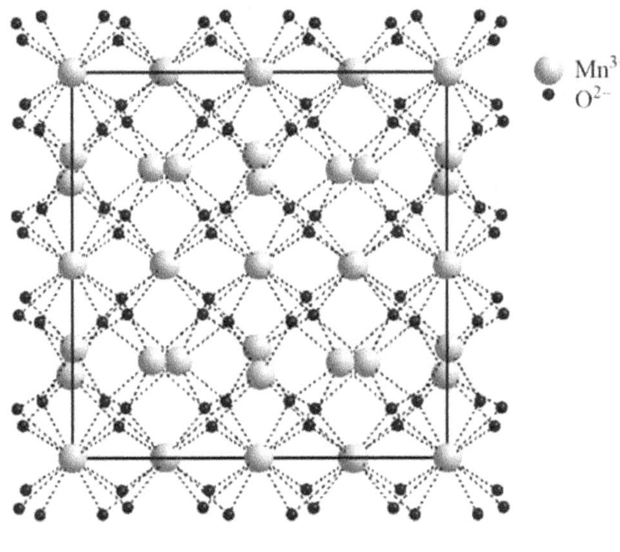

Unit Cell of α–Mn₂O₃

Figure 16. The unit cell view of α-Mn_2O_3 along *a*–axis.

The main features of this crystal structure are as follows:

1. The crystallization of α-Mn_2O_3 takes place in an orthorhombic lattice with five different octahedral coordinative environments around manganese ions.

2. Each Mn^{3+} ion is surrounded by six oxide ions while each O^{2-} ion is coordinated by four metal ions. The coordinating geometry of all five types of manganese ions is distorted octahedra and of all the six types of oxide ions are distorted tetrahedral in nature. Therefore, the coordination number ratio of Mn^{3+} and O^{2-} is 6:4.

3. The number of formula units present per unit cell is 16.

4. α-Mn_2O_3 has the cubic bixbyite structure, which is an example of a C-type rare earth sesquioxide.

The bandgap emission of nanocrystalline α–Mn_2O_3 is important due to their applications as ultra-violet (UV) emitters.

2. γ-Mn₂O₃: The unit cell of γ-Mn₂O₃ form is shown below.

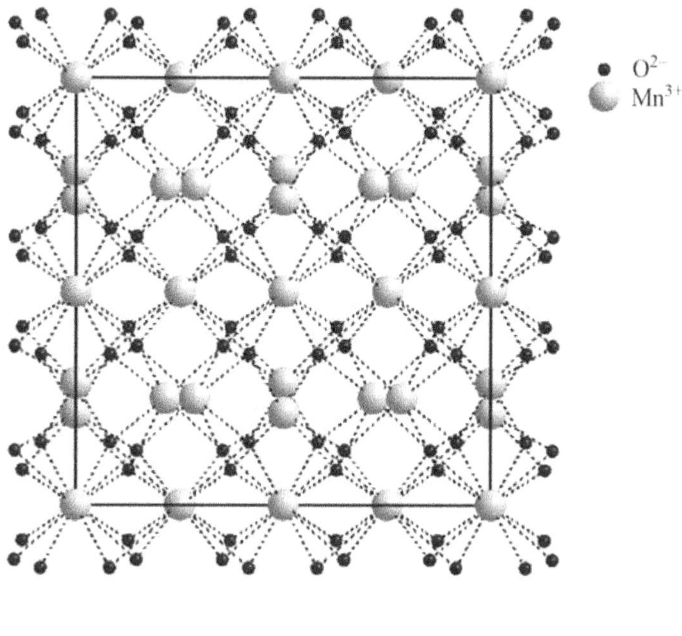

Unit Cell of γ–Mn₂O₃

Figure 17. The unit cell view of γ-Mn₂O₃ along *a*–axis.

The main features of this crystal structure are as follows:

1. The crystallization of γ-Mn₂O₃ takes place in a cubic lattice with two different octahedral coordinative environments around manganese ions.

2. Each Mn^{3+} ion is surrounded by six oxide ions while each O^{2-} ion is coordinated by four metal ions. The first type coordinating geometry around manganese ions is distorted octahedra while the second type coordinating geometry around manganese ion is a perfect octahedron. However, only one kind of coordinative environment around oxide ions is observed with a distorted tetrahedral symmetry in nature. Therefore, the coordination number ratio of Mn^{3+} and O^{2-} is 6:4.

3. The number of formula units present per unit cell is 16.

4. γ-Mn₂O₃ has a structure related to the spinel structure of Mn_3O_4 where the oxide ions are cubic close-packed. This is similar to the relationship between γ-Fe₂O₃ and Fe_3O_4.

γ-Mn₂O₃ is ferrimagnetic with a Neel temperature of 39 K and nanoparticle form has found a great deal of interest for applications in biology and material science.

➢ *Corundum*

Corundum is a crystalline state of Al_2O_3 typically having traces of titanium, vanadium, iron, and chromium in general. The corundum is a rock-forming mineral. It is a transparent material naturally but can show many colors when the impurities are present. The transparent specimens are used as gems, called padparadscha if pink-orange and ruby if red; while the remaining colors are called sapphire in common, e.g., green sapphire for a greenish specimen.

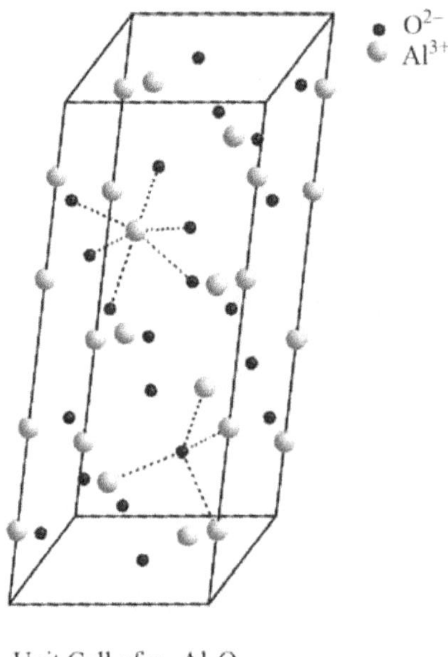

Figure 18. The unit cell of α–Al_2O_3 crystal system.

The main features of this crystal structure are as follows:

1. α–Al_2O_3 crystallizes in the trigonal lattice in which oxide ions form a slightly distorted hcp, in which two-thirds of the voids between the octahedral sites are occupied by aluminum ions.

2. Each Al^{3+} ion is surrounded by six oxide ions while each O^{2-} ion is coordinated by four aluminum ions. The coordinating geometry of Al^{3+} is distorted octahedral and of oxide, the ion is distorted tetrahedral in nature. Therefore, the coordination number ratio of Al^{3+} and O^{2-} is 6:4.

3. The number of formula units present per unit cell is six.

Besides its use as a valuable gem, corundum mineral also finds some use as an abrasive due to the extreme hardness of the material (9 on Mohs hardness scale). It is used for grinding optical glass and for polishing metals and has also been made into sandpapers and grinding wheels.

> ➢ *Perovskite*

The perovskite is an oxide mineral of calcium titanium with formula $CaTiO_3$. The mineral was found by Gustav Rose in the Ural Mountains of Russia in 1839 and is also named after a Russian mineralogist Lev Perovskite. The unit cell of perovskite is shown below.

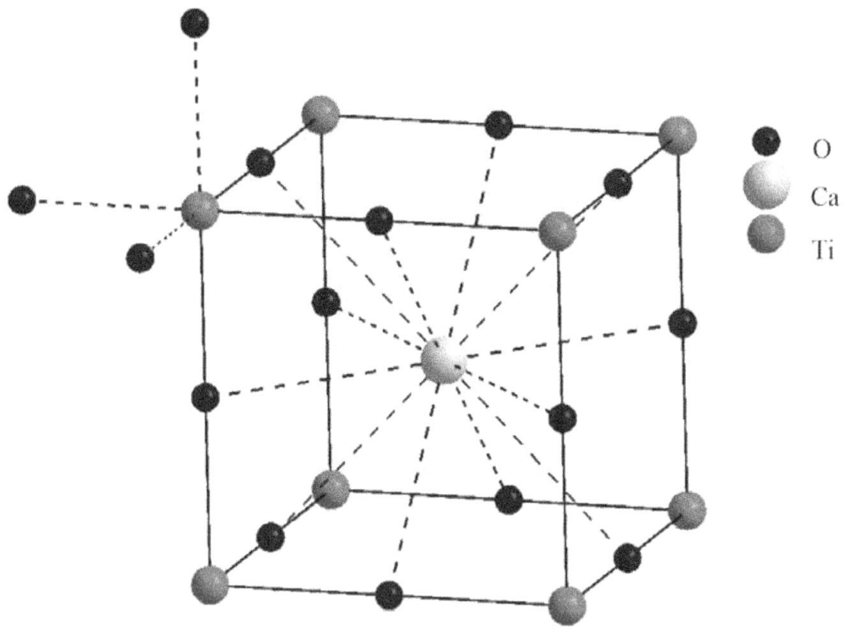

Unit Cell of Perovskite ($CaTiO_3$)

Figure 19. The unit cell of the perovskite crystal system.

The main features of this crystal structure are as follows:

1. $CaTiO_3$ crystallizes with a cubic unit cell in which Ti atoms are present at all the corners while the oxygen atoms are situated center of at all the edges. The large Ca^{2+} cation is present at the center of the unit cell.

2. Each Ti atom is surrounded by six oxygen atoms while each O atom is coordinated by two titanium atoms. The coordinating geometry of rhenium atom is octahedral and of oxygen atom is linear in nature. Therefore, the coordination number ratio of Ti and O is 6:2. Furthermore, the calcium cation is surrounded by twelve oxygens.

3. The octahedron units, TiO_6, share corners to form the 3-dimensional structure.

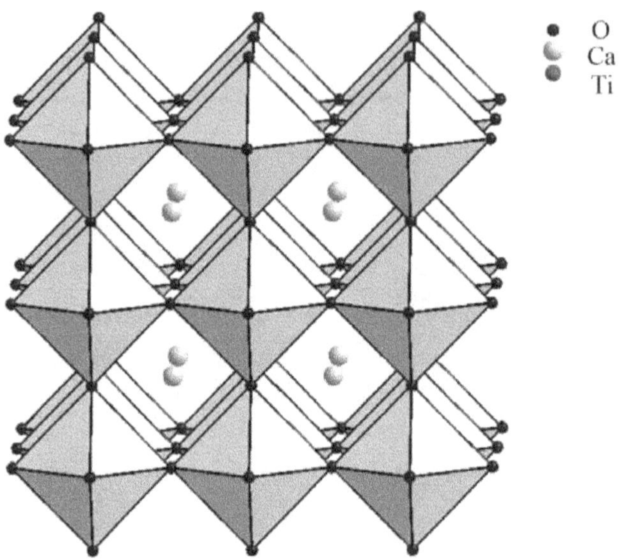

Three Dimensional structure of CaTiO₃

Figure 20. Three-dimensional structure of perovskite crystal system.

4. The number of formula units present per unit cell is one.

5. The structure of perovskite is adopted by many other oxides which have the ABO_3 formula.

6. The relative size different ions, required for stability of the cubic lattice are pretty much inflexible; therefore, very slight distortion and buckling can generate many lower-symmetry distorted versions, with coordination numbers of B cations, A cations or both are decreased. The Tilting of BO_6 polyhedra reduces the coordination of an undersized cation A from twelve to as low as eight. Conversely, off-centering of an undersized cation B within its polyhedral unit allows it to attain a stable bonding pattern. The resulting electric dipole is accountable for the ferroelectricity possessed and shown by some perovskites like $BaTiO_3$ which are distorted in profile.

7. The tetragonal and orthorhombic crystal phases are the most common non-cubic types.

8. The possibility of disordered and ordered variants results in some complex perovskite kinds which contain two dissimilar B-site cations.

Perovskite structures possess many intriguing and interesting properties from theoretical as well as the application viewpoint. Massive magnetoresistance, superconductivity, ferroelectricity, charge ordering, high thermos-power, spin-dependent transport and the interplay of magnetic, structural and transport properties are the most commonly observed features in this class. These compounds are used as catalyst electrodes and sensors in some types of fuel cells and are good candidates for spintronics applications and memory devices.

> ### *Ilmenite*

Ilmenite is an oxide mineral of titanium and iron with formula $FeTiO_3$. It is a weakly magnetic steel-gray or black solid material. The unit cell of Ilmenite is shown below.

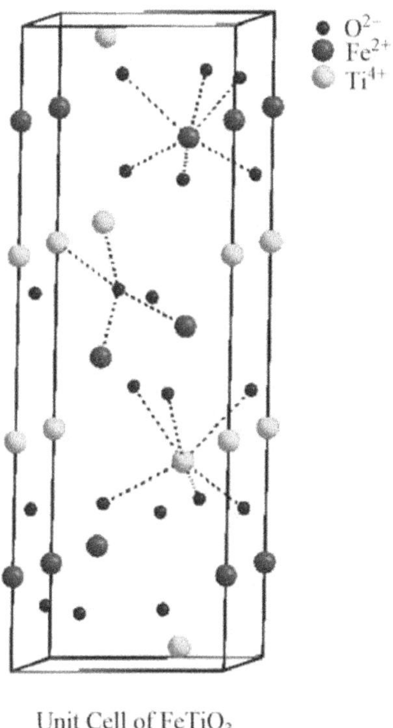

Unit Cell of FeTiO$_3$

Figure 21. The unit cell of the Ilmenite crystal system.

The main features of this crystal structure are as follows:

1. $FeTiO_3$ crystallizes in a trigonal crystal system which can be considered as an ordered imitative of the corundum structure. In corundum, all cations are the same but in $FeTiO_3$, the Fe^{2+} and Ti^{4+} ions occupy alternating layers, perpendicular to the trigonal c-axis. Ilmenite is paramagnetic due to high spin ferrous centers.

2. Each Ti^{4+} and Fe^{2+} ion is surrounded by six oxide ions while each O^{2-} atom is coordinated by two Ti^{4+} and two Fe^{2+} ions. The coordinating geometry of titanium ion and ferrous ion is distorted octahedral and of oxide ion is distorted tetrahedral in nature. Hence, the coordination number ratio of Ti^{4+}, Fe^{2+} and O^{2-} ions is 6:6:4.

As far as the commercial importance is concerned, the ilmenite is the best ore for Ti. It is the primary source of TiO_2, which is used in fabrics, paints, paper, food, plastics, cosmetics and sunscreen.

DALAL INSTITUTE

> ➤ *Calcite*

The carbonate mineral calcite is the most stable polymorph of $CaCO_3$ (calcium carbonate). The word calcite is a derivative of German word calcite, a term coined in the nineteenth century from the Latin for lime, and the suffix -ite for minerals. The unit cell of Ilmenite is shown below.

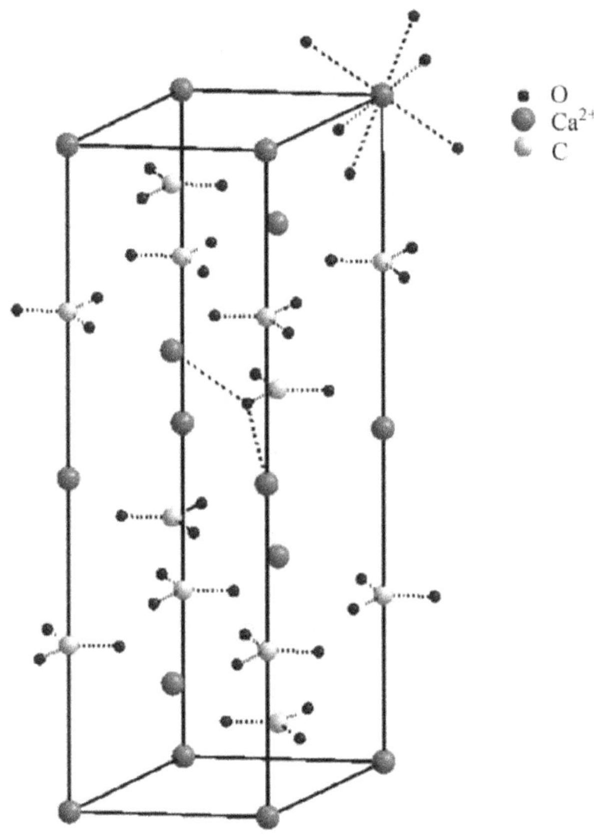

Unit Cell Of Calcite ($CaCO_3$)

Figure 21. The unit cell of calcite crystal.

The main features of this crystal structure are as follows:

1. $CaCO_3$ crystallizes in a trigonal (hexagonal) crystal system in which layers of Ca^{2+} metal ions are in alternation with stacks of carbonate layers. The carbonate layers are made-up of flat triangular-shaped CO_3^{-} ions, with a central carbon surrounded by three oxygens at each corner of a triangle. This triangular structural unit is the key component in the trigonal symmetry of this mineral group.

2. Each Ca^{2+} ion is surrounded by six oxygen atoms while the carbon atom of carbonate ion is coordinated by the oxygens in a trigonal fashion. However, each oxygen is coordinated by two Ca^{2+} ions and one C atom. The coordinating geometry of calcium ion is octahedral and of carbon is trigonal in nature. Hence, the coordination number ratio of Ca^{2+}, C and O ions is 6:3:3.

The optical calcite of very high-grade was used in the second world war (WWII) for gun sights, specifically in bomb sights and anti-aircraft artillery. Also, various experimental studies have been carried out to use calcite for a cloak of invisibility. The calcite precipitated microbiologically, has a wide range of applications, like soil-stabilization, soil remediation and concrete repair. Calcite, obtained from an 80-kg sample of Carrara marble, is used as the IAEA-603 isotopic standard in mass spectrometry for the calibration of C^{13} and O^{18}.

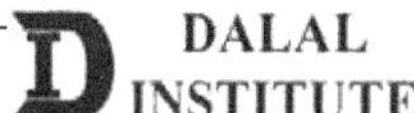

❖ Problems

Q 1. How many tetrahedral and octahedral voids would you expect from the cubic close packing of N identical spheres? Also discuss the same if hcp, in place of fcc, is used.

Q 2. Draw and discuss the crystal structure of fluorite.

Q 3. How does the Na_2O structure differ from the crystal structure of CaF_2?

Q 4. What is rutile? Explain its structure.

Q 5. Give five points differentiating the antirutile structure from the rutile one.

Q 6. Define the term layered lattice. Draw and discuss two examples in detail.

Q 7. Draw and discuss the crystal structure of β–cristobalite.

Q 8. How does the crystal structure of ReO_3 differ from the perovskite? Explain with suitable diagram.

Q 9. Compare the crystal structures of ilmenite and corundum.

Q 10. What is calcite? Draw and discuss its crystal structure.

❖ Bibliography

[1] J. D. Lee, *Concise Inorganic Chemistry*, Chapman & Hall, New York, USA, 1994.

[2] B. R. Puri, L. R. Sharma, K. C. Kalia, *Principals of Inorganic Chemistry*, Milestone Publishers, Delhi, India, 2012.

[3] J. E. Huheey, E. A. Keiter, R. L. Keiter, *Inorganic Chemistry: Principals of Structure and Reactivity*, HarperCollins College Publishers, New York, USA, 1993.

[4] A. F. Wells, *Structural Inorganic Chemistry*, Oxford University Press, London, UK, 1975.

[5] F. A. Cotton, G. Wilkinson, C. A. Murillo, M. Bochmann, *Advanced Inorganic Chemistry*, John Wiley & Sons, New Jersey, USA, 1999.

[6] G. Raj, *Advanced Inorganic Chemistry Vol-1*, Krishna Prakashan Media, Uttar Pradesh, India, 2008.

[7] G. S. Girolami, *X-ray Crystallography*, University Science Books, California, USA, 2016.

CHAPTER 7

Metal-Ligand Bonding:

❖ Limitation of Crystal Field Theory

The main drawback of the crystal field theory is that it does not consider the covalent character in metal-ligand bonding at all. It treats the metal-ligand interaction in a purely electrostatic framework which is pretty far from reality. All the effects which originate from covalence cannot be explained by this theory. Therefore, the main limitations of crystal field theory can be concluded only after knowing the causes and magnitude of the covalence in the metal-ligand bonds.

➢ *Evidences for the Covalent Character in Metal–Ligand Bond*

The crystal field theory considers the metal center as well as surrounding ligands as point charges and assumes that the interaction between them is 100% ionic. However, quite strong experimental evidences have proved that there is some covalent character too which cannot be ignored. Some of those experimental evidences are as follows:

1. The nephelauxetic effect: The electrons present in the partially filled d-orbitals of the metal center repel each other to produce a number of energy levels. The placement of these levels on the energy scale depends upon the arrangement of filled electrons. The energy of these levels can be given in terms of "Racah parameters" B and C (a measure of interelectronic repulsion). The energy difference between same multiplicity states is expressed in B and Dq while between different multiplicity states is given in term of B, Dq and C. It has been observed that the complexation of metal center always results in a decrease in interelectronic repulsion parameters which in turn also advocates a decrease in the repulsion between d-electron density. Now, as the magnitude of this interelectronic repulsion is dependent upon the distance between the areas of maximum charge density, the decrease in its value is expected only when the d-orbital lobes extend in space. This is called as the nephelauxetic effect and measured as the nephelauxetic parameter (β).

This extension or nephelauxetic effect may be attributed to the larger orbital overlap between metal and ligand resulting in greater stabilization due to covalent bonding. Hence, the direct effect of covalent bonding between metal and ligand is to decrease the interelectronic repulsion parameters. In other words, greater the decrease in "Racah parameters", larger is the extent of covalent bonding between metal and ligand. Different ligands have different capacity to extend their d-orbital and are arranged in ascending order, known as the nephelauxetic series as:

$$F^- < H_2O < NH_3 < Ox^{2-} < en < NCS^- < Cl^- < CN^- < Br^- < I^-$$

It has been observed that the nephelauxetic ratio always follows a certain trend with respect to the nature of the ligands present. However, there are many ligands that do not form complexes with a particular metal ion; the Racah parameter and for these complexes cannot be calculated empirical rather experimentally.

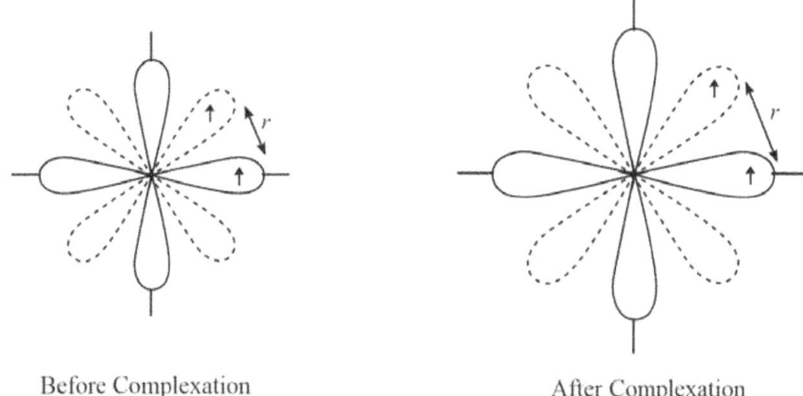

Figure 1. The expansion of d-electron cloud in transition metal centre after complexation due to nephelauxetic effect.

Hence, the observed decrease in inter-electronic repulsion parameters after complexation proves that there is always a somewhat more or less covalent character in the metal-ligand bond.

2. Lande's splitting factor: The value of Lande's splitting factor or simply "g-factor" for transition metal complexes was found to be different from what was expected from a pure ionic bonding perspective. The experimental value of g-factor showed that the electron from metal ion is always somewhat more or less delocalized to the ligand orbitals. This is possible only if metal orbitals are in overlap with the ligand orbitals via covalent interaction. Therefore, in view of the observed value of g-factor, we can claim that the partial covalence does exist in the metal-ligand bond.

3. Electron spin resonance spectra: It has been observed that the electron paramagnetic resonance or EPR spectra of some metal complexes is not that simple as expected. The fine lines were further split into hyperfine lines due to NMR active nuclei of ligands. This result can be explained only if the unpaired electrons of the metal center are delocalized to the nuclei centers of the ligands and as a consequence, electron-nucleus coupling occurs. Therefore, it quite obvious that we consider the metal-ligand as somewhat more or less covalent in nature. For example, the electron-spin resonance spectra of $[Ir(Cl_6)]^{2-}$ revealed that the unpaired $5d$-electron of Ir^{4+} spends 70% of its time on Ir^{4+} while 30% on chloride ligands.

4. Nuclear magnetic resonance spectra: The nuclear magnetic resonance of some ligands in metal complexes is markedly affected by the unpaired electrons of the metal center. This is possible only if metal orbitals are in overlap with the ligand orbitals. This proves a covalence in metal-ligand bond. For example, the chemical shift of tris-(acetylacetonato) vanadium(III) is shifted towards a position as if there is no paramagnetic contribution from the metal ion. This anomaly in the chemical shift may be attributed to the transfer of unpaired electron density from vanadium ion to the attached ligand.

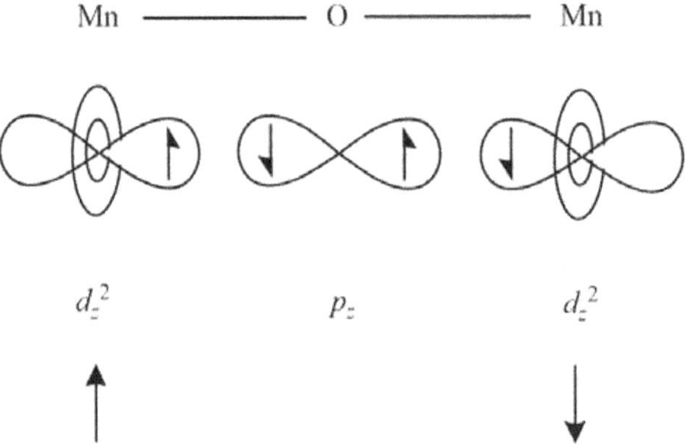

Furthermore, the F^{19} NMR of metal-fluoride complexes shows a significant effect arising from the unpaired electron from the metal centers, which obviously suggests that the electron on transition metal center does spend a non-zero time on the ligand too.

5. Nuclear quadrupole resonance: The nuclear quadrupole resonance spectra of some metal complexes having halide ions as ligands showed that the metal-halogen bond is not 100% ionic but has some covalent character too. In general, higher the NQR frequencies are, the larger will be the covalent character in metal-ligand bond. For example, one of the orders of the covalent character provided by nuclear quadrupole studies is Hg–X >> Cd–X > Zn–X, which is also supported by the Hard-Soft Acid-Base principal.

6. Kramers-Anderson superexchange: The phenomenon of superexchange was first suggested by H. Kramers in 1934 when he observed that in some crystals such as MnO, the Mn atoms interact with each another despite having diamagnetic intermediates like the oxide ions between them. Later on, Phillip Anderson modified the Kramers' model in the years of 1950. The superexchange phenomenon is the strong antiferromagnetic interaction between two nearest neighboring cations via a non-magnetic intermediate. Therefore, it is different from the direct-exchange in which the interaction is between nearest-neighbor cations without involving an intermediate anion.

Figure 2. The superexchange phenomenon in MnO.

However, it may also be worthy to mention that if two neighboring cations are joined at 90 degrees to the diamagnetic bridge, then the coupling can be ferromagnetic in nature.

The phenomenon of superexchange can be rationalized with the help of Pauli exclusion principle which governs that the superexchange would lead to anti-ferromagnetic behavior if the coupling occurs between two ions each of which has half-filled orbital; and would yield ferromagnetic behavior one metal ion is having half-filled orbital while the other ion has fully-filled orbitals; provided that in both cases, they interact via a diamagnetic intermediate like O^{2-}. Moreover, the interaction between an ion with either filled or a half-filled orbital and one with an empty orbital can be either ferromagnetic or antiferromagnetic, however, it usually favors ferromagnetism. However, if multiple kinds of interactions are simultaneously present, the antiferromagnetic ordering is the one that usually dominates because of its independence of the intra-atomic exchange term. The whole of the above explanation is based on the fact that the bonding between the metal center and ligand is not 100% ionic but does have a significant extant of orbital-overlap which confirms the presence of covalent character.

> ➢ *Limitations of CFT*

Now as we have discussed the covalence in metal-ligand bond, the main limitations of the crystal field theory can be summarized as follows:

1. The predictions of crystal field theory deviate from experimental results more and more as the extent of covalence between the metal center and ligands increases.

2. This theory does not give any satisfactory explanation for the relative strength of different ligands and hence is unable to explain the trends in the spectrochemical series. For example, ionic ligands are expected to show greater splitting effect due to the assumption of ligands as point charges but the neutral ligands like NH_3 or H_2O are actually stronger ligands than that of halide ions.

3. The crystal field theory gives no insight of back bonding between the metal and ligand and fails to explain the π-bonding or multiple bonds.

4. It does not explain the charge-transfer bands observed in the UV-visible spectra of transition complexes.

5. The crystal field theory considers only the d-orbitals of central metal ion but takes no account of the s and p-orbitals for its calculations.

6. The crystal field theory does not consider the orbitals of ligands at all and hence does not explain any properties associated with ligand orbitals and their interaction with orbitals of the metal center.

7. The assumption of the interaction between metal and ligand as purely electrostatic in nature is an idealistic one and is pretty far from reality.

8. This theory does not explain the effect of π-bonding on crystal field splitting Δ and therefore cannot compare the π-acid characters.

9. It could not explain the color the metal complexes having full or empty d-orbitals.

10. The CFT overlooks the attractive forces acting between the d-electrons of the metal center and nuclear charge on the ligands attached, which results in an unclear image of the properties dependent upon the same.

❖ Molecular Orbital Theory – Octahedral, Tetrahedral or Square Planar Complexes

The crystal field theory fails to explain many physical properties of the transition metal complexes because it does not consider the interaction between the metal and ligand orbitals. The molecular orbital theory can be very well applied to transition metal complexes to rationalize the covalent as well as the ionic character in the metal-ligand bond. A transition metal ion has nine valence atomic orbitals which are consisted of five nd, three $(n+1)p$, and one $(n+1)s$ orbitals. These orbitals are of appropriate energy to form bonding interaction with ligands. The molecular orbital theory is highly dependent on the geometry of the complex and can successfully be used for describing octahedral complexes, tetrahedral and square-planar complexes. The main features of molecular orbital theory for metal complexes are as follows:

1. The atomic orbital of the metal center and of surrounding ligands combine to form new orbitals, known as molecular orbitals.

2. The number of molecular orbitals formed is the same as that of the number of atomic orbitals combined.

3. The additive overlap results in the bonding molecular orbital while the subtractive overlap results in the antibonding overlap.

4. The energy of bonding molecular orbitals is lower than their nonbonding counterparts while the energy of antibonding molecular orbitals is higher than that of nonbonding orbitals.

5. The energy of nonbonding orbitals remains the same.

6. The ionic character of the covalent bond arises from the difference in the energy of combining orbitals.

7. If the energy of a molecular orbital is comparable to an atomic orbital, it will not be very much different in nature from atomic orbital.

The polarity of the bond can be explained by considering the overlap of two atomic orbitals of different energies. Suppose ϕ_A and ϕ_B are two atomic orbitals of atoms A and B, respectively. These two atomic orbitals have one electron in each of them and combine to form one bonding (σ) and one antibonding (σ^*) molecular orbital. After the formation of molecular orbitals, both electrons occupy σ-orbital. Now, if the energy of σ-orbital is closer to ϕ_A, it will have more ϕ_A character and hence the electron density of both of the electrons will be concentrated more on atom A than B. Similarly if the energy of σ-orbital is closer to ϕ_B, it will have more ϕ_B character and the electron density of both of the electrons will be concentrated more on atom B than A. This same explanation holds for the ionic character in metal-ligand bond.

The main concern here is that the total number of the molecular orbitals formed as a result of bonding and antibonding interactions is quite large; and therefore, many complications related to the understanding of symmetry-energy may arise, all of which needs a very sophisticated treatment of chemical bonding. Hence, without a comprehensive knowledge of the chemical applications of group theory, it is quite difficult to explain the whole concept. However, a primitive explanation for the σ-bonding in transition metal complexes of different geometry can still be given.

> ### *Octahedral Complexes*

In octahedral complexes, the molecular orbitals created by the coordination of metal center can be seen as resulting from the donation of two electrons by each of six σ-donor ligands to the *d*-orbitals on the metal. The metal orbitals taking part in this type of bonding are *nd*, $(n+1)p$ and $(n+1)s$. It should be noted down that not all *nd*-orbitals but only d_z^2 and $d_{x^2-y^2}$ orbitals are capable of participating in the σ-overlap. The d_{xy}, d_{xz} and d_{yz} orbitals remain non-bonding orbitals. The ligands approach the metal center along the *x*, *y* and *z*-axes in such a way that their σ-symmetry orbitals form bonding and anti-bonding combinations with metal's *s*, p_x, p_y, p_z, d_z^2 and $d_{x^2-y^2}$ orbitals. A total of six bonding and six anti-bonding molecular orbitals formed. The symmetry designations of different metal orbitals taking part in octahedral overlap are:

$d_z^2,\ d_{x^2-y^2}$	–	e_g
s	–	a_{1g}
p_x, p_y, p_z	–	t_{1u}
d_{xy}, d_{xz}, d_{yz}	–	t_{2g}

The exact nature of the Mulliken symbols, like e_g or $a_{1g,}$ etc., can be understood only after the core level understanding of group theory. However, it still can be noted that *a*, *e* and *t* represents singly, doubly and triply degenerate orbitals, respectively. The subscripts *g* and *u* represent gerade and ungerade nature of the orbitals. An orbital is said to be gerade if it has the same sign in all opposite directions from the center. These are also said to be having the centre of symmetry. On the other hand, an orbital is said to be ungerade if it has the opposite sign in all the opposite directions from the center. These are also said to be lacking the centre of symmetry.

The symmetry adapted linear combinations of atomic orbitals (SALCs) for metal-ligand direct overlap can be obtained just by resolving the reducible representation based on the bond vectors along the axis of σ-overlap. As there are total six bond vectors for the metal-ligand σ-bonding, hexa-dimensional reducible representation will be obtained. As the six ligands in an octahedral complex approach the central metal ion along the three Cartesian axes (two along *x*-axis, two along *y*-axis and two along *z*-axis), the six σ-basis bond vectors must also be selected along *x*, *y* and *z*-axis.

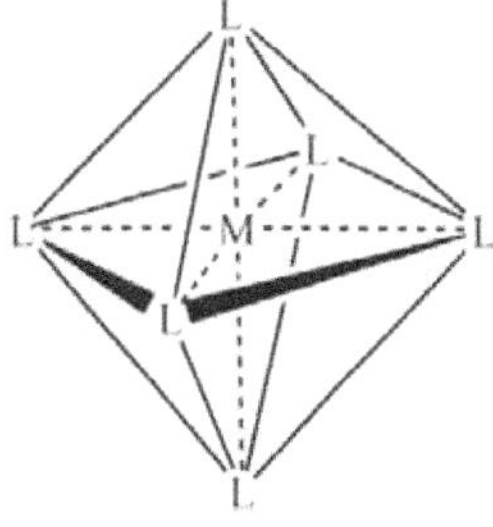

Figure 3. Continued on the next page…

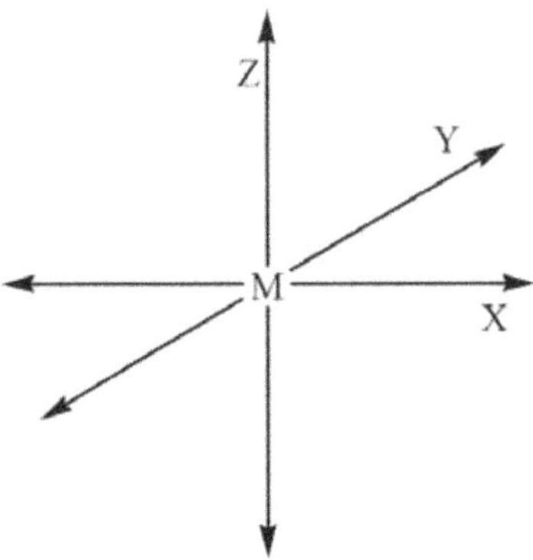

Figure 3. The octahedral coordination and corresponding σ-basis set for ligand orbitals in octahedral complexes.

The symmetry adapted linear combinations of these fall into three irreducible representations labeled as a_{1g}, e_g, and t_{1u}. The symmetry designations of different ligand orbitals taking part in octahedral overlap are:

Table 1. Reducible representation based on bond vectors in the octahedral geometry.

O_h	E	$8C_3$	$6C_2$	$6C_4$	$3C_2$	i	$6S_4$	$8S_6$	$3\sigma_h$	$6\sigma_d$	Irreducible components
Γ_π	6	0	0	2	2	0	0	0	4	2	$a_{1g} + e_g + t_{1u}$

The metal-orbital sets of a_{1g}, e_g and t_{1u} symmetry participate in σ-bonding by interacting with the same symmetry SALCs sets to produce σ-bonding and antibonding molecular orbitals. However, the triply degenerate t_{2g} orbitals set on the metal remains nonbonding as there no ligand orbitals of this symmetry. The exact nature of the "symmetry adapted linear combination of atomic orbitals" or simply SALCs is quite complex and is beyond the scope of this volume but still we can simply give a superficial explanation for these ligand orbitals. The pictorial representation of various SALC orbitals in octahedral complexes, capable of σ-overlap with metal orbitals, can be given as:

1. If the six atomic orbitals of six ligands are pointing their positive lobes towards metal's s-orbital, then the composite orbital becomes:

$$\varphi_1 = \sigma_1 + \sigma_2 + \sigma_3 + \sigma_4 + \sigma_5 + \sigma_6 \tag{1}$$

The pictorial representation is given below.

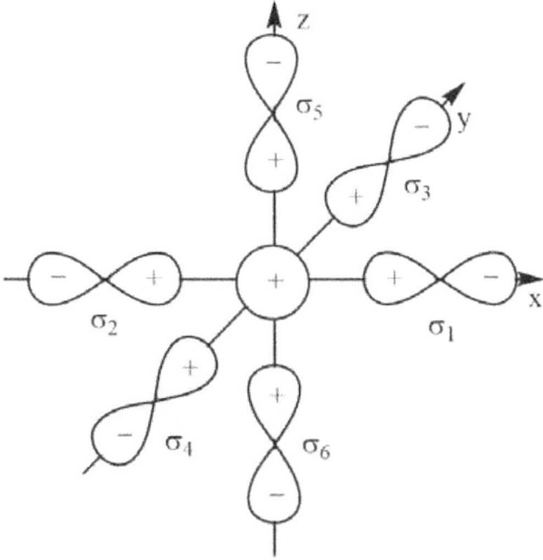

Figure 4. The combination of ligand orbitals overlapping with *s*-orbital of the metal center.

2. The p_x-orbital of metal center can overlap with ligands atomic orbital approaching along the *x*-axis. Hence, the composite orbital becomes:

$$\varphi_2 = \sigma_1 - \sigma_2 \tag{2}$$

The pictorial representation is given below.

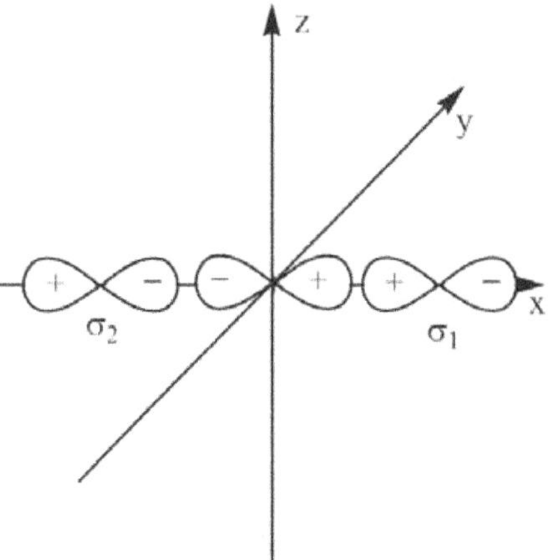

Figure 5. The combination of ligand orbitals overlapping with p_x-orbital of the metal center.

3. The p_y-orbital of metal center can overlap with ligands atomic orbital approaching along y-axis. Hence, the composite orbital becomes:

$$\varphi_3 = \sigma_3 - \sigma_4 \tag{3}$$

The pictorial representation is given below.

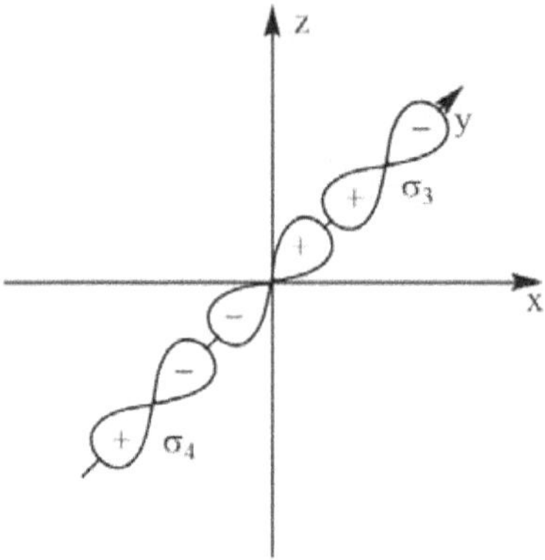

Figure 6. The combination of ligand orbitals overlapping with p_y-orbital of the metal center.

4. The p_z-orbital of metal center can overlap with ligands atomic orbital approaching along z-axis. Hence, the composite orbital becomes:

$$\varphi_4 = \sigma_5 - \sigma_6 \tag{4}$$

The pictorial representation is given below.

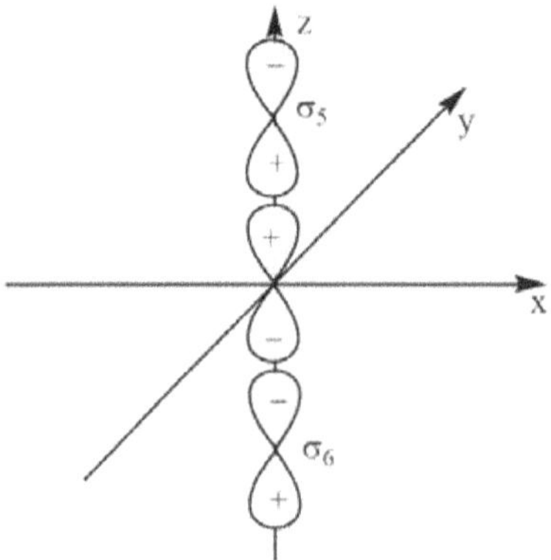

Figure 7. The combination of ligand orbitals overlapping with p_z-orbital of the metal center.

5. The $d_{x^2-y^2}$ orbital of metal center can overlap with ligands atomic orbital approaching along x and y-axis. Hence, the composite orbital becomes:

$$\varphi_5 = \sigma_1 + \sigma_2 - \sigma_3 - \sigma_4 \tag{5}$$

The pictorial representation is given below.

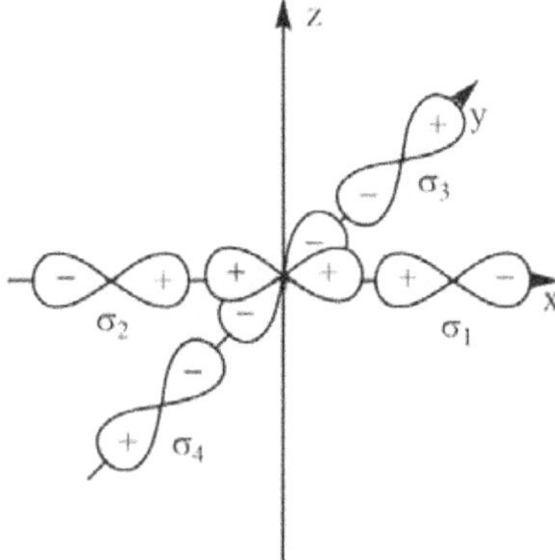

Figure 8. The combination of ligand orbitals overlapping with $d_{x^2-y^2}$ orbital of the metal center.

6. The d_z^2 orbital of metal center can overlap with ligands atomic orbital approaching along x, y and z-axis. Hence, the composite orbital becomes:

$$\varphi_6 = \sigma_5 + \sigma_6 - \sigma_1 - \sigma_2 - \sigma_3 - \sigma_4 \tag{6}$$

The pictorial representation is given below.

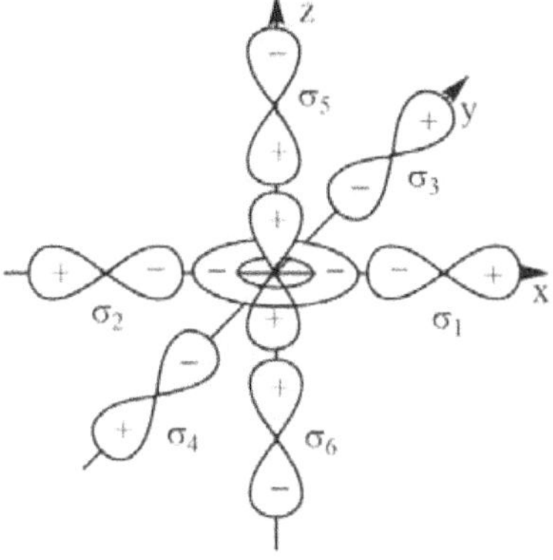

Figure 9. The combination of ligand orbitals overlapping with d_z^2 orbital of the metal center.

The overall molecular orbital energy level diagram for σ-bonding in octahedral complexes can be shown as:

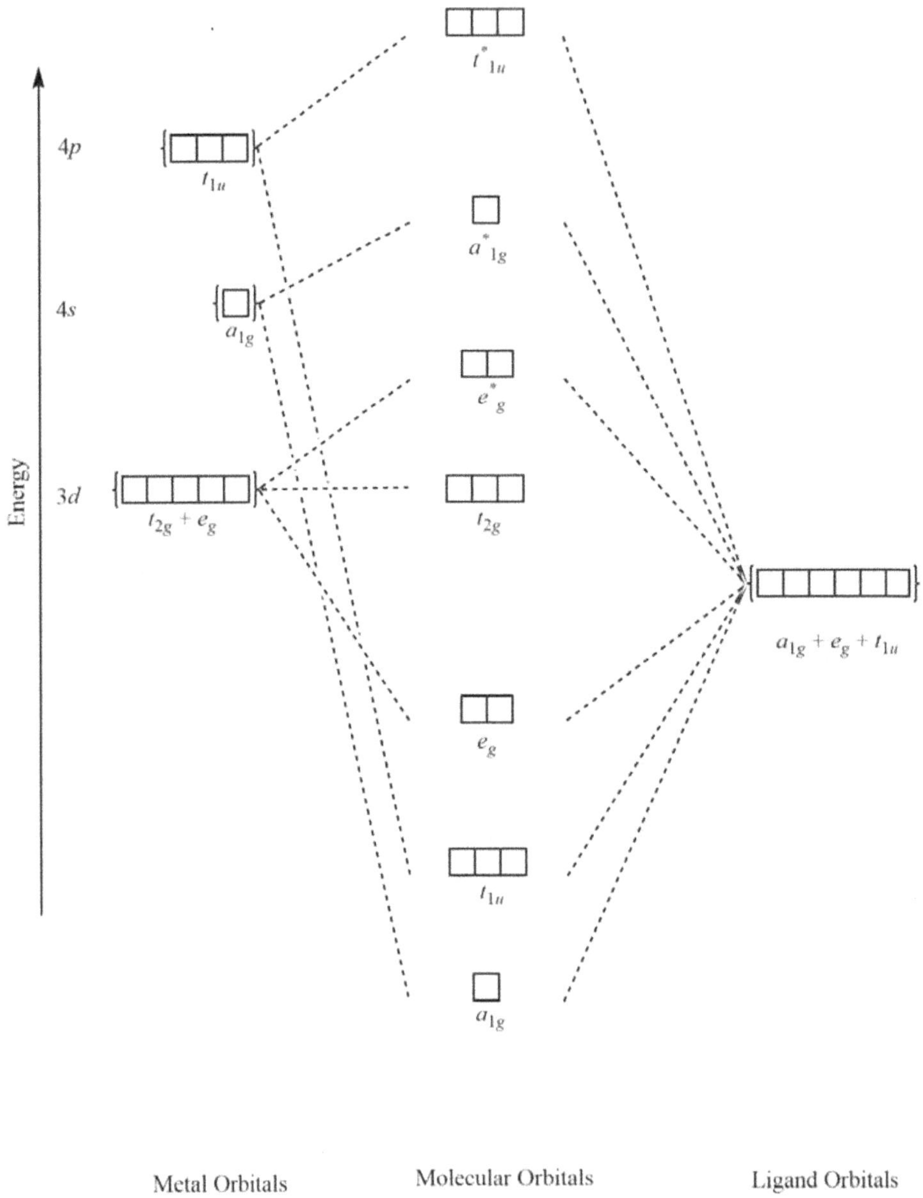

Figure 10. The formation of σ-molecular orbitals (bonding, antibonding and non-bonding) in octahedral complexes of transition metals.

The φ_1 composite orbital of a_{1g} symmetry interacts with a_{1g} orbital set of metal ions and produces one bonding (a_{1g}) and one antibonding molecular orbital (a^*_{1g}). The φ_2, φ_3 and φ_4 composite orbital set of t_{1u} symmetry interacts with t_{1u} orbital set of metal ions and produces triply-degenerate bonding (t_{1u}) and triply-degenerate antibonding molecular orbital (t^*_{1u}) set. The φ_5 and φ_6 composite orbital set of e_g symmetry interacts with e_g orbital set of metal ions and produces doubly-degenerate bonding (e_g) and doubly-degenerate antibonding molecular orbital (e^*_g) set. The t_{2g} orbital set of the metal center remains non-bonding in nature. The electron-filling in various orbitals is done in accordance with the Aufbau principle, Hund's rule and Pauli exclusion principle.

Some typical explanations in the view of MO-theory are:

i) Spin only magnetic moments: Just like crystal field theory, MO-theory can also be used to rationalize the spin only magnetic moments of different transition metal complexes. For instance, 5.9 B.M. magnetic moment of $[Mn(H_2O)_6]^{2+}$ is generated when seventeen electrons (five from Mn^{2+} and twelve electrons from six aquo ligands) are filled in different molecular orbitals. Twelve out of seventeen electrons occupy six bonding molecular orbitals of a_{1g}, t_{1u} and e_g. Three out of the remaining five electrons are filled in nonbonding t_{2g} and the last two electrons occupy e^*_g giving a total of five unpaired electrons.

ii) 18- electron rule: It is a well-known fact that metal carbonyls and many other complexes follow 18-electron rule quite well. This is also analyzed in terms of effective atomic number which suggests that a total of 18-electrons in the valence shell generally makes overall electron-count to resemble a noble gas core. However, a strong basis for this rule can found in molecular orbital theory. Six bonding molecular orbitals can accommodate a total of 12 electrons and only six electrons can be filled in the non-bonding t_{2g}. Any extra electron is bound to occupy the antibonding e^*_g orbital and hence affects the stability.

iii) The splitting of *d*-orbital: According to crystal field theory, the splitting, Δ_o of d-orbital in t_{2g} and e_g is produced by the difference in the Coulombic repulsion experienced various metal orbitals. However, the molecular orbital theory sees the splitting of d-orbital as a result of the interaction of metal orbitals with ligand orbitals and labels it as splitting between nonbonding t_{2g} and antibonding e^*_g.

iv) High spin – low spin complexes: Just like in the case of crystal field theory, the magnitude of Δ_o decides the multiplicity nature of the complexes. If the separation between nonbonding t_{2g} and antibonding e^*_g is pretty large, the electrons prefer to pair up with the electrons already present in nonbonding t_{2g} and the complex becomes low-spin. On the other side, If the separation between nonbonding t_{2g} and antibonding e^*_g is small, the electrons prefer to occupy antibonding e^*_g and the complex becomes high-spin. For example, in $[Co(NH_3)_6]^{3+}$, the magnitude of Δ_o is very large and the configuration becomes t_{2g}^6, e^{*0}_g. Furthermore, the comparatively lower stability and weaker metal-ligand bond in high spin complexes may be attributed to the presence of electrons in a molecular orbital of antibonding nature.

v) Jahn-Teller distortions: The concept of molecular orbital theory can also be used to rationalize the z-out or z-in distortion of the octahedral metal complexes. For example, octahedral complexes of Cu^{2+} ion have been known to undergo Jahn-Teller distortion due to d^9 configuration. According to molecular orbital theory, the electronic configuration octahedral complexes of Cu^{2+} will be t_{2g}^6, e^{*3}_g. Now, as the energy of e_g^* molecular

orbital is close to the pure atomic orbitals of metal ion, e_g^* would resemble more to the metal orbitals than the ligand's. Therefore, we can approximate e_g^* to pure $d_{x^2-y^2}$ and d_z^2. Now if the configuration is $(d_{x^2-y^2})^1 (d_z^2)^2$, higher antibonding electron density along z-axis would result in a weaker bond and an elongation of bonds would take place along z-axis, giving a z-out distortion. On the other hand, if the configuration is $(d_{x^2-y^2})^2 (d_z^2)^1$, higher antibonding electron density along x *and* y-axis would result in a weaker bond and an elongation would take place along x and y-axis, giving a z-in distortion.

vi) Variation of ionic radii: Experimental values of ionic radii of central metal ions in bivalent salts of first transition series were very well explained by the crystal field theory. The same can be done in the frame of molecular orbital theory. The decrease in ionic radii, when one moves from Ca^{2+} to V^{2+}, may be attributed to the increasing magnitude of effective nuclear charge while the filling of valence electrons takes place in nonbonding t_{2g}. However, after V^{2+}, the fourth electron in Cr^{2+} is filled in antibonding e^* and this effect makes the metal-ligand bond to increase, which in turn results in an increment in effective ionic radius. This trend continues up to Mn^{2+}, but as we move from Fe^{2+} to Ni^{2+}, the electron filling again goes to nonbonding t_{2g} and thus thereby decreases the ionic radii. The ninth and tenth electron in Cu^{2+} and Zn^{2+} are filled in antibonding e^* and therefore, the effective ionic radius increases again. The whole situation creates two minima in effective ionic radii, one for V^{2+} and other for Ni^{2+}.

It is worth to note down that the more sophisticated rationalization of the MO-diagram for octahedral complexes can be given in the terms of molecular symmetry combined with detailed quantum mechanics involved. In general practice, six symmetry-adapted linear combinations (SALCs) of atomic orbitals, corresponding to the irreducible components of the bond-vector based reducible representation, are created. The irreducible representations that these SALCs span to are a_{1g}, t_{1u} and e_g. The metal also has six valence orbitals that span the same irreducible representations (s-orbital is labeled as a_{1g}, a set of three p-orbitals is labeled t_{1u}, and the d_z^2 and $d_{x^2-y^2}$ orbitals are labeled e_g). The six σ-bonding molecular orbitals result from the combinations of ligand SALC's with metal orbitals of the same symmetry.

> ### *Tetrahedral Complexes*

In tetrahedral complexes, the molecular orbitals created by the coordination profile can be seen as resulting from the donation of one electron pair by each of four σ-donor ligands to the d-orbitals on the transition metal. The visualization of the composite orbital of a tetrahedral complex in the molecular orbital framework is quite difficult because of the absence of the centre of symmetry. However, the symmetry designations of different metal orbitals taking part in this type of overlap can still be given as:

s	—	a_1
p_x, p_y, p_z	—	t_2
d_{xy}, d_{xz}, d_{yz}	—	t_2
$d_z^2, d_{x^2-y^2}$	—	e

The symmetry adapted linear combinations of atomic orbitals (SALCs) for metal-ligand direct overlap can be obtained just by resolving the reducible representation based on the bond vectors along the axis of σ-overlap. As there is a total of four bond vectors for the metal-ligand σ-bonding, the four-dimensional reducible representation will be obtained.

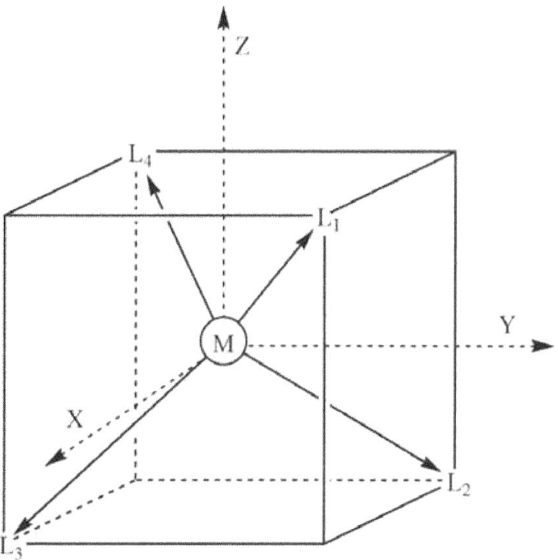

Figure 11. The σ-basis set for ligand orbitals in tetrahedral complexes.

The symmetry adapted linear combinations of these fall into two (one singly and one triply degenerate) irreducible representations labeled as a_1 and t_2. The symmetry designations of different ligand orbitals taking part in tetrahedral overlap are:

Table 2. Reducible representation based on perpendicular vectors in a tetrahedral geometry.

T_d	E	$8C_3$	$3C_2$	$6S_4$	$6\sigma_d$	Irreducible components
Γ_π	4	1	0	0	2	$a_1 + t_2$

The d_{xy}, d_{xz} and d_{yz} orbitals set on the metal also have t_2-symmetry. Similarly, the p_x, p_y and p_z-orbital set of the metal also has t_2 symmetry which resembles with the symmetry of one of SALCs set on ligands. Therefore, these same-symmetry (t_2) sets from metal and ligand interact to create three bonding and antibonding molecular orbitals. Moreover, s-orbital on metal has a_1-symmetry and hence mix-up with one of ligand SALC with a_1-symmetry. This also results in one bonding and one antibonding molecular orbital. One metal-orbital set of a_1 symmetry and two sets of t_2 symmetry participate in σ-bonding while the doubly degenerate e-symmetry set remains nonbonding. The overall molecular orbital energy level diagram for σ-bonding in tetrahedral complexes can be shown as:

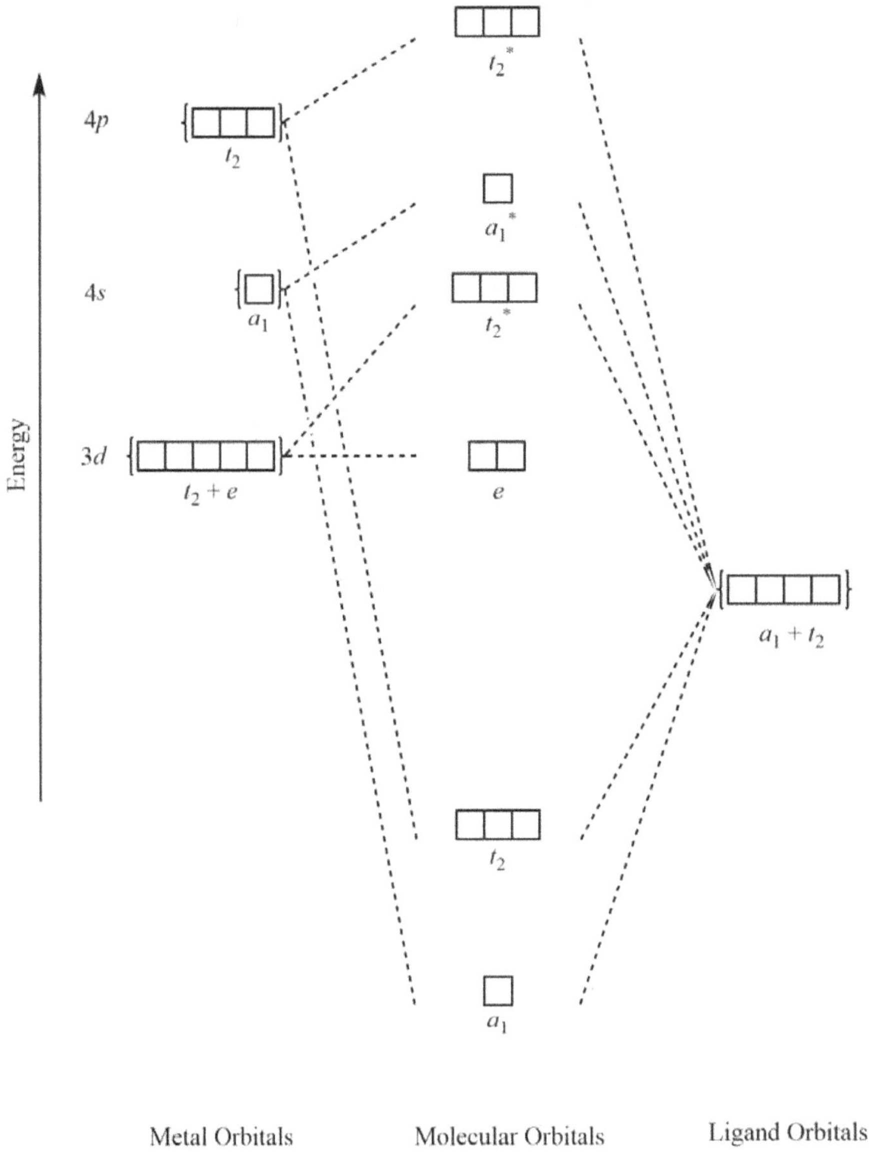

Figure 12. The formation of σ-molecular orbitals (bonding, antibonding and non-bonding) in tetrahedral complexes of transition metals.

> ### Square Planar Complexes

In square-planar complexes, the molecular orbitals created by coordination can be seen as resulting from the donation of two electrons by each of four σ-donor ligands to the d-orbitals on the metal. The metal orbitals taking part in this type of bonding are nd, $(n+1)p$ and $(n+1)s$. It should be noted down that not all nd or $(n+1)p$ orbitals but only d_z^2, $d_{x^2-y^2}$, p_x and p_y-orbitals are capable of participating in the σ-overlap. The d_{xy}, d_{xz}, d_{yz} and p_z-orbitals remain non-bonding orbitals. The symmetry designations of different metal orbitals taking part in square-planar overlap are:

s	$-$	a_{1g}
p_x, p_y	$-$	e_u
d_z^2	$-$	a_{1g}
$d_{x^2-y^2}$	$-$	b_{1g}
p_z	$-$	a_{2u}
d_{xy}	$-$	b_{2g}
d_{yz}, d_{xz}	$-$	e_g

The symmetry adapted linear combinations of ligand atomic orbitals (SALCs) for metal-ligand sidewise overlap in square planar complexes can be obtained just by resolving the reducible representation based on the bond vectors along to the axis of σ-overlap. As there are total four bond vectors for the metal-ligand σ-bonding, the four-dimensional reducible representation will be obtained.

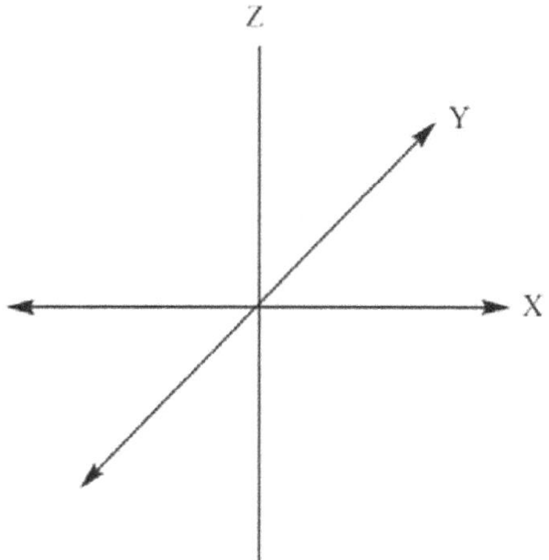

Figure 13. Continued on the next page…

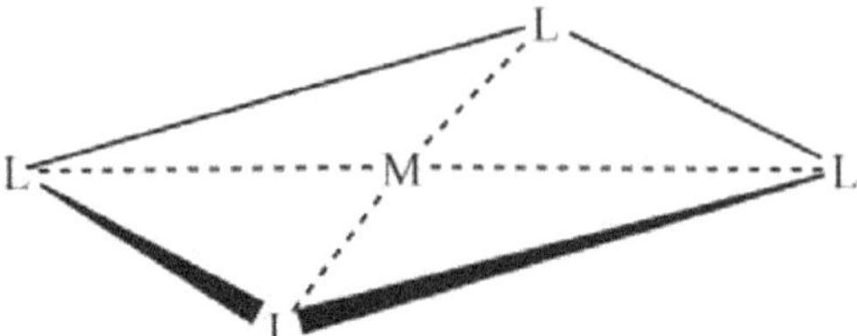

Figure 13. The σ-basis set for ligand orbitals and corresponding coordination in square-planar complexes of transition metals.

The symmetry adapted linear combinations of these fall into three irreducible representations labeled as a_{1g}, b_{1g} and e_u. The symmetry designations of different ligand SALC-orbitals taking part in square-planar overlap are:

Table 3. Reducible representation based on perpendicular vectors in square planar geometry.

D_{4h}	E	$2C_4$	C_2	$2C_2'$	$2C_2''$	i	$2S_4$	σ_h	$2\sigma_v$	$2\sigma_d$	Irreducible components
Γ_π	4	0	0	2	0	0	0	4	2	0	$a_{1g} + b_{1g} + e_u$

The irreducible components that these SALCs span to are a_{1g}, b_{1g} and e_u. The molecular orbitals of σ-bonding in square-planar complexes result from the mixing of ligand SALCs with metal orbitals of the same symmetry. Two metal-orbital sets of a_{1g} symmetry, one set of b_{1g} symmetry and one set on e_u symmetry participate in σ-bonding while the doubly degenerate e_g and singly degenerate b_{2g} sets remain nonbonding. This is quite logical because there are no ligand orbitals with these symmetry properties and hence no orbital overlap is possible. The ligands approach the metal center along the x and y-axes in such a way that their σ-symmetry orbitals form bonding and anti-bonding combinations with metal's s, p_x, p_y, d_{z^2} and $d_{x^2-y^2}$ orbitals. Two a_{1g}-symmetry sets of metal orbitals (s and d_{z^2}) interact with the ligands SALC of the same symmetry and form three molecular orbitals. The b_{1g}-symmetry set of metal orbital ($d_{x^2-y^2}$) interact with the ligands SALC of same symmetry and form two molecular orbitals, one bonding and one of antibonding nature.

Similarly, e_u-symmetry set of metal orbitals (p_x, p_y) interacts with the ligands SALC of the same symmetry and form bonding and antibonding molecular orbital sets. A total of nine sets of molecular orbitals are formed in which electron filling has to occur. Furthermore, it should also be noted that bonding molecular orbitals are predominantly associated with ligands which can be attributed to the lower energy of ligand SALCs while the antibonding molecular orbitals are primarily associated with the metal center owing to the higher energy of its atomic orbitals. In other words, the simplistic approach can treat bonding molecular orbitals as ligand orbitals and antibonding molecular orbitals as metal orbitals.

The molecular orbital energy level diagram for σ-bonding in square-planar complexes can be shown as:

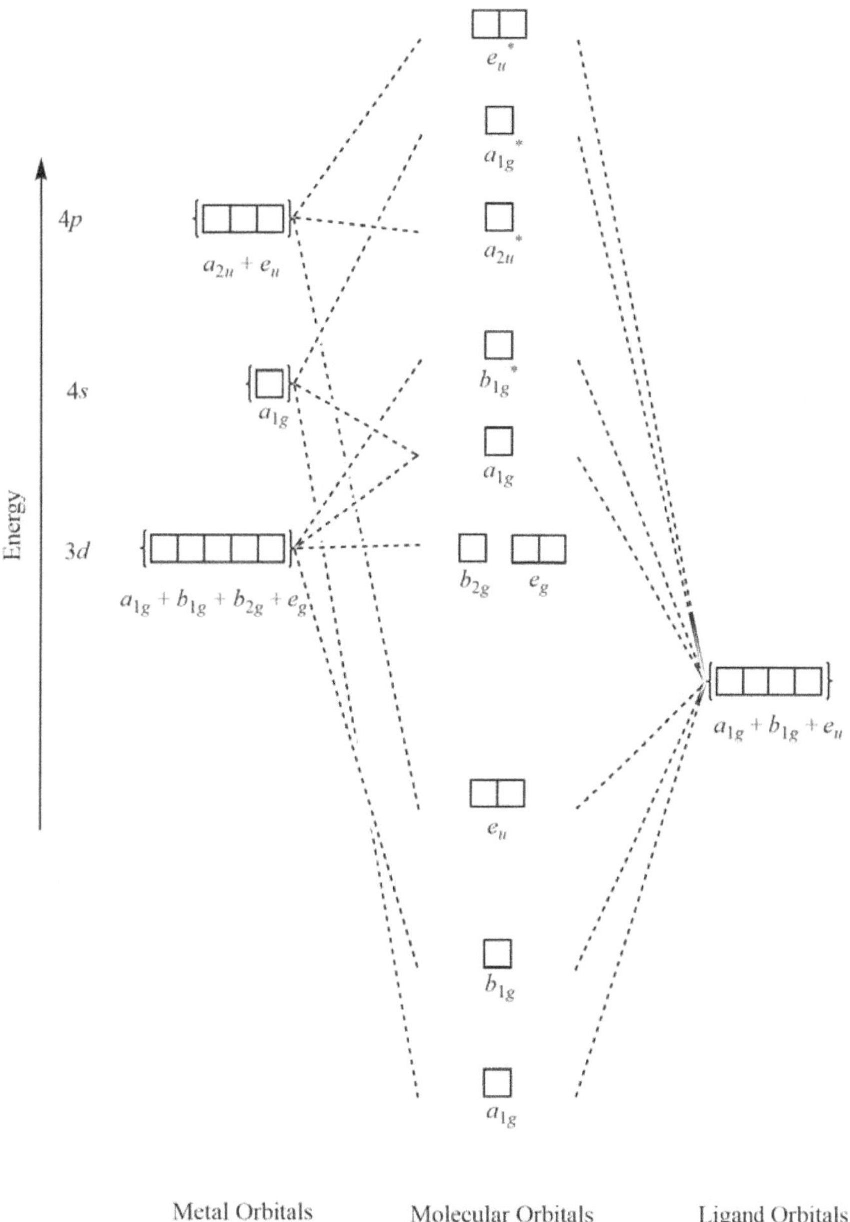

Figure 14. The generation of σ-molecular orbitals in square-planar complexes.

❖ π-Bonding and Molecular Orbital Theory

We have already studied the metal-ligand σ-overlap in the framework of molecular orbitals theory. Furthermore, this theory is also very useful for providing a rational explanation for the π-bonding in different metal-complex geometries. The basic approach remains the same except the fact that the orbital overlap in this situation is not along the internuclear axis but is sidewise in nature. The symmetry criteria governing these overlaps are absolutely clear but the extant up to which the overlap for different ligands takes place is still a matter of debate. In other words, the matching of metal and ligands orbital symmetry is not sufficient to assure the formation of π-bond as there are also some other factors (like size and energy) that signify the extent of overlap. It is quite possible that two orbital sets with the same symmetry may not be able to form molecular orbital due to a large difference in their energies.

There are four different types of metal-ligand interaction which can be resulted from the sidewise overlap of the orbitals.

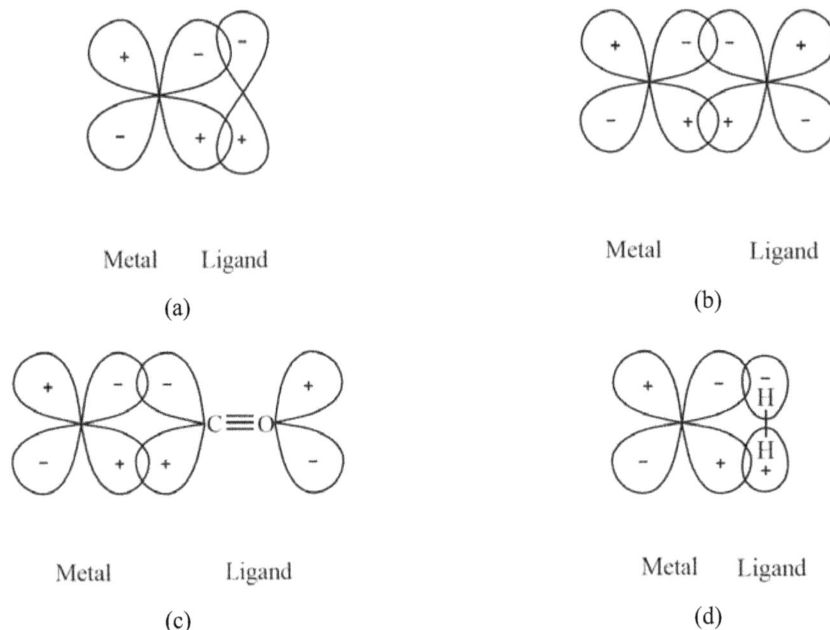

Figure 15. The sidewise overlap (π-bonding) of the d-orbital metal with different types of ligands orbital (a) d_π-p_π, (b) d_π-d_π, (c) d_π-π^*, (d) d_π-σ^*.

Electron density can be transferred from filled ligand orbital to the empty d-orbital of the metal center, or from the filled d-orbital of the metal to the empty orbital of the ligand.

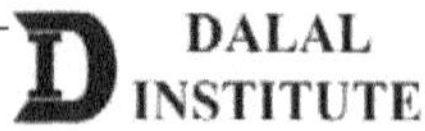

Table 4. Different types of π-bonding and the compatibility of various ligands.

Type	Explanation	Examples of the ligands involved
$d_\pi\text{-}p_\pi$	Transfer of electron density from filled p-orbital of the ligand to the empty d-orbital of the metal.	RS^-, RO^-, O^{2-}, F^-, Cl^-, Br^-, I^-, R_2N^-
$d_\pi\text{-}d_\pi$	Transfer of electron density from filled d-orbital of the metal to the empty d-orbital of the ligand.	R_2S, R_3P, R_3As
$d_\pi\text{-}\pi^*$	Transfer of electron density from filled d-orbital of the metal to the empty π^*-orbital of the ligand.	CN^-, CO, RNC, N_2, NO_2, ethylene, pyridine
$d_\pi\text{-}\sigma^*$	Transfer of electron density from filled d-orbital of the metal to the empty σ^*-orbital of the ligand.	R_3P, H_2, alkanes

It can be seen from the above table that some ligands belong to more than one category and hence can use more than one type orbitals for π-bonding. However, it is observed that the contribution from one type of orbitals dominates the other in many cases. For instance, R_3P can accept d-electron density from metal in its empty d-orbital, or in the antibonding σ^* which is also greater in magnitude. Similarly, I^- also has the ability to donate electron from its filled p-orbital, or to accept electron density in its low lying empty d-orbitals.

The generally accepted explanation for the π-bonding in transition metal complexes of different geometry can be given as.

> ➤ **π-Bonding in Octahedral Complexes**

The π-bonding in octahedral complexes can happen in two ways; one through ligand p-orbitals that are not being used in σ bonding and other via π or π^* molecular orbitals present on the ligand. The symmetry designations of different metal orbitals taking part in octahedral overlap are:

$$d_z^2,\ d_{x^2-y^2} \qquad\qquad - \qquad\qquad e_g$$

$$s \qquad\qquad - \qquad\qquad a_{1g}$$

$$p_x,\ p_y,\ p_z \qquad\qquad - \qquad\qquad t_{1u}$$

$$d_{xy},\ d_{xz},\ d_{yz} \qquad\qquad - \qquad\qquad t_{2g}$$

The symmetry adapted linear combinations of atomic orbitals (SALCs) for metal-ligand sidewise overlap can be obtained just by resolving the reducible representation based on the displacement vectors perpendicular to the axis of σ overlap. As each of the six ligands has two basis-vectors for π-symmetry, there are twelve in total.

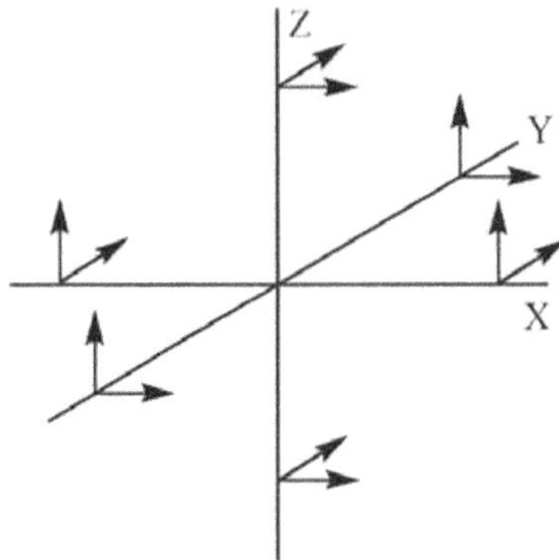

Figure 16. The π-basis set for ligand orbitals in octahedral complexes.

The symmetry adapted linear combinations of these fall into four triply degenerate irreducible representations labeled as t_{1g}, t_{2g}, t_{1u} and t_{2u}. The symmetry designations of different metal orbitals taking part in octahedral overlap are:

Table 5. Reducible representation based on perpendicular vectors in the octahedral geometry.

O_h	E	$8C_3$	$6C_2$	$6C_4$	$3C_2$	i	$6S_4$	$8S_6$	$3\sigma_h$	$6\sigma_d$	Irreducible components
Γ_π	12	0	0	0	−4	0	0	0	0	0	$t_{1g} + t_{2g} + t_{1u} + t_{2u}$

Two of these aforementioned sets are of t_{2g} and t_{1u} symmetry. The d_{xy}, d_{xz} and d_{yz} orbitals set on the metal also have t_{2g} symmetry, and therefore the π-bonds formed between a central metal and six ligands also have it as these π-bonds are just formed by the overlap of two sets of orbitals with t_{2g} symmetry. Similarly, the p_x, p_y and p_z-orbital set of the metal has t_{1u} symmetry which resembles with the symmetry of ligands SALCs. Therefore, these same-symmetry (t_{1u}) sets from metal and ligand may also interact to create bonding and antibonding molecular orbitals. However, the participation of t_{1u} set of the metal in π-overlap is highly disliked because the orbitals corresponding to this set are already being used in stronger σ-bonding and any deviation from this state is bound to destabilize the complex.

Hence, the SALC sets of t_{1g}, t_{2u} and also t_{1u} remain nonbonding. The pictorial representation of different SALC-orbitals in octahedral complexes, capable of π-overlap with metal orbitals, can be given as:

1. If four perpendicular p-orbitals of four ligands approaching along x and y-axis are pointing their lobes towards metal's d_{xy}-orbital in a sidewise manner, then the composite orbital becomes:

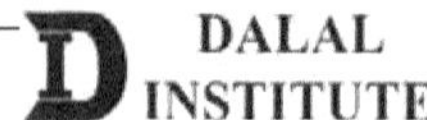

$$\varphi_1 = \pi_1 - \pi_2 + \pi_3 - \pi_4 \tag{7}$$

The pictorial representation is given below.

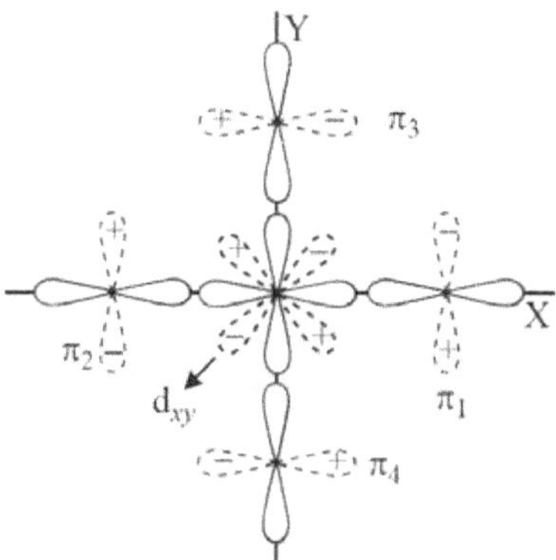

Figure 17. The combination of ligand orbitals overlapping with d_{xy}-orbital of the metal centre.

2. If four perpendicular p-orbitals of four ligands approaching along y and z-axis and are pointing their lobes towards metal's d_{yz}-orbital in a sidewise manner, then the composite orbital becomes:

$$\varphi_2 = \pi_3 - \pi_4 + \pi_5 - \pi_6 \tag{8}$$

The pictorial representation is given below.

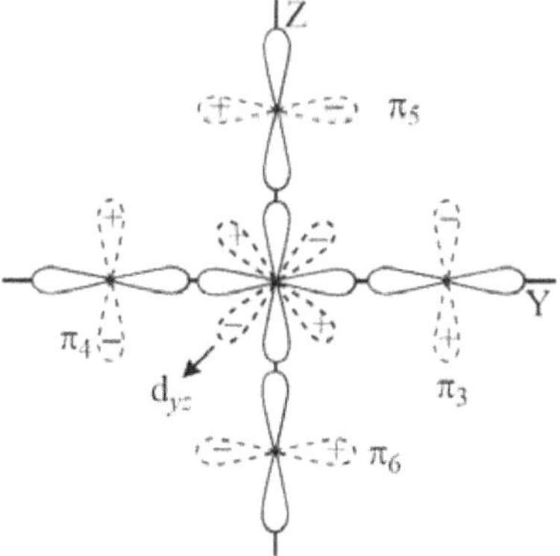

Figure 18. The combination of ligand orbitals overlapping with d_{yz}-orbital of the metal center.

3. If four perpendicular p-orbitals of four ligands approaching along x and z-axis are pointing their lobes towards metal's d_{xz}-orbital in a sidewise manner, then the composite orbital becomes:

$$\varphi_3 = \pi_1 - \pi_2 + \pi_5 - \pi_6 \tag{9}$$

The pictorial representation is given below.

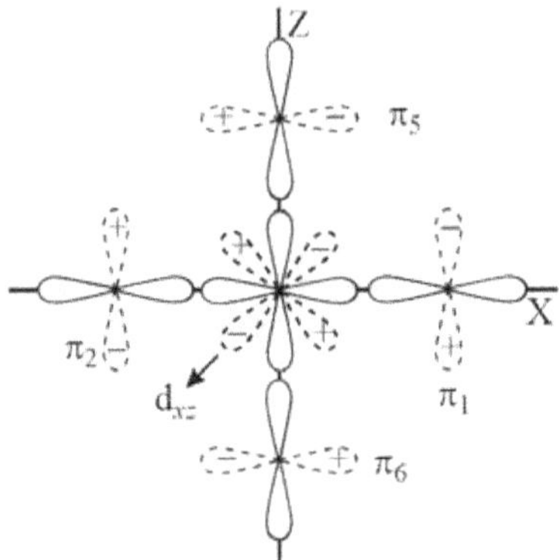

Figure 19. The combination of ligand orbitals overlapping with d_{xz}-orbital of the metal center.

Molecular orbitals and the effect of π-bonding on crystal field splitting can be categorized in three ways as follows:

1. When ligand π-orbitals are filled and are of lower energy: The one form of coordinative π-bonding is ligand-to-metal bonding. This situation arises when the t_{2g}-symmetry p or π-orbitals on the ligands are filled and are low in energy. They combine with the d_{xy}, d_{xz} and d_{yz} orbitals on the metal and donate electrons to the resulting π-symmetry bonding orbital between them and the metal.

The metal-ligand bond is somewhat strengthened by this interaction, but the complementary anti-bonding molecular orbital from metal-ligand overlap is not higher in energy than the antibonding molecular orbital from the σ-bonding. The molecular orbital of t_{2g}^{*} are greater in energy than nonbonding sets of t_{1g}, t_{2u} and t_{1u}. On the other hand, the bonding molecular orbitals of t_{2g} are lower in energy than that of nonbonding SALCs of the ligands.

Therefore, when 36 electrons (12×2 from π-bonding and 6×2 from σ-bonding) from ligands are filled in the molecular orbitals, they will completely saturate the a_{1g}-bonding, t_{1u}-bonding, e_g-bonding, t_{2g}-bonding and the nonbonding sets of t_{1g}, t_{2u} and t_{1u}. The electron previously filled in d-orbital of the free metal ion can now be considered as distributed between t_{2g}^{*} and e_g^{*}. Hence, the crystal field splitting Δ_o decreases when ligand to metal bonding takes place.

The overall molecular orbital energy level diagram for this type of π-bonding in octahedral complexes can be shown as:

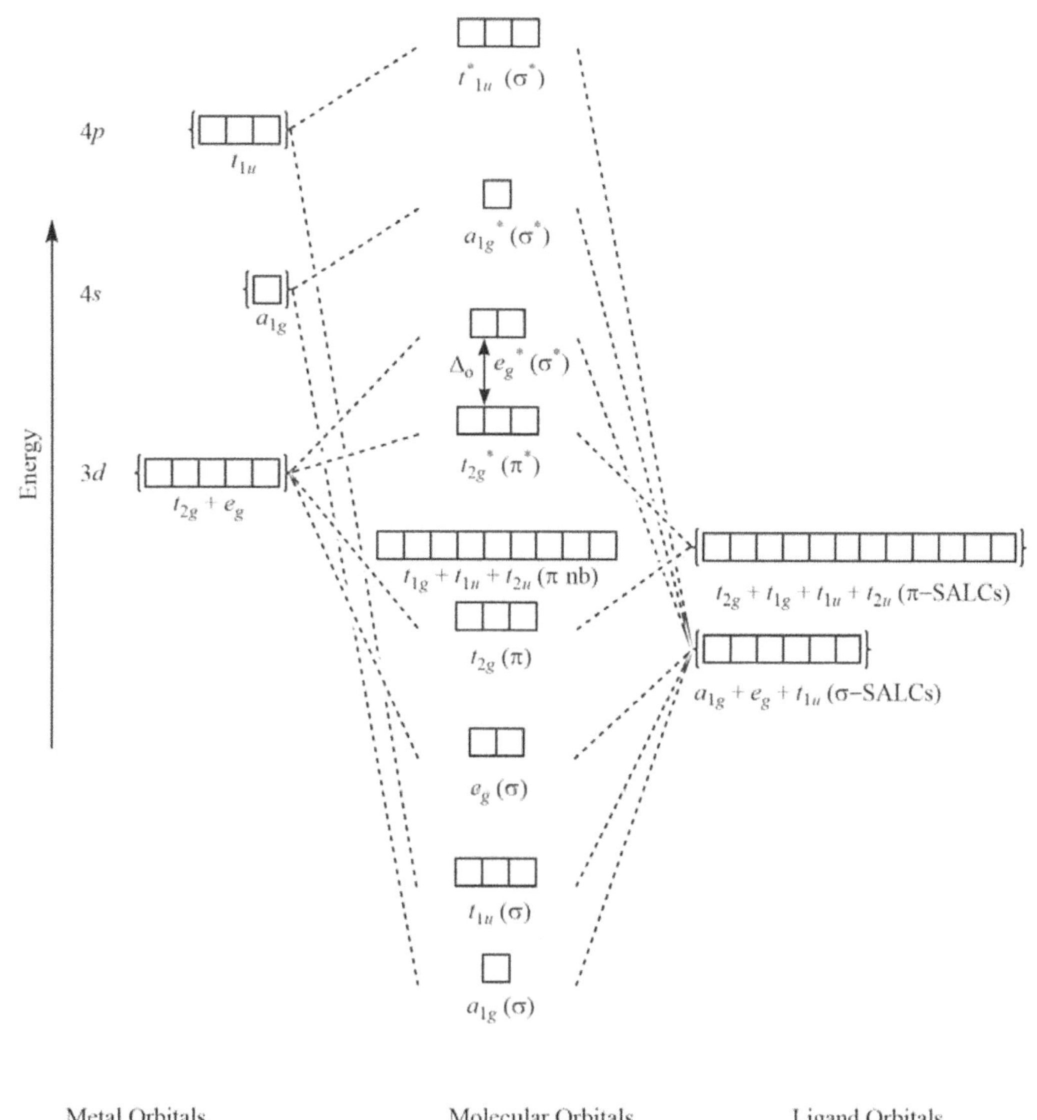

Figure 20. The generation of π and σ-molecular orbitals in octahedral complexes.

2. When ligand π-orbitals are empty and are of higher energy: The second important form of π-bonding in coordination complexes is metal-to-ligand π bonding, also called π-backbonding. It occurs when the LUMOs (lowest unoccupied molecular orbitals) of the ligand are anti-bonding π^*-orbitals and are high in energy. The ligands end up with electrons in their π^* molecular orbital, so the corresponding π-bond within the ligand weakens. The complementary anti-bonding molecular orbital from metal-ligand overlap is higher in energy than both, the first antibonding molecular orbital from the σ-bonding (e_g^*), and the nonbonding sets of t_{1g}, t_{2u} and t_{1u}. On the other hand, the bonding molecular orbitals of t_{2g} are higher in energy than all σ bonding molecular orbitals. The overall molecular orbital energy level diagram for this type of π-bonding in octahedral complexes can be shown as:

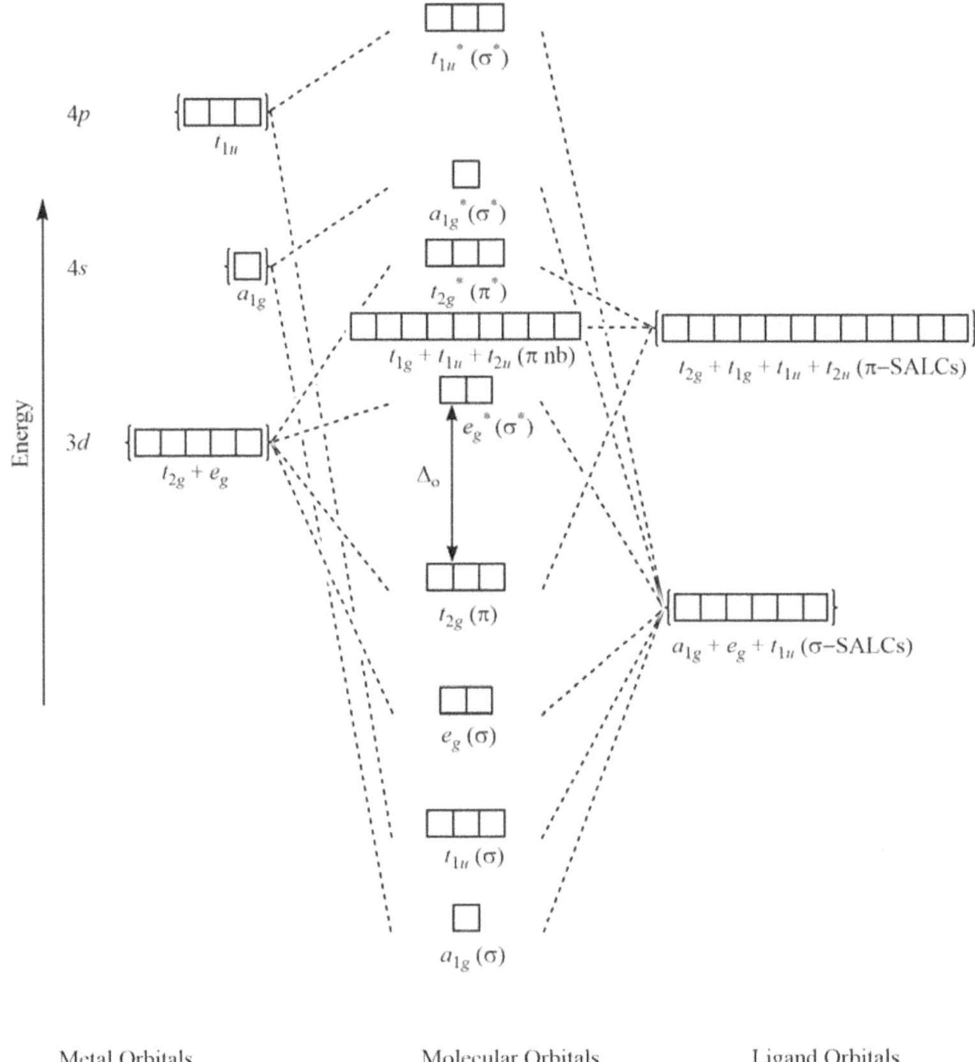

Figure 21. The generation of π and σ-molecular orbitals in octahedral complexes.

Therefore, when twelve electrons (12×0 from π-bonding and 6×2 from σ-bonding) from ligands are filled in the molecular orbitals, they will completely saturate the a_{1g}-bonding, t_{1u}-bonding and e_g-bonding. The electron previously filled in d-orbital of the free metal ion can now be considered as distributed between t_{2g} and e_g^*. Hence, the crystal field splitting Δ_o increases and the bond between the ligand and the metal strengthens when metal to ligand bonding takes place.

3. When ligand π-orbitals are empty as well as filled: There are many ligands, like I^-, which have both filled p-orbitals as well as vacant d-orbitals of π symmetry. In these cases, the prediction of crystal field splitting energy is quite difficult as the two effects counterbalance each other to different extents for different ligands. The six bonding MOs which are formed are actually filled by the electrons coming from the ligands, and electrons from the d-orbitals of the metal ion occupy the nonbonding and, sometimes, antibonding molecular orbitals. The energy difference between the latter two types of MOs is called Δ_o and is determined by the nature of the π-interaction between the ligand orbitals with the d-orbitals on the central atom. As described above, π-donor ligands lead to a small Δ_o and are called weak- or low-field ligands, whereas π-acceptor ligands lead to a large value of Δ_o and are called strong or high-field ligands. Ligands that are neither π-donor nor π-acceptor give a value of Δ_o somewhere in-between.

The magnitude of Δ_o determines the electronic structure of d^4-d^7 metal ions., The nonbonding and antibonding MOs in types of metals complexes can be filled in two ways; the first one in which maximum possible electrons are put in the nonbonding orbitals before the antibonding filling, and second one in which maximum possible unpaired electrons are put in. The first case is labeled as low-spin, while the second is labeled as high-spin complexes. A small Δ_o can be outranked by the energetic stabilization from avoiding the electron pairing, leading to the spin-free case. On the other hand, if Δ_o is large, the spin-pairing energy would become negligible in comparison and a spin-paired state arises. An empirically obtained list of various ligands which are arranged by the magnitude of splitting Δ they produce is called as the spectrochemical series. It can be seen that the low-field ligands are all π-donors (such as I^-), the high field ligands are π-acceptors (such as CN^- and CO), and ligands such as H_2O and NH_3, which are neither, are in the middle.

$I^- < Br^- < S^{2-} < SCN^- < Cl^- < NO_3^- < N^{3-} < F^- < OH^- < C_2O_4^{2-} < H_2O < NCS^- < CH_3CN <$ py (pyridine) $< NH_3 <$ en (ethylenediamine) $<$ bipy (2,2'-bipyridine) $<$ phen (1,10-phenanthroline) $< NO_2^- <$ $PPh_3 < CN^- < CO$

It is also worthy to note that the electrons transfer from metal d-orbital to the antibonding molecular orbitals of carbonyl and its analogs is actually partial in nature. The electron-transfer increases metal-carbon bond strength but weakens the carbon–oxygen bond. The increased M–CO bond strength is obvious in the increases of vibrational frequencies for the metal–carbon bond and usually lie outside the range of normal IR spectrophotometers. Moreover, the bond length of M–CO bond is also shortened. The decreased bond strength of C–O bond is obvious from the decreasing wavenumber of v_{CO} bands. The high stabilization which results from metal-to-ligand bonding (back-bonding) is caused by the relaxation of the negative charge from the metal center. This permits the metal ion to accept the σ-donation more effectively. The combination of ligand-to-metal σ-bonding and metal-to-ligand π-bonding is a synergic effect, as each enhances the other.

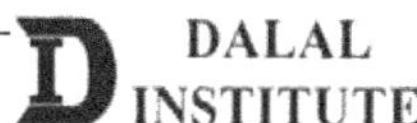

> ➤ *π-Bonding in Tetrahedral Complexes*

The π-bonding in tetrahedral complexes can happen in two ways; one through ligand *p*-orbitals that are not being used in σ bonding and other via π or π* molecular orbitals present on the ligand. The symmetry designations of different metal orbitals taking part in tetrahedral overlap are:

s	–	a_1
p_x, p_y, p_z	–	t_2
d_{xy}, d_{xz}, d_{yz}	–	t_2
$d_{z^2}, d_{x^2-y^2}$	–	e

The symmetry adapted linear combinations of atomic orbitals (SALCs) for metal-ligand sidewise overlap can be obtained just by resolving the reducible representation based on the displacement vectors perpendicular to the axis of σ overlap. As each of the four ligands has two basis-vectors for π-symmetry, there are eight in total. Each pair of perpendicular vectors (p_x and p_y) is attached about its own *z*-axis so that all the *y*-vectors are parallel to the *xy*-plane of the tetrahedral complex.

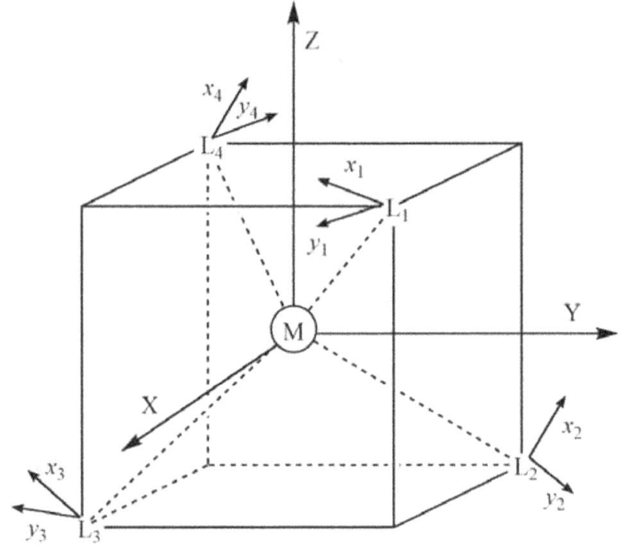

Figure 22. The π-basis set for ligand orbitals in tetrahedral complexes.

The symmetry adapted linear combinations of these fall into two doubly and on singly degenerate irreducible representations labeled as e, t_1 and t_2. The symmetry designations of different ligand orbitals taking part in tetrahedral overlap are:

Table 6. Reducible representation based on perpendicular vectors in the tetrahedral geometry.

T_d	E	$8C_3$	$3C_2$	$6S_4$	$6\sigma_d$	Irreducible components
Γ_π	8	-1	0	0	0	$e + t_1 + t_2$

Two of these aforementioned sets are of e and t_2 symmetry. The $d_{x^2-y^2}$ and d_{z^2} orbitals set on the metal also have e-symmetry, and therefore the π-overlap between a central metal and four ligands is possible as far as the generation of molecular orbitals with e-symmetry is concerned.

Similarly, the p_x, p_y and p_z-orbital set of the metal has t_2 symmetry which resembles with the symmetry of one of SALCs set on ligands. Therefore, these same-symmetry (t_2) sets from metal and ligand may also interact to create bonding and antibonding molecular orbitals. However, the π-overlap in tetrahedral complexes is not independent of the σ-bonding because the ligand SALCs and metals atomic orbital, of both types of e and t_2 symmetry, participate in σ as well as π-bonding.

Hence, when twenty-four electrons (8×2 from π-bonding and 4×2 from σ-bonding) from ligands are filled in the molecular orbitals, they will completely saturate the a_1 (σ), t_2 (σ), t_2 (π), e (π) and t_1 (π nb). The electron previously filled in d-orbital of the free metal ion can now be considered as distributed in e^* (π^*) and t_2^* (σ^*, π^*) which are predominantly $d_{x^2-y^2} = d_{z^2}$ and $d_{xz} = d_{yz} = d_{xy}$ in character, respectively.

Therefore, the pattern of crystal field splitting energy, provided by the molecular orbital theory, is the same as that of what it was given by crystal field theory. Moreover, the ligand to metal charge transfer spectral band can be considered as a result of the transition of an electron from nonbonding or bonding molecular orbitals to the antibonding molecular orbitals. These charge transfer peaks are very high in intensity they are spin as well as Laporte allowed in nature.

Tetrahedral complexes are almost always high-spin in nature because the energy separation between e^* (π^*) and t_2^* (σ^*, π^*) energy levels, Δ_t, is very small. That's why, during the filling of metal d-electrons, the promotion of electrons is always preferred over pairing. This is the same result as provided by the application of crystal field theory.

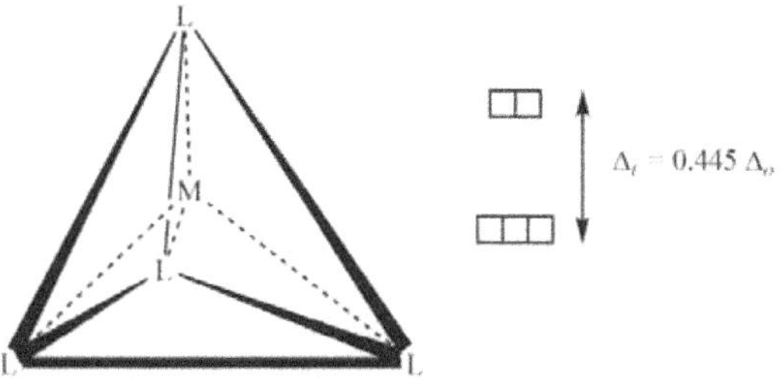

Figure 23. The crystal field splitting of d-subshell in tetrahedral complexes.

The overall molecular orbital energy level diagram for this type of π-bonding in tetrahedral complexes can be shown as:

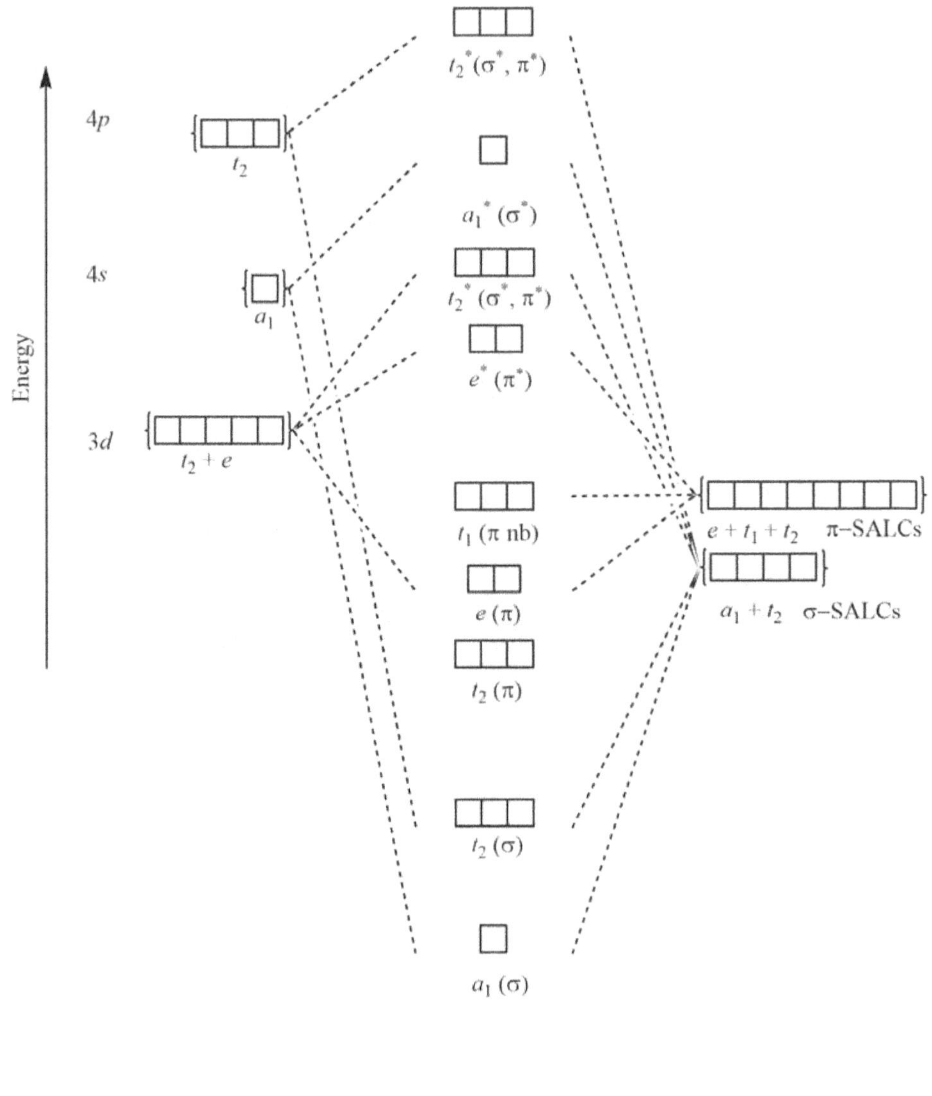

Figure 23. The generation of π and σ-molecular orbitals in four-coordinated tetrahedral complexes of transition metals.

> ➤ *π-Bonding in Square Planar Complexes*

The π-bonding in square planar complexes can be visualized in terms of the sidewise overlap of different kinds of orbitals on metal and ligands, having comparable energy and proper symmetry. The symmetry designations of different metal orbitals taking part in square-planar overlap are:

s	–	a_{1g}
p_x, p_y	–	e_u
d_{z^2}	–	a_{1g}
$d_{x^2-y^2}$	–	b_{1g}
p_z	–	a_{2u}
d_{xy}	–	b_{2g}
d_{yz}, d_{xz}	–	e_g

The symmetry adapted linear combinations of ligand atomic orbitals (SALCs) for metal-ligand sidewise overlap in square planar complexes can be obtained just by resolving the reducible representation based on the displacement vectors perpendicular to the axis of σ-overlap. As each of the four ligands has two basis-vectors for π-symmetry, there are eight in total.

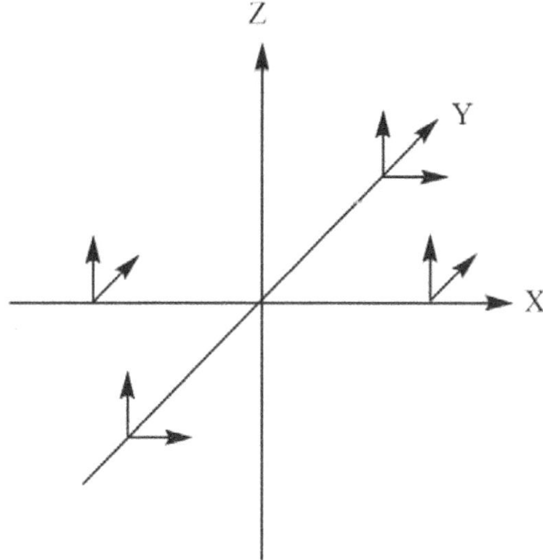

Figure 24. The complete π-basis set for ligand orbitals in square planar complexes of transition metal centers.

The symmetry adapted linear combinations of these fall into six irreducible representations labeled as a_{2g}, b_{2u}, e_u, a_{2u}, e_g and b_{2g}. The symmetry designations of different ligand SALC-orbitals taking part in square-planar overlap are:

Table 7. The complete reducible representation based on perpendicular vectors (usable in sidewise overlap) in square planar geometry.

D_{4h}	E	$2C_4$	C_2	$2C_2'$	$2C_2''$	i	$2S_4$	σ_h	$2\sigma_v$	$2\sigma_d$	Irreducible components
Γ_π	8	0	0	−4	0	0	0	0	0	0	$a_{2g} + b_{2u} + e_u + e_g +$ $b_{2g} + a_{2u}.$

Two of these aforementioned sets, with a_{2g} and b_{2u} symmetry, remain nonbonding and are localized on the ligands. This is quite logical because there are no metal orbitals with these symmetry properties and hence no orbital overlap is possible.

The metal orbitals of e_u-symmetry take part in σ as well as π-bonding. Now, as there are two sets of SLACs with e_u-symmetry, the number of molecular orbital sets formed with e_u-symmetry is also three. The lowest energy molecular orbitals set of e_u-symmetry is primarily of σ-bonding while the higher energy molecular orbitals set is mainly π-bonding in nature. The highest doubly degenerate molecular orbitals set of e_u-symmetry is of both σ^* and π^*-antibonding character.

The d_{yz} and d_{xz} orbitals set on the metal has e_g-symmetry, which resembles with the symmetry of one of SALCs set on ligands. Therefore, these same-symmetry (e_g) sets from metal and ligand also interact to create bonding and antibonding molecular orbitals. Furthermore, the b_{2g}-symmetry d_{xy} orbital and a_{2u}-symmetry p_z orbitals on the metal interacts with b_{2g}-symmetry and a_{2u}-symmetry SALC, respectively, to produce π-bonding and antibonding molecular orbitals.

Hence, when twenty-four electrons (8×2 from π-bonding and 4×2 from σ-bonding) from ligands are filled in the molecular orbitals, they will completely saturate the $a_{1g}(\sigma)$, $b_{1g}(\sigma)$, $e_u(\sigma)$, $b_{2g}(\pi)$, $e_g(\pi)$, $a_{2u}(\pi)$, e_u (π), a_{2g} (π nb) and b_{2u} (π nb). The electron previously filled in d-orbital of the free metal ion can now be considered as distributed in e_g^* (π^*), $a_{1g}(\sigma^*)$, b_{2g}^* (π^*) and b_{1g}^* (σ^*) which are predominantly $d_{xz}=d_{yz}$, d_z^2, d_{xy} and $d_{x^2-y^2}$ in character, respectively. Hence, the pattern of crystal field splitting energy, provided by the molecular orbital theory, is the same as that of what it was given by crystal field theory. In other words, the formal electronic configuration of tetrahedral metal complexes is the same from both theories.

Moreover, the ligand to metal charge transfer spectral band can be considered as a result of the transition of an electron from nonbonding or bonding molecular orbitals to b_{1g}^* (σ^*). These charge transfer peaks are pretty high in intensity due to their spin and Laporte allowance.

The overall molecular orbital energy level diagram for the π-bonding in square-planar complexes can be shown as:

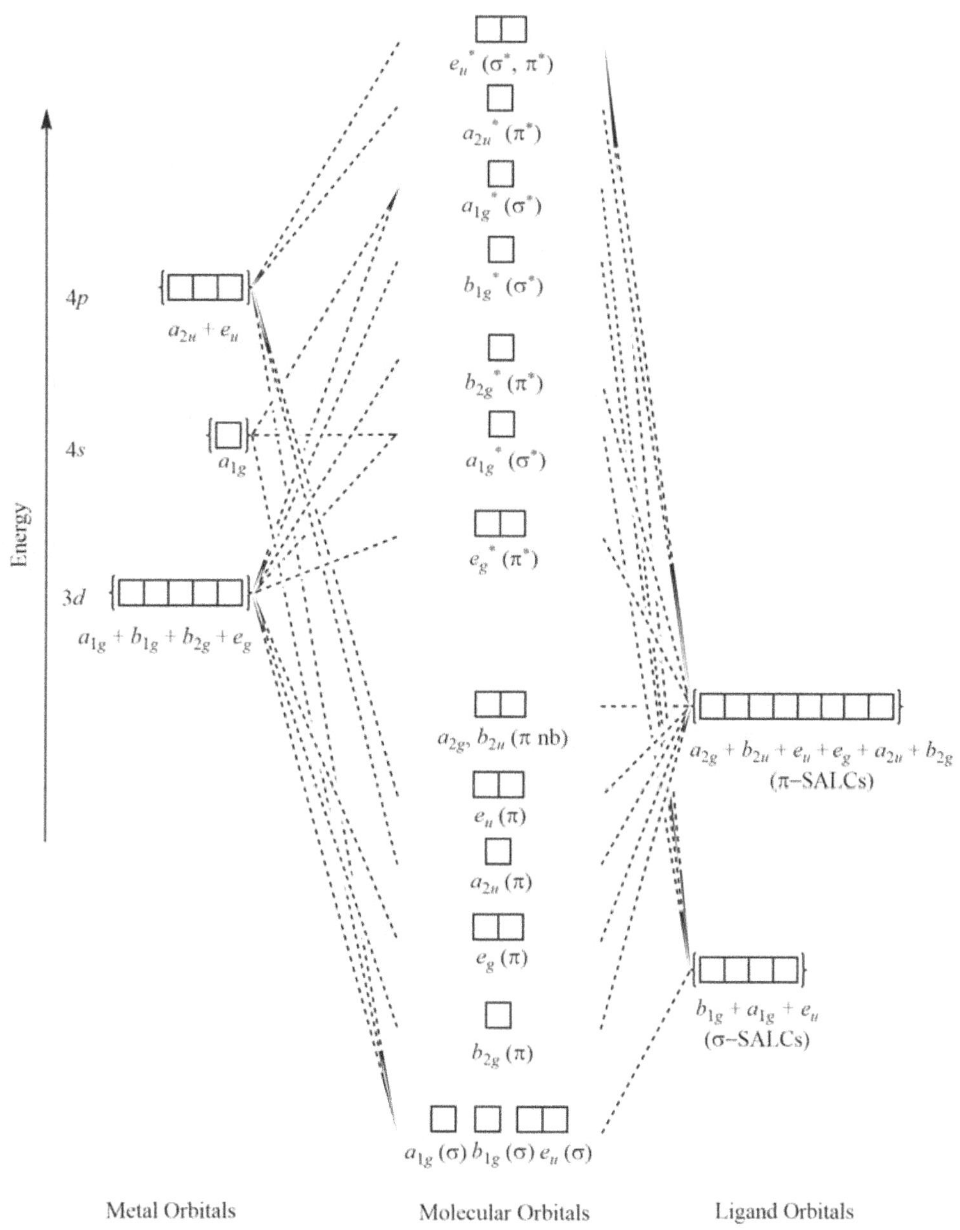

Figure 25. The generation of π and σ-molecular orbitals in square-planar complexes.

❖ Problems

Q 1. Give at least seven limitations of crystal field theory.

Q 2. Why does the crystal field theory fail to provide any rational explanation for the nephelauxetic effect?

Q 3. Discuss the "superexchange phenomenon" as an evidence for the presence of covalent-character in metal-ligand bonds.

Q 4. Write down the main postulates of molecular orbital theory for metal complexes.

Q 5. Explain the metal-ligand σ-bonding for octahedral complexes in the molecular orbital framework.

Q 6. Draw and discuss the molecular orbital energy level diagram involving σ-bonding for tetrahedral complexes.

Q 7. How does the π-bonding affect the magnitude of crystal field splitting in octahedral complexes when ligand π-orbitals are empty and are high in energy?

Q 8. Explain the metal-ligand π-bonding for octahedral complexes in the molecular orbital framework.

Q 9. Draw and discuss the molecular orbital energy level diagram involving π-bonding for tetrahedral complexes.

Q 10. Draw the molecular orbital energy level diagram for π-bonding for square-planar complexes. Also, explain ligand to metal charge transfer in brief.

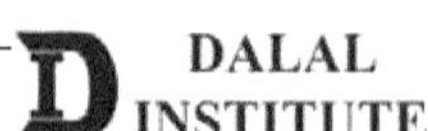

❖ Bibliography

[1] B. R. Puri, L. R. Sharma, K. C. Kalia, *Principals of Inorganic Chemistry*, Milestone Publishers, Delhi, India, 2012.

[2] J. E. Huheey, E. A. Keiter, R. L. Keiter, *Inorganic Chemistry: Principals of Structure and Reactivity*, HarperCollins College Publishers, New York, USA, 1993.

[3] B. W. Pfennig, *Principles of Inorganic Chemistry*, John Wiley & Sons, New Jersey, USA, 2015.

[4] F. A. Cotton, *Chemical Applications of Group Theory*, John Wiley & Sons, New Jersey, USA, 2006.

[5] N. N. Greenwood, A. Earnshaw, *Chemistry of the Elements*, Butterworth-Heinemann, Oxford, Britain, 1998.

[6] J. E. House, *Inorganic Chemistry*, Academic Press, California, USA, 2008.

[7] D. Shriver, M. Weller, T. Overton, J. Rourke, F. Armstrong, *Inorganic Chemistry*, W. H. Freeman and Company, New York, USA, 2014.

CHAPTER 8

Electronic Spectra of Transition Metal Complexes:

❖ Spectroscopic Ground States

The spectra of transition metal complexes is not as simple as it appears from just the splitting of d-orbitals with electrons get promoted from the lower energy orbital set to a higher energy orbital set. Actually, energy levels of a transition metal atom or ion with a particular electronic configuration are described not only by the electronic configuration itself but also by different types of electronic interactions such as spin-spin, orbital-orbital or spin-orbital which can be categorized by some special symbols, called as term symbols. The ground state term symbol is predicted by Hund's rule. In other words, the term symbol in quantum mechanics is an abbreviated description of the total angular momentum quantum numbers in a multi-electron atom. However, the quantum mechanical states of a single electron can also be described by a term symbol.

➢ *Calculation of Microstates in a Particular Electronic Configuration*

The various overall-arrangements of electronic cloud around the nucleus for a particular configuration are not the same as far as the energy and angular momentum are concerned. These different electronic arrangements can be classified on the basis of overall spin, orbital or total angular momentum.

The term microstates may be defined as the overall electronic arrangements of a particular electronic configuration which can be differentiated in terms of energy or angular momentum.

The two types of electronic configurations for which the microstate calculation has to be carried out are discussed below.

1. When unpaired electrons are present in the same subshell:

i) p^1-configuration: Let us consider that we want to study the number of ways in which a single electron can be filled in any p-subshell.

	p_x	p_y	p_z
1	↑		
2	↓		
3		↑	
4		↓	
5			↑
6			↓

ii) p^2-configuration: Let us consider that we want to study the number of ways in which two electrons can be filled in any p-subshell.

	p_x	p_y	p_z
1	↑	↑	
2		↑	↑
3	↑		↑
4	↓	↓	
5		↓	↑
6	↓		↓
7	↑	↓	
8		↑	↓
9	↑		↓
10	↓	↑	
11		↓	↑
12	↓		↑
13	↑↓		
14		↑↓	
15			↑↓

Hence, the total number of ways in which one and two electrons can be arranged in the p-subshell of an atom or ion are six and fifteen, respectively. However, the calculation of the number of microstates for the configurations like d^2 or f^2 using the abovementioned method is quite lengthy and difficult. Moreover, as we are interested only in the number of these electronic arrangements and not in the nature; permutation and combination can be used to find out all these numbers as follows:

$$\text{No. of microstates} = \binom{n}{r} = \frac{n!}{r!\,(n-r)!}$$

Where n is twice the number of orbitals present in the subshell under consideration and r is the number of unpaired electrons in them. The formulation related to the calculation of the number of microstate in particular term symbol will be discussed later in this chapter.

Now, the number of microstates for different electronic configurations can be calculated easily.

i) For p^1-configuration, $n = 6$ and $r = 1$. Therefore

$$\text{No. of microstates} = \binom{6}{1} = \frac{6!}{1!\,(6-1)!}$$

$$\binom{6}{1} = \frac{6 \times 5 \times 4 \times 3 \times 2 \times 1}{1(5 \times 4 \times 3 \times 2 \times 1)}$$

$$= 6 \text{ microstates}$$

ii) For p^2-configuration, $n = 6$ and $r = 2$. Therefore

$$\text{No. of microstates} = \binom{6}{2} = \frac{6!}{2!\,(6-2)!}$$

$$\binom{6}{2} = \frac{6 \times 5 \times 4 \times 3 \times 2 \times 1}{2 \times 1(4 \times 3 \times 2 \times 1)}$$

$$= 15 \text{ microstates}$$

iii) For d^1-configuration, $n = 10$ and $r = 1$. Therefore

$$\text{No. of microstates} = \binom{10}{1} = \frac{10!}{1!\,(10-1)!}$$

$$\binom{10}{1} = \frac{10 \times 9 \times 8 \times 7 \times 6 \times 5 \times 4 \times 3 \times 2 \times 1}{1(9 \times 8 \times 7 \times 6 \times 5 \times 4 \times 3 \times 2 \times 1)}$$

$$= 10 \text{ microstates}$$

iv) For d^2-configuration, $n = 10$ and $r = 2$. Therefore

$$\text{No. of microstates} = \binom{10}{2} = \frac{10!}{2!\,(10-2)!}$$

$$\binom{10}{2} = \frac{10 \times 9 \times 8 \times 7 \times 6 \times 5 \times 4 \times 3 \times 2 \times 1}{2 \times 1(8 \times 7 \times 6 \times 5 \times 4 \times 3 \times 2 \times 1)}$$

$$= 45 \text{ microstates}$$

The general form of atomic term symbols considers spin-spin (S-S coupling), orbital-orbital (L-L coupling) and spin-orbital (L-S or Russell-Saunders coupling) interactions; therefore, before categorizing all the microstates in these term symbols, it is extremely important to find out all possible microstates for different electronic configurations. The total number of microstates for other different electronic configurations can be calculated using the same method and are listed below.

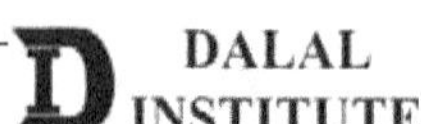

Table 1. The calculated microstates for different electronic configurations.

Electronic configuration	No. of microstates	Electronic configuration	No. of microstates
p^1	6	d^3	120
p^2	15	d^4	210
p^3	20	d^5	252
p^4	15	d^6	210
p^5	6	d^7	120
p^6	1	d^8	45
d^1	10	d^9	10
d^2	45	d^{10}	1

It can be seen that the number of microstates for the same number of unpaired electrons is equal. For example, there are six microstates for p^1 as well as p^5 and forty-five microstates for both d^2 as well as d^8. It can be explained in terms of electron-hole formalism which will be discussed later in this chapter.

2. When unpaired electrons are present in different subshell:

The same approach can also be applied to calculate the microstates in the case when the electrons are present in two different subshells. Therefore, permutation and combination can be used to find out all these numbers as follows:

$$\text{No. of microstates} = \binom{n}{r} \times \binom{m}{s} = \frac{n!}{r!\,(n-r)!} \times \frac{m!}{s!\,(m-s)!}$$

Where, n is twice the number of orbitals and r is the number of unpaired electrons in one subshell while m is twice the number of orbitals and s is the number of unpaired electrons in the other subshell. Now, the number of microstates for different electronic configurations can be calculated easily.

i) p^1p^1-configuration, $n = 6$, $r = 1$, $m = 6$ and $s = 1$.

$$\text{No. of microstates} = \binom{6}{1} \times \binom{6}{1} = \frac{6!}{1!\,(6-1)!} \times \frac{6!}{1!\,(6-1)!}$$

$$\binom{6}{1} \times \binom{6}{1} = \frac{6 \times 5 \times 4 \times 3 \times 2 \times 1}{1(5 \times 4 \times 3 \times 2 \times 1)} \times \frac{6 \times 5 \times 4 \times 3 \times 2 \times 1}{1(5 \times 4 \times 3 \times 2 \times 1)}$$

$$= 36 \text{ microstates}$$

ii) d^1d^1-configuration, $n = 10$, $r = 1$, $m = 10$ and $s = 1$.

$$\text{No. of microstates} = \left(\frac{10}{1}\right) \times \left(\frac{10}{1}\right) = \frac{10!}{1!\,(10-1)!} \times \frac{10!}{1!\,(10-1)!}$$

$$\left(\frac{10}{1}\right) \times \left(\frac{10}{1}\right) = \frac{10 \times 9 \times 8 \times 7 \times 6 \times 5 \times 4 \times 3 \times 2 \times 1}{1(9 \times 8 \times 7 \times 6 \times 5 \times 4 \times 3 \times 2 \times 1)} \times \frac{10 \times 9 \times 8 \times 7 \times 6 \times 5 \times 4 \times 3 \times 2 \times 1}{1(9 \times 8 \times 7 \times 6 \times 5 \times 4 \times 3 \times 2 \times 1)}$$

$$= 100 \text{ microstates}$$

iii) p^1d^1-configuration, $n = 6$, $r = 1$, $m = 10$ and $s = 1$.

$$\text{No. of microstates} = \left(\frac{6}{1}\right) \times \left(\frac{10}{1}\right) = \frac{6!}{1!\,(6-1)!} \times \frac{10!}{1!\,(10-1)!}$$

$$\left(\frac{6}{1}\right) \times \left(\frac{10}{1}\right) = \frac{6 \times 5 \times 4 \times 3 \times 2 \times 1}{1(5 \times 4 \times 3 \times 2 \times 1)} \times \frac{10 \times 9 \times 8 \times 7 \times 6 \times 5 \times 4 \times 3 \times 2 \times 1}{1(9 \times 8 \times 7 \times 6 \times 5 \times 4 \times 3 \times 2 \times 1)}$$

$$= 60 \text{ microstates}$$

The total number of microstates for different electronic configurations can be calculated using the same method and are listed below.

Table 2. The total number of microstates calculated for the different electronic configurations when unpaired electrons are present in two different subshells.

Electronic configuration	No. of microstates	Electronic configuration	No. of microstates
p^1p^1	36	d^2d^2	2025
p^1p^2	90	d^2d^3	5400
p^2p^2	225	p^1d^1	60
p^2p^3	300	p^1d^2	270
d^1d^1	100	p^2d^2	675
d^1d^2	450	p^2d^3	1800

This table can further be extended for the remaining combination of p-p, d-d, p-d or their combinations with f-subshell. The number of microstates in particular configuration can be distributed to various electronic states, represented by atomic term symbols. The distribution of these microstates in different term symbols will be discussed later in this section.

➢ *Atomic Term Symbols*

Atomic term symbols may be defined as the symbolic representations of various electronic states having different resultant angular momentums resulting from spin-spin, orbital-orbital or spin-orbital interactions and the transitions between two different atomic states may also be represented using their term symbols, to which certain rules apply.

The general form of any atomic term symbol that is used to represent any electronic state resulting from inter-electronic repulsion is:

$$^{2S+1}L_J$$

Where,

$2S+1$ = spin multiplicity

S = resultant spin angular momentum quantum number

L = resultant orbital angular momentum quantum number

J = resultant total angular momentum quantum number

Just like in the case of atomic orbitals, where l represents the individual orbital angular momentums; L represents the resultants orbital angular momentum of an electronic state and gives the base designation of any atomic term symbol.

L	=	0	1	2	3	4	5	6	7
State	=	S	P	D	F	G	H	I	K

The calculation of resultant spin and orbital angular momentum involves the concepts of space quantization and vector interactions. A somewhat simplified approach for the calculation of resultant orbital angular momentum quantum number (L), resultant spin angular momentum quantum number S and resultant total angular momentum quantum number (J) can be given by understanding the spin-spin, orbital-orbital and spin-orbital couplings schemes.

1. Orbital-orbital coupling (*l-l* interaction): Consider a multielectron system, then the resultant orbital angular momentum quantum number can be deduced as:

$$L = (l_1 + l_2),\ (l_1 + l_2 - 1)\ \ldots\ldots\ |l_1 - l_2|$$

Where l_1 and l_2 are the individual orbital angular momentum quantum numbers for electrons and modulus sign shows that the value of resultant orbital angular momentum quantum number is always positive. Given the eigenstates of l_1 and l_2, the construction of eigenstates of L (which still is conserved) is the coupling of the angular momenta of electrons 1 and 2.

i) For p^1p^1-configuration, $l_1 = 1$ and $l_2 = 1$, therefore

$$L = (1 + 1),\ (1 + 1 - 1),\ (1 - 1)$$

$$L = 2, 1, 0$$

$$\text{States} = D, P, S$$

Similarly,

ii) For d^1d^1-configuration, $l_1 = 2$ and $l_2 = 2$, therefore

$$L = (2 + 2),\ (2 + 2 - 1) \dots\dots\ (2 - 2)$$

$$L = 4, 3, 2, 1, 0$$

$$\text{States} = G, F, D, P, S$$

Although the above-mentioned procedure provides the resultant orbital angular momentum quantum number (L) quite easily, the exact concept of orbital-orbital coupling can be understood only after knowing the concepts of space quantization. One thing that is totally clear is that these symbols are nothing but the mathematical shorthand of the electronic arrangements around the nucleus. The quantization of individual orbital angular momentums can be used to calculate the resultant value as follows:

i) For p^1p^1-configuration, $l_1 = 1$ and $l_2 = 1$, therefore orbital angular momentum for each of the electron is $\sqrt{2}$ and it is a well-known fact from the quantum mechanics that $\sqrt{2}$ angular momentum can be oriented in space with three different ways (+1, 0 and −1). The different combinations of orbital angular momentum can be calculated as

$l_1(z)$	$=$	+1	+1	+1	0	0	0	−1	−1	−1
$l_2(z)$	$=$	+1	0	−1	+1	0	−1	+1	0	−1
L_z	$=$	+2	+1	0	+1	0	−1	0	−1	−2

Hence, the orbital angular momentums of two p-electrons can interact in nine ways, creating nine combinations; out of which, three quantum-mechanically allowed series can be fashioned.

$$L_z = (+2, +1, 0, -1, -2),\ (+1, 0, -1),\ (0)$$

or

$$L = 2, 1, 0$$

Hence

$$\text{States} = D, P, S$$

Therefore, we can say that there are nine ways in which the orbital motion can interact.

Similarly,

ii) For d^1d^1-configuration, $l_1 = 2$ and $l_2 = 2$, therefore orbital angular momentum for each of the electron is $\sqrt{6}$ and it is a well-known fact from the quantum mechanics that $\sqrt{6}$ angular momentum can be oriented in space with five different ways (+2, +1, 0, −1 and −2). The different combinations of orbital angular momentum can be calculated as:

$l_1(z)$	=	+2	+2	+2	+2	+2	+1	+1	+1	+1	+1
$l_2(z)$	=	+2	+1	0	−1	−2	+2	+1	0	−1	−2
L_z	=	+4	+3	+2	+1	0	+3	+2	+1	0	−1

and

$l_1(z)$	=	0	0	0	0	0	−1	−1	−1	−1	−1
$l_2(z)$	=	+2	+1	0	−1	−2	+2	+1	0	−1	−2
L_z	=	+2	+1	0	−1	−2	+1	0	−1	−2	−3

and

$l_1(z)$	=	−2	−2	−2	−2	−2
$l_2(z)$	=	+2	+1	0	−1	−2
L_z	=	0	−1	−2	−3	−4

Hence, the orbital angular momentums of two d-electrons can interact in twenty-five ways, creating twenty-five combinations; out of which, five quantum-mechanically allowed series can be fashioned.

$$L_z = (+4, +3, +2, +1, 0, -1, -2, -3, -4), (+3, +2, +1, 0, -1, -2, -3), (+2, +1, 0, -1, -2), (+1, 0, -1), (0)$$

or

$$L = 4, 3, 2, 1, 0$$

Which means

$$\text{States} = G, F, D, P, S$$

2. Spin-spin coupling (s-s interaction): Consider a multielectron system, then the resultant spin angular momentum quantum number can be deduced as:

$$S = (s_1 + s_2), \ (s_1 + s_2 - 1) \ \ldots\ldots \ |s_1 - s_2|$$

Where s_1 and s_2 are the individual spin angular momentum quantum numbers for electrons and modulus sign shows that the value of resultant spin angular momentum quantum number is always positive.

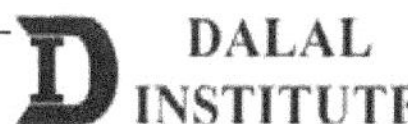

i) For p^1p^1 or d^1d^1-configuration, $s_1 = 1/2$ and $s_2 = 1/2$, therefore

$$S = (1/2 + 1/2), (1/2 - 1/2)$$

or

$$S = 1, 0$$

Which means

$$\text{Multiplicity} = (2S + 1) = 3 \text{ and } 1$$

The aforementioned procedure offers the resultant spin angular momentum quantum number (S) quite easily but the exact concept of spin-spin interaction can be understood only after knowing the concepts of space quantization. The quantization of individual spin angular momentums can be used to calculate the resultant value as follows:

i) For p^1p^1 or d^1d^1-configuration, $s_1 = 1/2$ and $s_2 = 1/2$, therefore spin angular momentum for each of the electrons is $\sqrt{0.75}$ and it is a well-known fact from the quantum mechanics that $\sqrt{0.75}$ angular momentum can be oriented in space with two different ways (+1/2 and −1/2). The different combinations of spin angular momentum can be calculated as:

$s_1(z)$	=	+1/2	+1/2	−1/2	−1/2
$s_2(z)$	=	+1/2	−1/2	+1/2	−1/2
S_z	=	+1	0	0	−1

Hence, the spin angular momentums of two p or d-electrons can interact in four ways, creating four combinations; out of which, two quantum-mechanically allowed series can be fashioned.

$$S_z = (+1, 0, -1), (0)$$

or

$$S = 1, 0$$

$$\text{Multiplicity} = (2S + 1) = 3 \text{ and } 1$$

The multiplicity actually represents the number of orientations possible for the total spin relative to the total orbital angular momentum L, and thus to the number of near-degenerate levels that differ only in their spin-orbit coupling energy. For example, the ground state of the carbon atom is a ^3P state. The superscript of three specifies that the multiplicity 2S+1 = 3 i.e. triplet, so that the total spin S = 1. This spin is due to two unpaired electrons, as a result of Hund's rule which favors the single filling of degenerate orbitals. The spin multiplicity is a primary factor in governing the overall energy of an electronic state and maybe summarised for different electron combination.

Table 3. Spin multiplicities for the different number of unpaired electron in an electronic configuration.

Unpaired Electrons	S	2S+1	State
0	0	1	Singlet
1	1/2	2	Doublet
2	1	3	Triplet
3	3/2	4	Quartet
4	2	5	Quintet

3. Spin-orbital coupling (L-S interaction): In a multi-electron system, the resultant orbital angular momentum (L) and resultant spin angular momentum (S) interact with each other to give total angular momentum which is defined by the quantum number J.

$$J = (L + S),\ (L + S - 1) \ldots\ldots\ |L - S|$$

Where L and S are the quantum numbers for resultant orbital angular momentum and resultant spin angular momentum, respectively. The modulus sign shows that the value of the resultant total angular momentum quantum number is always positive. The value of J is assigned as the subscripts of the overall term symbol.

i) For p^1p^1-configuration, 9 combinations given by orbital-orbital coupling (L = 2, 1, 0) and 4 combinations given by spin-spin coupling (S = 1, 0) combine to create a total of 36 microstates. therefore

$$L = 0, 1, 2 \text{ and } S = 1, 0$$

$$\text{States} = {}^3S,\ {}^3P,\ {}^3D,\ {}^1S,\ {}^1P,\ {}^1D$$

The summarization of spin-orbital coupling for p^1p^1-configurations is given below.

Table 4. Splitting of the term symbols for p^1p^1-configuration due to L-S coupling.

State	Value L and S	Value of J	States after L-S coupling
${}^3S$	L = 0 and S =1	J = (0 + 1) = 1	${}^3S_1$
${}^3P$	L = 1 and S = 1	J = (1 + 1) ....(1 − 1) = 2, 1, 0	${}^3P_2,\ {}^3P_1,\ {}^3P_0$
${}^3D$	L = 2 and S = 1	J = (2 + 1) ....(2 − 1) = 3, 2, 1	${}^3D_3,\ {}^3D_2,\ {}^3D_1$

Table 4. Continued on the next page…

^1S	$L = 0$ and $S = 0$	$J = (0 + 0) = 0$	$^1S_0$
^1P state:	$L = 1$ and $S = 0$	$J = (1 + 0) = 1$	$^1P_1$
^1D state:	$L = 2$ and $S = 0$	$J = (2 + 0) = 2$	$^1D_2$

ii) For d^1d^1-configuration, 25 combinations given by orbital-orbital coupling ($L = 4, 3, 2, 1, 0$) and 4 combinations given by spin-spin coupling ($S = 1, 0$) couple to create a total of 100 microstates. therefore

$$L = 4, 3, 2, 1, 0$$

$$S = 1, 0$$

$$\text{States} = {}^3S, {}^3P, {}^3D, {}^3F, {}^3G, {}^1S, {}^1P, {}^1D, {}^1F, {}^1G$$

The summarization of spin-orbital coupling for d^1d^1-configurations is given below.

Table 5. Splitting of the term symbols for d^1d^1-configuration due to L-S coupling.

State	Value L and S	Value of J	States after L-S coupling
^3S	$L = 0$ and $S = 1$	$J = (0 + 1) = 1$	$^3S_1$
^3P	$L = 1$ and $S = 1$	$J = (1 + 1) \ldots (1 - 1) =$ $2, 1, 0$	$^3P_2, {}^3P_1, {}^3P_0$
^3D	$L = 2$ and $S = 1$	$J = (2 + 1) \ldots (2 - 1) =$ $3, 2, 1$	$^3D_3, {}^3D_2, {}^3D_1$
^3F	$L = 3$ and $S = 1$	$J = (3 + 1) \ldots (3 - 1) =$ $4, 3, 2$	$^3F_4, {}^3F_3, {}^3F_2$
^3G	$L = 4$ and $S = 1$	$J = (4 + 1) \ldots (4 - 1) =$ $5, 4, 3$	$^3G_5, {}^3G_4, {}^3G_3$
^1S	$L = 0$ and $S = 0$	$J = (0 + 0) = 0$	$^1S_0$
^1P	$L = 1$ and $S = 0$	$J = (1 + 0) = 1$	$^1P_1$
^1D	$L = 2$ and $S = 0$	$J = (2 + 0) = 2$	$^1D_2$
^1F	$L = 3$ and $S = 0$	$J = (3 + 0) = 3$	$^1F_3$
^1G	$L = 4$ and $S = 0$	$J = (4 + 0) = 4$	$^1G_4$

The quantization of individual orbital angular momentums can be used to calculate the resultant value as follows: For ^3P state of p^1p^1-configuration, L = 1 and S = 1. Therefore, the different combinations of resultant orbital angular momentum and resultant spin angular momentum can be calculated:

L_z	=	+1	+1	+1	0	0	0	−1	−1	−1
S_z	=	+1	0	−1	+1	0	−1	+1	0	−1
J_z	=	+2	+1	0	+1	0	−1	0	−1	−2

Hence, the resultant orbital angular momentums and resultant spin angular momentums of ^3P can interact in nine ways, creating nine microstates; out of which, three quantum-mechanically allowed series can be fashioned.

$$J_z = (+2, +1, 0, -1, -2), (+1, 0, -1), (0)$$

or

$$J = 2, 1, 0$$

Which means

$$\text{States} = {}^3P_2, {}^3P_1 \text{ and } {}^3P_0$$

> ➢ ***Derivation of the Term Symbols for Unpaired Electrons in the Same Subshell***

In the previous section, we have distributed all the 36 microstates for p^1p^1-configuration in six electronic states labeled by ^3S, ^3P, ^3D, ^1S, ^1P and ^1D term symbols. However, for p^2-configuration (both of the unpaired electrons in same subshell), many microstates that were possible in p^1p^1-configuration cannot exist if they violate the Pauli exclusion principle. Therefore, there are only 15 microstates (complying with Pauli principle) for p^2-configuration which can be distributed in three electronic states labeled by ^1S, ^3P and ^1D term symbols.

Similarly, all the 100 microstates for d^1d^1-configuration (^3S, ^3P, ^3D, ^3F, ^3G, ^1S, ^1P, ^1D, ^1F and ^1G) cannot exist if they both the electrons are present in the same d-subshell. Therefore, there are only 45 microstates (complying with Pauli principle) for d^2-configuration which can be distributed in five electronic states labeled by ^1S, ^3P, ^1D, ^3F and ^1G term symbols.

Term symbols for electronic configurations with unpaired electrons in the same subshell can be derived using "pigeon hole" diagrams. The principal steps for such operations are:

1. Create vertical columns for all allowed orientation or effect of individual orbital angular momentum in the reference direction. For example, +1, 0 and −1 are the allowed orientations of the orbital angular momentum corresponding to $l = 1$ value.

2. Fill up the electrons in these columns by exhausting all the possibilities of parallel, paired and opposite orientations.

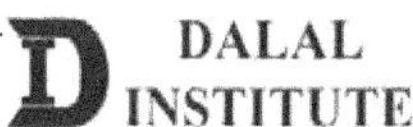

3. Sum-up all the l_z values of every column to find out the resultant L_z values.

4. Quantum mechanically allowed series set-up should be carried out for different parallel, paired and opposite orientation, which in turn can be used to provide the resultant orbital angular momentum quantum number L for particular resultant spin quantum number S.

5. Combine the values of L and S to find out the term symbols for allowed electronic states.

1. s^1-configuration:

$$l_z$$

$$0 \quad \boxed{\uparrow}$$

$$L_z \quad 0$$

As $L_z = 0$, the value of resultant orbital angular momentum quantum number L = 0. There is only one unpaired electron, therefore S = 1/2.

From L = 0, the state is S; and from S = 1/2, the multiplicity is 2. Thus, the overall term symbol is ^{2}S.

2. s^2-configuration:

$$l_z$$

$$0 \quad \boxed{\uparrow\downarrow}$$

$$L_z \quad 0$$

As $L_z = 0$, the value of resultant orbital angular momentum quantum number L = 0. There are no unpaired electrons, therefore S = 0.

From L = 0, the state is S; and from S = 0, the multiplicity is 1. Thus, the overall term symbol is ^{1}S.

3. p^1 and p^5-configuration:

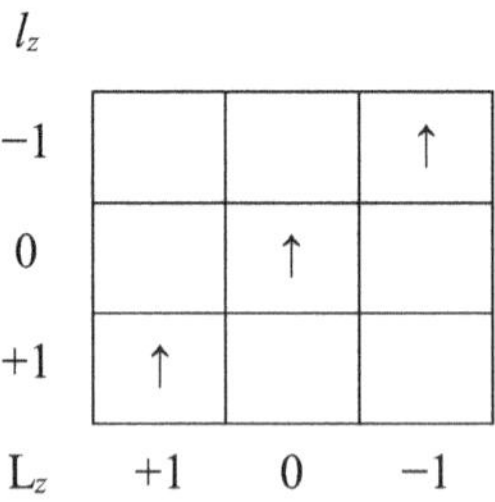

As $L_z = +1, 0, -1$; the value of resultant orbital angular momentum quantum number L = 1. There is one unpaired electron, therefore S = 1/2.

From L = 1, the state is P; and from S = 1/2, the multiplicity is 2. Thus, the overall term symbol is ^{2}P. Hence, all the 6 microstates for p^1 and p^5-configurations are distributed in ^{2}P term symbol.

4. p^2 and p^4-configuration:

For parallel arrangements,

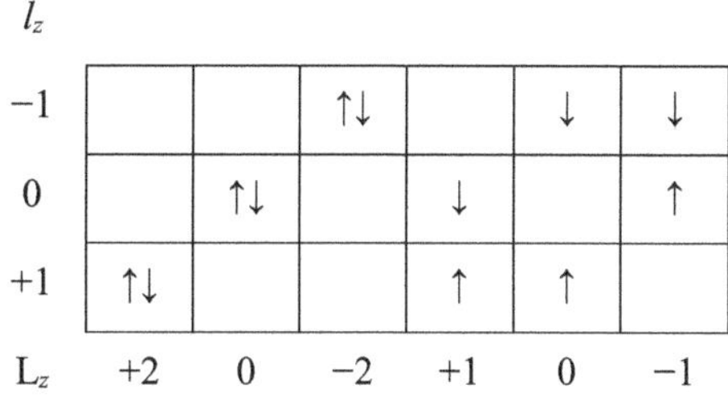

As $L_z = +1, 0, -1$; the value of resultant orbital angular momentum quantum number $L = 1$.

There are two unpaired electrons, therefore $S = 1$.

From $L = 1$, the state is P.

From $S = 1$, the multiplicity is 3.

Thus, the overall term symbol is ^3P.

Similarly,

For paired and opposite arrangements,

Out of six values of L_z (resultant orbital angular momentum in reference direction), two quantum-mechanically allowed series can be setup. One with $L_z = +2, +1, 0, -1, -2$; giving resultant orbital angular momentum quantum number $L = 2$. The second series with $L_z = 0$; giving resultant orbital angular momentum quantum number $L = 0$.

There are zero unpaired electrons, therefore $S = 0$.

From $L = 2$ and 0; the states are D and S, respectively.

From $S = 0$, the multiplicity is 1.

Thus, the overall term symbols are ^1D and ^1S.

Hence, all the 15 microstates for p^2 and p^4 electronic configurations which can be distributed in ^1S, ^3P and ^1D term symbols.

5. p^3-configuration:

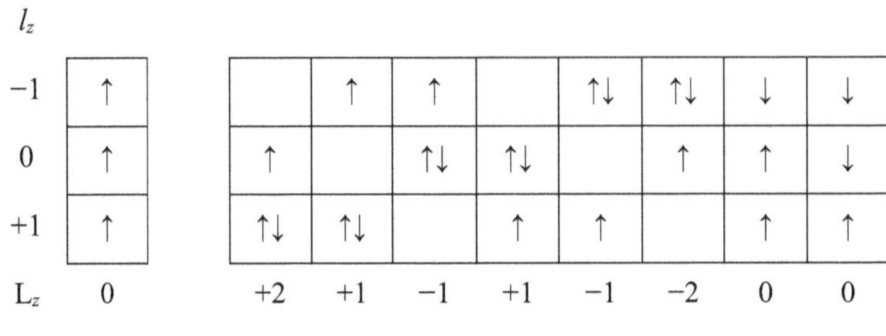

Parallel   Paired and opposite

For parallel arrangements,

As $L_z = 0$; the value of resultant orbital angular momentum quantum number $L = 0$. There are three unpaired electrons, therefore $S = 3/2$.

From $L = 0$, the state is S; and from $S = 3/2$, the multiplicity is 4. Thus, the overall term symbol is ^{4}S.

For paired and opposite arrangements,

Out of eight values of L_z, two quantum-mechanically allowed series can be setup. One with $L_z = +2, +1, 0, -1, -2$; giving resultant orbital angular momentum quantum number $L = 2$. The second series with $L_z = +1, 0, -1$; giving resultant orbital angular momentum quantum number $L = 1$. There is one unpaired electron, therefore $S = 1/2$.

From $L = 2$ and 1 the states are D and P, respectively.

From $S = 1/2$, the multiplicity is 2.

Thus, the overall term symbols are ^{2}D and ^{2}P.

Hence, all the 20 microstates for p^3-configurations are distributed in ^{4}S, ^{2}P and ^{2}D term symbols.

6. p^6-configuration:

l_z

-1	↑↓
0	↑↓
$+1$	↑↓

L_z 0

As $L_z = 0$, the value of resultant orbital angular momentum quantum number $L = 0$; and zero unpaired electrons, therefore $S = 0$. Thus, the overall term symbol is ^{1}S which contains the one and only microstate of p^6-electronic configuration.

7. d^1 and d^9-configuration:

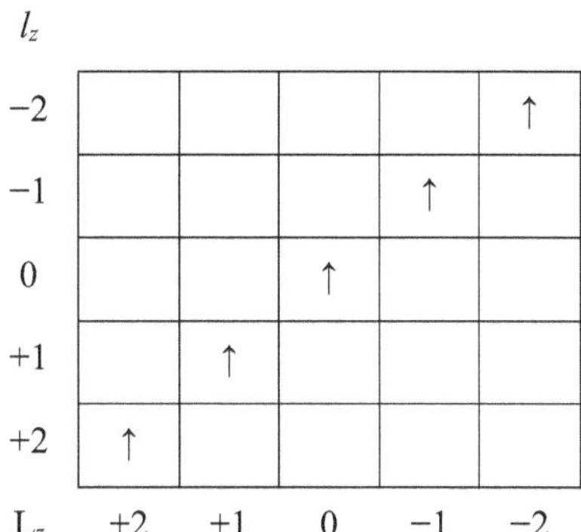

As $L_z = +2, +1, 0, -1, -2$; the value of resultant orbital angular momentum quantum number $L = 2$.

There is only one unpaired electron, therefore $S = 1/2$.

From $L = 2$, the state is D; and from $S = 1/2$, the multiplicity is also 2.

Thus, the overall term symbol is ^2D.

Hence, all the 10 microstates for d^1 and d^9-configurations are distributed in ^2D term symbols.

8. d^2 and d^8-configuration:

For parallel arrangements,

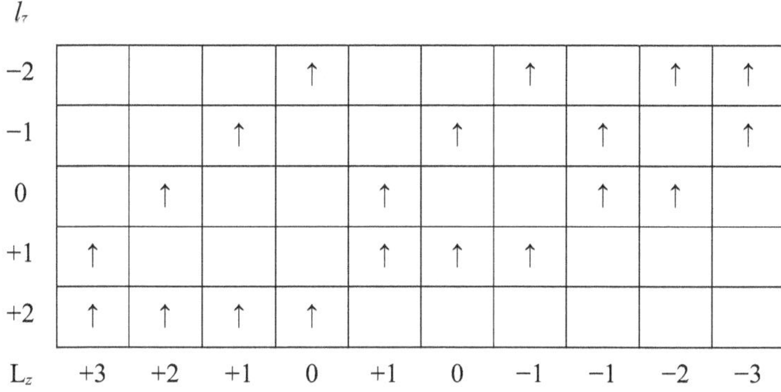

Out of ten values of L_z, two quantum-mechanically allowed series can be setup. One with $L_z = +3, +2, +1, 0, -1, -2, -3$; giving resultant orbital angular momentum quantum number $L = 3$. The second series with $L_z = +1, 0, -1$; giving resultant orbital angular momentum quantum number $L = 1$. There are two unpaired electrons, therefore $S = 1$.

From $L = 3$ and 1, the states are F and P, respectively. From $S = 1$, the multiplicity is 3. Thus, the overall term symbols are ^3F and ^3P.

For paired and opposite arrangements,

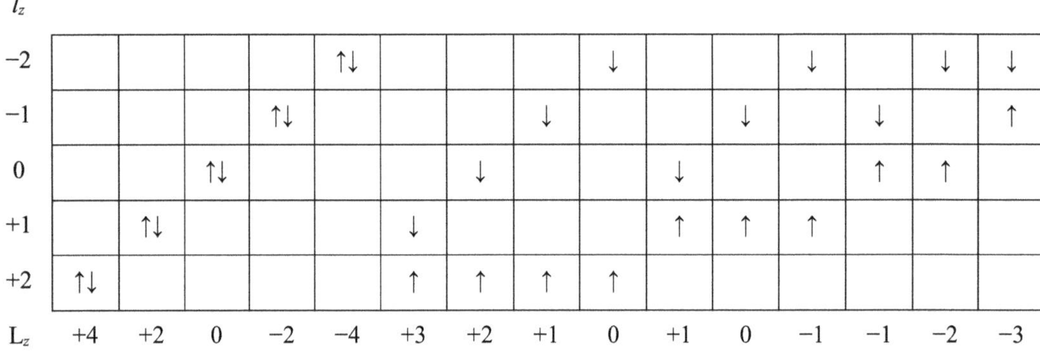

l_z															
−2					↑↓				↓			↓		↓	↓
−1				↑↓				↓			↓		↓		↑
0			↑↓				↓			↓			↑	↑	
+1		↑↓				↓				↑	↑	↑			
+2	↑↓					↑	↑	↑	↑						
L_z	+4	+2	0	−2	−4	+3	+2	+1	0	+1	0	−1	−1	−2	−3

Out of fifteen values of L_z, three quantum-mechanically allowed series can be setup. First with L_z = +4, +3, +2, +1, 0, −1, −2, −3, −4; giving resultant orbital angular momentum quantum number L = 4. Second series with L_z = +2, +1, 0, −1, −2; giving resultant orbital angular momentum quantum number L = 2. The third series with L_z = 0; giving resultant orbital angular momentum quantum number L = 0. There are zero unpaired electrons, therefore S = 0.

From L = 4, 2 and 0 the states are G, D and S, respectively. From S = 0, the multiplicity is 1. Thus, the overall term symbols are ^1G, ^1D and ^1S.

Hence, all the 45 microstates for d^2 and d^8-configurations are distributed in ^1S, ^3P, ^1D, ^3F and ^1G term symbols.

9. d^3 and d^7-configuration:

For parallel arrangements,

l_z										
−2			↑		↑	↑		↑	↑	↑
−1		↑		↑		↑	↑	↑	↑	
0	↑			↑	↑	↑	↑			↑
+1	↑	↑	↑	↑	↑				↑	
+2	↑	↑	↑				↑	↑		↑
L_z	+3	+2	+1	0	−1	−3	+1	−1	−2	0

Out of ten values of L_z, two quantum-mechanically allowed series can be setup.

One with L_z = +3, +2, +1, 0, −1, −2, −3; giving resultant orbital angular momentum quantum number L = 3.

The second series with L_z = +1, 0, −1; giving resultant orbital angular momentum quantum number L = 1. There are three unpaired electrons, therefore S = 3/2.

From L = 3 and 1; the states are F and P, respectively. From S = 3/2, the multiplicity is 4. Thus, the overall term symbols are ^4F and ^4P.

For paired and opposite arrangements,

l_z

−2				↑			↑			↑			↑		
−1			↑			↑			↑				↑↓	↑↓	↑↓
0		↑			↑				↑↓	↑↓	↑↓	↑↓		↑	
+1	↑				↑↓	↑↓	↑↓	↑↓			↑				↑
+2	↑↓	↑↓	↑↓	↑↓				↑				↑			
L_z	+5	+4	+3	+2	+2	+1	0	+4	−1	−2	+1	+2	−4	−2	−1

and

l_z

−2		↑↓	↑↓	↑↓	↑↓			↓		↓	↓		↓	↓	↓
−1	↑↓	↑					↓		↓		↑	↓	↑	↑	
0			↑			↓			↑	↑	↑	↑			↑
+1				↑		↑	↑	↑	↑	↑				↑	
+2	↑				↑	↑	↑	↑				↑	↑		↑
L_z	0	−5	−4	−3	−2	+3	+2	+1	0	−1	−3	+1	−1	−2	0

Out of thirty values of L_z, six quantum-mechanically allowed series can be setup. First with L_z = +5, +4, +3, +2, +1, 0, −1, −2, −3, −4, −5; giving resultant orbital angular momentum quantum number L = 5. Second series with L_z = +4, +3, +2, +1, 0, −1, −2, −3, −4; giving resultant orbital angular momentum quantum number L = 4. The third series with L_z = +3, +2, +1, 0, −1, −2, −3; giving resultant orbital angular momentum quantum number L = 3. Fourth and fifth series with L_z = +2, +1, 0, −1, −2; giving resultant orbital angular momentum quantum number L = 2 and 2, respectively. The sixth series is consisted of L_z = +1, 0, −1; giving resultant orbital angular momentum quantum number L = 1.

There is one unpaired electron, therefore S = 1/2.

From L = 5, 4, 3, 2, 2 and 1; the states are H, G, F, D, D and P, respectively.

From S = 1/2, the multiplicity is 2.

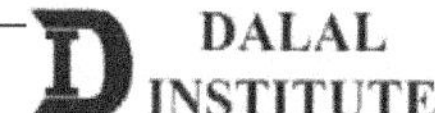

Thus, the overall term symbols are ^2H, ^2G, ^2F, ^2D, ^2D and ^2P.

Hence, all the 120 microstates for d^3 and d^7-configurations are distributed in ^4F, ^4P, ^2H, ^2G, ^2F, ^2D, ^2D and ^2P term symbols.

10. d^4 and d^6-configuration:

For parallel arrangements,

l_z

−2		↑	↑	↑	↑
−1	↑		↑	↑	↑
0	↑	↑		↑	↑
+1	↑	↑	↑		↑
+2	↑	↑	↑	↑	
L_z	+2	+1	0	−1	−2

Out of five values of L_z, only one quantum-mechanically allowed series can be set up with $L_z = +2, +1, 0, -1, -2$; giving resultant orbital angular momentum quantum number $L = 2$.

There are four unpaired electrons, therefore $S = 2$.

From $L = 2$, the state is D.

From $S = 2$, the multiplicity is 5.

Thus, the overall term symbol is ^5D.

For two electrons paired or opposite arrangement,

l_z

−2			↑		↑	↑		↑	↑			↑	↑		
−1		↑	↑	↑		↑	↑		↑		↑		↑		↑
0	↑			↑	↑		↑	↑		↑			↑↓	↑↓	↑↓
+1	↑	↑	↑				↑↓	↑↓	↑↓	↑↓	↑↓	↑↓		↑	
+2	↑↓	↑↓	↑↓	↑↓	↑↓	↑↓				↑	↑	↑		↑	↑
L_z	+5	+4	+2	+3	+2	+1	+1	0	−1	+4	+3	+2	−3	+3	+1

and

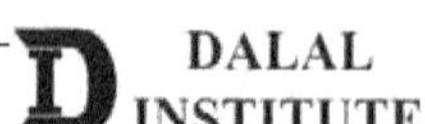

l_z

l_z															
−2	↑		↑			↑		↑	↑	↑↓	↑↓	↑↓	↑↓	↑↓	↑↓
−1		↑		↑↓	↑↓	↑↓	↑↓	↑↓	↑↓			↑		↑	↑
0	↑↓	↑↓	↑↓		↑		↑		↑		↑		↑		↑
+1		↑	↑	↑			↑	↑		↑			↑	↑	
+2	↑			↑	↑	↑				↑	↑	↑			
L_z	0	0	−1	+1	0	−2	−1	−3	−4	−1	−2	−3	−3	−4	−5

and

l_z

l_z															
−2		↓	↓	↓	↓		↑	↑	↑	↑		↑	↑	↑	↑
−1	↓		↑	↑	↑	↑		↑	↓	↓	↑		↑	↑	↑
0	↑	↑		↑	↑	↓	↓		↑	↑	↑	↑		↓	↓
+1	↑	↑	↑		↑	↑	↑	↓		↑	↓	↓	↓		↑
+2	↑	↑	↑	↑		↑	↑	↑	↑		↑	↑	↑	↑	
L_z	+2	+1	0	−1	−2	+2	+1	0	−1	−2	+2	+1	0	−1	−2

Out of forty-five values of L_z, seven quantum-mechanically allowed series can be setup as. First with L_z = +5, +4, +3, +2, +1, 0, −1, −2, −3, −4, −5; giving resultant orbital angular momentum quantum number L = 5. Second series with L_z = +4, +3, +2, +1, 0, −1, −2, −3, −4; giving resultant orbital angular momentum quantum number L = 4. The third and fourth series with L_z = +3, +2, +1, 0, −1, −2, −3; giving resultant orbital angular momentum quantum number L = 3 and 3. fifth series with L_z = +2, +1, 0, −1, −2; giving resultant orbital angular momentum quantum number L = 2. The sixth and seventh series is consisted of L_z = +1, 0, −1; giving resultant orbital angular momentum quantum number L = 1 and 1, respectively.

There are two unpaired electrons, therefore S = 1.

From L = 5, 4, 3, 3, 2, 1 and 1; the states are H, G, F, F, D, P and P, respectively.

From S = 1, the multiplicity is 3.

Thus, the overall term symbols are ^3H, ^3G, ^3F, ^3F, ^3D, ^3P and ^3P.

Similarly,

For all electrons paired or opposite arrangement,

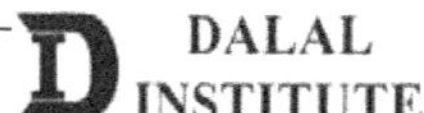

l_z															
−2				↑↓			↑↓		↑↓	↑↓			↓		↓
−1			↑↓			↑↓		↑↓		↑↓		↓		↓	
0		↑↓			↑↓			↑↓	↑↓		↓			↑	↑
+1	↑↓				↑↓	↑↓	↑↓				↑	↑	↑		
+2	↑↓	↑↓	↑↓	↑↓							↑↓	↑↓	↑↓	↑↓	↑↓
L_z	+6	+4	+2	0	+1	0	−2	−2	−4	−6	+5	+4	+2	+3	+2

and

l_z															
−2	↓		↓	↓			↓	↓			↓		↓		
−1	↑	↓		↑		↓		↑		↓		↓		↑↓	↑↓
0		↑	↑		↓			↑↓	↑↓	↑↓	↑↓	↑↓	↑↓		↓
+1		↑↓	↑↓	↑↓	↑↓	↑↓	↑↓		↓			↑	↑	↓	
+2	↑↓				↑	↑	↑		↑	↑	↑			↑	↑
L_z	+1	+1	0	−1	+4	+3	+2	−3	+3	+1	0	0	−1	+1	0

and

l_z															
−2	↓		↓	↓	↑↓	↑↓	↑↓	↑↓	↑↓	↑↓		↓	↓	↓	↓
−1	↑↓	↑↓	↑↓	↑↓			↓		↓	↓	↓		↓	↓	↓
0		↓		↑		↓		↓		↑	↓	↓		↑	↑
+1		↑	↑		↓			↑	↑		↑	↑	↑		↑
+2	↑				↑	↑	↑				↑	↑	↑	↑	
L_z	−2	−1	−3	−4	−1	−2	−3	−3	−4	−5	+2	+1	0	−1	−2

and

l_z

−2		↓	↓	↓	↓
−1	↓		↑	↑	↑
0	↑	↑		↓	↓
+1	↓	↓	↓		↑
+2	↑	↑	↑	↑	

L_z +2   +1   0   −1   −2

Out of fifty values of L_z, eight quantum-mechanically allowed series can be setup as. First with L_z = +6, +5, +4, +3, +2, +1, 0, −1, −2, −3, −4, −5, −6; giving resultant orbital angular momentum quantum number L = 6. Second and third series with L_z = +4, +3, +2, +1, 0, −1, −2, −3, −4; giving resultant orbital angular momentum quantum number L = 4 and 4, respectively. The fourth series with L_z = +3, +2, +1, 0, −1, −2, −3; giving resultant orbital angular momentum quantum number L = 3. Fifth and sixth series with L_z = +2, +1, 0, −1, −2; giving resultant orbital angular momentum quantum number L = 2 and 2, respectively. The seventh and eighth series is consisted of L_z = 0; giving resultant orbital angular momentum quantum number L = 0 and 0, respectively. There are zero unpaired electrons, therefore S = 0.

From L = 6, 4, 4, 3, 2, 2, 0 and 0; the states are I, G, G, F, D, D, S and S, respectively. From S = 0, the multiplicity is 1. Thus, the overall term symbols are ^1I, ^1G, ^1G, ^1F, ^1D, ^1D, ^1S and ^1S.

Hence, all the 210 microstates for d^4 and d^6-configurations are distributed in ^5D, ^3H, ^3G, ^3F, ^3F, ^3D, ^3P, ^3P, ^1I, ^1G, ^1G, ^1F, ^1D, ^1D, ^1S and ^1S term symbols.

11. d^5-configuration:

For parallel arrangements,

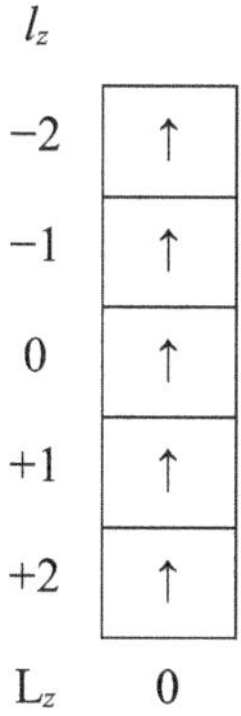

l_z

−2	↑
−1	↑
0	↑
+1	↑
+2	↑

L_z 0

As $L_z = 0$; the value of resultant orbital angular momentum quantum number $L = 0$. There are five unpaired electrons, therefore $S = 5/2$.

From $L = 0$, the state is S. From $S = 5/2$, the multiplicity is 6. Thus, the overall term symbol is ^6S.

For two electrons paired or opposite arrangement,

l_z

−2		↑	↑	↑		↑	↑	↑		↑	↑	↑		↑	↑
−1	↑		↑	↑	↑		↑	↑	↑		↑	↑	↑↓	↑↓	↑↓
0	↑	↑		↑	↑	↑		↑	↑↓	↑↓	↑↓	↑↓	↑		↑
+1	↑	↑	↑		↑↓	↑↓	↑↓	↑↓	↑	↑		↑	↑	↑	
+2	↑↓	↑↓	↑↓	↑↓	↑	↑	↑		↑	↑	↑		↑	↑	↑
L_z	+4	+3	+2	+1	+3	+2	+1	−1	+2	+1	−1	−2	+1	−1	−2

and

l_z

−2	↑	↑↓	↑↓	↑↓	↑↓	↓	↑	↑	↑
−1	↑↓		↑	↑	↑	↑	↓	↑	↑
0	↑	↑		↑	↑	↑	↑	↓	↑
+1	↑	↑	↑		↑	↑	↑	↑	↓
+2		↑	↑	↑		↑	↑	↑	↑
L_z	−3	−1	−2	−3	−4	0	0	0	0

Out of twenty-four values of L_z, four quantum-mechanically allowed series can be setup as. First with $L_z = +4$, $+3, +2, +1, 0, -1, -2, -3, -4$; giving resultant orbital angular momentum quantum number $L = 4$. Second series with $L_z = +3, +2, +1, 0, -1, -2, -3$; giving resultant orbital angular momentum quantum number $L = 3$. The third series with $L_z = +2, +1, 0, -1, -2$; giving resultant orbital angular momentum quantum number $L = 3$. The fourth series is consisted of $L_z = +1, 0, -1$; giving resultant orbital angular momentum quantum number $L = 1$.

There are three unpaired electrons, therefore $S = 3/2$.

From $L = 4, 3, 2$ and 1; the states are G, F, D and P, respectively.

From $S = 3/2$, the multiplicity is 4.

Thus, the overall term symbols are ^4G, ^4F, ^4D and ^4P.

For four electrons paired or opposite arrangement,

l_z															
−2			↑			↑			↑	↑↓	↑↓	↑↓			↑
−1		↑			↑		↑↓	↑↓	↑↓			↑		↑	
0	↑			↑↓	↑↓	↑↓		↑			↑		↑↓	↑↓	↑↓
+1	↑↓	↑↓	↑↓	↑			↑			↑			↑↓	↑↓	↑↓
+2	↑↓	↑↓	↑↓	↑↓	↑↓	↑↓	↑↓	↑↓	↑↓	↑↓	↑↓	↑↓	↑		
L_z	+6	+5	+4	+5	+3	+2	+3	+2	0	+1	0	−1	+4	+1	0

and

l_z															
−2			↑	↑↓	↑↓	↑↓			↑	↑↓	↑↓	↑↓	↑↓	↑↓	↑↓
−1	↑↓	↑↓	↑↓			↑	↑↓	↑↓	↑↓			↑	↑↓	↑↓	↑↓
0		↑			↑		↑↓	↑↓	↑↓	↑↓	↑↓	↑↓			↑
+1	↑↓	↑↓	↑↓	↑↓	↑↓	↑↓		↑			↑			↑	
+2	↑			↑			↑			↑			↑		
L_z	+2	0	−2	0	−2	−3	0	−1	−4	−2	−3	−5	−4	−5	−6

and

l_z															
−2		↓	↓	↓		↓	↓	↓		↓	↓	↓		↓	↓
−1	↓		↑	↑	↓		↑	↑	↓		↑	↑	↑↓	↑↓	↑↓
0	↑	↑		↑	↑	↑		↑	↑↓	↑↓	↑↓	↑↓	↓		↑
+1	↑	↑	↑		↑↓	↑↓	↑↓	↑↓	↑	↑		↑	↑	↑	
+2	↑↓	↑↓	↑↓	↑↓	↑	↑	↑		↑	↑	↑		↑	↑	↑
L_z	+4	+3	+2	+1	+3	+2	+1	−1	+2	+1	−1	−2	+1	−1	−2

and

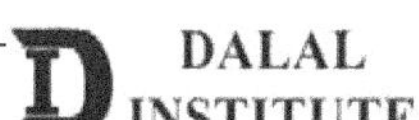

l_z

l_z															
−2	↓	↑↓	↑↓	↑↓	↑↓		↑	↑	↑		↑	↑	↑		↑
−1	↑↓		↓	↓	↓	↑		↓	↓	↑		↓	↓	↑	
0	↑	↓		↑	↑	↓	↓		↑	↓	↓		↑	↑↓	↑↓
+1	↑	↑	↑		↑	↑	↑	↑		↑↓	↑↓	↑↓	↑↓	↓	↓
+2		↑	↑	↑		↑↓	↑↓	↑↓	↑↓	↑	↑	↑		↑	↑
L_z	−3	−1	−2	−3	−4	+4	+3	+2	+1	+3	+2	+1	−1	+2	+1

and

l_z

l_z															
−2	↑	↑		↑	↑	↑	↑↓	↑↓	↑↓	↑↓	↓	↑	↑	↑	↑
−1	↓	↓	↑↓	↑↓	↑↓	↑↓		↑	↑	↑	↓	↓	↑	↑	↓
0	↑↓	↑↓	↑		↓	↓	↑		↓	↓	↑	↓	↓	↑	↑
+1		↑	↓	↓		↑	↓	↓		↑	↑	↑	↓	↓	↓
+2	↑		↑	↑	↑		↑	↑	↑		↑	↑	↑	↓	↑
L_z	−1	−2	+1	−1	−2	−3	−1	−2	−3	−4	0	0	0	0	0

Out of seventy-five values of L_z, eleven quantum-mechanically allowed series can be setup. First with $L_z = +6$, +5, +4, +3, +2, +1, 0, −1, −2, −3, −4, −5, −6; giving resultant orbital angular momentum quantum number $L = 6$. Second series with $L_z = +5$, +4, +3, +2, +1, 0, −1, −2, −3, −4 −5; giving resultant orbital angular momentum quantum number $L = 5$. The third and fourth series with $L_z = +4$, +3, +2, +1, 0, −1, −2, −3, −4; giving resultant orbital angular momentum quantum number $L = 4$. Fifth and sixth series with $L_z = +3$, +2, +1, 0, −1, −2, −3; giving resultant orbital angular momentum quantum number $L = 3$ and 3, respectively. The seventh, eighth and ninth series are consisted of $L_z = +2$, +1, 0, −1, −2; giving resultant orbital angular momentum quantum number $L = 2$, 2 and 2, respectively. The tenth and eleventh series are consisted of $L_z = +1$, 0, −1 and $L_z = 0$; giving resultant orbital angular momentum quantum number $L = 1$ and 0, respectively.

There is one unpaired electron which $S = 1/2$.

From $L = 6$, 5, 4, 4, 3, 3, 2, 2, 2, 1 and 0; the states are I, H, G, G, F, F, D, D, D, P and S, respectively. From $S = 1/2$, the multiplicity is 2. Thus, the overall term symbols are ^2I, ^2H, ^2G, ^2G, ^2F, ^2F, ^2D, ^2D, ^2D, ^2P and ^2S.

Hence, all the 252 microstates for d^5-configurations are distributed in ^2I, ^2H, ^2G, ^2G, ^2F, ^2F, ^2D, ^2D, ^2D, ^2P, ^2S, ^4G, ^4F, ^4D, ^4P and ^6S term symbols.

12. d^{10}-configuration:

$$l_z$$

-2	↑↓
-1	↑↓
0	↑↓
$+1$	↑↓
$+2$	↑↓

$$L_z \quad 0$$

As $L_z = 0$; the value of resultant orbital angular momentum quantum number $L = 0$. There are no unpaired electrons, therefore $S = 0$.

From $L = 0$, the state is S. From $S = 0$, the multiplicity is 1. Thus, the overall term symbol is ^1S which contains the only microstate of d^{10}-configuration.

It can clearly be seen that the number of microstates, as well as the term symbols for dn and d10−n configurations, are the same. This is due to the fact that the number of unpaired electrons is the same for both of the configurations. In other words, the possible arrangements for unpaired electrons in less than half-filled or for holes in more than half-filled configurations are the same. The same analogy is true for s, p or f-subshell.

Moreover, the number of microstates distributed in any term symbol can be calculated using the following relations:

1. Term symbols without J-value: $(2L+1) \times (2S+1)$

2. Term symbols without J-value: $(2J+1)$

Let us tally the number of microstates for p^2 electronic configuration with term symbols distribution.

		$(2L+1) \times (2S+1)$	$(2J+1)$
Distribution of 15 microstates of p^2-configuration in ^1S, ^3P, ^1D term symbols	→	$^1S = (2\times0+1) \times (2\times0+1) = 1$	$^1S_0 = 2\times0 + 1 = 1$
		$^3P = (2\times1+1) \times (2\times1+1) = 9$	$^3P_2 = 2\times2 + 1 = 5$
		$^1D = (2\times2+1) \times (2\times0+1) = 5$	$^3P_1 = 2\times1 + 1 = 3$
			$^3P_0 = 2\times0 + 1 = 1$
			$^1D_2 = 2\times2 + 1 = 5$

> ➤ *Determination of Spectroscopic Ground State Term*

In a particular configuration, the classification of various microstates in different term symbols also distinguishes them energetically. It can be explained in terms of the effect of spin-spin, orbital-orbital and spin-orbital coupling. There are two different approaches, both are based upon some conclusive results from quantum mechanics, to find out the term symbol for ground electronic state.

1. From correlation diagram: This approach is based on the correlation of all the free ion term of a particular atom or ion and then the ground state term is calculated with the help of certain rules.

i) The energy of different terms (electronic states) depends primarily on the Hund's rule, which states that the most stable state should have the highest multiplicity. Hence, triplet states are more stable than the singlet one. In other words, the higher the value of 2S+1, the greater will be the stability.

ii) If the value of spin multiplicities for two different states is the same, the value of resultant angular momentum will the deciding factor in the determination of the lower energy term. A higher value of L gives the lower energy state and the vice-versa is also true.

iii) The energy dependence of different terms upon total angular momentum is configuration-specific in nature. The perturbed Hamiltonian of spin-orbital interaction can have both types of effect, stabilization or destabilization, over the energy states and depends upon the magnitude of J-value. It has been proved that the interaction of spin-orbital motion destabilizes the less than half filed configuration and stabilizes the more than half-filled. Hence, a lower J-value for less than half-filled and a higher J-value of more than half will give the lower energy.

Let us apply this procedure to find out the ground state term symbol for p^2-configuration:

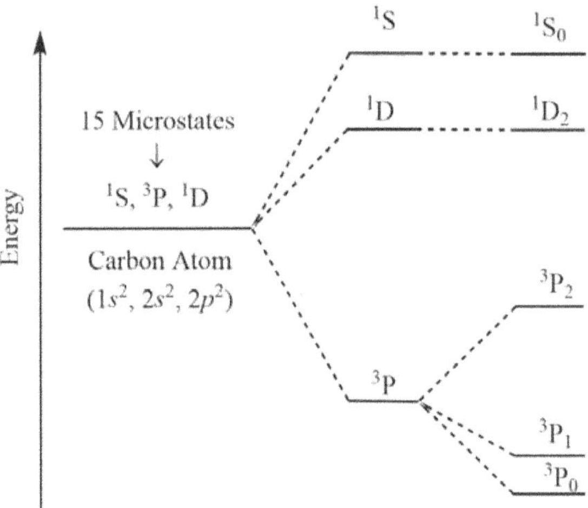

Figure 1. Splitting of free ion terms in the carbon atom.

The effect of spin-spin interaction makes the triplet (^{3}P) state more stable in comparison to the singlet ^{1}D or ^{1}S. Furthermore, the stabilization of ^{1}D state after the consideration of orbital-orbital interaction can be attributed to the higher value of orbital angular momentum. Finally, as the p^2-configuration is less than half-filled; the splitting of ^{3}P state goes with ^{3}P$_0$ as the ground state term symbol.

2. Using L-S coupling scheme: This procedure is applicable only to find out the ground state term symbol. We do not need to calculate all the microstates but only present in the ground electronic state. Various steps involved in this procedure are:

1. The electron filling in the valence subshell should be carried out in such a way that a maximum spin multiplicity (2S+1) is produced.

2. Maximize the value of resultant orbital angular momentum component by filing electrons first in the orbitals with the positive component of individual orbital angular momentum.

3. Select maximum J-value for more than half-filled and minimum J-value for less than half-filled configuration.

Let us determine the ground state term symbol for carbon atom. The electronic configuration for carbon is $1s^2, 2s^2, 2p^2$.

$$l_z \quad +1 \quad 0 \quad -1$$

$$2p \quad \boxed{\uparrow \;|\; \uparrow \;|\; }$$

From the total component of orbital and spin angular momentum, L = (+1) + (0) = 1 and S = 1/2 + 1/2 = 1.

From L-S coupling, |L+S| ...... |L−S|; (1+1) .... (1 −1) = 2, 1, 0.

Now, as the configuration is less than half-filled, the lower J-value is selected for lower energy. Thus, this gives rise to a ^{3}P$_0$ state.

For nitrogen atom, the determination of the ground state term symbol can be calculated through the same route. The electronic configuration for nitrogen is $1s^2, 2s^2, 2p^3$.

$$l_z \quad +1 \quad 0 \quad -1$$

$$2p \quad \boxed{\uparrow \;|\; \uparrow \;|\; \uparrow}$$

From the total component of orbital and spin angular momentum, L = (+1) + (0) + (−1) = 0 and S = 1/2 + 1/2 + 1/2 = 3/2.

From L-S coupling, |L+S| ...... |L−S|; (0 + 3/2) = 3/2.

Now, as the configuration is half-filled, the only J-value is bound to be selected for lower energy. Thus, this gives rise to a ^{4}S$_{3/2}$ state.

For trivalent vanadium ion (V^{3+}), the electronic configuration can be given as is $1s^2, 2s^2, 2p^6, 3s^2, 3p^6, 4s^0, 3d^2$.

l_z	+2	+1	0	−1	−2
$3d$	↑	↑			

From the total component of orbital and spin angular momentum, L = (+2) + (+1) = 3 and S = 1/2 + 1/2 = 1.

From L-S coupling, |L+S| ...... |L−S|; (3 + 1) .... (3 − 1) = 4, 3, 2. Now, as the configuration is less than half-filled, the lower J-value is selected for lower energy. Thus, this gives rise to a $^3F_2$ state. However, the ground state term symbol for d^8-configuration will be $^3F_4$.

All the free ions terms (including ground states) for different electronic configurations are summarized in the following table. However, it must be kept in mind that the ground state term symbol after L-S coupling will be different for more than half-filled and less than half-filled counterparts.

Table 6. The free ion terms and the corresponding ground electronic states for different types of electronic configuration.

Electronic configuration	Free ion terms	Ground State
s^1	^2S	^2S
s^2	^1S	^1S
p^1, p^5	^2P	^2P
p^2, p^4	$^1S, ^3P, ^1D$	^3P
p^3	$^4S, ^2P, ^2D$	^4S
p^6	^1S	^1S
d^1, d^9	^2D	^2D
d^2, d^8	$^1S, ^3P, ^1D, ^3F, ^1G$	^3F
d^3, d^7	$^4F, ^4P, ^2H, ^2G, ^2F, ^2D, ^2D, ^2P$	^4F
d^4, d^6	$^5D, ^3H, ^3G, ^3F, ^3F, ^3D, ^3P, ^3P, ^1I, ^1G, ^1G, ^1F, ^1D, ^1D, ^1S, ^1S$	^5D
d^5	$^2I, ^2H, ^2G, ^2G, ^2F, ^2F, ^2D, ^2D, ^2D, ^2P, ^2S, ^4G, ^4F, ^4D, ^4P, ^6S$	^6S
d^{10}	^1S	^1S

It is also worthy to note that the ground state term symbol for all fully-filled subshells like s^2, p^6, or d^{10} is ^1S always which also includes the one and only microstate available.

❖ Correlation and Spin-Orbit Coupling in Free Ions for 1st Series of Transition Metals

The energy of different free ion terms primarily depends on the spin-spin interaction, yielding the most stable state with the highest multiplicity. Furthermore, if the values of spin multiplicity for two different states are equivalent, the value of resultant angular momentum will the deciding factor in the determination of the lower energy term. A higher value of L gives the lower energy state and the vice-versa is also true. The energy dependence of different terms upon total angular momentum is configuration-specific in nature. The perturbed Hamiltonian of spin-orbital interaction can have both types of effect, stabilization or destabilization, over the energy states and depends upon the magnitude of J-value. It has been proved that the interaction of spin-orbital motion destabilizes the less than half filed configuration and stabilizes the more than half-filled.

Hence, a lower J-value for less than half-filled and a higher J-value for more than half will give the lower energy. However, the actual value of the energies for different terms, and hence their relative ordering, must be determined from the analysis of spectroscopic data. The correlation diagram of different free ion terms and thereafter the effect of L-S interaction for different ions can be given as:

1. Sc^{2+} and Cu^{2+} (d^1, d^9):

The number of unpaired electrons and hence microstates in Sc^{2+} (d^1) and Cu^{2+} (d^9) are the same. The one and only term symbol containing all the 10 microstates is ^2D, which also represents the ground electronic state. However, the splitting pattern of ^2D term due to L-S coupling for Cu^{2+} (d^9) is just the reverse of what is for Sc^{2+} (d^1).

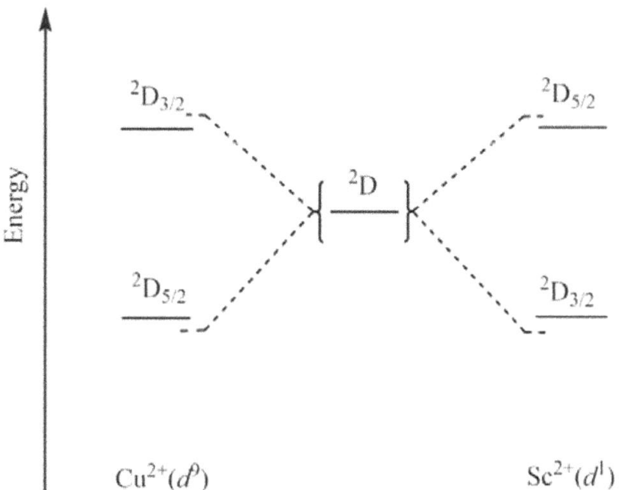

Figure 2. Correlation and spin-orbital coupling of free ion terms of Sc^{2+} and Cu^{2+}.

2. V^{3+} and Ni^{2+} (d^2, d^8):

The number of unpaired electrons and hence microstates in V^{3+} (d^2) and Ni^{2+} (d^8) are the same. All the 45 microstates for d^2 and d^8-configurations are distributed in ^1S, ^3P, ^1D, ^3F and ^1G term symbols with ^3F as the ground electronic state. However, the splitting pattern of ^3F term due to L-S coupling for Ni^{2+} (d^8) is just the reverse of what is for V^{3+} (d^2).

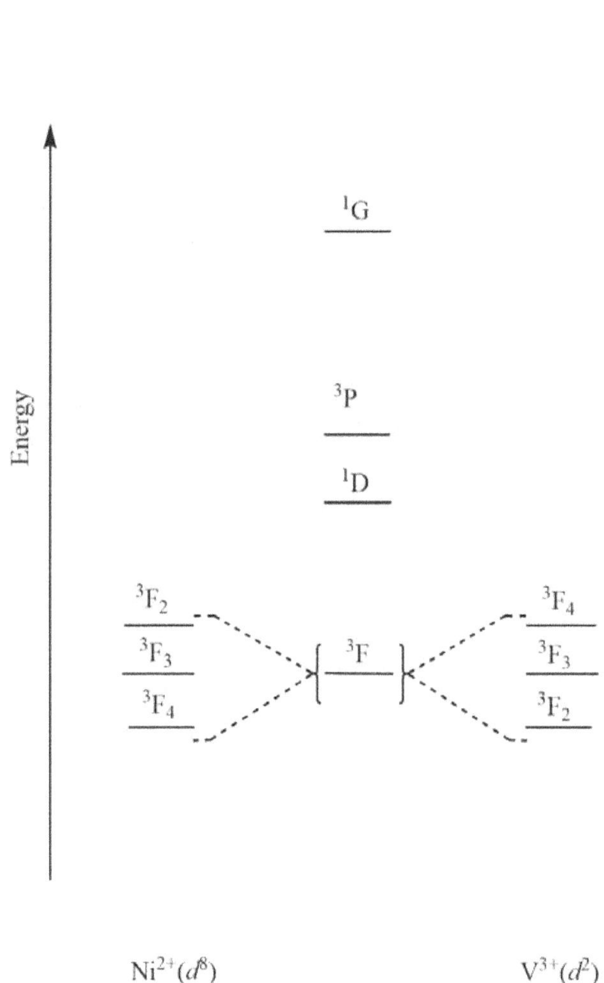

Figure 3. Correlation and spin-orbital coupling of free ion terms of V^{3+} and Ni^{2+}.

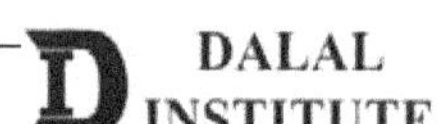

3. Cr^{3+} and Co^{2+} (d^3, d^7):

The number of unpaired electrons and hence microstates in Cr^{3+} (d^3) and Co^{2+} (d^7) are the same. All the 120 microstates for d^3 and d^7-configurations are distributed in ^4F, ^4P, ^2H, ^2G, ^2F, ^2D, ^2D and ^2P term symbols with ^4F as the ground electronic state. However, the splitting pattern of ^4F term due to L-S coupling for Co^{2+} (d^7) is just the reverse of what is for Cr^{3+} (d^3).

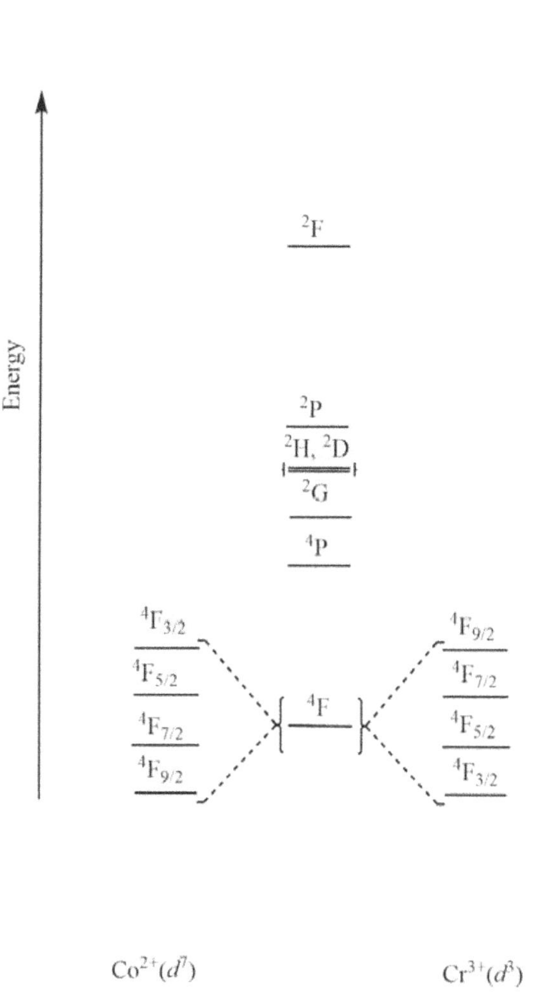

Figure 4. Correlation and spin orbital coupling of free ion terms of Cr^{3+} and Co^{2+}.

4. Mn^{3+} and Fe^{2+} (d^4, d^6):

The number of unpaired electrons and hence microstates in Mn^{3+} (d^4) and Fe^{2+} (d^6) are the same. All the 210 microstates for d^4 and d^6-configurations are distributed in ^{5}D, ^{3}H, ^{3}G, ^{3}F, ^{3}F, ^{3}D, ^{3}P, ^{3}P, ^{1}I, ^{1}G, ^{1}G, ^{1}F, ^{1}D, ^{1}D, ^{1}S and ^{1}S term symbols with ^{5}D as the ground electronic state. However, the splitting pattern of ^{5}D term due to L-S coupling for Fe^{2+} (d^6) is just the reverse of what is for Mn^{3+} (d^4).

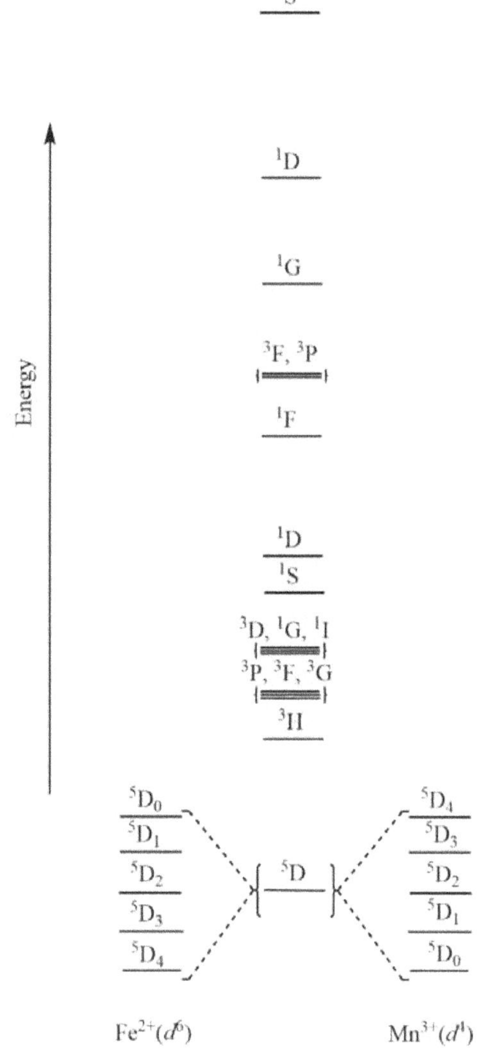

Figure 5. Correlation and spin-orbital coupling of free ion terms of Mn^{3+} and Fe^{2+}.

5. Fe^{3+} and Mn^{2+} (d^5):

The number of unpaired electrons and hence microstates in Fe^{3+} (d^5) and Mn^{2+} (d^5) are the same. All the 252 microstates for d^5-configurations are distributed in ^2I, ^2H, ^2G, ^2G, ^2F, ^2F, ^2D, ^2D, ^2D, ^2P, ^2S, ^4G, ^4F, ^4D, ^4P and ^6S term symbols with ^6S as the ground electronic state. Moreover, the splitting of ^6S term due to L-S coupling is not possible because of the absence of resultant orbital motion. However, higher-energy free ion terms with 4 or two multiplicities will get split.

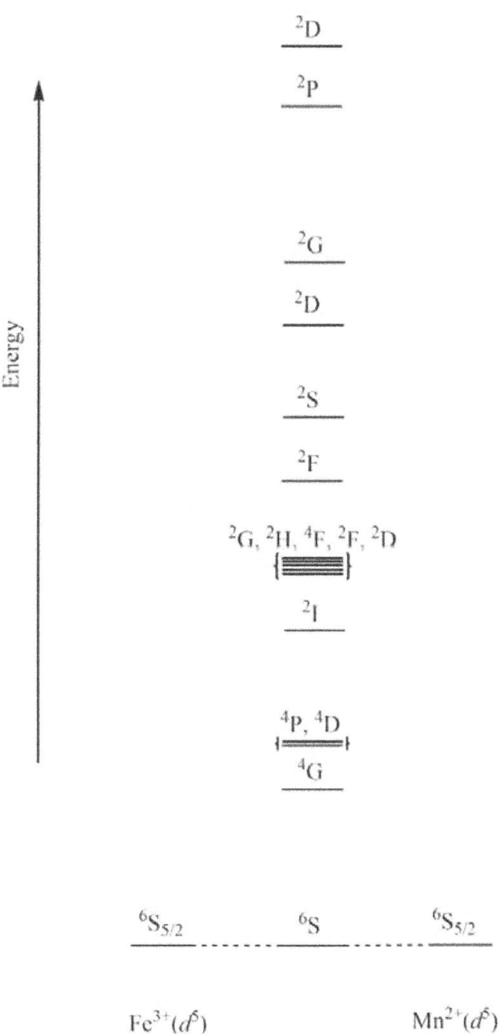

Figure 6. Correlation and spin-orbital coupling of free ion terms of Fe^{3+} and Mn^{2+}.

❖ Orgel and Tanabe-Sugano Diagrams for Transition Metal Complexes (d^1 – d^9 States)

It is a well-known fact that electronic transitions are always accompanied by vibrational as well as rotational changes which results in a considerable broadening of the bands in the UV-visible spectra of transition metal complexes too. The nature of these transitions is quite complex to understand and requires some basic knowledge of quantum mechanics and chemical applications of group theory.

The selection rules:

The selection rules governing the electronic spectra of transition metal complexes are given below.

1. There should be no change in the number of unpaired electrons. In other words, for all the multiplicity allowed transitions, $\Delta S = 0$. Hence, a triplet-triplet or singlet-singlet transitions will be multiplicity allowed while a singlet-triplet or triplet-singlet will be multiplicity forbidden.

2. If the complex possesses the center of symmetry, all the transitions which do not involve a change of ±1 in angular momentum quantum number are Laporte forbidden. In other words, in molecular geometries with the centre of symmetry, electronic transitions with $\Delta l = \pm1$ are Laporte allowed. Hence, are the d-d transitions in free ions as well as in perfectly octahedral environment are Laporte forbidden.

Relaxation of the abovementioned rules can occur through two mechanisms. The first one is the spin-orbit coupling which gives rise to weak spin forbidden bands. The second one is the absence of perfect octahedral geometry due to the presence of a different ligand or vibronic coupling in which a perfect octahedral complex may have some allowance of d-d transitions. vibrations where the molecule is asymmetric and the absorption of light at that moment is then possible. Both of these effects (vibronic coupling or six dissimilar ligands) generally mixes the d and p-orbitals of the transition metal so that the transitions are no longer purely d-d in nature. For example, the tetrahedral $[MnCl_4]^{2-}$ is colored because it does not possess the centre of symmetry and $[Co(NH_3)_5Cl]^{2+}$ also lacking the center of symmetry due to the presence of one Cl^- ligand, and therefore is colored. However, the perfect octahedral geometries like $[Mn(H_2O)_6]^{2+}$ with a d^5-electronic configuration is expected to be colorless (spin forbidden and Laporte forbidden) but does show a pale-pink color which can be explained in term of slight vibronic allowance.

The splitting of free ion terms:

It is pretty interesting to note that the degeneracy of free-ion terms like ^2D or ^3F can be removed not only by L-S coupling alone but can also be removed by the perturbation produced by the ligands. Moreover, the wavefunctions for S, P, D, F or G states have the same symmetry as that s, p, d, f or g orbitals sets; which means that the splitting pattern of D and F states will same as d and f-subshell, respectively.

The s-orbital is spherically symmetric in nature and is not affected by any crystal field. Hence, S state also does not get split in any type of ligand field. The p-orbitals are directional in nature and are affected by different types of crystal field differently. Hence, P state may or may not get split in the presence of ligand field. For example, P state does not get split in octahedral or tetrahedral field but does get split in square planar

crystal field. The d-orbitals are also directional in nature and are affected by different types of crystal field differently. Hence, D state does get split in the presence of the ligand field. For example, D state does get split in the octahedral or tetrahedral field with triply and doubly degenerate sets but the splitting pattern and degeneracy are totally different in pentagonal bipyramidal crystal field. The splitting profile of different electronic states in octahedral and tetrahedral crystal fields in the following tables, which will be used very frequently in the further text of this chapter.

Table 7. Splitting of free ion terms in the tetrahedral crystal field.

Electronic state	Symmetry designation in the tetrahedral field (Mulliken symbols)
S	A_1
P	T_1
D	$E + T_2$
F	$A_2 + T_1 + T_2$
G	$A_1 + E + T_1 + T_2$
H	$E + T_1 + T_1 + T_2$
I	$A_1 + A_2 + E + T_1 + T_2 + T_2$

Table 8. Splitting of free ion terms in the octahedral crystal field.

Electronic state	Symmetry designation in the octahedral field (Mulliken symbols)
S	A_{1g}
P	T_{1g}
D	$E_g + T_{2g}$
F	$A_{2g} + T_{1g} + T_{2g}$
G	$A_{1g} + E_g + T_{1g} + T_{2g}$
H	$E_g + T_{1g} + T_{1g} + T_{2g}$
I	$A_{1g} + A_{2g} + E_g + T_{1g} + T_{2g} + T_{2g}$

The A, E and T represent singly, doubly and triply degenerate states, respectively. The presence of "g" in symmetry designations of the octahedral field is for the gerade or centrosymmetric environment.

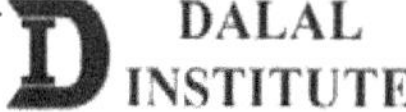

Terms correlation in the tetrahedral and octahedral field:

The qualitative description of different energy term for d^n and d^{10-n} configuration, from free ion to strong crystal field configurations ignoring inter-electronic repulsions, can be given as:

1. The total number of energy levels remain the same under the influence of weak and strong crystal fields.

2. The one to one correspondence of different energy levels in a strong crystal field may get stabilized or destabilized in comparison to the weak field case.

3. Energy levels of the same symmetry never cross each other and each level has a contribution in its energy from all other energy states of the same symmetry.

4. Term correlation for d^1, d^9, d^2, d^8 configuration are shown completely while for d^3, d^4, d^5, d^6, d^7 are shown partially by taking only lower energy levels.

5. According to hole formalism, the number of microstates and hence all free ion terms for d^n and d^{10-n} configuration are same. Now, as the magnitude of the crystal field experienced positive electrons is the same as what experienced by the negative electrons, but is of opposite sign. Therefore, the splitting pattern for d^n and d^{10-n} configurations are opposite of each other.

6. Owing to the hole formalism in quantum mechanics, strong field configuration of t_{2g}^4, t_{2g}^5, e_g^3 give rise to the same terms as given by the strong field configuration of t_{2g}^2, t_{2g}^1, e_g^1. However, weaker inter-electronic repulsion is considered as the perturbation over stronger V_o.

7. The splitting pattern of d^n tetrahedral is just the opposite of what is for d^n octahedral. However, no g or u are used in the tetrahedral case because there is no center of symmetry in a tetrahedral geometry.

It has already been discussed that the total number of microstates for electronic configuration without inter-electronic repulsion or with inter-electronic repulsion (free ion terms) remains the same. Furthermore, the number of microstates also remains same even in the presence of weak or strong crystal field; and when the inter-electronic repulsion is completed neglected in comparison to the ligand field strength, the calculation of microstates is carried out individually for t_{2g} and e_g set and multiplied afterward to give the total.

Furthermore, the strength of the crystal field does not alter the ground state Mulliken symbol in the case of d^1, d^9-octahedral or tetrahedral complexes. However, in the case of d^2-d^8 electronic configurations, the splitting pattern of free ion term at weak and strong crystal fields is quite different. Generally, the energy of some irreducible component of low multiplicity free ion term decreases so rapidly with the increase in the strength of the crystal field that it becomes the ground state symbol. In other words, the ground state term symbol of metal complexes with d^2-d^8 electronic configurations is different in weak and strong crystal fields. For example, the ground state term symbol for d^5-configuration with small ligand field is $^6A_{1g}$ (from ^6S) but as the magnitude of crystal field increases, the $^2T_{2g}$ Mulliken state (from ^2I) becomes highly stable and also make up the ground electronic state.

The correlation diagrams (with the corresponding microstates shown below each level) for different electronic configurations in transition metal complexes with four-coordinated tetrahedral and six-coordinated octahedral symmetry are given below.

1. d^1, d^9 systems:

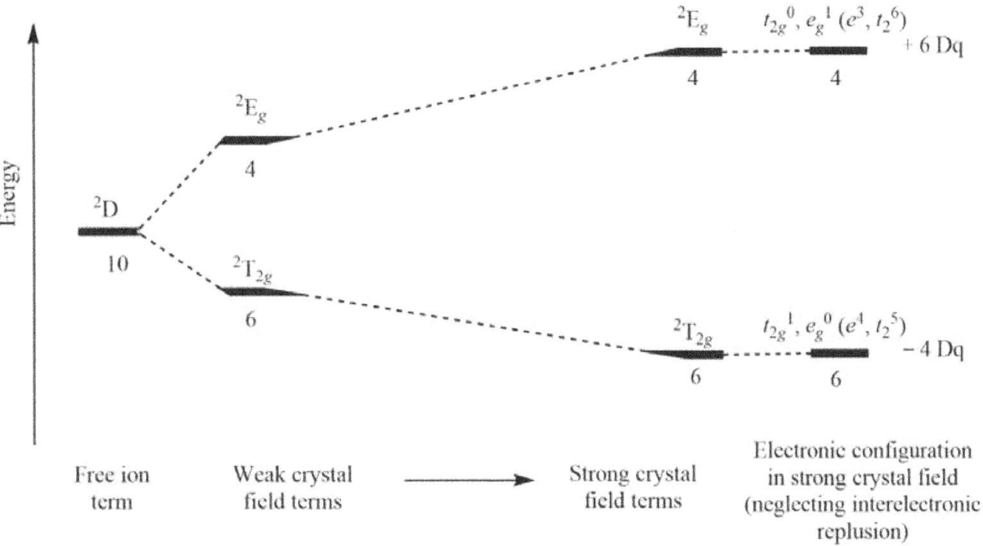

Figure 7. Qualitative correlation diagram for d^1 octahedral and d^9 tetrahedral complexes.

and

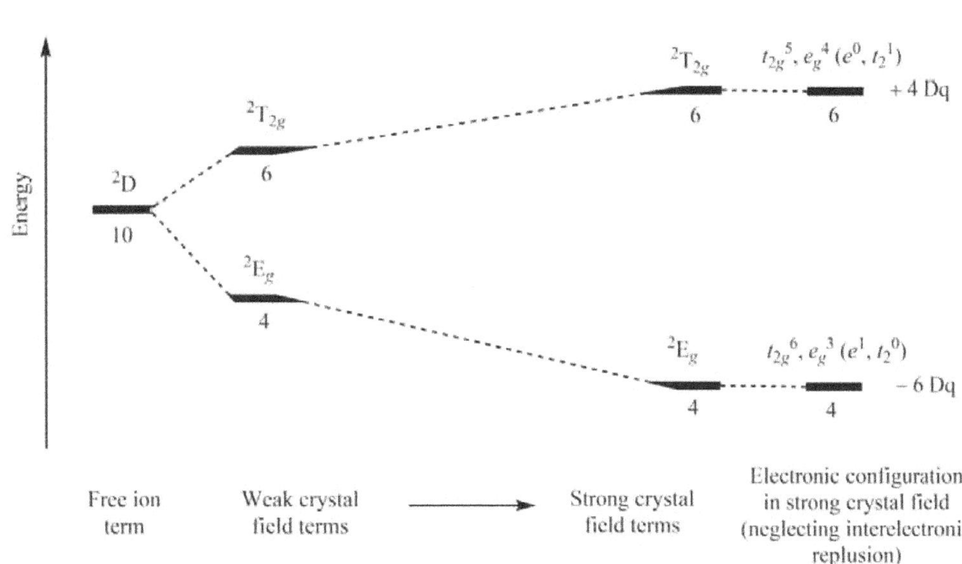

Figure 8. Qualitative correlation diagram for d^9 octahedral and d^1 tetrahedral complexes.

2. d^2, d^8 systems:

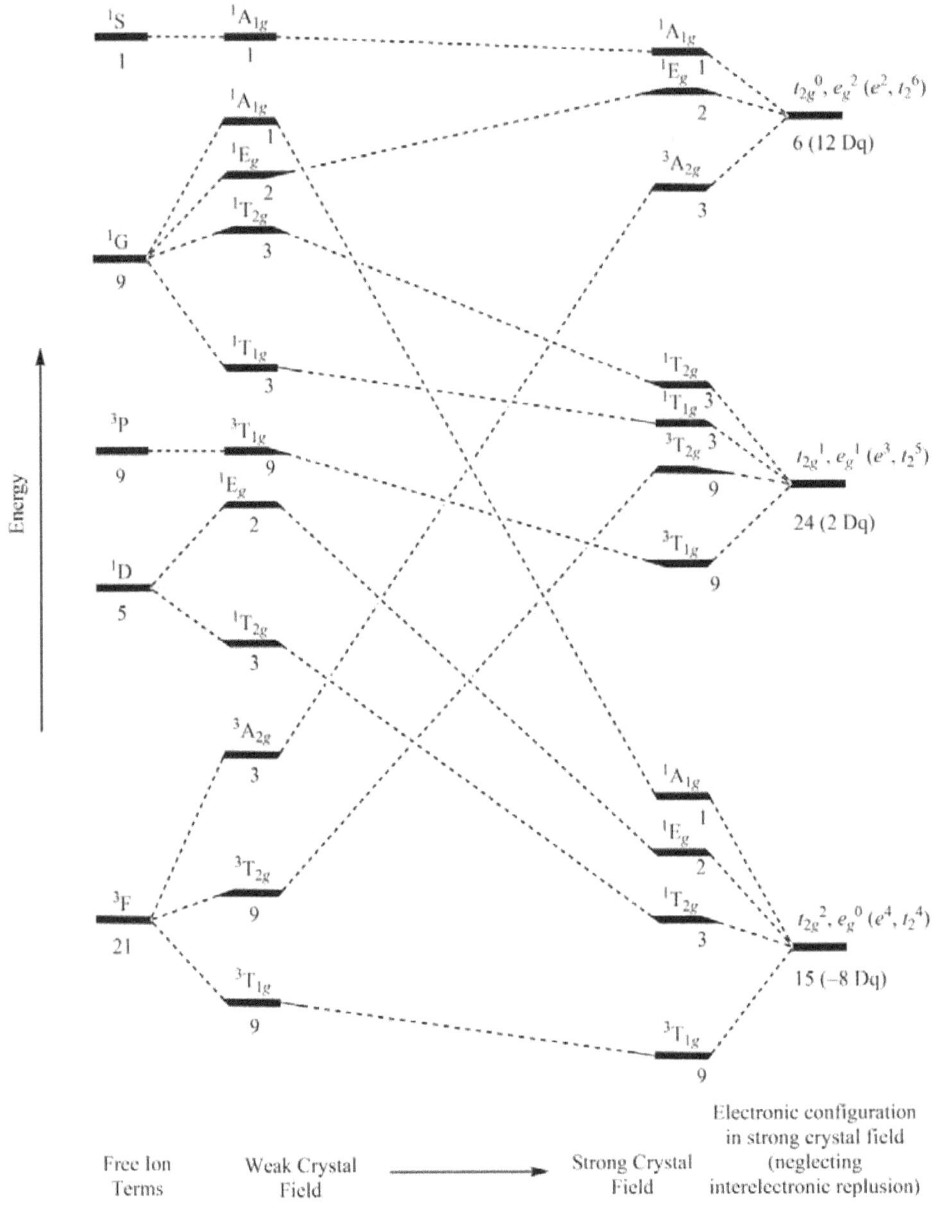

Figure 9. Qualitative correlation diagram for d^2 octahedral and d^8 tetrahedral complexes.

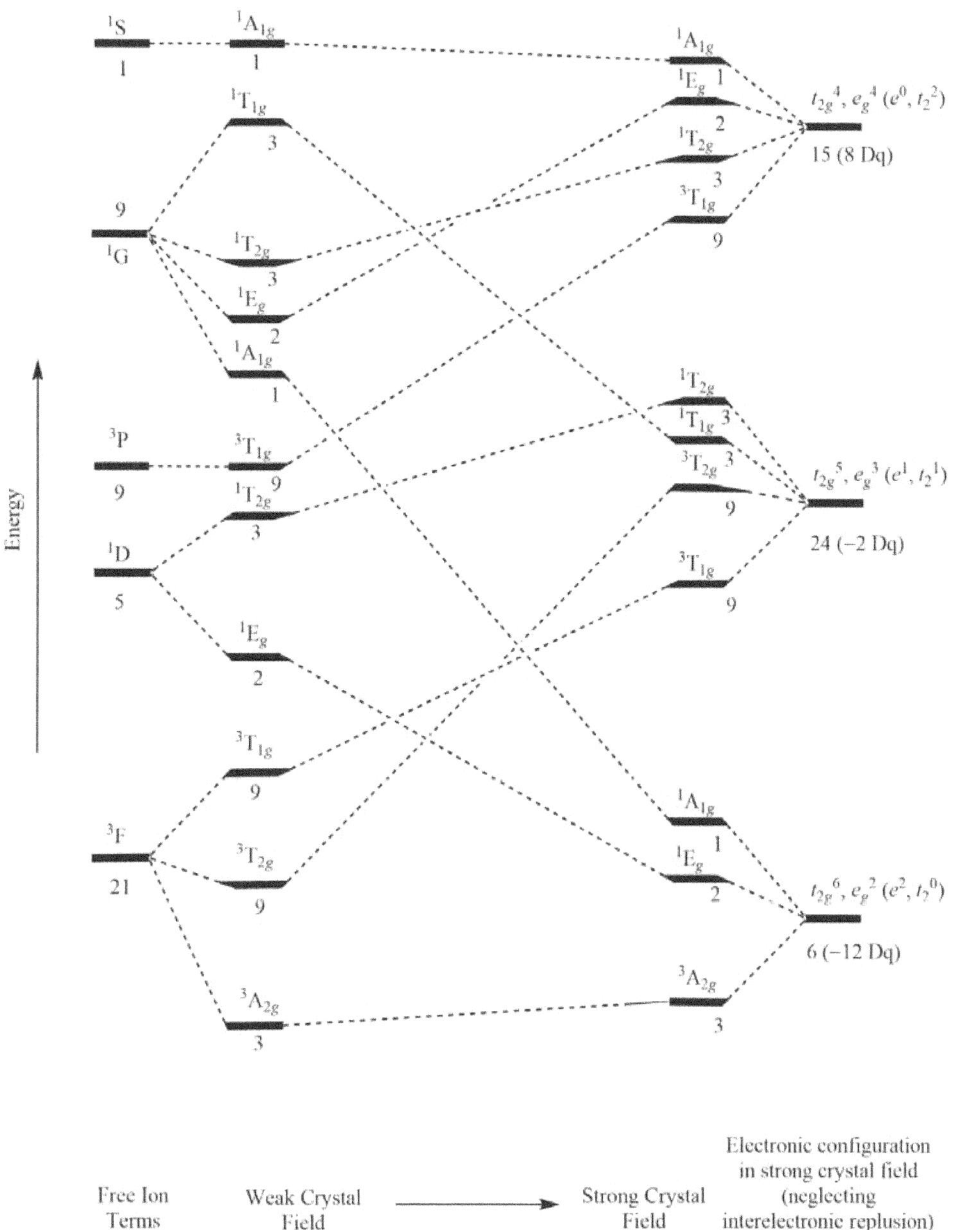

Figure 10. Qualitative correlation diagram for d^8 octahedral and d^2 tetrahedral complexes.

3. d^3, d^7 systems:

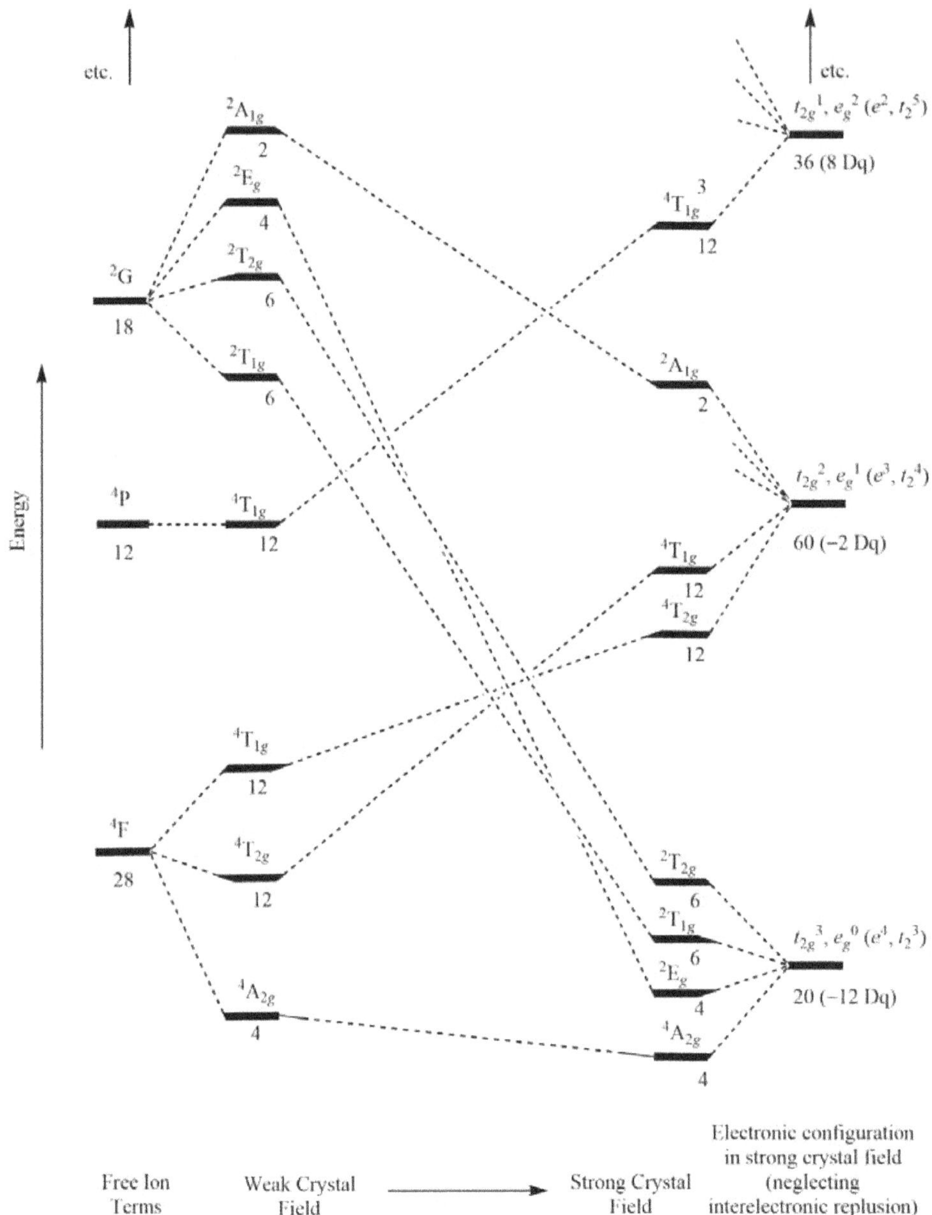

Figure 11. Qualitative correlation diagram for d^3 octahedral and d^7 tetrahedral complexes.

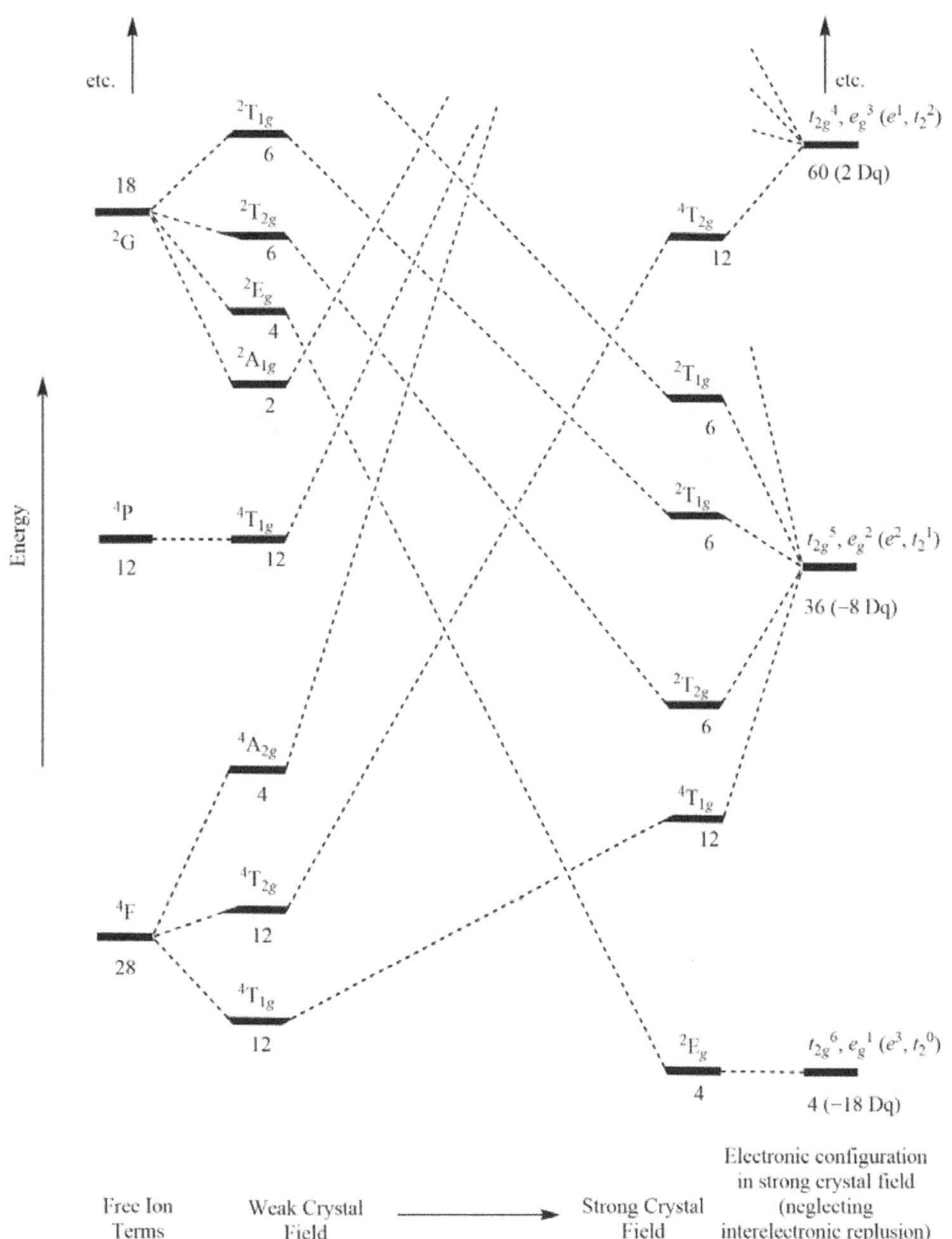

Figure 12. Qualitative correlation diagram for d^7 octahedral and d^3 tetrahedral complexes.

4. d^4, d^6 systems:

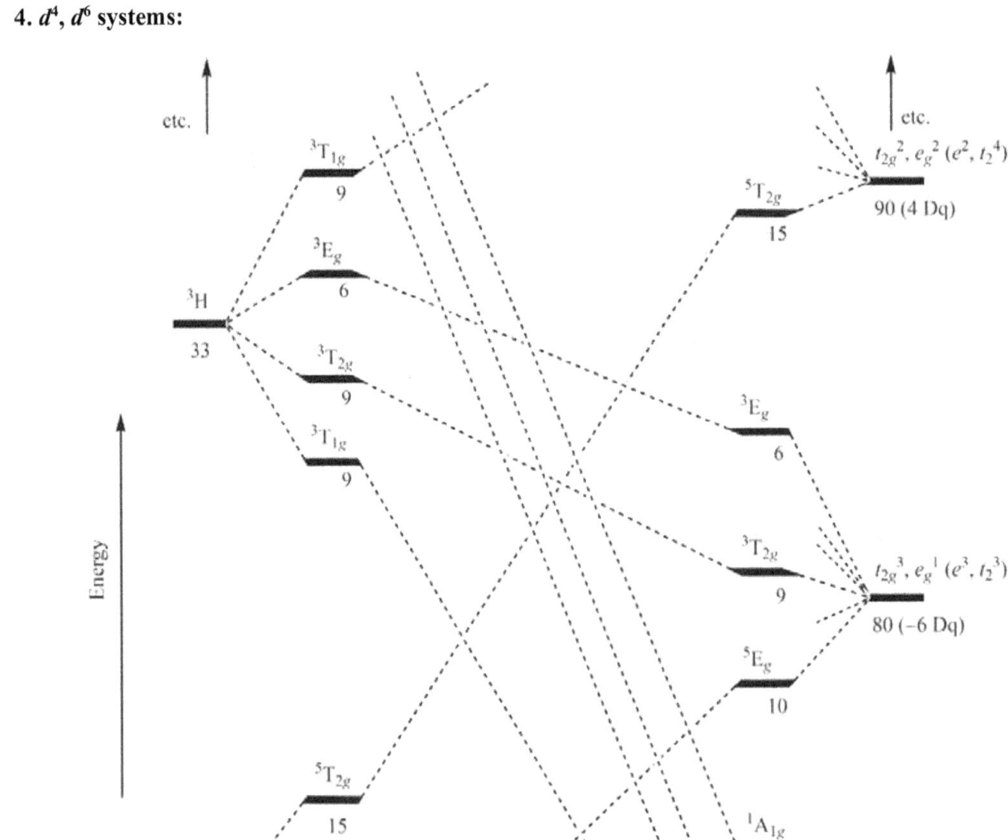
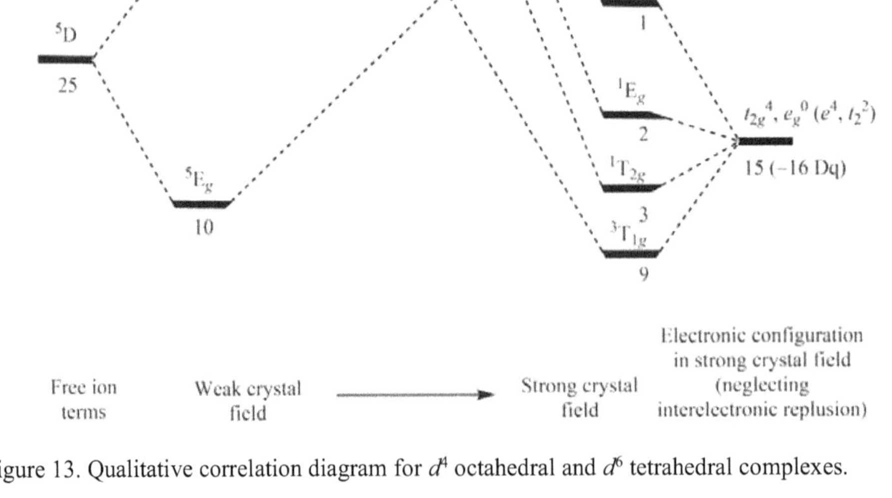

Figure 13. Qualitative correlation diagram for d^4 octahedral and d^6 tetrahedral complexes.

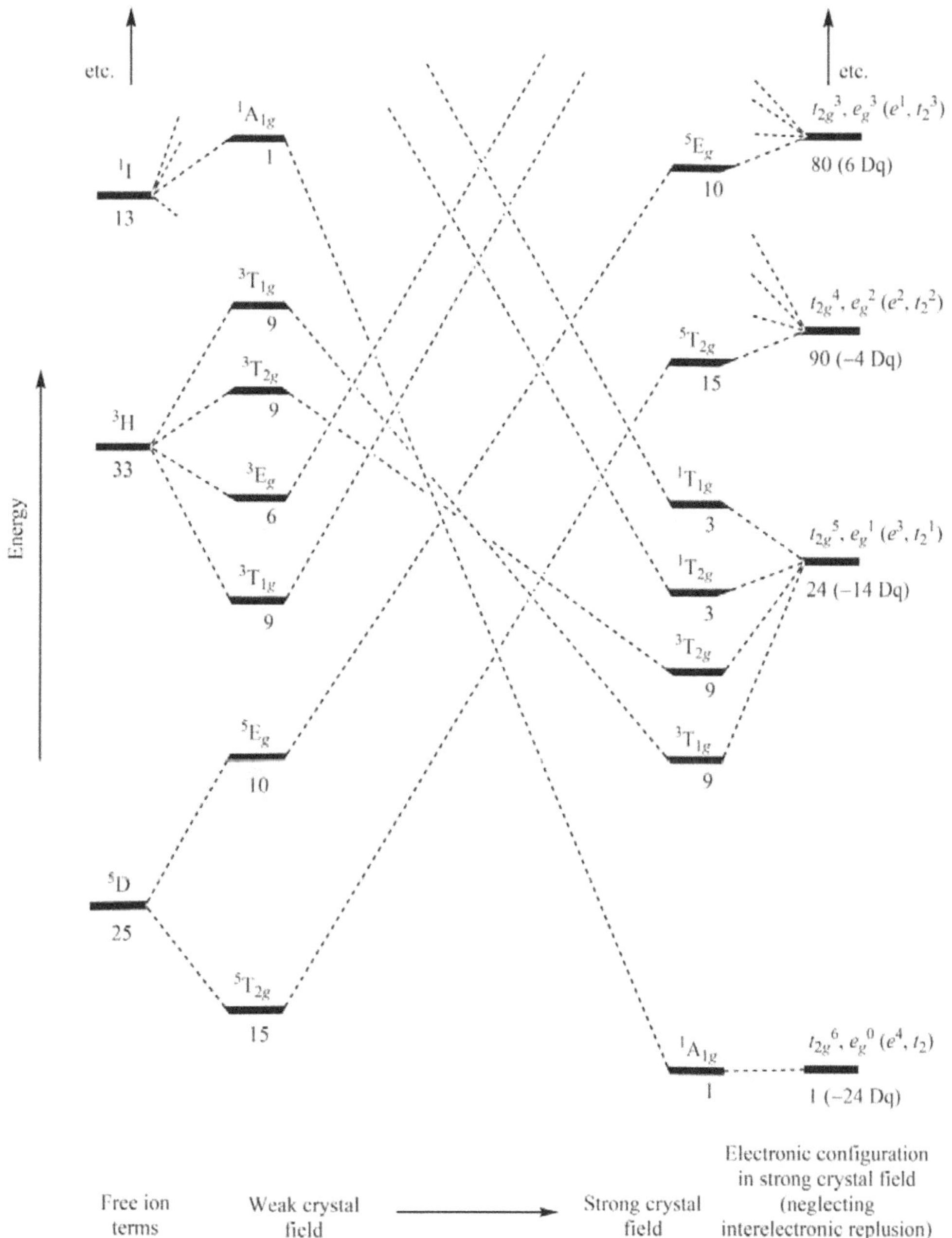

Figure 14. Qualitative correlation diagram for d^6 octahedral and d^4 tetrahedral complexes.

5. d^5 systems:

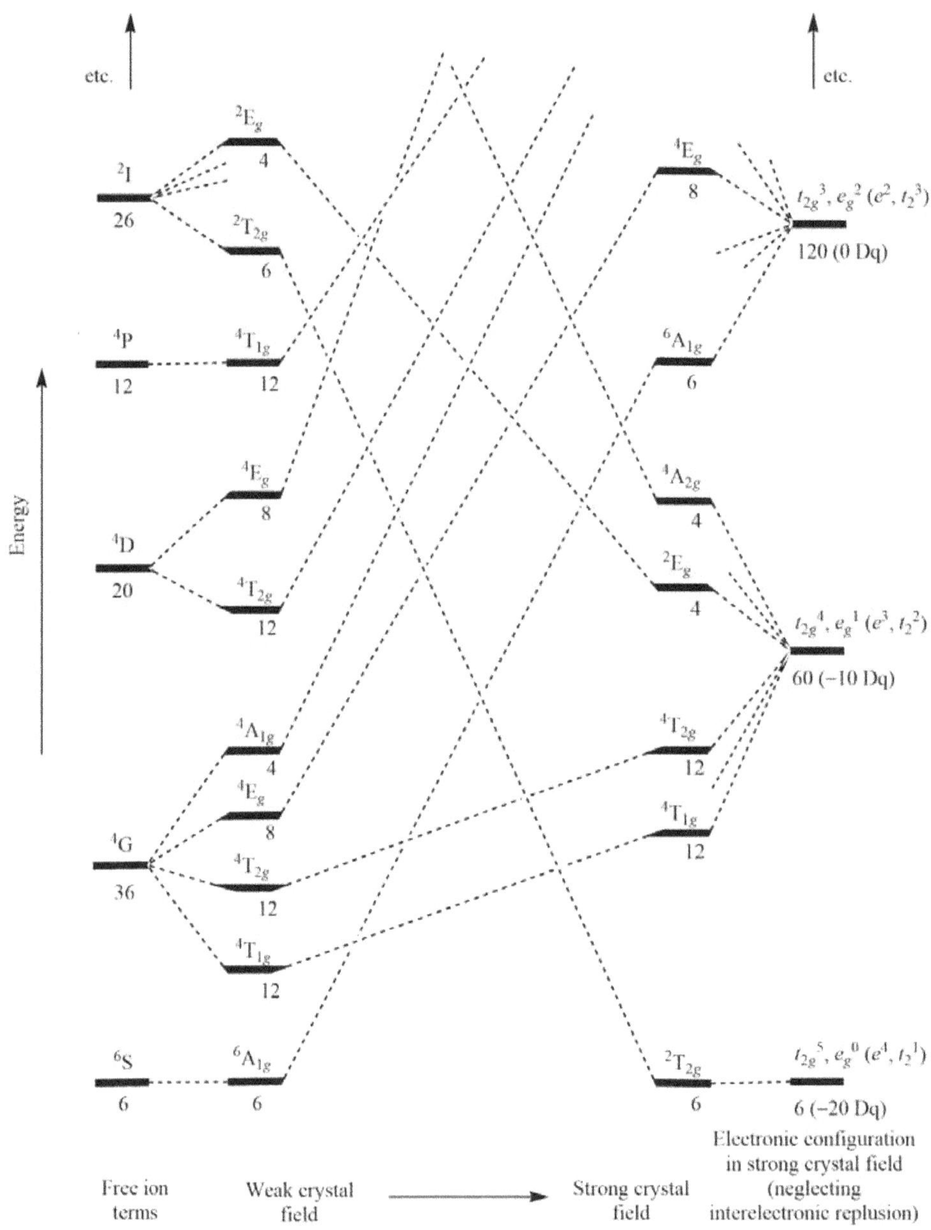

Figure 15. Qualitative correlation diagram for d^5 octahedral and tetrahedral complexes.

➤ *Orgel Diagrams*

Orgel diagrams are the oversimplified version of correlation diagrams that show the relative energies of electronic terms in transition metal complexes. They are named after their inventor, Leslie Orgel. These diagrams are restricted only to show weak field cases and offer no information about strong field cases. Because Orgel diagrams are qualitative, no energy calculations can be performed from these diagrams. Moreover, Orgel diagrams only show the symmetry states of the highest spin multiplicity instead of all possible terms, unlike a general correlation diagram. Thus, Orgel diagrams include only those transitions which are spin-allowed in nature, along with corresponding symmetry designations.

In an Orgel diagram, the parent term (P, D, or F) in the presence of no ligand field is located in the center of the diagram; and the Mulliken terms arising from different electronic configurations in a ligand field are represented at each side. There are two Orgel diagrams, one for d^1, d^4, d^6, and d^9 configurations and the other with d^2, d^3, d^7, and d^8 configurations. An Orgel diagram for d^5 has also been very popular which includes spin-forbidden transitions too. In the Orgel diagram, lines with the same Russell-Saunders terms will diverge due to the non-crossing rule, but all other lines will be linear.

1. d^1, d^9, d^4, d^6 systems:

For the "D" Orgel diagram, the left side contains d^1 and d^6 octahedral, and d^4 and d^9 tetrahedral complexes. The right side contains d^4 and d^9 octahedral, and d^1 and d^6 tetrahedral complexes. The lowest energy absorption band on the left side of the spectrum is $T_{2g} \rightarrow E_g$ while on the right side of the spectrum it is $E_g \rightarrow T_{2g}$ transition.

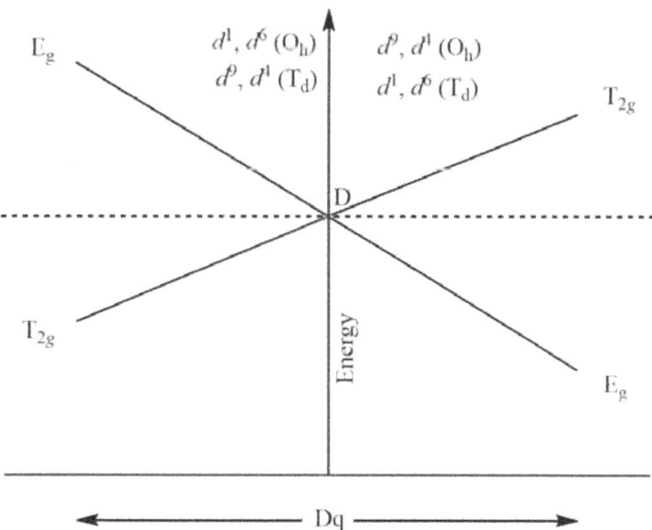

Figure 16. Orgel diagram for d^1, d^9, d^4, d^6 complexes in octahedral (O$_h$) and tetrahedral (T$_d$) crystal fields.

i) Spectra of d^1 and d^9 octahedral complexes:

The ground state term symbol for d^1 and d^9 complexes is ^2D; but the splitting pattern of their ^2D states is just the opposite of each other, which can be attributed to the electron-hole inverse relationship or simply the hole-formalism. In other words, a d^9 metal has an electron vacancy or "hole" in its d-subshell and thus can be considered as the inverse of the d^1 arrangement. Therefore, despite having identical ground state term symbol ^2D (split into $^2T_{2g}$ and ^2E_g in the octahedral field), the energy order of Mulliken states in d^9-configuration complexes will be just the inverse of what is in d^1 system.

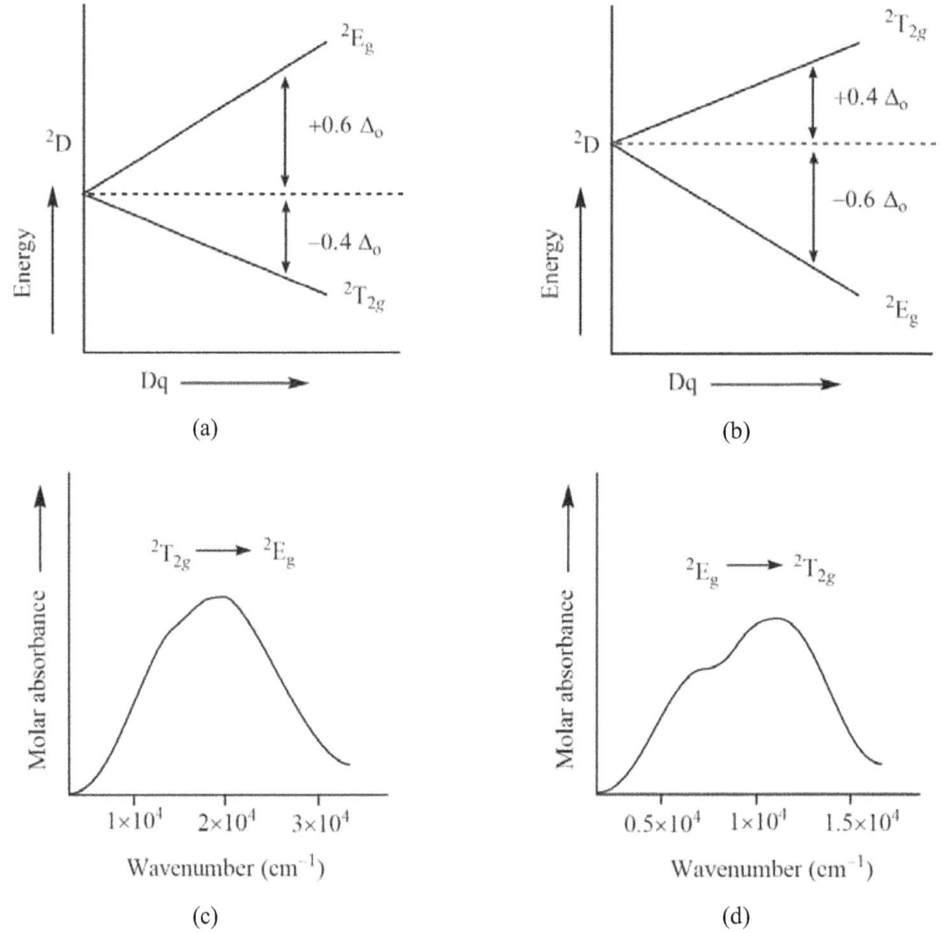

Figure 17. The splitting pattern of ^2D state in octahedral complexes with (a) d^1-configuration and (b) d^9-configuration; and the corresponding electronic spectra of (c) $[Ti(H_2O)_6]^{3+}$ and (d) $[Cu(H_2O)_6]^{2+}$.

ii) Spectra of d^1 and d^9 tetrahedral complexes:

In addition to the electron-hole inverse relationship for d^1 and d^9 octahedral complexes, an inverse relationship for octahedral-tetrahedral crystal field symmetries also exists. This is simply because the crystal fields of these two symmetries produce the inverse splitting patterns of the d-subshell. Therefore, despite having identical ground state term symbol ^2D (split into $^2T_2$ and ^2E in the tetrahedral field), the energy order of Mulliken states in the tetrahedral field will be just the inverse of what is in octahedral systems. However, it is worthy to note that the subscript "g" or "u" cannot be used due to the lack of the center of symmetry.

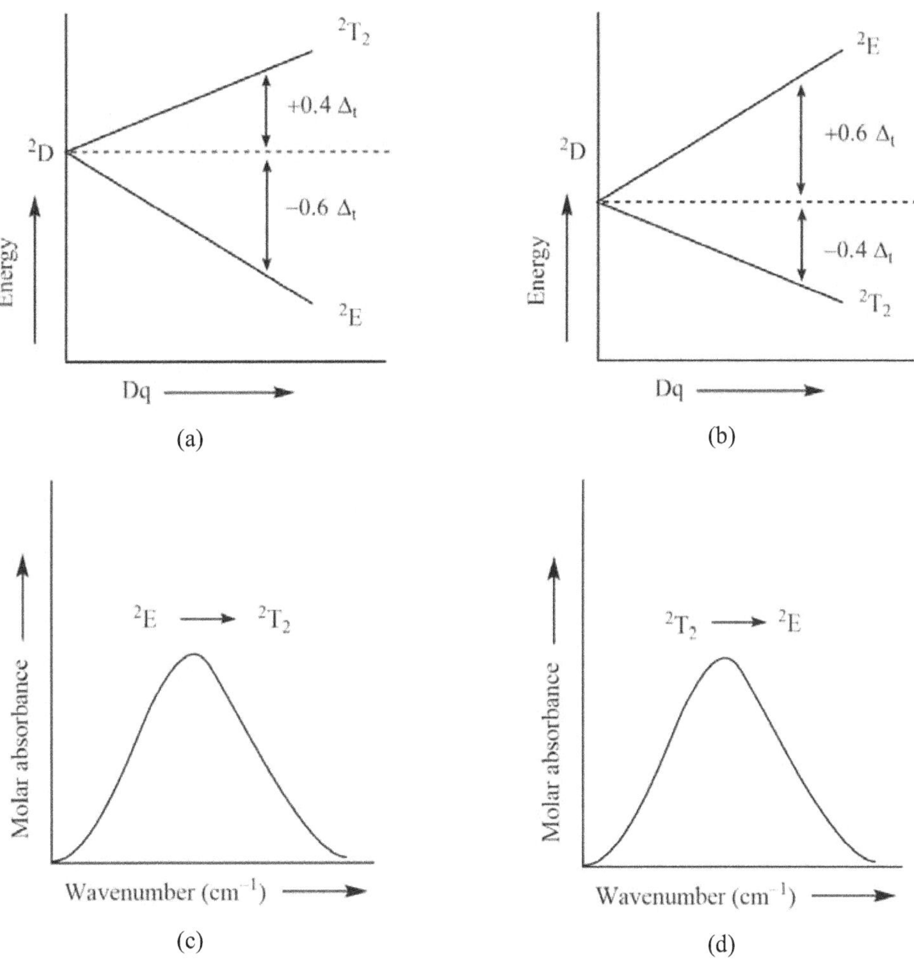

Figure 18. The splitting pattern of ^2D state in tetrahedral complexes with (a) d^1-configuration and (b) d^9-configuration; and the corresponding electronic spectra of (c) ML_4 (d^1) and (d) ML_4 (d^9).

iii) Spectra of d^4 and d^6 octahedral complexes:

The ground state term symbol for d^4 and d^6 complexes is ${}^5D$; but the splitting pattern of their ${}^5D$ states is just the opposite of each other, which can be attributed to the electron-hole inverse relationship or simply the hole formalism. In other words, a d^6 metal has an electron vacancy or "hole" in its d-subshell and thus can be considered as the inverse of the d^4 arrangement. Therefore, despite having identical ground state term symbol ${}^5D$ (split into ${}^5T_{2g}$ and ${}^5E_g$ in the octahedral field), the energy order of Mulliken states in d^6 configuration will be just the inverse of what is in d^4 system.

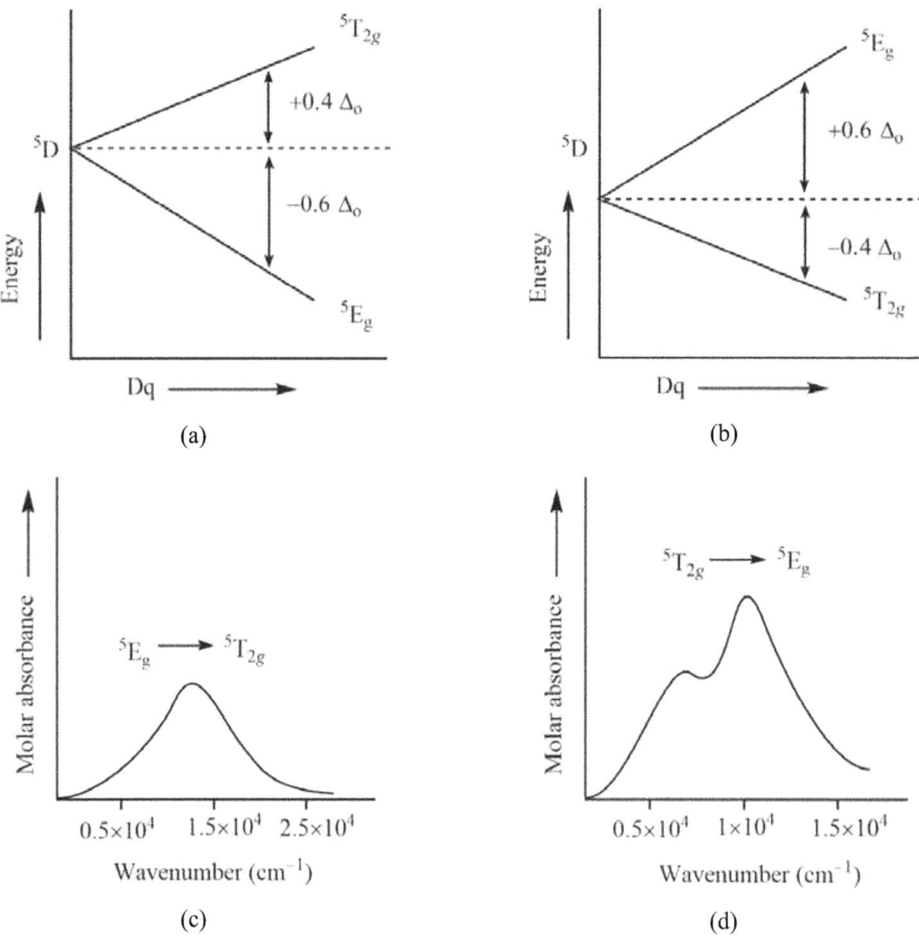

Figure 19. The splitting pattern of ${}^5D$ state in octahedral complexes with (a) d^4-configuration and (b) d^6-configuration; and the corresponding electronic spectra of (c) $[Cr(H_2O)_6]^{2+}$ and (d) $[Fe(H_2O)_6]^{2+}$.

iv) Spectra of d^4 and d^6 tetrahedral complexes:

In addition to the electron-hole inverse relationship for d^4 and d^6 octahedral complexes, an inverse relationship for octahedral-tetrahedral crystal field symmetries also exists. This is simply because the crystal fields of these two symmetries produce the inverse splitting pattern of the d-subshell. Therefore, despite having identical ground state term symbol ^5D (split into $^5T_2$ and ^5E in the tetrahedral field), the energy order of Mulliken states in the tetrahedral field will be just the inverse of what is in octahedral systems. However, it is worthy to note that the subscript "g" or "u" cannot be used due to the lack of the center of symmetry.

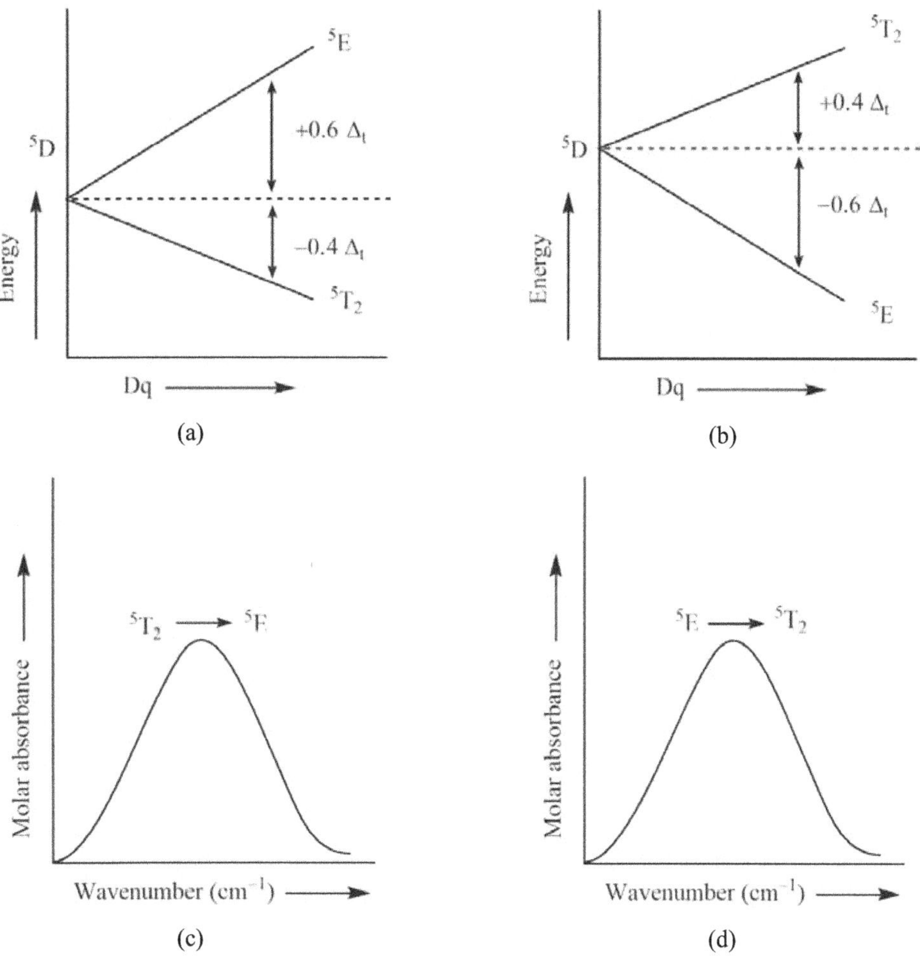

Figure 20. The splitting pattern of ^5D state in tetrahedral complexes with (a) d^4-configuration and (b) d^6-configuration; and the corresponding electronic spectra of (c) ML$_4$ (d^4) and (d) ML$_4$ (d^6).

2. d^2, d^8, d^3, d^7 systems:

For the "F" Orgel diagram, the left side contains d^2 and d^7 tetrahedral and d^3 and d^8 octahedral complexes. The right side contains d^3 and d^8 tetrahedral and d^2 and high spin d^7 octahedral complexes. The lowest energy absorption band on the left side of the spectrum is $A_{2g}(F) \rightarrow T_{2g}(F)$ while on the right side of the spectrum it is $T_{1g}(F) \rightarrow T_{2g}(F)$ transition. The difference in energy between these two states is solely attributable to electron-electron repulsions. The two free ion electronic states are separated by an energy difference 15B, where B is the Racah parameter that acts as a measure of electron-electron repulsions. The value of B can be calculated experimentally in a very similar manner as the value of ligand-field splitting Δ is obtained.

For instance, in an octahedral d^2 complex, there are three ways of arranging the two d electrons. One is t_{2g}^2, second is $t_{2g}^1 e_g^1$ and the last is e_g^2. These are the three electronic states under consideration and are one should use the right-hand side of the diagram. Moreover, the energy gap between each state is equal to Δ since it requires the promotion of one electron from t_{2g} to e_g. It is clear from the Orgel diagram that there are four states: two T_{1g} states, one T_{2g} state, and one A_{2g} state. The spin multiplicities are omitted in the diagram that allows it to be generalized for d^8 complexes. The subscript g and u should be omitted if the same diagram is to be used to generalize the spectra of tetrahedral counterparts. It is also worthy to note that the ordering of the second and third transition on the right-hand side is reversed after the crossover point.

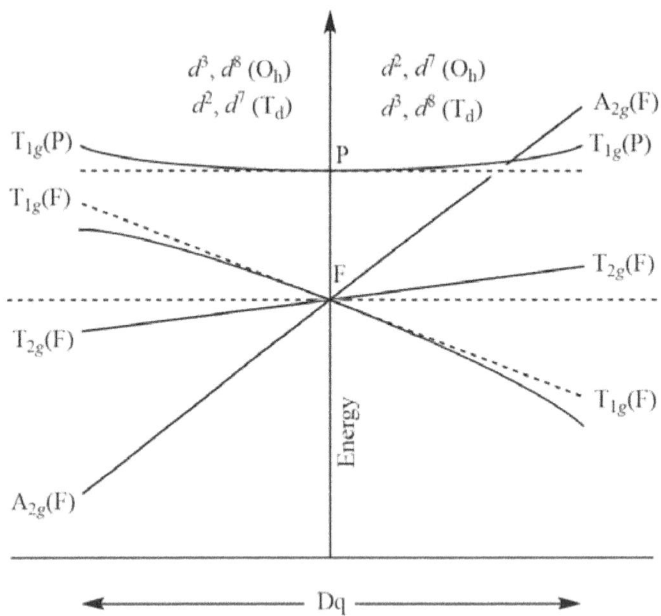

Figure 21. Orgel diagram for d^2, d^8, d^3, d^7 complexes in octahedral (O$_h$) and tetrahedral (T$_d$) crystal fields.

i) Spectra of d^2 and d^8 octahedral complexes:

The ground state term symbol for d^2 and d^8 complexes is ^3F; but the splitting pattern of their ^3F states is just the opposite of each other, which can be attributed to the electron-hole inverse relationship or simply the hole-formalism. In other words, a d^8 metal has two electron vacancies or "holes" in its d-subshell and thus can be considered as the inverse of the d^2 arrangement. Therefore, despite having identical ground state term symbol ^3F (split into $^3A_{2g}$, $^3T_{2g}$ and $^3T_{1g}$ in the octahedral field), the energy order of Mulliken states in d^8-configuration complexes will be just the inverse of what is in d^2 system.

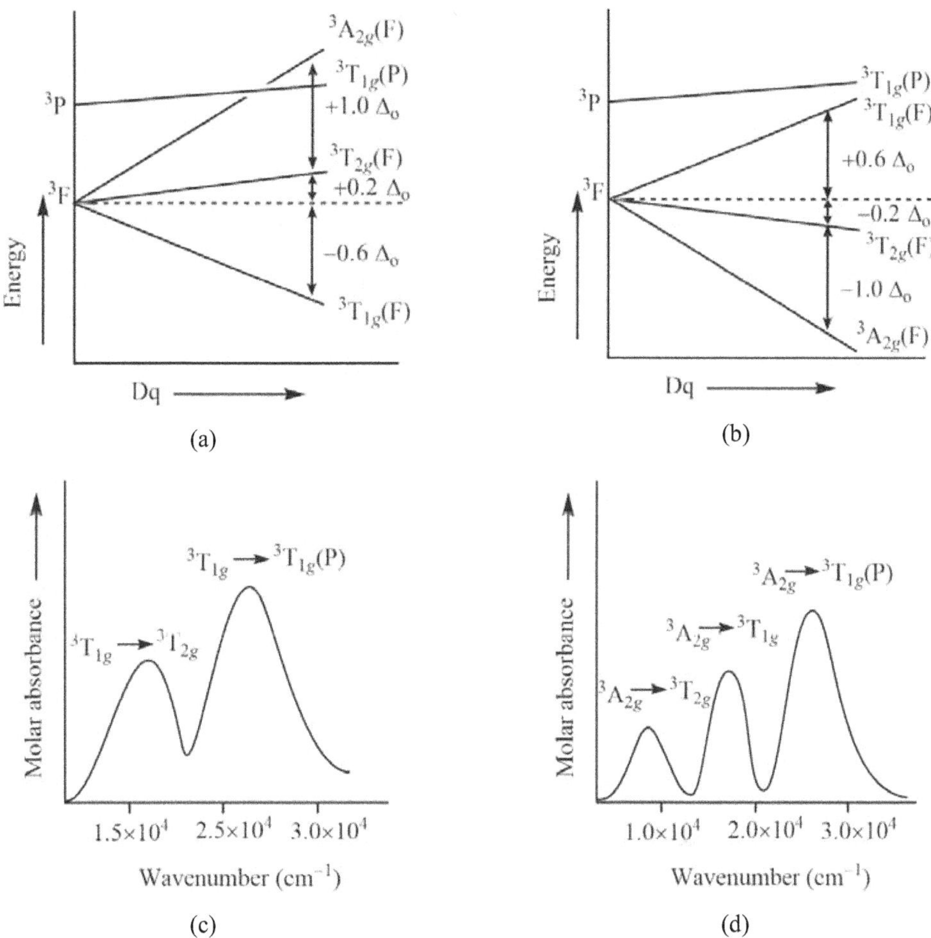

Figure 22. The splitting pattern of ^3F state in octahedral complexes with (a) d^2-configuration and (b) d^8-configuration; and the corresponding electronic spectra of (c) $[V(H_2O)_6]^{3+}$ and (d) $[Ni(H_2O)_6]^{2+}$.

ii) Spectra of d^2 and d^8 tetrahedral complexes:

In addition to the electron-hole inverse relationship for d^2 and d^8 octahedral complexes, an inverse relationship for octahedral-tetrahedral crystal field symmetries also exists. This is simply because the crystal fields of these two symmetries produce the inverse splitting pattern of the d-subshell. Therefore, despite having identical ground state term symbol ^3F (split into $^3A_2$, $^3T_2$ and $^3T_1$ tetrahedral field), the energy order of Mulliken states in the tetrahedral field will be just the inverse of what is in octahedral systems. However, it is worthy to note that the subscript "*g*" or "*u*" cannot be used due to the lack of the center of symmetry.

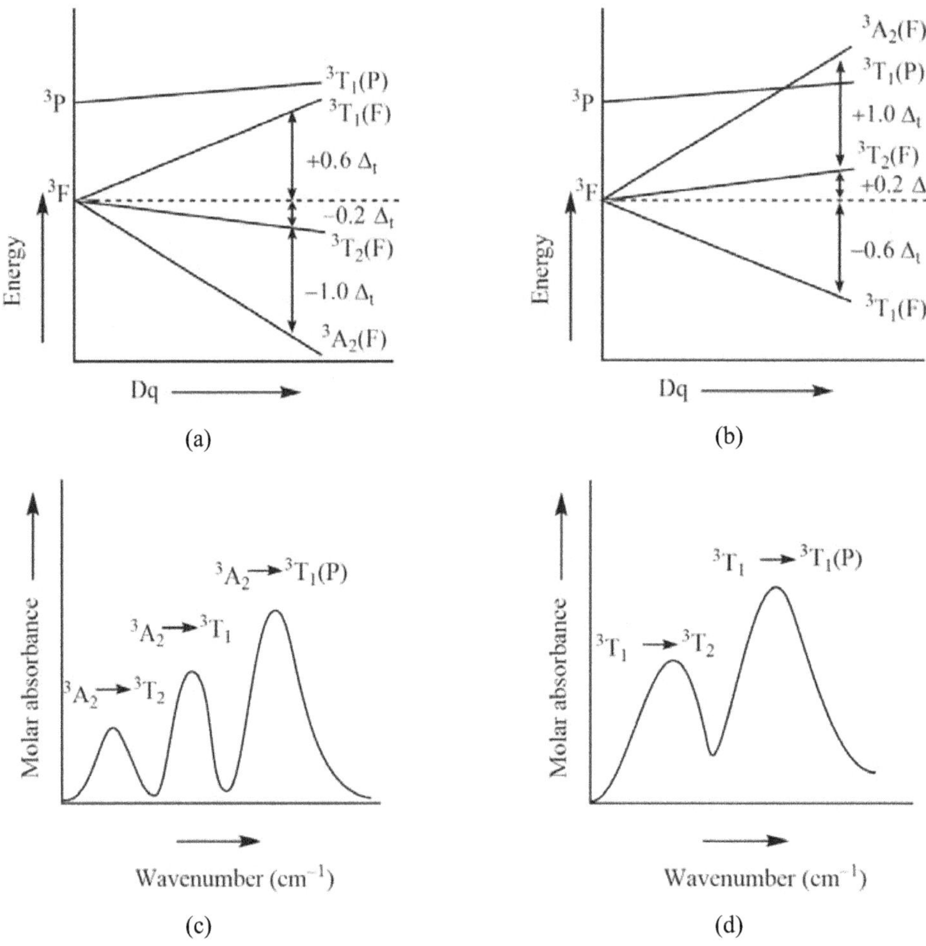

Figure 23. The splitting pattern of ^3F state in tetrahedral complexes with (a) d^2-configuration and (b) d^8-configuration; and the corresponding electronic spectra of (c) ML_4 (d^2) and (d) ML_4 (d^8).

iii) Spectra of d^3 and d^7 octahedral complexes:

The ground state term symbol for d^3 and d^7 complexes is ^4F; but the splitting pattern of their ^4F states is just the opposite of each other, which can be attributed to the electron-hole inverse relationship or simply the hole formalism. In other words, a d^7 metal has an electron vacancy or "hole" in its d-subshell and thus can be considered as the inverse of the d^3 arrangement. Therefore, despite having identical ground state term symbol ^4F (split into $^4A_{2g}$, $^4T_{2g}$ and $^4T_{1g}$ in the octahedral field), the energy order of Mulliken states in d^7 configuration will be just the inverse of what is in d^3 system.

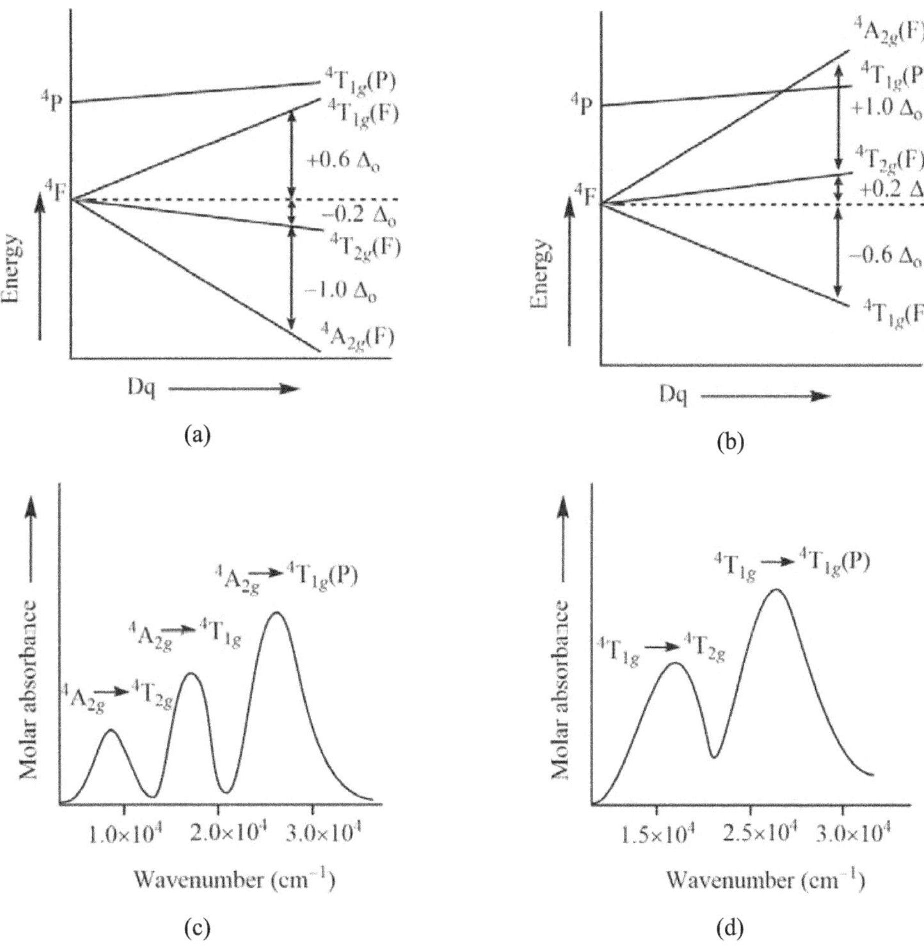

Figure 24. The splitting pattern of ^4F state in octahedral complexes with (a) d^3-configuration and (b) d^7-configuration; and the corresponding electronic spectra of (c) $[Cr(H_2O)_6]^{3+}$ and (d) $[Co(H_2O)_6]^{2+}$.

iv) Spectra of d^3 and d^7 tetrahedral complexes:

In addition to the electron-hole inverse relationship for d^3 and d^7 octahedral complexes, an inverse relationship for octahedral-tetrahedral crystal field symmetries also exists. This is simply because the crystal fields of these two symmetries produce the inverse splitting pattern of the d-subshell. Therefore, despite having identical ground state term symbol ^4F (split into $^4A_2$, $^4T_2$ and $^4T_1$ in the tetrahedral field), the energy order of Mulliken states in the tetrahedral field will be just the inverse of what is in octahedral systems. However, it is worthy to note that the subscript "g" or "u" cannot be used due to the lack of the center of symmetry.

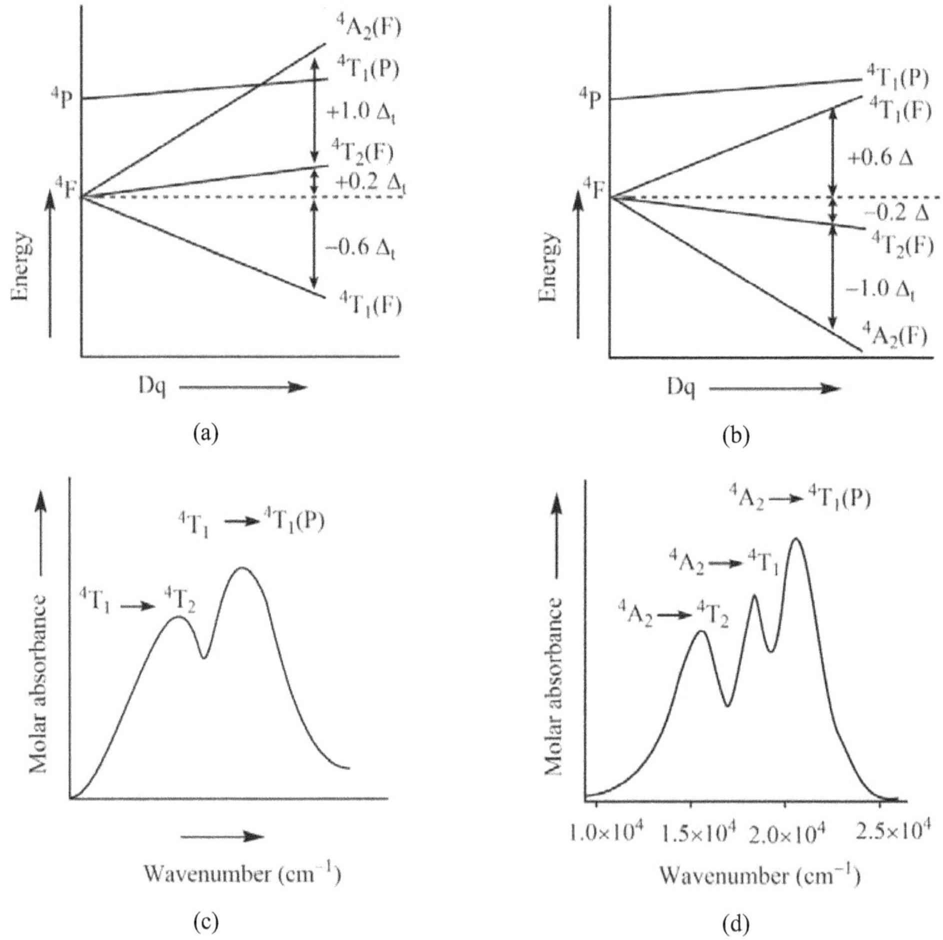

Figure 25. The splitting pattern of ^4F state in tetrahedral complexes with (a) d^3-configuration and (b) d^7-configuration; and the corresponding electronic spectra of (c) ML_4 (d^3) and (d) $[CoCl_4]^{2-}$ (d^7).

DALAL INSTITUTE

3. d^5 systems:

Orgel diagrams we have studied so far include only spin allowed transitions. However, in the case of d^5-configuration, this is not possible as there is only one electronic state with a multiplicity of six, $^6A_{1g}$. Therefore, all the transitions must occur with a change in the spin multiplicity and are spin forbidden for octahedral as well as tetrahedral complexes. The lowest energy absorption band of the spectrum is $^6A_{1g}(S) \rightarrow$ $^4T_{1g}(G)$ in octahedral; while in the tetrahedral complex, it is $^6A_1(S) \rightarrow {}^4T_1(G)$ transition. The difference in energy between these two states is also attributable to electron-electron repulsions and B, the Racah parameter that acts as a measure of electron-electron repulsions.

The electronic configuration for ground state term symbol is $t_{2g}^3 e_g^2$ with five unpaired electrons having parallel spins and any promotion or rearrangement of the electrons would lead to a lower multiplicity state. Moreover, if we combine this fact perfect octahedral complexes like $[Mn(H_2O)_6]^{2+}$, the absorption intensities become very weak due to the additional selection rule of Laporte forbiddance. This makes $[Mn(H_2O)_6]^{2+}$ pale pink in color but the tetrahedral complexes of Mn^{2+} are quite instance due to the absence of the centre of symmetry. The subscript g and u should be omitted if the same diagram is to be used to generalize the spectra of tetrahedral counterparts. It is also worthy to note that the electronic states with symmetry never cross each other but repel each other due to their quantum mechanical mixing.

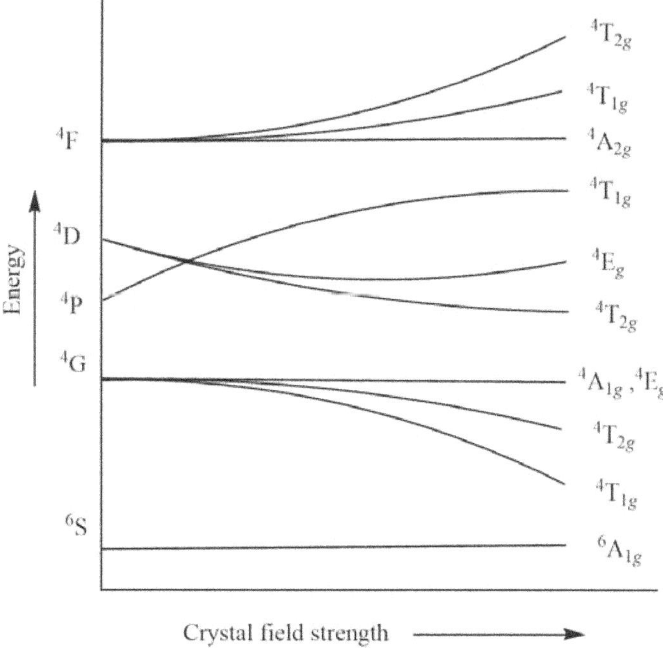

Figure 26. Orgel diagram for d^5 complexes in octahedral (O_h) and tetrahedral (T_d) crystal fields.

Unlike other electronic configurations, the ground state of d^5 system remains the same but the gerade subscript will be removed i.e. ^6A. The same is true for other Mulliken states. In other words, the subscript "g" or "u" cannot be used due to the lack of the center of symmetry.

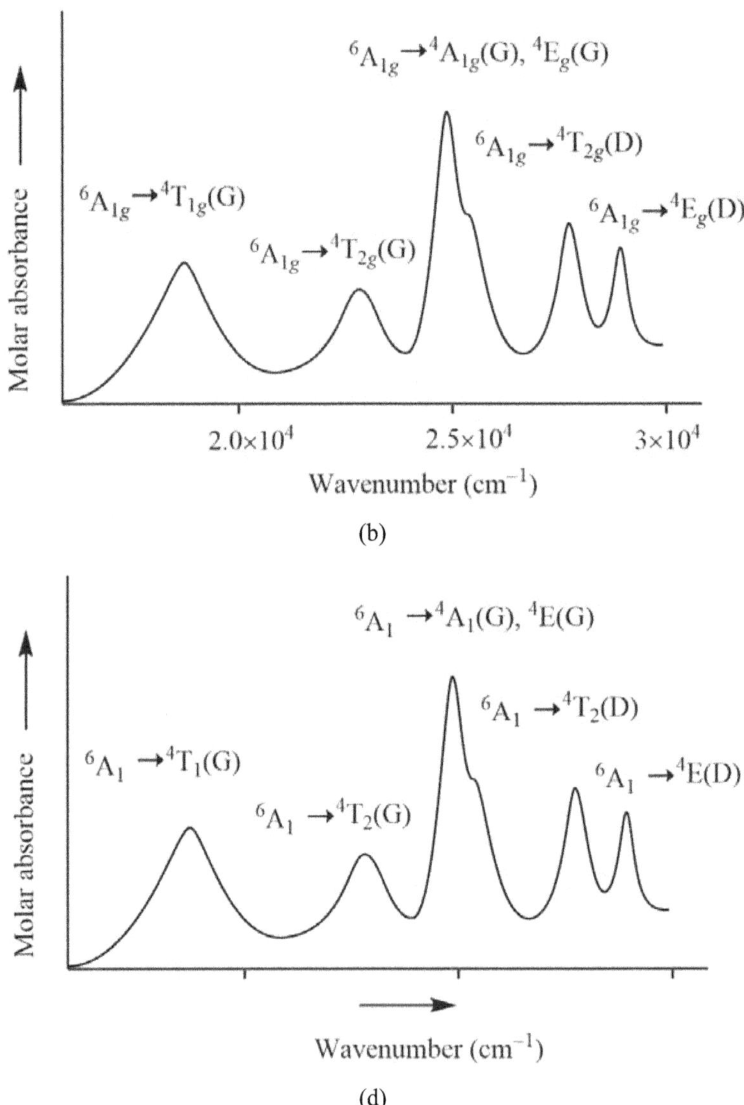

Figure 27. The electronic spectra of (a) $[Mn(H_2O)_6]^{2+}$ and (b) $[ML_4]$ (d^5).

> ### *Tanabe-Sugano Diagrams*

In 1954, two Japanese scientists, Yukito Tanabe and Satoru Sugano published a paper entitled "On the absorption spectra of complex ions". Earlier than that, very little was available regarding the excited electronic states of transition metal complexes. Tanabe and Sugano used the Hans Bethe's CFT concept and Giulio Racah's linear combinations of Slater integrals (Racah parameters) to rationalize the absorption spectra of the octahedral metal complex in a quantitative frame than what had been accomplished earlier. After a number of spectroscopic experiments, they estimated the values B and C (two of Racah's parameters) for every d-electronic configuration based on the trends in the UV visible absorption spectrum of isoelectronic transition metals of $3d$ series. The energy plots of various electronic states of every electron configuration are now called as Tanabe-Sugano or simply the T-S diagrams. In other words, Tanabe-Sugano diagrams are nothing but some special kind of correlation diagrams that can be used in weak as well as strong field complexes. These diagrams are used in transition metal chemistry to forecast the UV-visible absorption spectrum of coordination compounds and the results from these diagrams can also be compared to experimentally obtained spectroscopic data. T-S diagrams can be used qualitatively to approximate the value of crystal field splitting energy (Δ). Unlike Orgel diagrams (applicable to high-spin complexes only), these diagrams can be used for spin-free as well as spin-paired complexes. T-S diagrams can also be exploited to find the size of the ligand field required to cause high-spin–low-spin transitions. In contrast to the Orgel diagrams, the ground state is set as a reference in a T-S diagram. The ground-state-energy is taken to be zero for all ligand field strengths, and the energy of all other states along-with their components are plotted with respect to the ground state term.

In Tanabe-Sugano diagrams, the x-axis is expressed in terms of the crystal field splitting parameter (Δ or Dq) divided by the Racah parameter (B). Along y-axis, the energy, E, divided by B is taken. Out of three Racah parameters (A, B, and C), which describe various aspects of inter-electronic repulsion, A is an average total inter-electron repulsion whereas B and C correspond with individual d-electronic repulsions. The value of A is constant for d-electronic configuration, and therefore, is unnecessary for obtaining relative energies, and so is absent from T-S diagram studies of metal complexes. The parameter C is required only in some special cases. The parameter B is the most significant of all Racah's parameters in transition metals complexes. Moreover, certain lines bend due to the mixing of states with same symmetry. Although transitions between the same spin-multiplicity are allowed, energy profiling of spin-forbidden states is also included in the T-S diagrams, which were absent in Orgel diagrams. All states are usually Labelled on the right side of the diagram; nonetheless, some labels may be allotted on other locations for more clarity in case of complicated T-S diagrams like d^6. The free ion terms for a specific d^n configuration are shown in order of increasing energy on the y-axis of the T-S diagram. The relative of energies free ion terms are obtained using Hund's rules. The splitting of free ion terms has already been discussed in Table 7 and Table 8. Certain T-S diagrams like d^4, d^5, d^6, and d^7; also have a starlight vertical line at a certain Dq/B value, which represents a discontinuity in the slopes of the excited-states' terms. This occurs when the pairing energy (P) becomes equal to the crystal field splitting energy. Metal complexes to the left of this vertical line are spin-free, while the complexes to the right are spin-paired. There is no low-spin or high-spin designation for d^2, d^3, or d^8. The subscript "g" and "u" can be omitted as all designations are gerade.

1. d^1, d^9 systems:

i) d^1-configuration

Metal complexes with d^1-configuration do not have any inter electronic repulsion and the single electron resides in the t_{2g} orbital ground state. When t_{2g} orbital set holds the single electron, six microstates will have $^2T_{2g}$ state energy of -4 Dq; and when the electron is promoted to the e_g orbital, the four microstates will have ^2E_g state energy of $+6$ Dq. This is in accordance with the single absorption band in a UV-vis experiment; and thus, the transition from $^2T_{2g}$ to ^2E_g does not require a Tanabe–Sugano diagram.

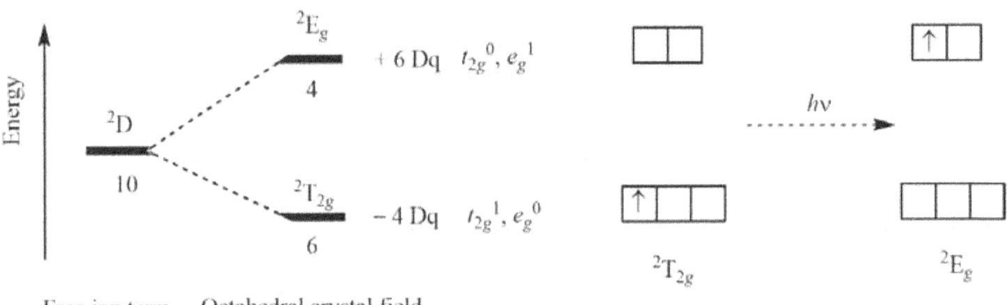

Figure 28. Splitting of free ion term for d^1 complexes in the octahedral crystal field.

ii) d^9-configuration

In d^9 octahedral metal complexes, the ground state filling of electrons ($t_{2g}^6\,e_g^3$) has only four microstates that have ^2E_g energy state with -6 Dq. When the electron from t_{2g} is promoted to the e_g orbital set; the new configuration will have six microstates that have $^2T_{2g}$ energy state with $+4$ Dq. This could also be described as a positive "hole" that moves from the e_g to the t_{2g} orbital set. The sign of Dq is opposite that for d^1, with a ^2E_g ground state and a $^2T_{2g}$ excited state. Like the d^1 case, d^9 octahedral complexes do not require the Tanabe–Sugano diagram to predict their absorption spectra.

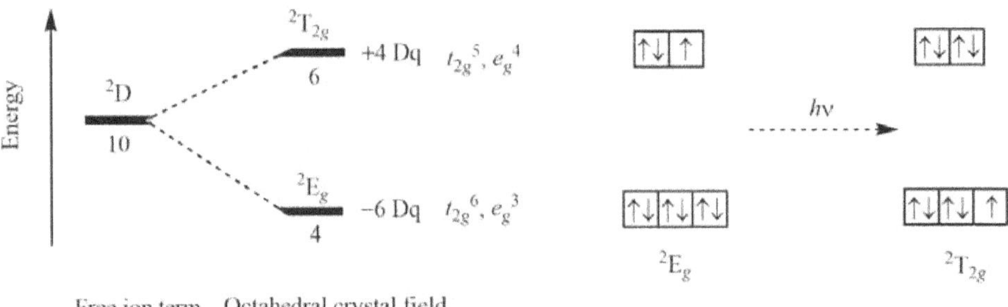

Figure 29. Splitting of free ion term for d^9 complexes in the octahedral crystal field.

2. d^2 systems:

Metal complexes with d^2-configuration have ^3F ground state term symbol in the absence of any crystal field. However, when six ligands approach in octahedral coordination, the ground state term symbol becomes $^3T_{1g}$ and remains as such in the weak field as well as strong ligand field.

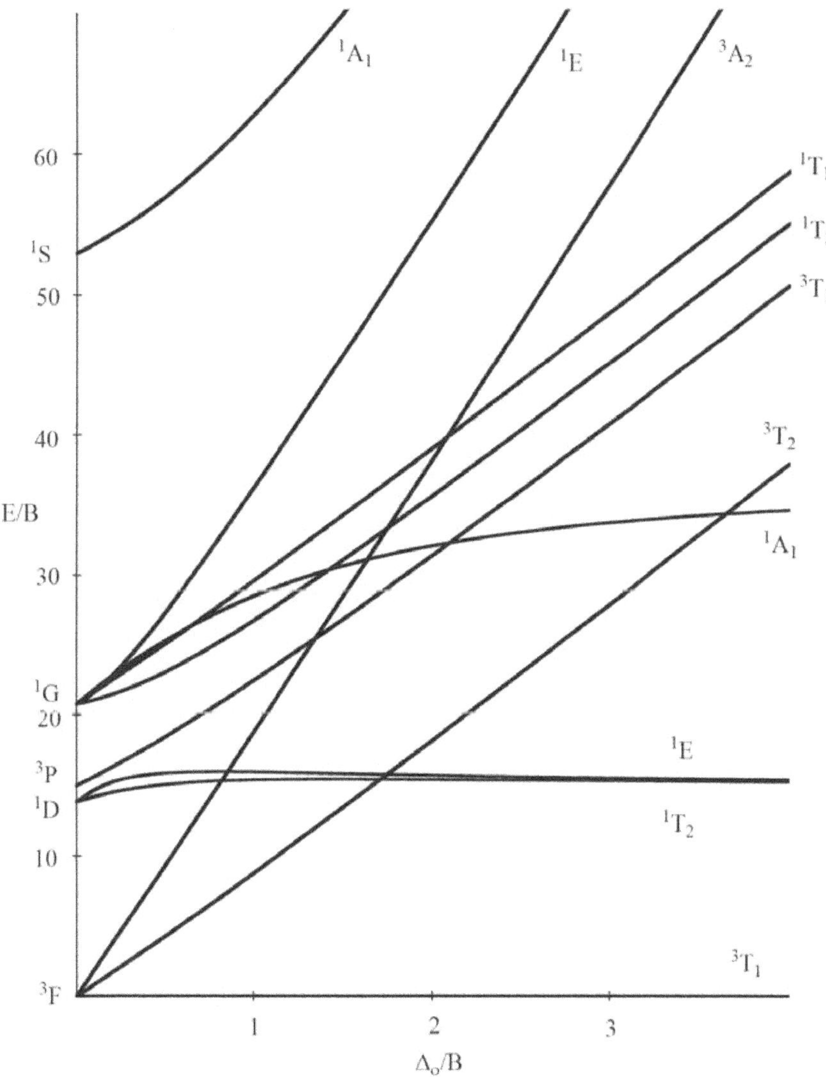

Figure 30. Splitting of free ion terms for d^2 complexes in the octahedral crystal field.

3. d^3 systems:

Metal complexes with d^3-configuration have ^4F ground state term symbol in the absence of any crystal field. However, when six ligands approach in octahedral coordination, the ground state term symbol becomes $^4A_{2g}$ and remains as such in the weak field as well as strong ligand field.

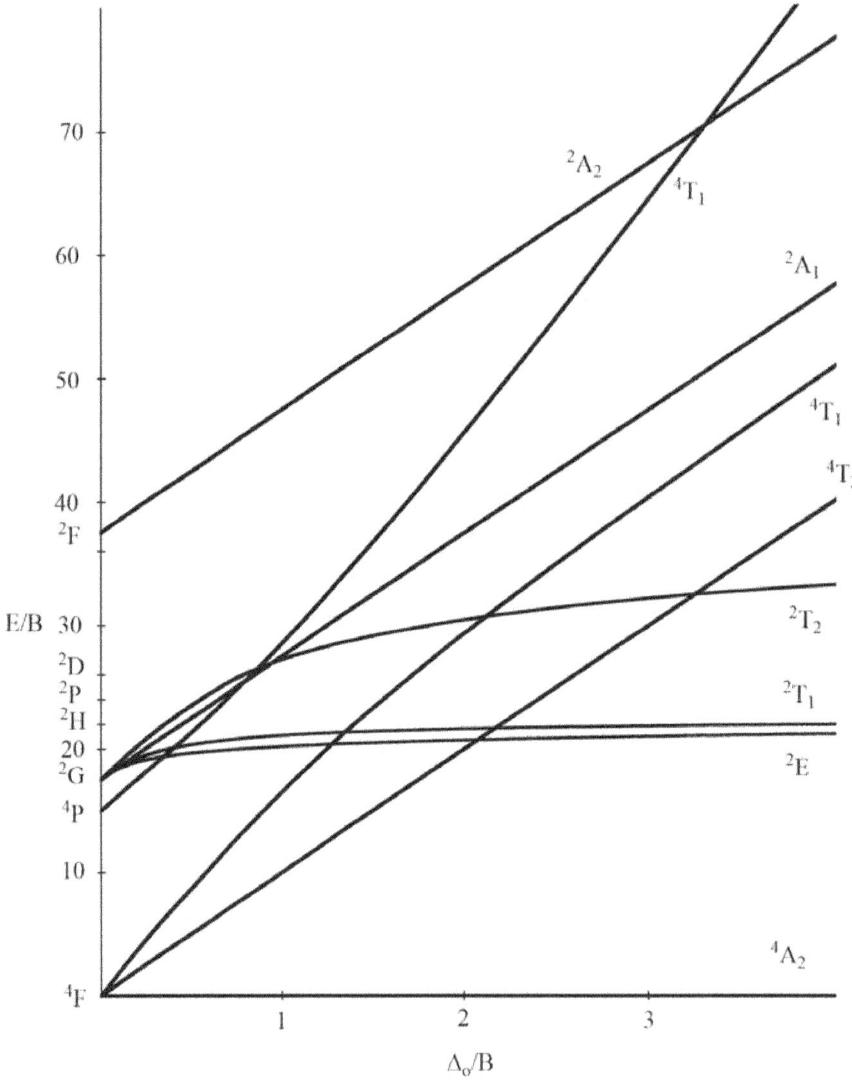

Figure 31. Splitting of free ion terms for d^3 complexes in the octahedral crystal field.

4. d^4 systems:

Metal complexes with d^4-configuration have ^5D ground state term symbol in the absence of any crystal field. However, when six ligands approach in octahedral coordination, the ground state term symbol becomes ^5E_g in the weak field and $^3T_{1g}$ in a strong ligand field.

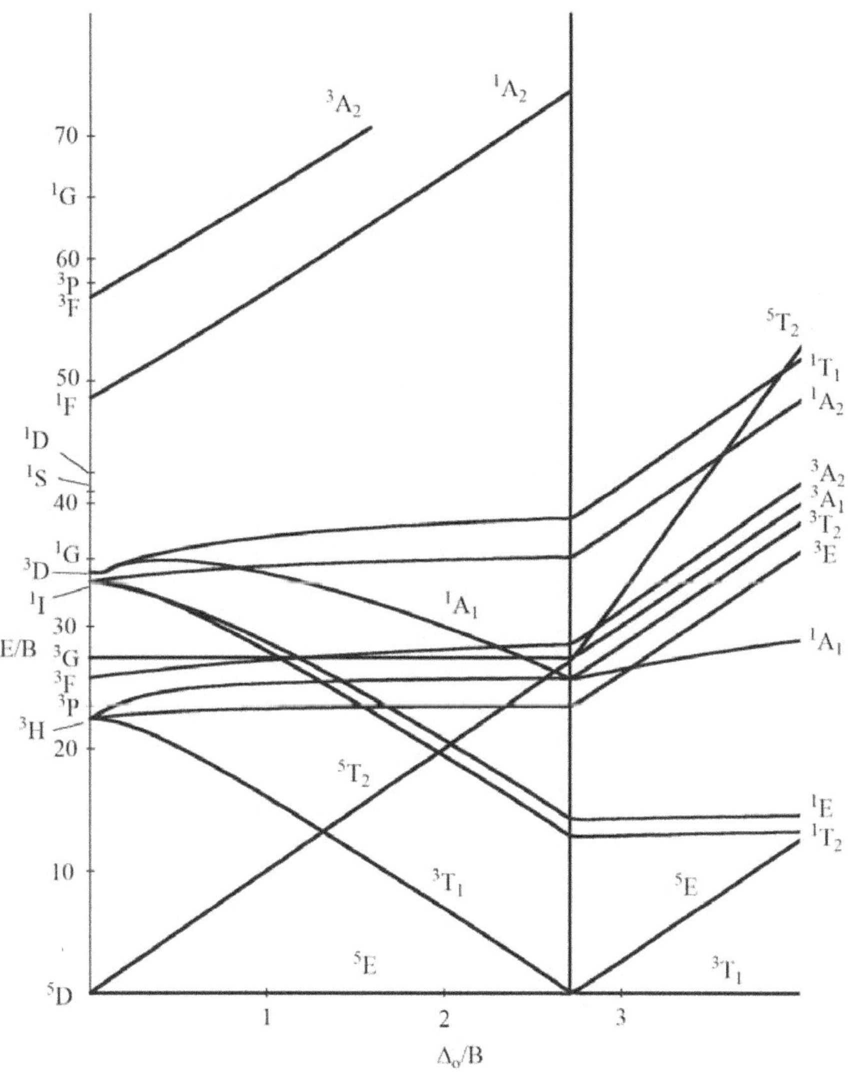

Figure 32. Splitting of free ion terms for d^4 complexes in the octahedral crystal field.

5. d^5 systems:

Metal complexes with d^5-configuration have ^6S ground state term symbol in the absence of any crystal field. However, when six ligands approach in octahedral coordination, the ground state term symbol becomes $^6A_{1g}$ in weak field and $^2T_{2g}$ in strong ligand field.

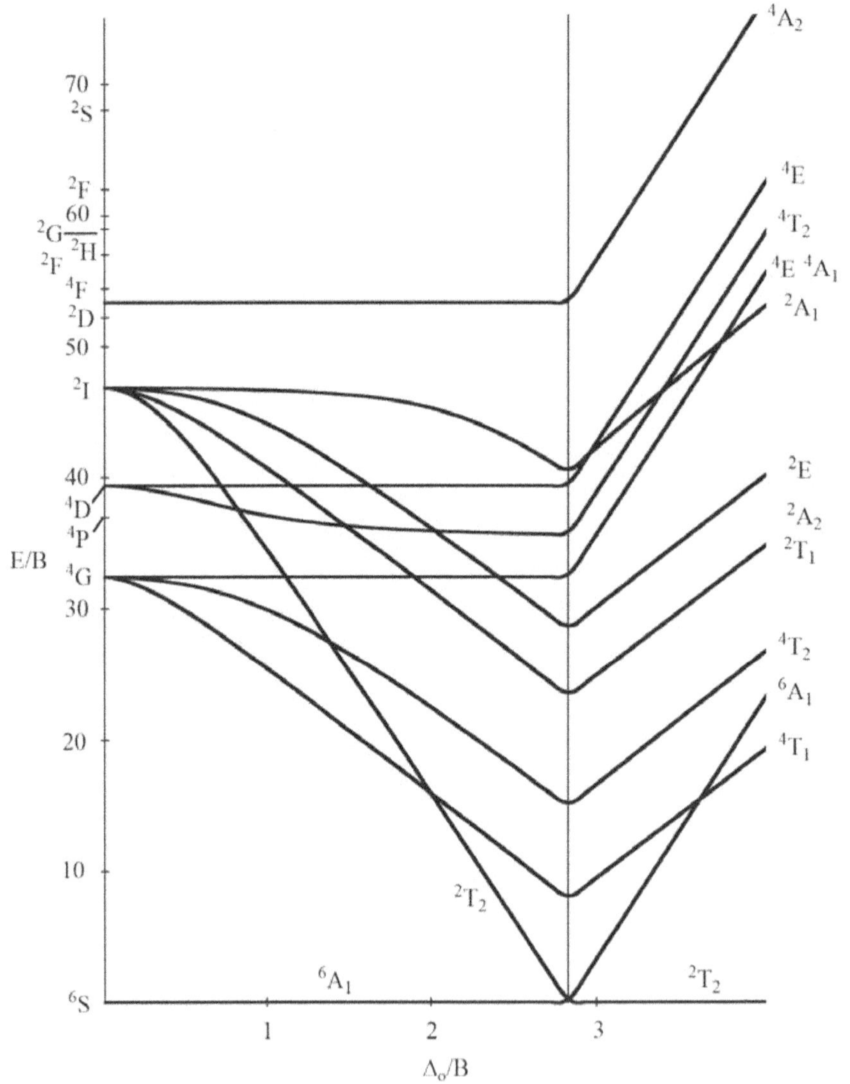

Figure 33. Splitting of free ion terms for d^5 complexes in the octahedral crystal field.

6. d^6 systems:

Metal complexes with d^6-configuration have ^5D ground state term symbol in the absence of any crystal field. However, when six ligands approach in octahedral coordination, the ground state term symbol becomes $^1T_{2g}$ in weak and $^1A_{1g}$ in strong ligand fields.

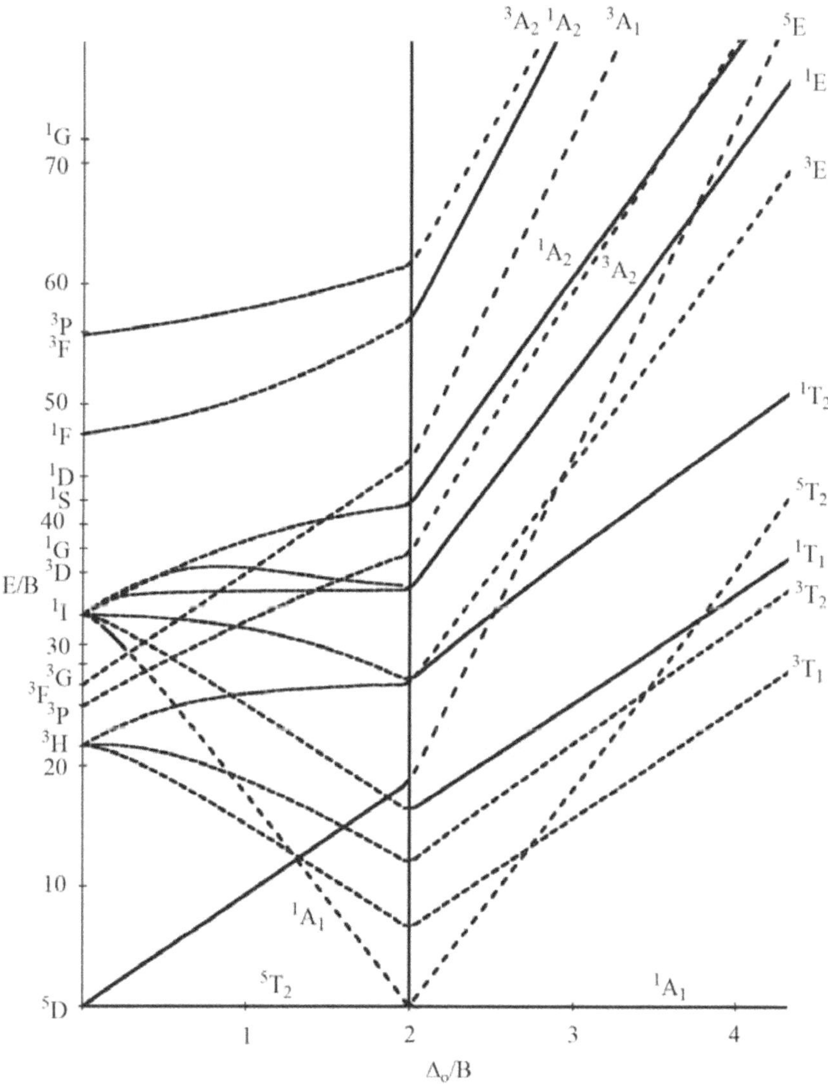

Figure 34. Splitting of free ion terms for d^6 complexes in the octahedral crystal field.

7. d^7 systems:

Metal complexes with d^7-configuration have ^4F ground state term symbol in the absence of any crystal field. However, when six ligands approach in octahedral coordination, the ground state term symbol becomes $^4T_{1g}$ in weak ^2E_g in strong ligand fields.

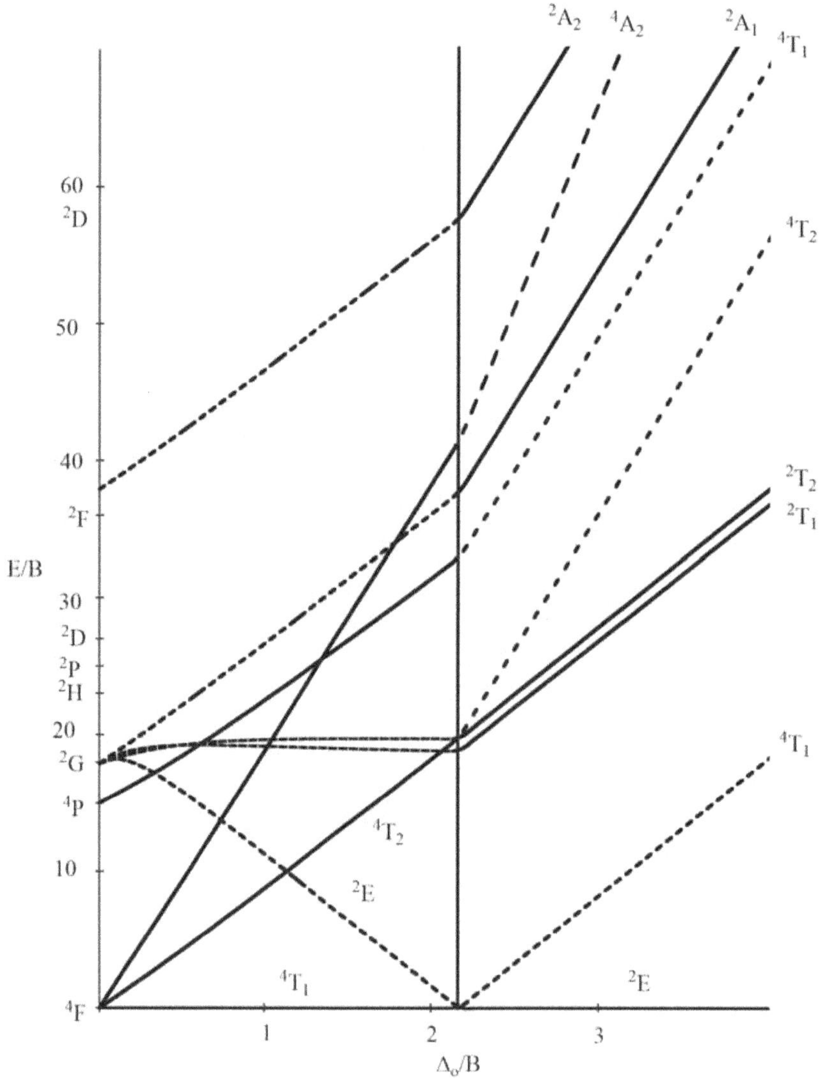

Figure 35. Splitting of free ion terms for d^7 complexes in the octahedral crystal field.

8. d^8 systems:

Metal complexes with d^8-configuration have ^3F ground state term symbol in the absence of any crystal field. However, when six ligands approach in octahedral coordination, the ground state term symbol becomes $^3A_{2g}$ and remains as such in weak as well as strong ligand fields.

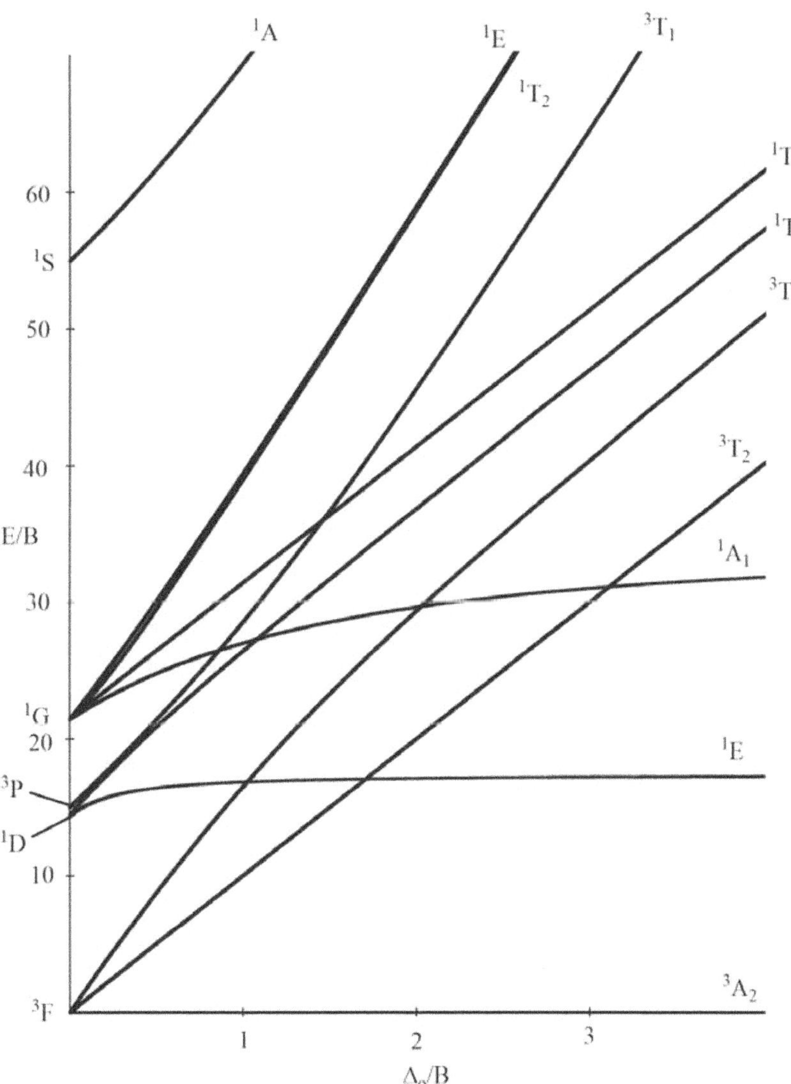

Figure 36. Splitting of free ion terms for d^8 complexes in the octahedral crystal field.

❖ Calculation of Dq, B and β Parameters

The Orgel diagrams explain how the magnitude of the splitting energy exerted by the ligands on *d*-orbitals very when a free metal ion is approached by a ligand field; and can also act as deciding factor for governing the placement of electrons, just like the inter-electronic repulsion energy. However, if the ligand field splitting energy is greater than the inter-electronic repulsion energy, then Orgel diagrams fail in determining the placement of electrons. In that case, Orgel diagrams are restricted only to the high-spin complexes. Tanabe-Sugano diagrams do not have this restriction and can be applied to the situations when Δ is significantly greater than inter-electronic repulsion. Thus, the Tanabe-Sugano diagrams can be utilized in determining electron placements for high-spin and low-spin metal complexes. However, they are limited in the sense that they have only qualitative significance.

Despite that, Tanabe-Sugano and Orgel diagrams are fairly valuable in interpreting UV-vis spectra and can be used to determine the value of crystal field splitting energy (Dq), Racah parameter (B) and also the nephelauxetic ratio (β).

➤ *d¹ Complexes*

Metal complexes with d^1-configuration do not have any inter electronic repulsion and the single electron resides in the t_{2g} orbital ground state. When t_{2g} orbital set holds the single electron, six microstates will have $^2T_{2g}$ state energy of -4 Dq; and when the electron is promoted to the e_g orbital, the four microstates will have ^2E_g state energy of $+6$ Dq. Thus, the only parameter that is needed to be calculated is the magnitude of crystal field splitting energy (10 Dq); and the single absorption band in a UV-vis experiment is exactly what we are looking for. Hence, the energy of the transition $^2T_{2g} \rightarrow {}^2E_g$ gives the value of Δ directly.

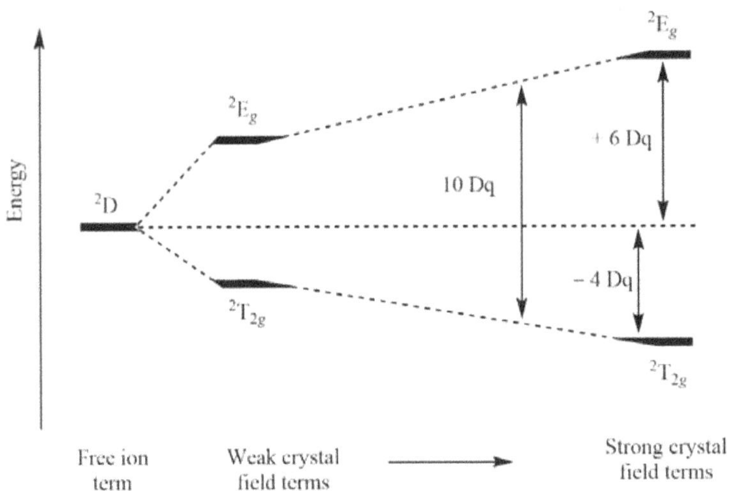

Figure 36. The splitting pattern of free ion term for d^1 complexes in the octahedral crystal field.

Consider the example of $[Ti(H_2O)_6]^{3+}$

1. Calculation of B: No need to calculate the Racah parameter.

2. Calculation of Δ_o: The purple color of the complex ion $[Ti(H_2O)_6]^{3+}$ is due to a broad absorption band at 20300 cm^{-1} arising from $^2T_{2g} \rightarrow {}^2E_g$ transition. Hence, 10 Dq for this complex is 20300 cm^{-1}.

3. Calculation of β: No need to calculate the nephelauxetic ratio.

> ➤ *d^9 Complexes:*

In d^9 octahedral metal complexes, the ground state filling of electrons ($t_{2g}^6 e_g^3$) has only four microstates that have ^2E_g energy state with -6 Dq. When the electron from t_{2g} is promoted to the e_g orbital set; the new configuration will have six microstates that have $^2T_{2g}$ energy state with $+4$ Dq. This could also be described as a positive "hole" that moves from the e_g to the t_{2g} orbital set. The sign of Dq is opposite that for d^1, with a ^2E_g ground state and a $^2T_{2g}$ excited state. Like the d^1 case, the only parameter that is needed to be calculated in d^9 complexes is the magnitude of crystal field splitting energy (10 Dq); and the single absorption band in a UV-vis experiment is exactly what we are looking for. Hence, the energy of the transition $^2E_g \rightarrow {}^2T_{2g}$ gives the value of Δ directly.

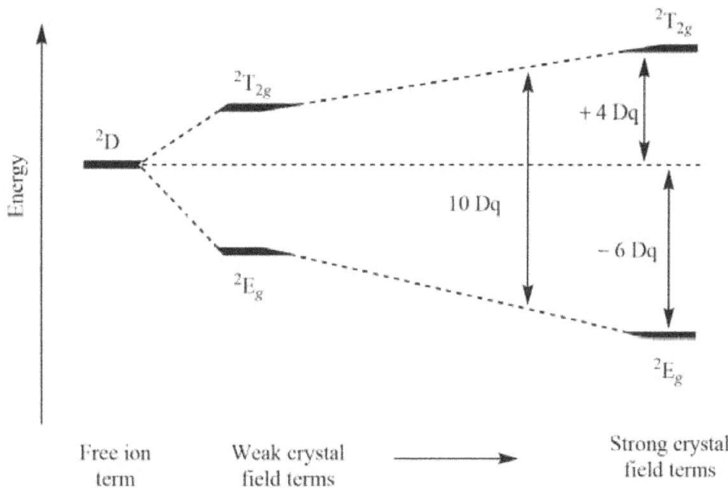

Figure 37. The splitting pattern of free ion term for d^9 complexes in the octahedral crystal field.

Consider the example of $[Cu(H_2O)_6]^{2+}$.

1. Calculation of B: No need to calculate the Racah parameter.

2. Calculation of Δ_o: In the UV-visible spectra of $[Cu(H_2O)_6]^{2+}$, the broad band at 12000 cm^{-1} is due to spin-allowed $^2E_g \rightarrow {}^2T_{2g}$ transition; and hence, 10 Dq for this complex is 12000 cm^{-1}.

3. Calculation of β: No need to calculate the nephelauxetic ratio.

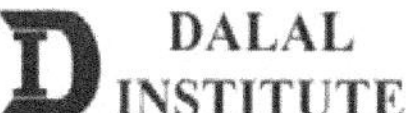

> ➢ *d^2 Complexes*

Metal complexes with d^2-configuration have ^3F ground state term symbol in the absence of any crystal field. However, when six ligands approach in octahedral coordination, the ground state term symbol becomes $^3T_{1g}$ and remains as such in weak as well as in strong ligand fields. The Orgel and Tanabe-Sugano diagram for d^2-configuration can be used to estimate the value of crystal field splitting energy for these transition metal complexes.

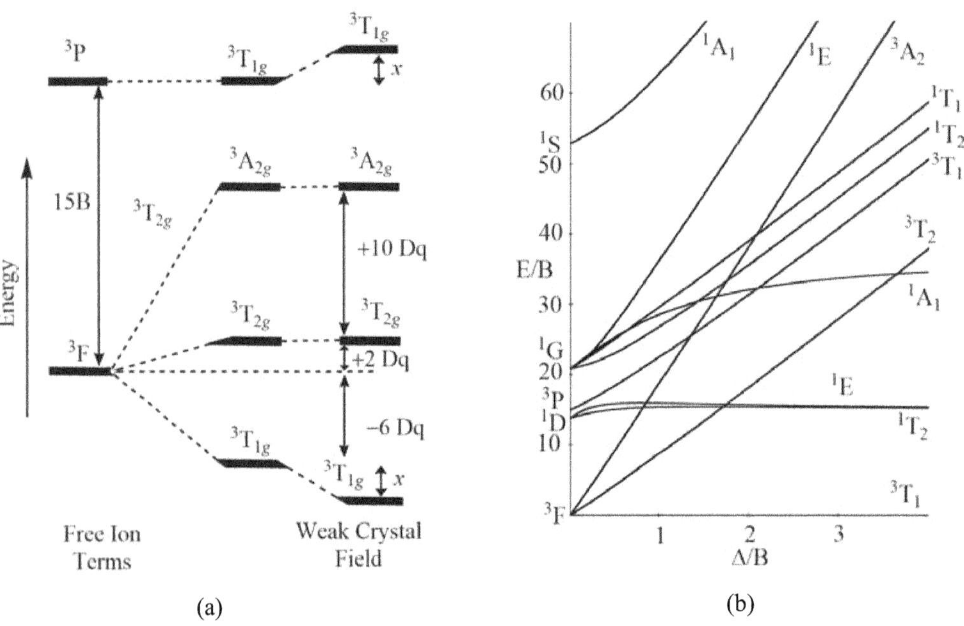

Figure 38. The (a) Orgel and (b) Tanabe-Sugano diagrams for d^2 complexes in the octahedral crystal field.

Consider the example of $[V(H_2O)_6]^{3+}$.

1. Calculation of B: From the Orgel diagram, it can be clearly seen that the ground state for d^2-octahedral complexes is $^3T_{1g}$ and there are three main transitions before the crossover point. Moreover, it is worthy to note down that the order of second and third transitions is reversed after the crossover point and only two bands will be observed at or near the crossover point. As the magnitude of the crystal field splitting energy increases, the $^3T_{1g}(F)$ and $^3T_{1g}(P)$ states repel each other more and more with a magnitude of x energy value.

$$\nu_1 = {}^3T_{1g} \rightarrow {}^3T_{2g}$$

$$\nu_2 = {}^3T_{1g} \rightarrow {}^3A_{2g}$$

$$\nu_3 = {}^3T_{1g} \rightarrow {}^3T_{1g}(P)$$

Which gives

$$v_1 = 8\ Dq + x \tag{1}$$

$$v_2 = 18\ Dq + x \tag{2}$$

$$v_3 = 15\ B + 6\ Dq + 2x \tag{3}$$

Adding equation (1) and (2), we get

$$v_2 + v_1 = 18\ Dq + x + 8\ Dq + x$$

$$v_2 + v_1 = 26\ Dq + 2x \tag{4}$$

Subtracting equation (1) and (2), we get

$$v_2 - v_1 = 18\ Dq + x - 8\ Dq - x$$

$$v_2 - v_1 = 10\ Dq \tag{5}$$

Putting the value of $2x$ from equation (4) in equation (3), we get

$$v_3 = 15\ B + 6\ Dq + v_2 + v_1 - 26\ Dq$$

$$v_3 = 15\ B + v_2 + v_1 - 20\ Dq \tag{6}$$

Multiplying equation (5) by 2 and putting the value of 20 Dq from equation (5) in equation (6), we get

$$v_3 = 15\ B + v_2 + v_1 - 20\ Dq$$

$$v_3 = 15\ B + v_2 + v_1 - 2v_2 + 2v_1$$

$$15\ B = v_3 + v_2 - 3v_1$$

$$B = \frac{v_3 + v_2 - 3v_1}{15} \tag{7}$$

However, only two transitions are observed, this method is difficult to apply in a precise manner and only gives approximations.

From the Tanabe-Sugano diagram, in the UV-visible spectra of $[V(H_2O)_6]^{3+}$, two bands are observed with maxima at around 17500 and 26000 cm^{-1}. There are three possible transitions expected, which include: $v_1 = {}^3T_{1g} \rightarrow {}^3T_{2g}$, $v_2 = {}^3T_{1g} \rightarrow {}^3T_{1g}(P)$, and $v_3 = {}^3T_{1g} \rightarrow {}^3A_{2g}$; but only two are observed. The ratio of experimental band energies is:

$$\frac{v_2}{v_1} = \frac{E_2}{E_1} = \frac{E_2/B}{E_1/B} = \frac{26000}{17500} = 1.49$$

Now slide a ruler across the printed diagram (perpendicular to the abscissa) until the ratio of E_2/B to E_1/B between lines becomes equivalent to 1.49. In this particular example, this ratio becomes 1.49 when Δ_o/B = 31. Stop the ruler movement and find out the values of E_2/B and E_1/B

$$\frac{E_2}{B} = 43; \quad \frac{E_1}{B} = 27$$

Thus, on the T-S diagram, where Δ_o/B = 31; the value of $^3T_{1g} \rightarrow {}^3T_{2g}$ and $^3T_{1g} \rightarrow {}^3T_{1g}(P)$ i.e. E_1/B and E_2/B, are 27 and 43, respectively. The Racah parameter can be found by calculating B from both v_2 and v_1.

$$\frac{26000}{B} = 43; \quad \frac{17500}{B} = 27$$

$$B = \frac{26000}{43} = 604 \text{ cm}^{-1}; \quad B = \frac{17500}{27} = 648 \text{ cm}^{-1}$$

$$\textit{Average value of Racah parameter (B)} = \frac{604 + 648}{2} = 626 \text{ cm}^{-1}$$

2. Calculation of Δ_o: Being a weak-complex, the theoretical value of lowest-energy absorption band given by the Orgel diagram is 8 Dq ($^3T_{1g} \rightarrow {}^3T_{2g}$); and the experimental value for lowest-energy absorption band is 17500 cm^{-1}. Hence, the value of 10 Dq or Δ_o can be calculated as:

$$8 \text{ Dq} = 0.8\, \Delta_o = 17500 \text{ cm}^{-1}$$

$$\Delta_o = \frac{17500 \text{ cm}^{-1}}{0.8}$$

$$\Delta_o = 10 \text{ Dq} = 21875 \text{ cm}^{-1}$$

However, this is just the approximation and a more precise and refined calculation should be carried out using the Tanabe-Sugano diagram. From the average value of the Racah parameter, the ligand field splitting parameter can be found as follows.

$$\frac{\Delta_o}{B} = 31; \quad \frac{\Delta_o}{626 \text{ cm}^{-1}} = 31; \quad \Delta_o = 19406 \text{ cm}^{-1}$$

3. Calculation of β: In order to calculate the nephelauxetic ratio, we must have the value of Racah parameter for a free metal ion in its gaseous state. For free d^2 ion like V^{3+}, it has been observed that ^3P state lies 12925 cm^{-1} above to the ^3F state. Hence, 15B = 12925 cm^{-1} or B = 862 cm^{-1}. Now, the value of nephelauxetic ratio can be calculated as

$$\text{Nephelauxetic ratio} = \beta = \frac{B_{\text{complex}}}{B_{\text{free ion}}} = \frac{626 \text{ cm}^{-1}}{862 \text{ cm}^{-1}} = 0.726$$

Hence, inter-electronic repulsion has been decreased during the process of complexation.

> ### d^8 *Complexes*

Metal complexes with d^8-configuration have ^3F ground state term symbol in the absence of any crystal field. However, when six ligands approach in octahedral coordination, the ground state term symbol becomes $^3A_{2g}$ and remains as such in weak as well as in strong ligand fields. The Orgel and Tanabe-Sugano diagram for d^8-configuration can be used to estimate the value of crystal field splitting energy for these transition metal complexes.

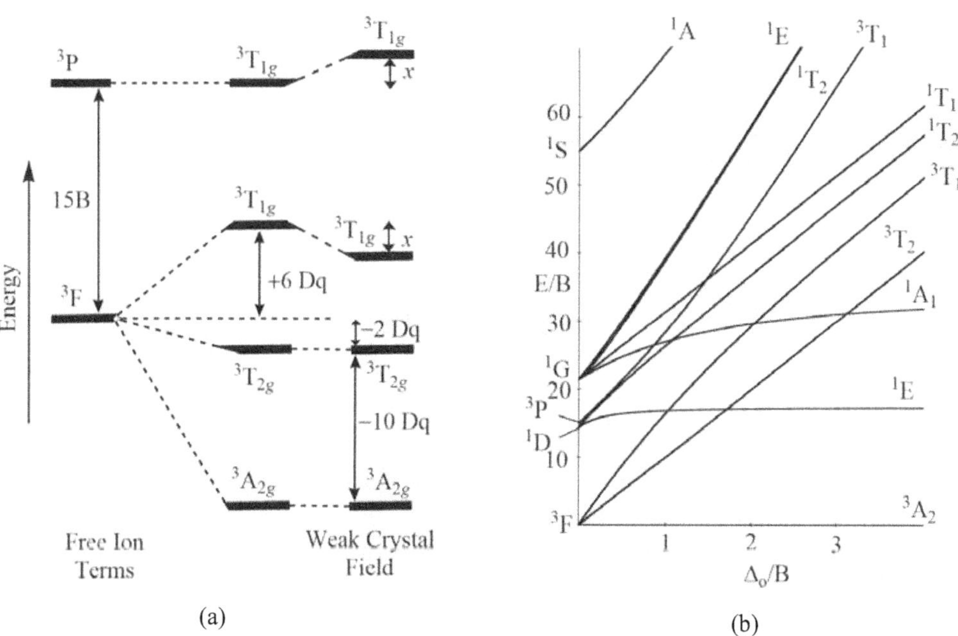

Figure 39. The (a) Orgel and (b) Tanabe-Sugano diagrams for d^8 complexes in the octahedral crystal field.

Consider the example of $[Ni(H_2O)_6]^{2+}$.

1. Calculation of B: From the Orgel diagram, it can be clearly seen that the ground state for d^8-octahedral complexes is $^3A_{2g}$ and there are three main transitions. As the magnitude of the crystal field splitting energy increases, the $^3T_{1g}(F)$ and $^3T_{1g}(P)$ states repel each other more and more with a magnitude of x energy value owing to the non-crossing rule of the same symmetry states.

$$\nu_1 = {}^3A_{2g} \rightarrow {}^3T_{2g}$$

$$\nu_2 = {}^3A_{2g} \rightarrow {}^3T_{1g}$$

$$\nu_3 = {}^3A_{2g} \rightarrow {}^3T_{1g}(P)$$

Which gives

$$\nu_1 = 10\ Dq \tag{1}$$

$$\nu_2 = 18\ Dq - x \tag{2}$$

$$\nu_3 = 15\ B + 12\ Dq + x \tag{3}$$

Putting value of x from equation (2) in (3), we get

$$\nu_3 = 15\ B + 12\ Dq + 18\ Dq - \nu_2$$

$$\nu_3 = 15\ B + 30\ Dq - \nu_2 \tag{4}$$

Multiplying equation (1) by 3 and putting the value of 30 Dq from equation (1) in (4), we get

$$\nu_3 = 15\ B + 3\nu_1 - \nu_2$$

$$15\ B = \nu_3 + \nu_2 - 3\nu_1$$

$$B = \frac{\nu_3 + \nu_2 - 3\nu_1}{15} \tag{5}$$

However, this method is applicable only when three transitions are observed. Moreover, this method is difficult to apply in a precise manner and only gives approximations.

From the Tanabe-Sugano diagram, in the UV-visible spectra of $[Ni(H_2O)_6]^{2+}$, three bands are observed with maxima at around 8500, 14500 and 25300 cm^{-1}. There are three possible transitions expected, which include: $\nu_1 = {}^3A_{2g} \rightarrow {}^3T_{2g}$, $\nu_2 = {}^3A_{2g} \rightarrow {}^3T_{1g}$, and $\nu_3 = {}^3A_{2g} \rightarrow {}^3T_{1g}(P)$. The ratio of experimental band energies of ν_3 to ν_2 is:

$$\frac{\nu_3}{\nu_2} = \frac{E_3}{E_2} = \frac{E_3/B}{E_2/B} = \frac{25300}{14500} = 1.74$$

Now slide a ruler across the printed diagram (perpendicular to the abscissa) until the ratio of E_2/B to E_1/B between lines becomes equivalent to 1.74. In this particular example, this ratio becomes 1.74 when Δ_o/B = 10. Stop the ruler movement and find out the values of E_3/B and E_2/B as:

$$\frac{E_3}{B} = 28; \quad \frac{E_2}{B} = 16$$

Thus, on the T-S diagram, where Δ_o/B = 10; the value of ${}^3A_{2g} \rightarrow {}^3T_{1g}$ and ${}^3A_{2g} \rightarrow {}^3T_{1g}(P)$ i.e. E_2/B and E_3/B, are 28 and 16, respectively. The Racah parameter can be found by calculating B from second and third i.e. from ν_3 and ν_2 transitions.

From ν_3, we get

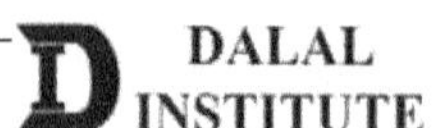

$$\frac{25300}{B} = 28$$

$$B = \frac{25300}{28} = 904 \text{ cm}^{-1}$$

Similarly

$$\frac{14500}{B} = 16$$

$$B = \frac{14500}{16} = 906 \text{ cm}^{-1}$$

Therefore,

$$\text{Average value of Racah parameter (B)} = \frac{904 + 906}{2} = 905 \text{ cm}^{-1}$$

2. Calculation of Δ_o: Being a weak-complex, the theoretical value of lowest-energy absorption band given by the Orgel diagram is 10 Dq ($^3A_{2g} \rightarrow {}^3T_{2g}$); and the experimental value for lowest-energy absorption band is 8500 cm^{-1}. Hence, the value of 10 Dq or Δ_o can be calculated as

$$10 \text{ Dq} = 8500 \text{ cm}^{-1}$$

$$\Delta_o = 8500 \text{ cm}^{-1}$$

However, this is just the approximation and a more precise and refined calculation should be carried out using the Tanabe-Sugano diagram. From the average value of the Racah parameter, the ligand field splitting parameter can be found as follows.

$$\frac{\Delta_o}{B} = 10$$

$$\frac{\Delta_o}{905 \text{ cm}^{-1}} = 10$$

$$\Delta_o = 9050 \text{ cm}^{-1}$$

3. Calculation of β: In order to calculate the nephelauxetic ratio, we must have the value of the Racah parameter for a free metal ion in its gaseous state. For free d^8 ion like Ni^{2+}, it has been observed that ^3P state lies 16200 cm^{-1} above to the ^3F state. Hence, 15B = 16200 cm^{-1} or B = 1080 cm^{-1}. Now, the value of nephelauxetic ratio can be calculated as

$$\text{Nephelauxetic ratio} = \beta = \frac{B_{complex}}{B_{free\ ion}} = \frac{905 \text{ cm}^{-1}}{1080 \text{ cm}^{-1}} = 0.838$$

Hence, inter-electronic repulsion has been decreased during the process of complexation.

➢ *d^3 Complexes*

Metal complexes with d^3-configuration have ^4F ground state term symbol in the absence of any crystal field. However, when six ligands approach in octahedral coordination, the ground state term symbol becomes $^4A_{2g}$ and remains as such in weak as well as in strong ligand fields. The Orgel and Tanabe-Sugano diagram for d^3-configuration can be used to estimate the value of crystal field splitting energy for these transition metal complexes.

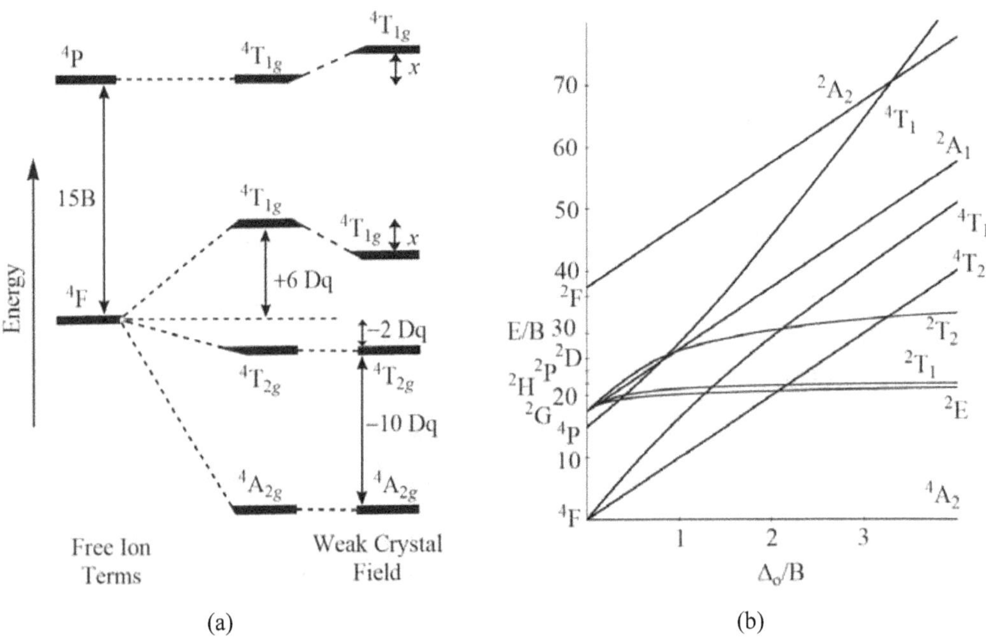

Figure 40. The (a) Orgel and (b) Tanabe-Sugano diagrams for d^3 complexes in the octahedral crystal field.

Consider the example of $[Cr(H_2O)_6]^{3+}$.

1. Calculation of B: From the Orgel diagram, it can be clearly seen that the ground state for d^3-octahedral complexes is $^4A_{2g}$ and there are three main transitions. As the magnitude of the crystal field splitting energy increases, the $^4T_{1g}(F)$ and $^4T_{1g}(P)$ states repel each other more and more with a magnitude of x energy value owing to the non-crossing rule of the same symmetry states.

$$\nu_1 = {}^4A_{2g} \rightarrow {}^4T_{2g}$$

$$\nu_2 = {}^4A_{2g} \rightarrow {}^4T_{1g}$$

$$\nu_3 = {}^4A_{2g} \rightarrow {}^4T_{1g}(P)$$

Which gives

$$\nu_1 = 10 \, Dq \tag{1}$$

$$\nu_2 = 18 \, Dq - x \tag{2}$$

$$\nu_3 = 15 \, B + 12 \, Dq + x \tag{3}$$

Putting the value of x from equation (2) in (3), we get

$$\nu_3 = 15 \, B + 12 \, Dq + 18 \, Dq - \nu_2$$

$$\nu_3 = 15 \, B + 30 \, Dq - \nu_2 \tag{4}$$

Multiplying equation (1) by 3 and putting the value of 30 Dq from equation (1) in (4), we get

$$\nu_3 = 15 \, B + 3\nu_1 - \nu_2$$

$$15 \, B = \nu_3 + \nu_2 - 3\nu_1$$

$$B = \frac{\nu_3 + \nu_2 - 3\nu_1}{15} \tag{5}$$

However, this method is applicable only when three transitions are observed. Moreover, this method is difficult to apply in a precise manner and only gives approximations.

From the Tanabe-Sugano diagram, in the UV-visible spectra of $[Cr(H_2O)_6]^{3+}$, three bands are observed with maxima at around 17000, 24000 and 37000 cm^{-1}. There are three possible transitions expected, which include: $\nu_1 = {}^4A_{2g} \longrightarrow {}^4T_{2g}$, $\nu_2 = {}^4A_{2g} \longrightarrow {}^4T_{1g}$, and $\nu_3 = {}^4A_{2g} \longrightarrow {}^4T_{1g}(P)$. The ratio of experimental band energies of ν_2 to ν_1 is:

$$\frac{\nu_2}{\nu_1} = \frac{E_2}{E_1} = \frac{E_2/B}{E_1/B} = \frac{24000}{17000} = 1.41$$

Now slide a ruler across the printed diagram (perpendicular to the abscissa) until the ratio of E_2/B to E_1/B between lines becomes equivalent to 1.41. In this particular example, this ratio becomes 1.41 when Δ_o/B = 24. Stop the ruler movement and find out the values of E_2/B and E_1/B as:

$$\frac{E_2}{B} = 33.90; \quad \frac{E_1}{B} = 24$$

Thus, on the T-S diagram, where Δ_o/B = 24; the value of ${}^4A_{2g} \rightarrow {}^4T_{1g}$ and ${}^4A_{2g} \rightarrow {}^4T_{1g}(P)$ i.e. E_2/B and E_3/B, are 33.90 and 24, respectively. The Racah parameter can be found by calculating B from first and second i.e. from ν_2 and ν_1 transitions.

From ν_2, we get

$$\frac{24000}{B} = 33.90$$

$$B = \frac{24000}{33.90} = 708 \text{ cm}^{-1}$$

Similarly

$$\frac{17000}{B} = 24$$

$$B = \frac{17000}{24} = 708 \text{ cm}^{-1}$$

Therefore,

$$\text{Average value of Racah parameter (B)} = \frac{708 + 708}{2} = 708 \text{ cm}^{-1}$$

2. Calculation of Δ_o: Being a weak-complex, the theoretical value of lowest-energy absorption band given by the Orgel diagram is 10 Dq ($^4A_{2g} \rightarrow {}^4T_{2g}$); and the experimental value for lowest-energy absorption band is 17000 cm^{-1}. Hence, the value of 10 Dq or Δ_o can be calculated as

$$10 \text{ Dq} = 17000 \text{ cm}^{-1}$$

$$\Delta_o = 17000 \text{ cm}^{-1}$$

However, this is just the approximation and a more precise and refined calculation should be carried out using the Tanabe-Sugano diagram. From the average value of the Racah parameter, the ligand field splitting parameter can be found as follows.

$$\frac{\Delta_o}{B} = 24$$

$$\frac{\Delta_o}{708 \text{ cm}^{-1}} = 24$$

$$\Delta_o = 16992 \text{ cm}^{-1}$$

3. Calculation of β: In order to calculate the nephelauxetic ratio, we must have the value of the Racah parameter for a free metal ion in its gaseous state. For free d^3 ion like Cr^{3+}, it has been observed that ^3P state lies 15450 cm^{-1} above to the ^3F state. Hence, 15B = 15450 cm^{-1} or B = 1030 cm^{-1}. Now, the value of nephelauxetic ratio can be calculated as:

$$\text{Nephelauxetic ratio} = \beta = \frac{B_{complex}}{B_{free\ ion}} = \frac{708 \text{ cm}^{-1}}{1030 \text{ cm}^{-1}} = 0.687$$

Hence, inter-electronic repulsion has been decreased during the process of complexation.

> ➢ *d^7 Complexes*

Metal complexes with d^7-configuration have ^4F ground state term symbol in the absence of any crystal field. However, when six ligands approach in octahedral coordination, the ground state term symbol becomes $^4T_{1g}$ in the weak field; and becomes ^2E_g when the ligand field becomes sufficiently strong. The Orgel and Tanabe-Sugano diagram for d^7-configuration can be used to estimate the value of crystal field splitting energy for these complexes.

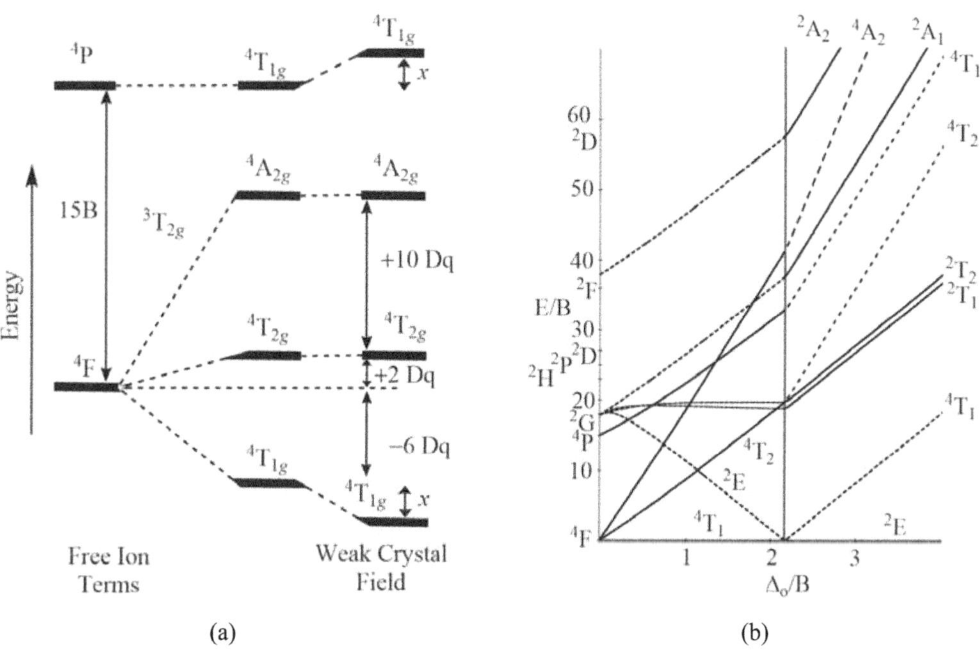

Figure 41. The (a) Orgel and (b) Tanabe-Sugano diagrams for d^7 complexes in the octahedral crystal field.

Consider the example of $[Co(H_2O)_6]^{2+}$

1. Calculation of B: From the Orgel diagram, it can be clearly seen that the ground state for d^7-octahedral complexes is $^4T_{1g}$ and there are three main transitions before the crossover point. Moreover, it is worthy to note down that the order of second and third transitions is reversed after the crossover point and only two bands will be observed at or near the crossover point. As the magnitude of the crystal field splitting energy increases, the $^4T_{1g}(F)$ and $^4T_{1g}(P)$ states repel each other more and more with a magnitude of x energy value.

$$\nu_1 = {}^4T_{1g} \rightarrow {}^4T_{2g}$$

$$\nu_2 = {}^4T_{1g} \rightarrow {}^4A_{2g}$$

$$v_3 = {}^4T_{1g} \rightarrow {}^4T_{1g}(P)$$

Which gives

$$v_1 = 8\ Dq + x \tag{1}$$

$$v_2 = 18\ Dq + x \tag{2}$$

$$v_3 = 15\ B + 6\ Dq + 2x \tag{3}$$

Adding equation (1) and (2), we get

$$v_2 + v_1 = 18\ Dq + x + 8\ Dq + x$$

$$v_2 + v_1 = 26\ Dq + 2x \tag{4}$$

Subtracting equation (1) and (2), we get

$$v_2 - v_1 = 18\ Dq + x - 8\ Dq - x$$

$$v_2 - v_1 = 10\ Dq \tag{5}$$

Putting the value of $2x$ from equation (4) in equation (3), we get

$$v_3 = 15\ B + 6\ Dq + v_2 + v_1 - 26\ Dq$$

$$v_3 = 15\ B + v_2 + v_1 - 20\ Dq \tag{6}$$

Multiplying equation (5) by 2 and putting the value of 20 Dq from equation (5) in equation (6), we get

$$v_3 = 15\ B + v_2 + v_1 - 20\ Dq$$

$$v_3 = 15\ B + v_2 + v_1 - 2v_2 + 2v_1$$

$$15\ B = v_3 + v_2 - 3v_1$$

$$B = \frac{v_3 + v_2 - 3v_1}{15} \tag{7}$$

However, only two transitions are observed, this method is difficult to apply in a precise manner and only gives approximations.

From the Tanabe-Sugano diagram, in the UV-visible spectra of $[Co(H_2O)_6]^{2+}$, two bands are observed with maxima at around 8000, 19600 and 21600 cm^{-1}. There are three possible transitions expected, which include: $v_1 = {}^4T_{1g} \rightarrow {}^4T_{2g}$, $v_2 = {}^4T_{1g} \rightarrow {}^4A_{2g}$ and $v_3 = {}^4T_{1g} \rightarrow {}^4T_{1g}(P)$. The ratio of experimental band energies is:

$$\frac{v_3}{v_1} = \frac{E_3}{E_1} = \frac{E_3/B}{E_1/B} = \frac{21600}{8000} = 2.70$$

Now slide a ruler across the printed diagram (perpendicular to the abscissa) until the ratio of E_3/B to E_1/B between lines becomes equivalent to 2.70. In this particular example, this ratio becomes 2.70 when Δ_o/B = 9.5. Stop the ruler movement and find out the values of E_2/B and E_1/B as:

$$\frac{E_3}{B} = 22; \quad \frac{E_1}{B} = 8.2$$

Thus, on the T-S diagram, where $\Delta_o/B = 31$; the value of $^3T_{1g} \rightarrow {}^3T_{2g}$ and $^3T_{1g} \rightarrow {}^3T_{1g}(P)$ i.e. E_1/B and E_3/B, are 8.2 and 22, respectively. The Racah parameter can be found by calculating B from both v_2 and v_1.

$$\frac{21600}{B} = 22; \quad \frac{8000}{B} = 8.2$$

$$B = \frac{21600}{22} = 982 \text{ cm}^{-1}; \quad B = \frac{8000}{8.2} = 976 \text{ cm}^{-1}$$

$$\textit{Average value of Racah parameter (B)} = \frac{982 + 976}{2} = 979 \text{ cm}^{-1}$$

2. Calculation of Δ_o: Being a weak-complex, the theoretical value of lowest-energy absorption band given by the Orgel diagram is 8 Dq $(^3T_{1g} \rightarrow {}^3T_{2g})$; and the experimental value for lowest-energy absorption band is 8000 cm^{-1}. Hence, the value of 10 Dq or Δ_o can be calculated as

$$8 \text{ Dq} = 0.8 \, \Delta_o = 8000 \text{ cm}^{-1}$$

$$\Delta_o = \frac{8000 \text{ cm}^{-1}}{0.8}$$

$$\Delta_o = 10 \text{ Dq} = 10000 \text{ cm}^{-1}$$

However, this is just the approximation and a more precise and refined calculation should be carried out using the Tanabe-Sugano diagram. From the average value of the Racah parameter, the ligand field splitting parameter can be found as follows.

$$\frac{\Delta_o}{B} = 9.5; \quad \frac{\Delta_o}{979 \text{ cm}^{-1}} = 9.5; \quad \Delta_o = 9300 \text{ cm}^{-1}$$

3. Calculation of β: In order to calculate the nephelauxetic ratio, we must have the value of the Racah parameter for a free metal ion in its gaseous state. For free d^7 ion like Co^{2+}, it has been observed that ^3P state lies 16755 cm^{-1} above to the ^3F state. Hence, $15B = 16755$ cm^{-1} or $B = 1117$ cm^{-1}. Now, the value of nephelauxetic ratio can be calculated as:

$$\textit{Nephelauxetic ratio} = \beta = \frac{B_{complex}}{B_{free \, ion}} = \frac{979 \text{ cm}^{-1}}{1117 \text{ cm}^{-1}} = 0.876$$

Hence, inter-electronic repulsion has been decreased during the process of complexation.

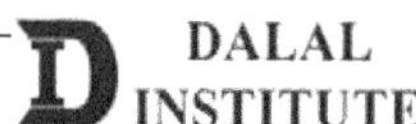

> ➤ *d^4 Complexes*

Metal complexes with d^4-configuration have ^5D ground state term symbol in the absence of any crystal field. However, when six ligands approach in octahedral coordination, the ground state term symbol becomes ^5E_g in the weak field; and becomes $^3T_{1g}$ when the ligand field becomes sufficiently strong. The Orgel and Tanabe-Sugano diagram for d^4-configuration can be used to estimate the value of crystal field splitting energy for these complexes.

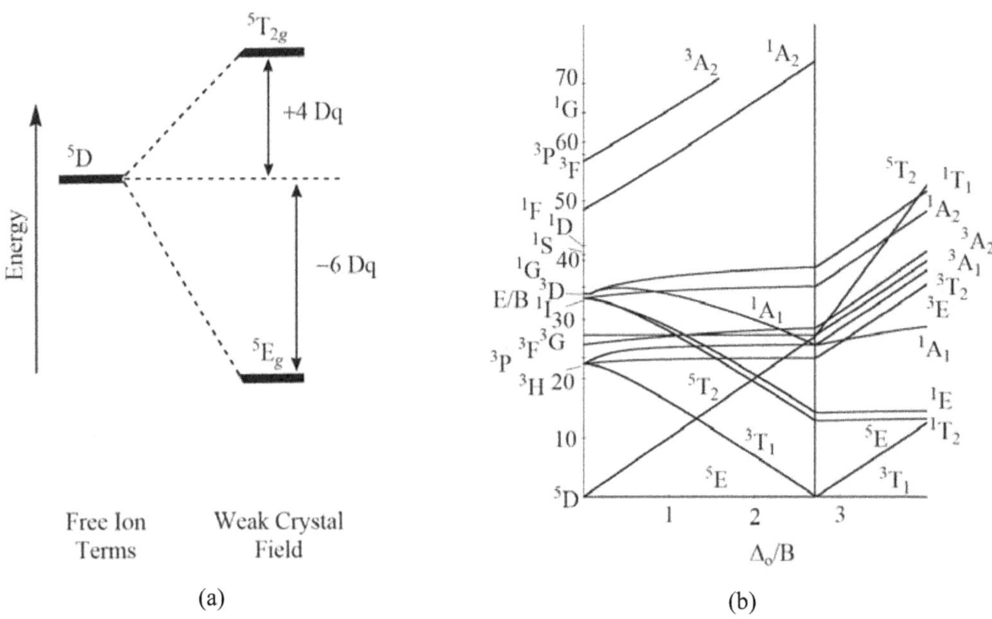

Figure 42. The (a) Orgel and (b) Tanabe-Sugano diagrams for d^4 complexes in the octahedral crystal field.

Consider the example of $[Mn(CN)_6]^{3-}$

1. Calculation of B: From the Tanabe-Sugano diagram, it can clearly be seen that the spin-allowed d-d transitions in low-spin d^4 metal complexes are $^3T_{1g} \rightarrow {}^3E_g$, $^3T_{1g} \rightarrow {}^3T_{2g}$, $^3T_{1g} \rightarrow {}^3A_{1g}$ and $^3T_{1g} \rightarrow {}^3A_{2g}$. In the UV-visible absorption spectra of $[Mn(CN)_6]^{3-}$, three bands are observed; one strong band with maxima at around 27000 and other two bands at 29000 and 34000 cm^{-1}. Moreover, the bands at 21800 and 43500 cm^{-1} can be ignored as they correspond to charge transfer transitions. Thus, the ratio of experimental energies is

$$\frac{v_2}{v_1} = \frac{E_2}{E_1} = \frac{E_2/B}{E_1/B} = \frac{29000 \text{ cm}^{-1}}{27000 \text{ cm}^{-1}} = 1.07$$

Now slide a ruler across the printed diagram (perpendicular to the abscissa) until the ratio of E_2/B to E_1/B i.e. the ratio between the lines corresponding to the first two spin-allowed transitions becomes equivalent

to 1.07. In this particular example, this ratio becomes 1.07 when $\Delta_o/B = 40$. Stop the ruler movement and find out the values of E_2/B and E_1/B.

$$\frac{E_2}{B} = 38; \quad \frac{E_1}{B} = 35$$

Thus, on the T-S diagram, where $\Delta_o/B = 40$; the value of $^3T_{1g} \rightarrow \,^3T_{2g}$ and $^3T_{1g} \rightarrow \,^3E_g$ i.e. E_2/B and E_1/B, are 38 and 35, respectively. The Racah parameter can be found by calculating B from both v_2 and v_1.

$$\frac{29000 \text{ cm}^{-1}}{B} = 38; \quad \frac{27000 \text{ cm}^{-1}}{B} = 35$$

$$B = \frac{29000 \text{ cm}^{-1}}{38} = 763 \text{ cm}^{-1}; \quad B = \frac{27000 \text{ cm}^{-1}}{35} = 771 \text{ cm}^{-1}$$

$$\textit{Average value of Racah parameter (B)} = \frac{763 \text{ cm}^{-1} + 771 \text{ cm}^{-1}}{2} = 767 \text{ cm}^{-1}$$

2. Calculation of Δ_o: The only parameter that is needed to be sought for the calculation of the magnitude of crystal field splitting energy (10 Dq) in weak field complexes is the single absorption band in a UV-vis experiment. Hence, the energy of the transition $^5E_g \rightarrow \,^5T_{2g}$ should give the value of Δ directly. In other words, the lowest energy absorption band in d^4 high-spin complexes is equal to the crystal field splitting energy. However, the magnitude of crystal field splitting energy for high-spin d^4 complexes cannot be obtained accurately from the Orgel diagram as the Jahn-Teller distortion reduces the symmetry from perfectly octahedral to a tetragonal geometry. The effect of Jahn-Teller distortion will be discussed later in this chapter. Furthermore, the practical applicability of the Tanabe-Sugano diagram in the high-spin region (before $\Delta_o/B = 27$) is strongly doubted because only one spin allowed transition is present, and it is a fact that two minimum spin-allowed transitions are required for the ratio calculation.

Being a strong-field complex, the theoretical value of crystal field splitting energy in $[Mn(CN)_6]^{3-}$ cannot be given by the Orgel diagram; hence, we are bound to use Tanabe-Sugano diagram. From the average value of the Racah parameter, what we have deduced earlier, the ligand field splitting parameter can be found as follows.

$$\frac{\Delta_o}{B} = 40; \quad \frac{\Delta_o}{767 \text{ cm}^{-1}} = 40; \quad \Delta_o = 30680 \text{ cm}^{-1}$$

3. Calculation of β: In order to calculate the nephelauxetic ratio, we must have the value of the Racah parameter for a free metal ion in its gaseous state. For free d^4 ion like Mn^{3+}, the value of B is found to be 1140 cm^{-1}. Now, the value of nephelauxetic ratio can be calculated as

$$\text{Nephelauxetic ratio} = \beta = \frac{B_{complex}}{B_{free\ ion}} = \frac{767 \text{ cm}^{-1}}{1140 \text{ cm}^{-1}} = 0.673$$

Hence, inter-electronic repulsion has been decreased during the process of complexation.

> ➤ *d^6 Complexes*

Metal complexes with d^6-configuration have ^5D ground state term symbol in the absence of any crystal field. However, when six ligands approach in octahedral coordination, the ground state term symbol becomes $^5T_{2g}$ in the weak field; and becomes $^1A_{1g}$ when the ligand field becomes sufficiently strong. The Orgel and Tanabe-Sugano diagram for d^6-configuration can be used to estimate the value of crystal field splitting energy for these complexes.

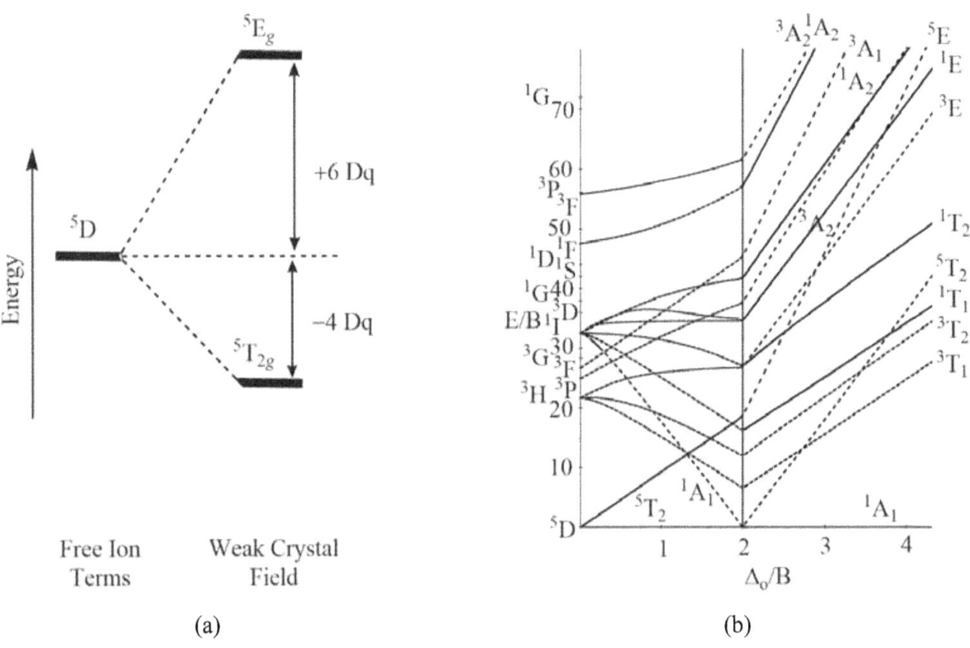

Figure 42. The (a) Orgel and (b) Tanabe-Sugano diagrams for d^6 complexes in octahedral crystal field.

Consider the example of $[Co(en)_3]^{3+}$

1. Calculation of B: From the Tanabe-Sugano diagram, it can clearly be seen that the spin-allowed *d-d* transitions in low-spin d^6 metal complexes are $^1A_{1g} \rightarrow {}^1T_{1g}$ and $^1A_{1g} \rightarrow {}^1T_{2g}$. In the UV-visible absorption spectra of $[Co(en)_3]^{3+}$, two bands are observed; one strong band with maxima at around 21450 and the other band at 29450 cm^{-1}. Therefore, the ratio of experimental band energies is

$$\frac{v_2}{v_1} = \frac{E_2}{E_1} = \frac{E_2/B}{E_1/B} = \frac{29450 \text{ cm}^{-1}}{21450 \text{ cm}^{-1}} = 1.37$$

Now slide a ruler across the printed diagram (perpendicular to the abscissa) until the ratio of E_2/B to E_1/B i.e. the ratio between the lines corresponding to the first two spin-allowed transitions becomes equivalent

to 1.37. In this particular example, this ratio becomes 1.37 when $\Delta_o/B = 40$. Stop the ruler movement and find out the values of E_2/B and E_1/B.

$$\frac{E_2}{B} = 52; \quad \frac{E_1}{B} = 38$$

Thus, on the T-S diagram, where $\Delta_o/B = 40$; the value of $^1A_{1g} \rightarrow {}^1T_{2g}$ and $^1A_{1g} \rightarrow {}^1T_{1g}$ i.e. E_2/B and E_1/B, are 52 and 38, respectively. The Racah parameter can be found by calculating B from both v_2 and v_1.

$$\frac{29450 \text{ cm}^{-1}}{B} = 52; \quad \frac{21450 \text{ cm}^{-1}}{B} = 38$$

$$B = \frac{29450 \text{ cm}^{-1}}{52} = 566 \text{ cm}^{-1}; \quad B = \frac{21450 \text{ cm}^{-1}}{38} = 564 \text{ cm}^{-1}$$

$$\textit{Average value of Racah parameter (B)} = \frac{566 \text{ cm}^{-1} + 564 \text{ cm}^{-1}}{2} = 565 \text{ cm}^{-1}$$

2. Calculation of Δ_o: The only parameter that is needed to be sought for the calculation of the magnitude of crystal field splitting energy (10 Dq) in weak-field d^6-complexes is the single absorption band in a UV-vis experiment. Hence, the energy of the transition $^5T_{2g} \rightarrow {}^5E_g$ should give the value of Δ directly. In other words, the lowest energy absorption band in d^6 high-spin complexes is equal to the crystal field splitting energy. However, the magnitude of crystal field splitting energy for high-spin d^6 complexes cannot be obtained accurately from the Orgel diagram as the Jahn-Teller distortion reduces the symmetry from perfectly octahedral to a tetragonal geometry. The effect of Jahn-Teller distortion will be discussed later in this chapter. Furthermore, the practical applicability of the Tanabe-Sugano diagram in the high-spin region (before $\Delta_o/B = 20$) is strongly doubted because only one spin allowed transition is present, and it is a fact that two minimum spin-allowed transitions are required for the ratio calculation.

     Being a strong-field complex, the theoretical value of crystal field splitting energy in $[\text{Co(en)}_3]^{3+}$ cannot be given by the Orgel diagram; hence, we are bound to use Tanabe-Sugano diagram. From the average value of the Racah parameter, what we have deduced earlier, the ligand field splitting parameter can be found as follows.

$$\frac{\Delta_o}{B} = 40; \quad \frac{\Delta_o}{565 \text{ cm}^{-1}} = 40; \quad \Delta_o = 22600 \text{ cm}^{-1}$$

3. Calculation of β: In order to calculate the nephelauxetic ratio, we must have the value of the Racah parameter for a free metal ion in its gaseous state. For free d^6 ion like Co^{3+}, the value of B is found to be 1100 cm^{-1}. Now, the value of nephelauxetic ratio can be calculated as

$$\text{Nephelauxetic ratio} = \beta = \frac{B_{complex}}{B_{free\ ion}} = \frac{565 \text{ cm}^{-1}}{1100 \text{ cm}^{-1}} = 0.514$$

Hence, inter-electronic repulsion has been decreased during the process of complexation.

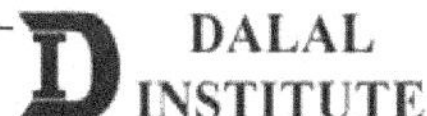

> ➤ *d^5 Complexes*

Metal complexes with d^5-configuration have ^6S ground state term symbol in the absence of any crystal field. However, when six ligands approach in octahedral coordination, the ground state term symbol becomes $^6A_{1g}$ in the weak field; and becomes $^2T_{2g}$ when the ligand field becomes sufficiently strong. The Tanabe-Sugano diagram for d^5-configuration can be used to estimate the value of crystal field splitting energy for these complexes.

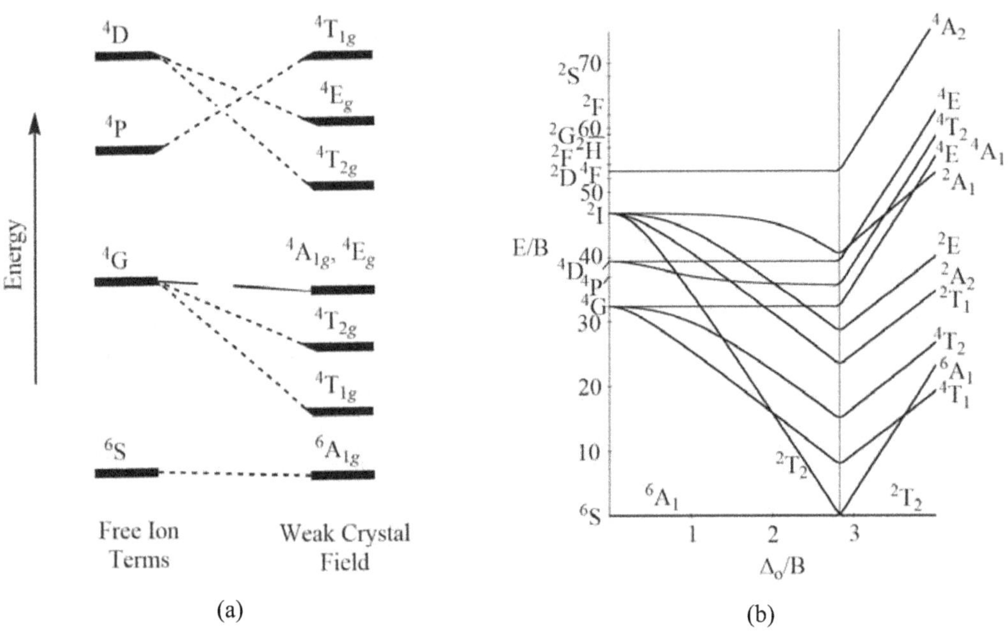

(a)   (b)

Figure 42. The (a) Orgel and (b) Tanabe-Sugano diagrams for d^5 complexes in the octahedral crystal field.

Consider the example of $[Mn(H_2O)_6]^{2+}$

1. Calculation of B: From the Tanabe-Sugano diagram, it can clearly be seen that there is no spin-allowed d-d transitions in high-spin d^5 metal complexes. However, main spin-forbidden transitions are $^6A_{1g} \rightarrow {}^4T_{1g}(G)$, $^6A_{1g} \rightarrow {}^4T_{2g}(G)$, $^6A_{1g} \rightarrow {}^4A_{1g}(G)$, $^6A_{1g} \rightarrow {}^4E_g(G)$, $^1A_{1g} \rightarrow {}^4T_{2g}(D)$ and $^1A_{1g} \rightarrow {}^4E_g(D)$. In the UV-visible absorption spectra of $[Mn(H_2O)_6]^{2+}$, the first two bands are observed at around 18600 and the other band at 22900 cm^{-1}. Therefore, the ratio of experimental band energies is

$$\frac{\nu_2}{\nu_1} = \frac{E_2}{E_1} = \frac{E_2/B}{E_1/B} = \frac{22900 \text{ cm}^{-1}}{18600 \text{ cm}^{-1}} = 1.23$$

Now slide a ruler across the printed diagram (perpendicular to the abscissa) until the ratio of E_2/B to E_1/B i.e. the ratio between the lines corresponding to the first two spin-allowed transitions becomes equivalent

to 1.23. In this particular example, this ratio becomes 1.23 when $\Delta_o/B = 11$. Stop the ruler movement and find out the values of E_2/B and E_1/B.

$$\frac{E_2}{B} = 29$$

$$\frac{E_1}{B} = 24$$

Thus, on the Tanabe-Sugano diagram, where $\Delta_o/B = 11$; the value of $^6A_{1g} \rightarrow {}^4T_{2g}(G)$ and $^6A_{1g} \rightarrow {}^4T_{1g}(G)$ i.e. E_2/B and E_1/B, are 29 and 24, respectively. The Racah parameter can be found by calculating B from both v_2 and v_1.

$$\frac{22900 \text{ cm}^{-1}}{B} = 29$$

$$\frac{18600 \text{ cm}^{-1}}{B} = 24$$

$$B = \frac{22900 \text{ cm}^{-1}}{29} = 789 \text{ cm}^{-1}; \quad B = \frac{18600 \text{ cm}^{-1}}{24} = 775 \text{ cm}^{-1}$$

$$\textit{Average value of Racah parameter (B)} = \frac{789 \text{ cm}^{-1} + 775 \text{ cm}^{-1}}{2} = 782 \text{ cm}^{-1}$$

2. Calculation of Δ_o: The magnitude of crystal field splitting energy for high-spin d^5 complexes cannot be obtained accurately from the Orgel diagram as the degeneracy of the ground state term is only one and does not split at all in the octahedral field. Therefore, we are bound to use the Tanabe-Sugano diagram. From the average value of the Racah parameter, what we have deduced earlier, the ligand field splitting parameter can be found as follows.

$$\frac{\Delta_o}{B} = 11$$

$$\frac{\Delta_o}{782 \text{ cm}^{-1}} = 11$$

$$\Delta_o = 8602 \text{ cm}^{-1}$$

3. Calculation of β: In order to calculate the nephelauxetic ratio, we must have the value of Racah parameter for a free metal ion in its gaseous state. For free d^5 ion like Mn^{2+}, the value of B is found to be 960 cm^{-1}. Now, the value of nephelauxetic ratio can be calculated as

$$\text{Nephelauxetic ratio} = \beta = \frac{B_{complex}}{B_{free \ ion}} = \frac{782 \text{ cm}^{-1}}{960 \text{ cm}^{-1}} = 0.814$$

Hence, the inter-electronic repulsion has been decreased during the process of complexation.

❖ Effect of Distortion on the *d*-Orbital Energy Levels

The splitting pattern of five *d*-orbitals for different geometries is also different; and therefore, this also induces an evident effect upon the *d-d* transitions in a metal ion complex. The UV-visible spectrum arising from these transitions provide information about the structure of these complexes. Hence, the effect of structural distortion upon the energies of the various free ion terms must be studied in detail.

➤ *Distortion in Six-Coordinated Complexes*

The hexa-coordination of ligands in different metal complexes can be categorized as octahedral, tetragonal, rhombic and trigonal anti-prismatic. If six ligands are supposed to approach the metal center along *x*, *y*, and *z*-axis (each passing through the two faces of a cube) then the octahedral, tetragonal and rhombic geometries can be represented by the bond length correlation of $x = y = z$, $x = y \neq z$ and $x \neq y \neq z$, respectively. The tetragonal distortions from perfect octahedron occur as the elongation or compression along only one four-fold axis while the rhombic distortions occur as the unequal amount of elongation or compression along two four-fold axes of rotation. Moreover, the trigonal distortions occur as the elongation or compression along one of the four three-fold symmetry axis.

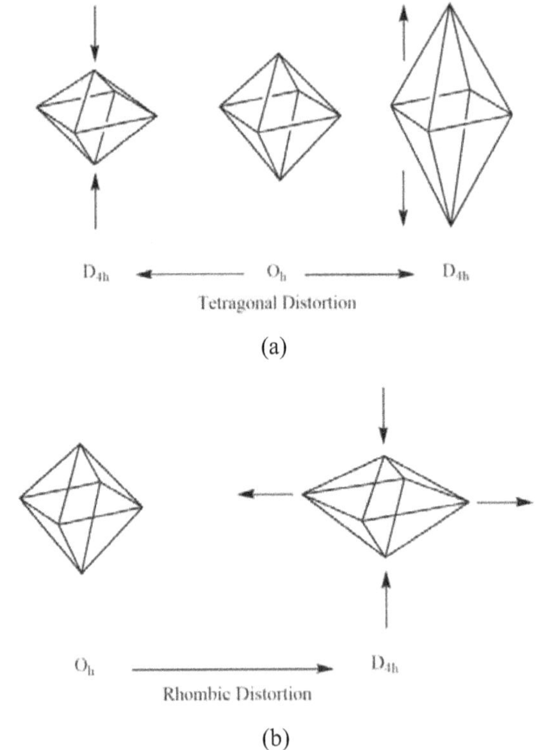

Figure 43. Continued on the next page...

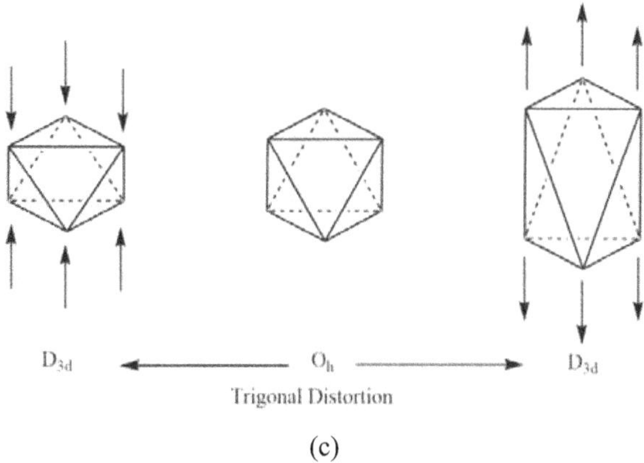

(c)

Figure 43. The (a) tetragonal (b) rhombic and (c) trigonal distortion in octahedral metal complexes.

1. Tetragonal distortion: As there is no inter-electronic repulsion in the d^1-complexes, the electronic energy states can be correlated with the splitting pattern of d-orbital set. Hence, in the case of z-out tetragonal distortion, the repulsion between metal electrons and ligand electrons will be less for the states comprising of electron density concentrated in orbitals directed toward z-axis. However, the tetragonal compression along z-axis will destabilize the electronic states having electron density distribution along or near z-axis. The effect of tetragonal distortion upon free ion terms of and d^1 and d^4-metal complexes can be shown as:

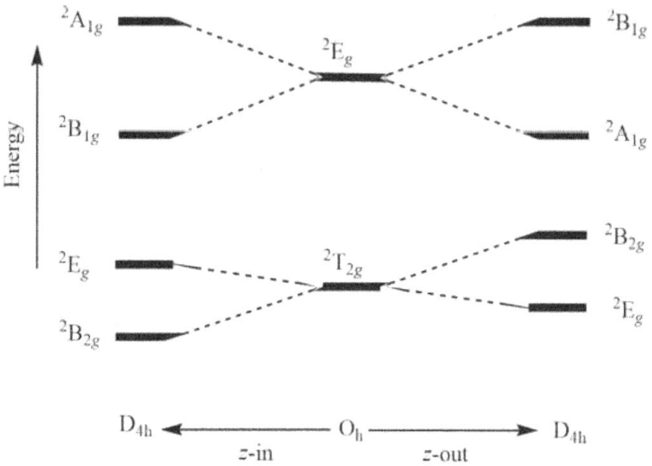

Figure 44. The splitting pattern of d-orbital energy levels of d^1-metal complexes undergoing tetragonal distortion.

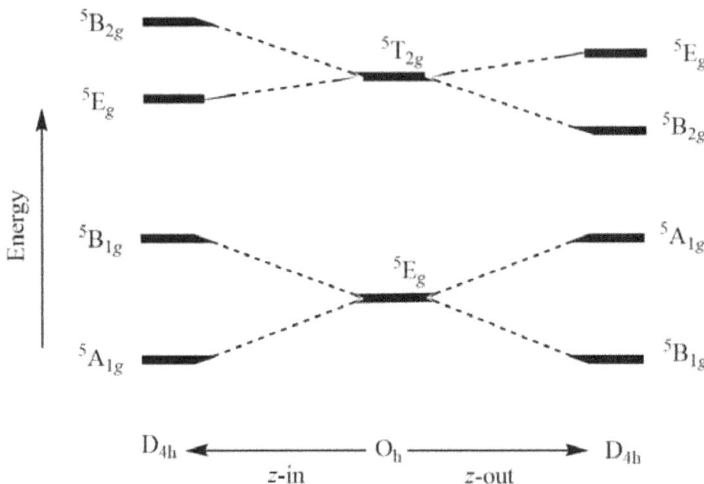

Figure 45. The splitting pattern of d-orbital energy levels of d^4-metal complexes undergoing tetragonal distortion.

It is worthy to mention that the homoleptic complexes of d^1-configuration prefer to undergo tetragonal compression due to Jahn-Teller distortion rather than the elongation. However, the heteroleptic complexes do show a splitting profile shown in the left side of Figure 44, like trans-$[TiA_4B_2]^{3+}$, if A-type ligands occupy a higher position in the spectrochemical series than B-type.

Apart from the Jahn-Teller distortions, the degeneracy of electronic states of octahedral complexes is also lifted when all the six ligands are not the same. This lowering of symmetry creates additional energy levels. For instance, Cr^{3+} is a d^3 system and does not show conventional Jahn-Teller distortion but the complexes like trans-$[CrA_4B_2]^{3+}$ show quite complex UV-visible spectrum which can only be explained in terms of energy levels of D_{4h} symmetry. Furthermore, chelate metal complexes like $[Cr(ox)_3]^{3-}$ and $[Cr(en)_3]^{3+}$ are no longer ideal octahedral geometries as their symmetry is lowered down to D_3 point group. However, the UV-visible spectrum of these complexes can successfully be rationalized as if they are arising from a perfectly octahedral complex like $[CrA_6]^{3+}$. This is obviously due to the fact that the extent of perturbation in these chelate complexes is quite small. Nevertheless, if we replace one ethylenediamine with two F^- ligands to form trans-$[Cr(en)_3]^{3+}$, the perturbation produced by the differentiation of the ligands is no more intolerable limit for ideal octahedral coordination.

The energy levels for a d^3 system in an octahedral complex and their splitting during tetragonal distortion is shown below.

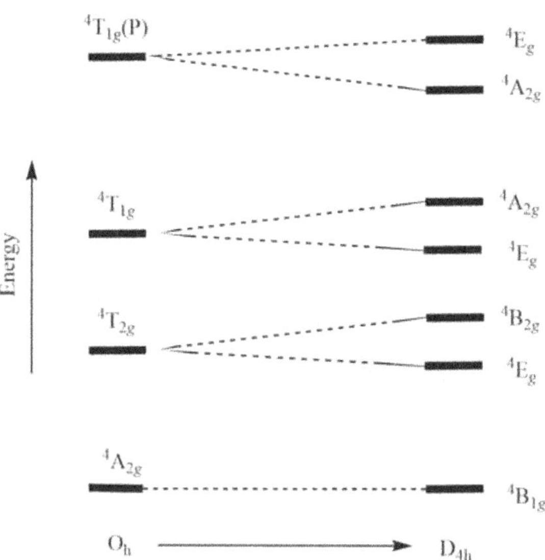

Figure 46. The splitting pattern of *d*-orbital energy levels of d^8-metal complexes in tetragonal distortion.

2. Rhombic distortion: The unequal amount of elongation or compression along two four-fold axes of rotation in octahedral complexes produces rhombic distortions. The common examples of rhombic distortion are high-spin Mn(III) and spin-paired Co(III) complex. Owing to more than one *d*-electron, we must consider ourselves with electronic states and not simply orbitals. The splitting pattern that occurs in a rhombic field is also shown below.

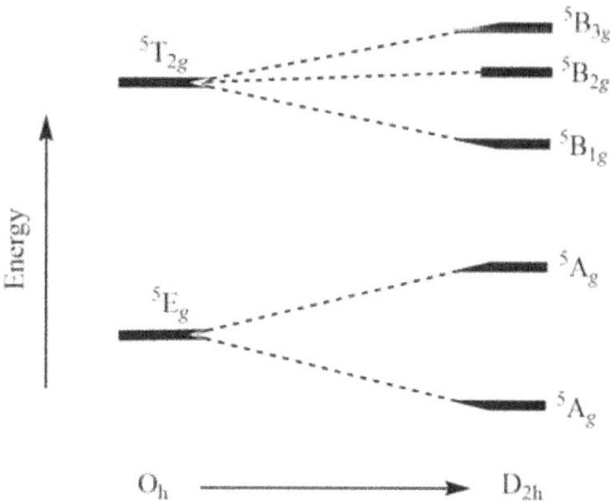

Figure 47. The splitting pattern of *d*-orbital energy levels of d^4-metal complexes in tetragonal distortion.

3. Trigonal distortion: The trigonal distortions occur as the elongation or compression along one of the four three-fold symmetry axis. The splitting pattern that occurs in a rhombic field is also shown below.

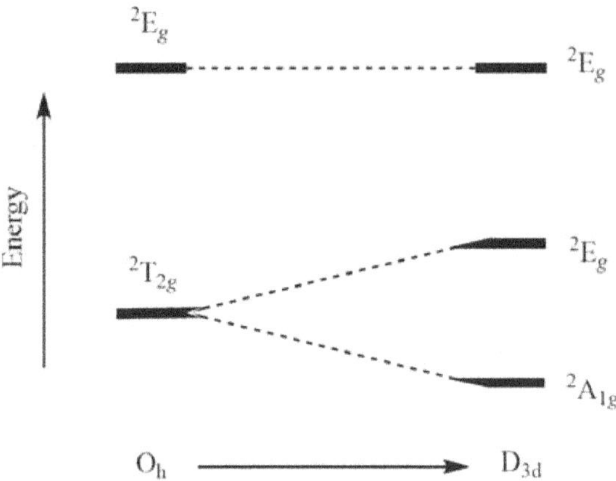

Figure 48. The splitting pattern of d-orbital energy levels of d^1-metal complexes in trigonal distortion.

> ➤ *Distortion in Four-Coordinated Complexes*

The tetra-coordination of ligands in different metal complexes can be categorized mainly as the tetrahedral and square planer. The tetrahedral geometry can be distorted in various ways to produce other four-coordinated structures. For instance, a perfect tetrahedral geometry can be compressed or elongated along two-fold and three-fold axis of symmetry to produces structures having D_{2d} and C_{3v} symmetry profiles, respectively. Moreover, the extreme case of compression along one two-fold axis is bound to yield a perfectly flat square-planar structure with D_{4h} point group.

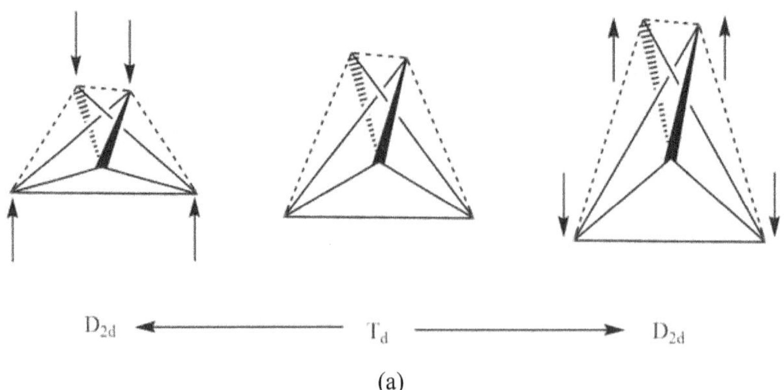

(a)

Figure 49. Continued on the next page…

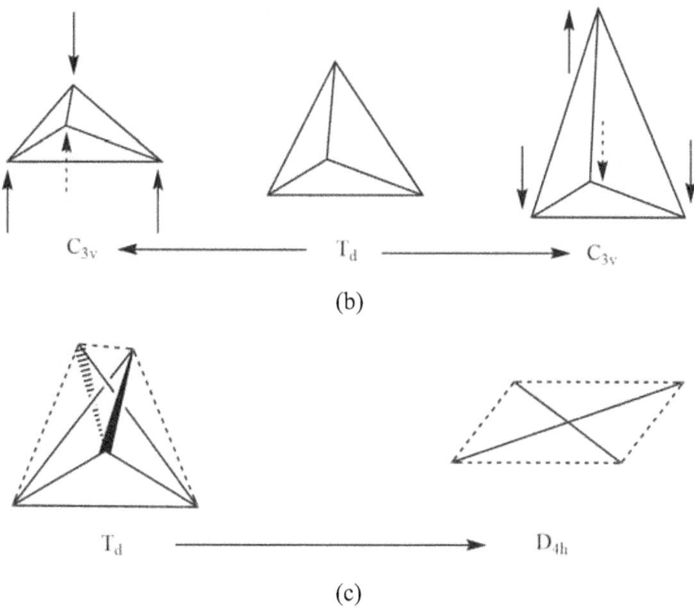

Figure 49. The (a) two-fold (b) three-fold and (c) square-planar distortion in tetrahedral metal complexes.

1. Two-fold distortion: The two-fold distortions occur as the elongation or compression along one of the six two-fold symmetry axis. The splitting pattern that occurs in a D_{2d} crystal field is also shown below.

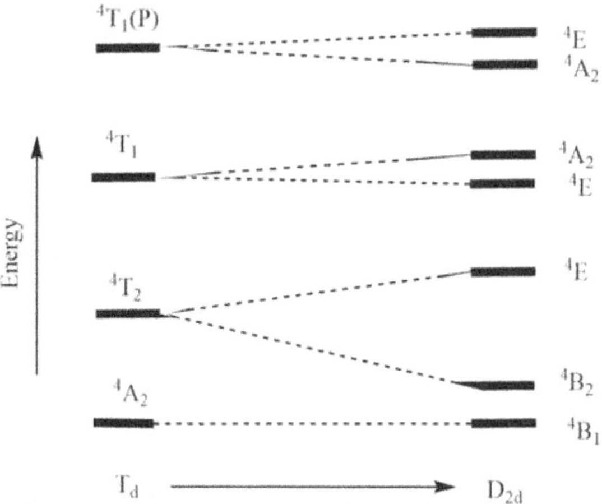

Figure 50. The splitting pattern of d-orbital energy levels of d^7-metal complexes in two-fold distortion.

2. Three-fold distortion: The three-fold distortions occur as the elongation or compression along one of the four three-fold symmetry axis. The splitting pattern that occurs in a C_{3v} crystal field is also shown below.

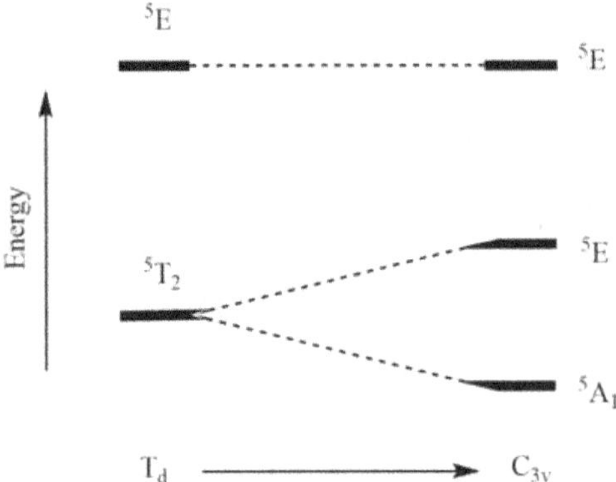

Figure 51. The splitting pattern of d-orbital energy levels of d^4-metal complexes in three-fold distortion.

3. Square-planer distortion: The square-planar distortions occur as the extreme compression along one of the six two-fold symmetry axis. The splitting pattern that occurs in a D_{4h} crystal field is shown below.

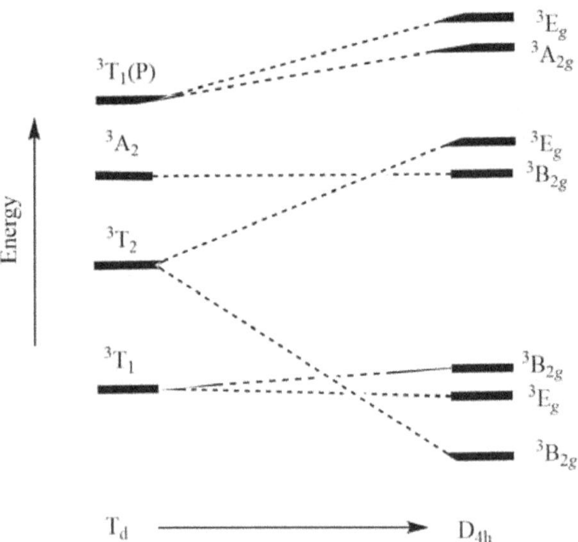

Figure 52. The splitting pattern of d-orbital energy levels of d^8-metal complexes in square-planer distortion.

❖ Structural Evidence from Electronic Spectrum

The electronic spectrum of transition metal complexes in the near-infrared, visible and ultraviolet region is quite useful in providing the information regarding their structural prototypes. The number of bands, their positions and molar absorptivity values must be considered for any conclusive remarks. Moreover, the spectra in the solution phase should also be checked against the solid-state spectral profile to be sure that no structural alteration has been taken place. These alterations could result in ligand displacement by solvent, ligand rearrangement, or an increase in the coordination number due to solvation. Very reliable information about the ligand environment in transition metal complexes can be found from their electronic spectrum. Some of the main sources of structural information from the UV-visible spectrum of transition metal complexes are given below.

➢ *Molar Absorptivity of assigned Bands*

The intensities of electronic transitions depend upon the wavefunctions of the ground and excited states; and provide important information about the electronic structure. Ligand field transitions that are forbidden in octahedral complexes may partly be allowed in tetrahedral complexes due to significant *d-p* mixing. The intensity of spin-allowed, as well as spin-forbidden d-d transitions in tetrahedral complexes, is 10-100 times more than in octahedral complexes. However, the values of molar extinction coefficients for *d-d* transitions in tetrahedral complexes are still lower than what we observe for charge transfer bands. This is because of the fact that the intensities in the former largely depend upon the ligand contribution to e and t_2 orbital sets, which is approximately 30% in magnitude. On the other hand, charge transfer transitions occur from an orbital which is largely p to an orbital with large *d*-character. The values for molar extinction coefficients for different types of transitions are given below.

Table 9. The order of magnitude for the molar extinction coefficient for different types of transitions.

Type	Extinction coefficient
d-d spin forbidden, Laporte forbidden (O_h)	0.1
d-d spin forbidden, Laporte allowed (T_d)	1
d-d spin allowed, Laporte forbidden (O_h)	10
d-d spin allowed, Laporte allowed (T_d)	100
Charge transfer, spin allowed and Laporte allowed	1000

Therefore, as the asymmetry around metal center increases, the intensity of various bands is also enhanced and a superficial shortlisting between tetrahedral and octahedral coordination can be done just on the basis of intensity and number of bands observed. It should also be noted that many complexes that ideally belong to a non-centrosymmetric point group usually have only a small asymmetric contribution to the crystal field. On the basis of the magnitude of this asymmetric contribution, the intensity lies between that of a truly centrosymmetric and a complex with a highly asymmetric crystal field, such as a square-pyramidal or distorted tetrahedral complex. If geometric distortion is so small that it is in the range of vibrational amplitudes, the intensities of the spectral band would not be much greater than those in typical centrosymmetric compounds. For example, peak intensities in tris-chelate complexes of ethylenediammine are not much more intense than their hexa-ammine counterparts. When asymmetry is the result of a difference in the ligand type on either side of the metal, the deviation depends on how different the ligands actually are. The bands in the spectrum of a complex such as the cis-Co[(NH$_3$)$_4$Cl$_2$]$^+$ ion are two to three times as intense as those in the corresponding trans complex; and Thus, the two isomers can be distinguished on the basis of intensity only. Tetrahedral complexes can easily be distinguished from hexa-coordinated counterparts on the basis of the intensity of the bands. The complex Ni{OP[N(CH$_3$)$_2$]$_3$}$_4$Cl$_2$ could have tetrahedral or distorted octahedral geometries but the similarity of its electronic spectrum to that of [NiCl$_4$]$^{2-}$ implied that this was the first cationic, tetrahedral nickel(II) complex ever synthesized. Further confirmation of the structure comes from the similarity of its x-ray powder diffraction patterns with tetrahedral zinc(II) complexes.

Consistent structural information can be obtained from the electronic spectra of Co(II) complexes. Hexa-coordinated complexes of Co(II) are usually high-spin and the Orgel diagram for d^7-configuration should be used to assign different peaks. The ground state is $^4T_{1g}$ and a considerable magnitude of spin-orbit coupling is expected in $^4T_{1g}(F) \rightarrow {}^4T_{2g}(F)$, $^4T_{1g}(F) \rightarrow {}^4A_{2g}(F)$ and $^4T_{1g}(F) \rightarrow {}^4T_{1g}(P)$ transitions. Being a two-electron transition $^4T_{1g}(F) \rightarrow {}^4A_{2g}(F)$, its intensity is pretty low and hardly observed. The electronic spectrum of octahedral [Co(H$_2$O)$_6$]$^{2+}$ and tetrahedral [CoCl$_4$]$^{2-}$ are shown in the following figure.

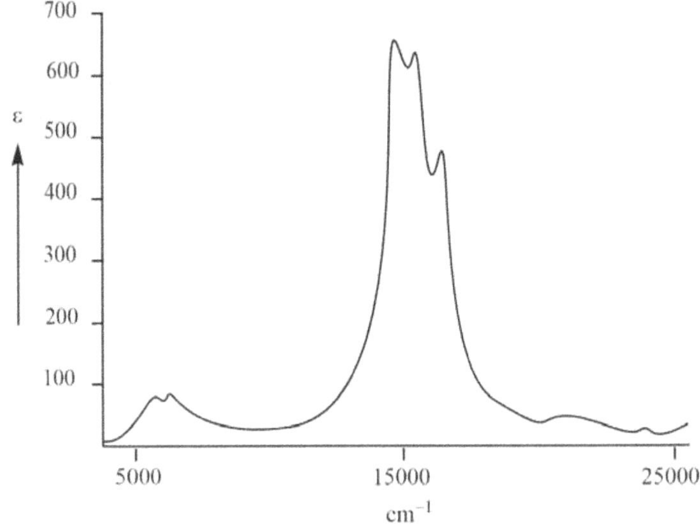

Figure 53. Electronic spectrum of tetrahedral [CoCl$_4$]$^{2-}$.

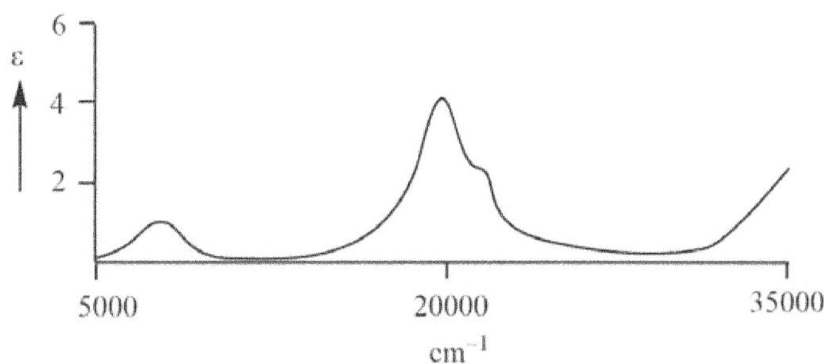

Figure 54. Electronic spectrum of octahedral $[Co(H_2O)_6]^{2+}$.

Tanabe and Sugano diagram for d^7-configuration has to be used to explain the electronic spectrum of tetrahedral complexes of Co(II) which are always high-spin in nature. The absorption at 15000 cm^{-1} may be attributed to $^4A_2(F) \rightarrow {}^4T_1(P)$ and fine structure can be explained in terms of to spin-orbital coupling of the T electronic state. The presence of spin-orbital interaction also permits some quartet $\rightarrow$ doublet transitions to happen. The other peak shown is assigned to $^4A_2(F) \rightarrow {}^4T_1(F)$ transition while the expected $^4A_2(F) \rightarrow {}^4T_2(F)$ transition is predicted to result in a peak at 3000-4500 cm^{-1}. However, most of the UV-visible spectrophotometers cannot record them as they are out of detectable range and usually overlap with the infrared bands of attached ligands. On the other hand, the main peak in the UV-visible spectrum of $[Co(H_2O)_6]^{2+}$ is at about 20000 cm^{-1} and is allocated to the $^4T_{1g}(F) \rightarrow {}^4T_{1g}(P)$ transition. The shoulder owes to the lifting of degeneracy due to spin-orbital coupling in the excited $^4T_{1g}(P)$ state. The second band at 8350 cm^{-1} is ascribed to $^4T_{1g}(F) \rightarrow {}^4T_{2g}(F)$ transition.

> ### *Splitting Pattern of Assigned Bands*

The splitting pattern of absorption peaks provides useful information not only on the symmetry but also on the relative σ and π bonding nature of the attached ligands. The effects of decreased ligand-field symmetry in a mixed ligand metal complex can easily be explained by the analysis of UV-visible spectra of the $[Co(en)_3]^{3+}$, cis-$[Co(en)_2F_2]^{1+}$ and trans-$[Co(en)_2F_2]^{1+}$. Although the complex $[Co(en)_3]^{3+}$ belongs to D_3 point group symmetry, its ligand field pretty close to O_h symmetry. Therefore, the Tanabe-Sugano diagram for d^6-configuration should be used to interpret the electronic spectrum which is comprised of two symmetric bands due to the $^1A_{1g} \rightarrow {}^1T_{1g}$ and $^1A_{1g} \rightarrow {}^1T_{2g}$ transitions, respectively. The spectrum of cis-$[Co(en)_2F_2]^{1+}$ is somewhat similar to what is observed for $[Co(en)_3]^{3+}$; however, both bands are broader and slightly shifted to lower energy value. In case of trans-$[Co(en)_2F_2]^{1+}$, the lower energy band is obviously split into two components; and the peak intensities are slightly lesser than those of cis-$[Co(en)_2F_2]^{1+}$ and $[Co(en)_3]^{3+}$ which obviously comply with the fact that the former complex does possess a center of symmetry so that the electronic transitions are formally Laporte forbidden.

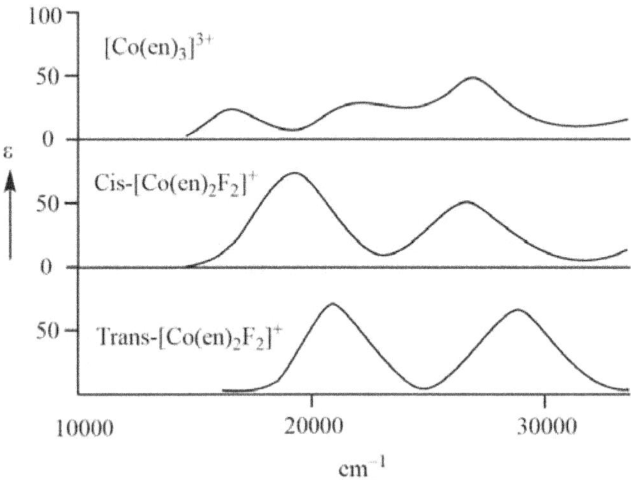

Figure 55. Electronic spectrum of trans-$[Co(en)_2F_2]^{1+}$, cis-$[Co(en)_2F_2]^{1+}$ and $[Co(en)_3]^{3+}$ in aqueous solution.

The main cause of the splitting of bands in trans-$[Co(en)_2F_2]^{1+}$ is the loss of degeneracy of excited states due to the lowering of ligand-field symmetry; which can be explained in terms of the effect of an increasing tetragonal component of the ligand field on the $^1T_{1g}$ and $^1T_{2g}$ states. The whole schematic is shown in the energy level diagram given below. The two lower energy peaks of trans-$[Co(en)_2F_2]^{1+}$ may be ascribed to the $^1A_{1g} \rightarrow {}^1E_g$ and $^1A_{1g} \rightarrow {}^1A_{2g}$ transitions, respectively. It is worthy to note that both excited states in the tetragonal field, ^1E_g and $^1A_{2g}$, are the components of $^1T_{1g}$ state of O_h-symmetry. Moreover, as the splitting of $^1T_{2g}$ state is quite insignificant, the two corresponding transitions ($^1A_{1g} \rightarrow {}^1E_g$ and $^1A_{1g} \rightarrow {}^1B_{2g}$) are not resolved even in the trans complex.

Now, the angular overlap method states that it is the "holohedrized" symmetry that governs the energy levels in a complex; or in other words, d-orbital energies depend on the sum of the effects of the ligands along each axis. Therefore, the magnitude of the tetragonal component in the ligand field of the trans-complex is twice that what is active in the cis counterpart. This gives rise to much more overlapping transitions in the spectrum of cis-$[Co(en)_2F_2]^{1+}$ and no clear splitting of the bands can be resolved. Even though it is not evident in the solution spectra, the sign of ligand field splitting in cis-isomer is different than the trans-isomers. The energy level diagram given in this section is actually for the situation in which the in-plane ligand field is stronger than the axial field, just like in the case of trans-$[Co(en)_2F_2]^{1+}$ complex. The degree of the splitting is correlated to the difference in σ and π-bonding ability of the in-plane and axial ligands. Thus, the peak splitting profile for mixed-ligand complexes offers a powerful means of obtaining metal-ligand bonding parameters.

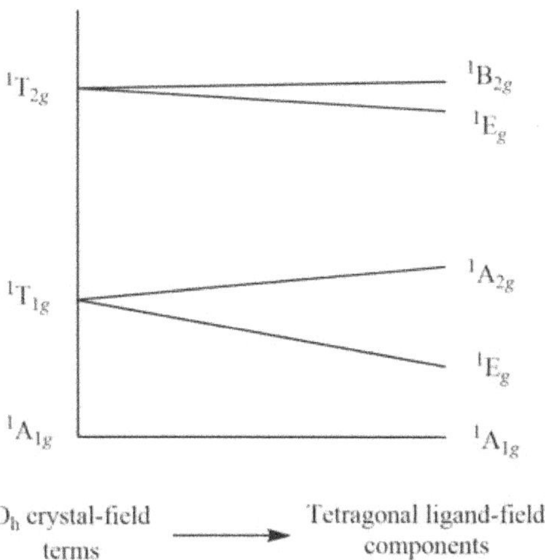

Figure 56. The splitting pattern of $^1T_{1g}$ and $^1T_{2g}$ terms of octahedral Co^{3+} complex in increasing the tetragonal ligand field.

In order to analyze the system in a precise manner, the assignment of different bands must be done properly; and if possible, this is best achieved by recording single-crystal polarized spectra. The average position of the split components of the $^1T_{1g}$ is approximately 18000 cm^{-1} and of $^1T_{2g}$ is 26000 cm^{-1} for trans-$[Co(en)_2F_2]^{1+}$. These values of band positions are rather similar to those of cis-$[Co(en)_2F_2]^{1+}$ and give a ligand field splitting 21000 cm^{-1}.

This result is also supported by "the rule of average environment" which states that the magnitude of ligand field splitting in a mixed-ligand complex is around the weighted average of the octahedral ligand fields associated with a complete set of each ligand. For example, the value of crystal field splitting in MA_nB_{6-n} complex can be given as:

$$\Delta = \frac{[n\Delta(MA_6) + (6-n)\Delta(MB_6)]}{6}$$

Putting the values of crystal field splitting values of $[Co(en)_3]^{3+}$ and $[CoF_6]^{3-}$ in the above equation, we get the expected magnitude of Δ for $[Co(en)_2F_2]^{1+}$.

$$\Delta = \frac{[4(23000) + 2(13000)]}{6} = 19667 \text{ cm}^{-1}$$

Which is pretty close to the experimentally observed value of 21000 cm^{-1} and therefore confirms the practical applications in structure determination.

❖ Jahn-Teller Effect

The Jahn-Teller theorem essentially states that any nonlinear molecule system possessing electronic degeneracy will be unstable and will undergo distortion to form a system of lower symmetry as well as lower energy and thus the degeneracy will be removed.

This effect describes the geometrical distortion of molecules and ions that is associated with electronically degenerate configurations. A configuration is said to be electronically degenerate if more than one sites are available for the filling of a single electron. The Jahn-Teller effect is generally encountered in octahedral transition metal complexes. The phenomenon is much more common in hexacoordinated complexes of bivalent copper. The d^9 configuration of Cu^{2+} ion yields three electrons in the doubly degenerate e_g orbitals set, leading to a doubly degenerate electronic state as well. Such complexes distort along one of the molecular four-fold axis (always labeled the z-axis), which has the effect of removing the orbital and electronic degeneracies and lowering the overall energy. The distortion usually occurs via the elongation the metal-ligand bonds along the z-axis, but sometimes also occurs as a shortening of the same bonds instead. Moreover, the Jahn-Teller theorem predicts the presence of an unstable geometry only and not the direction of the distortion. When distortion involving elongation occurs to decrease the electrostatic repulsion between the electron-pair on the ligand-attached and any extra electrons in metal orbitals with a z-component; and hence lowering the energy of the metal complex. Inversion center is retained after z-out as well as z-in the distortion. Symmetrical configurations possess electronic degeneracy while the unsymmetrical ones do not. Various symmetrical and unsymmetrical configurations are given below.

Table 10. Symmetrical and Unsymmetrical t_{2g} and e_g orbitals.

Symmetrical configurations	Unsymmetrical configurations
t_{2g}^0, t_{2g}^3, t_{2g}^6	t_{2g}^1, t_{2g}^2, t_{2g}^4, t_{2g}^5
e_g^0, e_g^4, e_g^2 (high-spin)	e_g^1, e_g^3, e_g^2 (low-spin)

Now, the conditions for different kinds of distortion can be summed up as:

Table 11. Conditions for Jahn-Teller distortion.

Type of distortion	Configuration required
No distortion	t_{2g} (symmetrical) + e_g (symmetrical)
Slight distortion	t_{2g} (unsymmetrical)
Strong distortion	e_g (unsymmetrical)

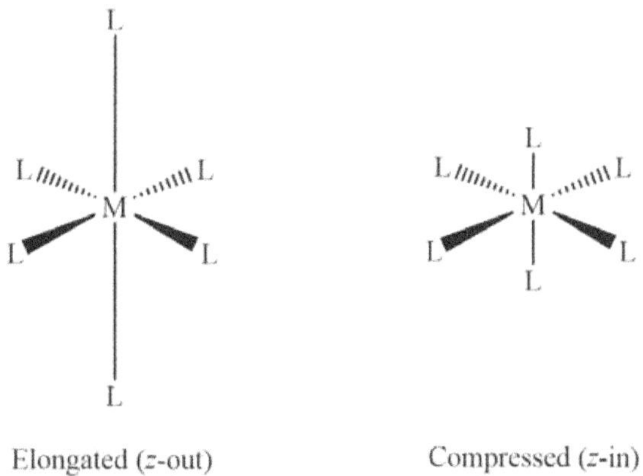

Figure 57. The Jahn-Teller distortions for an octahedral complex.

Let the case of the low-spin Co^{2+} octahedral complex. The corresponding electronic configuration is $t_{2g}^6 e_g^1$. Owing to the electronically degenerate state, the Jahn-Teller distortion is expected. Now suppose that the single electron of e_g set is present in d_z^2 orbital; the ligands approaching from z-axis will feel more repulsion than the ligands coming from x and y-axis. Therefore, the bonds along z-axis will be weaker in comparison to the bond along x and y-axis. This results in a tetragonal elongation about z-axis with two longer and four shorter bonds. This is formally called as z-out distortion. On the other hand, if the single electron of e_g set is present in $d_{x^2-y^2}$ orbital; the ligands approaching from x and y-axis will feel more repulsion than the ligands coming from z-axis. Therefore, the bonds along x and y-axis will be weaker in comparison to the bond along z-axis. This results in a tetragonally flattened octahedral geometry about z-axis with two shorter and four longer bonds. This is formally called as z-in distortion.

Consider the following examples

1. $[Co(CN)_6]^{4-}$: It is a low-spin complex with $t_{2g}^6 e_g^1$ electronic configuration and will undergo strong Jahn-Teller distortion.

2. $[Cr(NH_3)_6]^{3+}$: It is a high-spin complex with $t_{2g}^3 e_g^0$ electronic configuration which is completely symmetrical; and therefore, will not show any Jahn-Teller distortion.

3. $[FeF_6]^{4-}$: It is a high-spin complex with $t_{2g}^4 e_g^2$ electronic configuration and will undergo slight Jahn-Teller distortion.

> ### Energetics of Jahn-Teller Distortion

The Jahn-Teller distortion results in a system of lower symmetry and lower energy. This is actually the opposite of what is expected. Generally, symmetry leads to stability; but the Jahn-Teller effect is actually an exception to this statement. Therefore, it is necessary to discuss the driving force responsible for this behavior. The magnitude of Jahn-Teller effect is larger where the electron density associated with the degenerate set orbitals is more concentrated. Hence, Jahn-Teller effect plays a significant role in determining the structure of transition metal complexes with active $3d$-orbitals. The whole energetics of the Jahn-Teller can be understood by the case study of d^9 and d^1 complexes.

1. Cu^{2+} complexes: The electronic configuration of free Cu^{2+} ion is d^9; and in an octahedral environment, it is $t_{2g}^6\, e_g^3$. Before we put any conclusive remark on the direction or nature of the distortion, we shall find the crystal field stabilization energy for z-out as well as for z-in case.

i) Crystal field stabilization energy for z-out distortion:

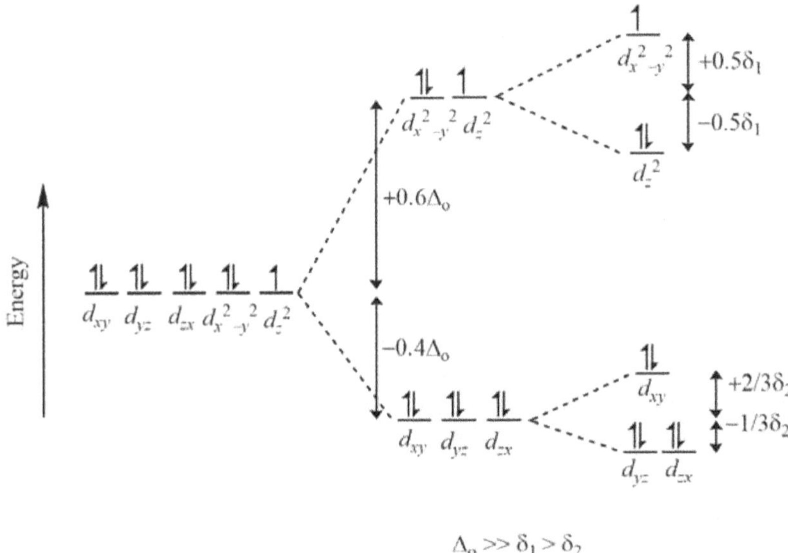

$$\Delta_o \gg \delta_1 > \delta_2$$

Figure 58. The splitting pattern and filling of d-orbital set of Cu^{2+} in octahedral and subsequently in the tetragonally elongated complex due to Jahn-Teller effect.

CFSE due to distortion = Energy of the distorted complex (E_2) − Energy of the complex without distortion (E_1)

$$E_1 = 6(-0.4\Delta_o) + 3(+0.6\Delta_o)$$

$$E_1 = -0.6\Delta_o$$

$$E_2 = 4(-0.4\Delta_o - \delta_2/3) + 2(-0.4\Delta_o + 2\delta_2/3) + 2(+0.6\Delta_o - \delta_1/2) + 1(+0.6\Delta_o + \delta_1/2)$$

$$E_2 = -0.6\Delta_o - \delta_1/2$$

$$\text{CFSE due to distortion} = E_2 - E_1 = -\delta_1/2$$

Hence, the crystal field stabilization due to z-out distortion is $-\delta_1/2$.

ii) Crystal field stabilization energy for z-in distortion:

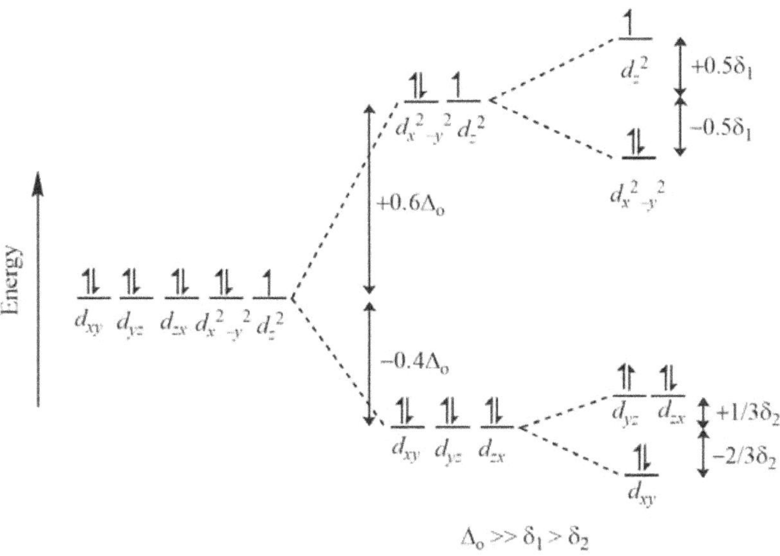

$$\Delta_o \gg \delta_1 > \delta_2$$

Figure 59. The splitting pattern and filling of d-orbital set of Cu^{2+} in octahedral and subsequently in the tetragonally compressed complex due to Jahn-Teller effect.

CFSE due to distortion = Energy of the distorted complex (E_2) − Energy of the complex without distortion (E_1)

$$E_1 = 6(-0.4\Delta_o) + 3(+0.6\Delta_o)$$

$$E_1 = -0.6\Delta_o$$

$$E_2 = 4(-0.4\Delta_o - 2\delta_2/3) + 2(-0.4\Delta_o + \delta_2/3) + 2(+0.6\Delta_o - \delta_1/2) + 1(+0.6\Delta_o + \delta_1/2)$$

$$E_2 = -0.6\Delta_o - \delta_1/2$$

$$\text{CFSE due to distortion} = E_2 - E_1 = -\delta_1/2$$

Hence, the crystal field stabilization due to z-in distortion is $-\delta_1/2$.

Hence, the magnitude of crystal field stabilization in z-out case is same as that is present in z-in complex. This implies that Jahn-Teller effect cannot predict the direction of the distortion. However, it has been observed that it is the z-out case that dominates in most of the cases. It may depend on the repulsive forces between the d-electrons and the ligands, so the odd electron will prefer $d_z{}^2$-orbital more than $d_{x^2-y^2}$ due to the lesser number of ligands it will repel with. Moreover, when a z-in distortion occurs, one can also view it terms of equatorial elongation while z-out will mean the weakening of two axial metal-ligand bonds. In other words, it is easier to weaken two bonds rather stretching four metal-ligand bonds.

2. Ti^{3+} complexes: The electronic configuration of free Ti^{3+} ion is d^1; and in an octahedral environment, it is $t_{2g}^1\ e_g^0$. Before we put any conclusive remark on the direction or nature of the distortion, we shall find the crystal field stabilization energy for z-out as well as for z-in case.

i) Crystal field stabilization energy for z-out distortion:

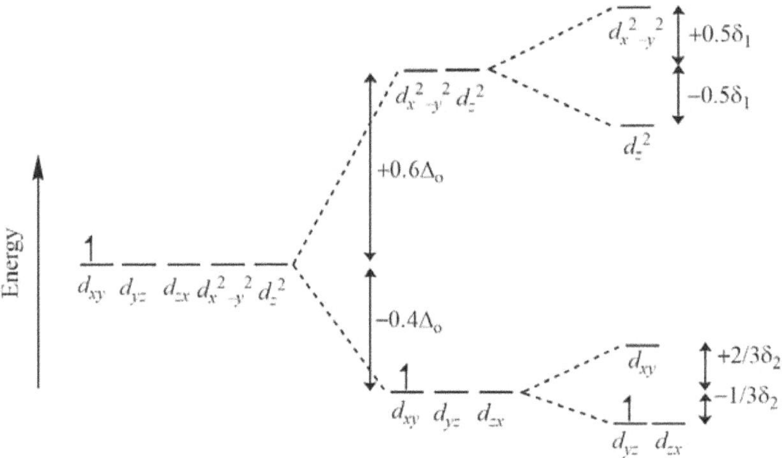

Figure 60. The splitting pattern and filling of d-orbital set of Ti^{3+} in octahedral and subsequently in the tetragonally elongated complex due to Jahn-Teller effect.

CFSE due to distortion = Energy of the distorted complex (E$_2$) − Energy of the complex without distortion (E$_1$)

$$E_1 = 1(-0.4\Delta_o)$$

$$E_1 = -0.4\Delta_o$$

$$E_2 = 1(-0.4\Delta_o - \delta_2/3)$$

$$E_2 = -0.4\Delta_o - \delta_2/3$$

$$\text{CFSE due to distortion} = E_2 - E_1 = -\delta_2/3$$

Hence, the crystal field stabilization due to z-out distortion is $-\delta_1/3$. Moreover, it is worthy to note that this type of Jahn-Teller distortion is unable to remove the electronic degeneracy completely (the single electron can still be filled in two degenerate orbitals namely d_{yz} and d_{zx}). Therefore, homoleptic octahedral complexes of d^1-configuration undergoing z-in distortion are quite rare.

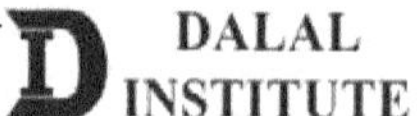

ii) Crystal field stabilization energy for z-in distortion:

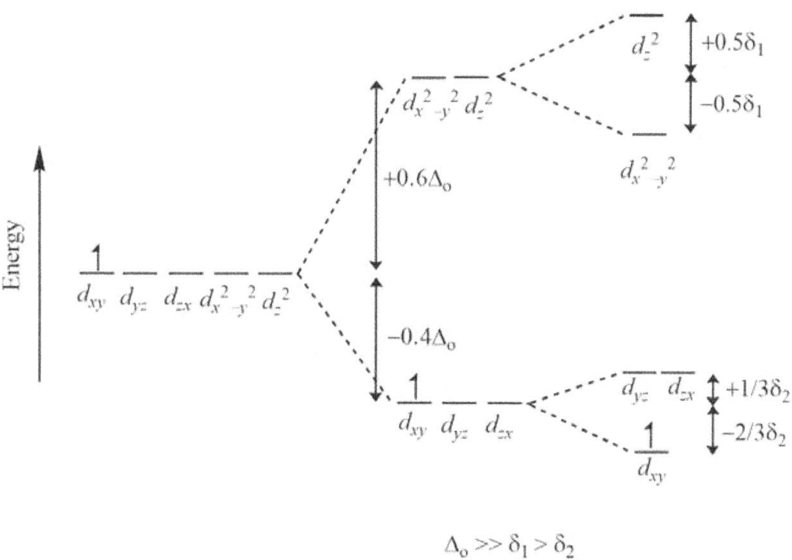

Figure 61. The splitting pattern and filling of d-orbital set of Ti^{3+} in octahedral and subsequently in the tetragonally compressed complex due to the Jahn-Teller effect.

CFSE due to distortion = Energy of the distorted complex (E_2) − Energy of the complex without distortion (E_1)

$$E_1 = 1(-0.4\Delta_o)$$

$$E_1 = -0.4\Delta_o$$

$$E_2 = 1(-0.4\Delta_o - 2\delta_2/3)$$

$$E_2 = -0.4\Delta_o - 2\delta_2/3$$

$$\text{CFSE due to distortion} = E_2 - E_1 = -2\delta_2/3$$

Hence, the crystal field stabilization due to z-in distortion is $-2\delta_2/3$.

Hence, the magnitude of crystal field stabilization in z-in case is half of what is observed in z-in complex. This implies that Jahn-Teller effect can predict the direction of the distortion. However, it is the d^1-configuration that has been observed to show z-in distortion; otherwise, the z-out case dominates in most of the cases. It may be explained in terms of the repulsive forces between the d-electrons and the ligands, so the odd electron will prefer d_z^2-orbital more than $d_{x^2-y^2}$ due to the lesser number of ligands it will repel with. Nevertheless, Ti^{3+} octahedral complexes prefer to undergo z-in due to the greater value of crystal field stabilization energy.

> ### *Effect of Jahn-Teller Distortion on Electronic Spectra*

Jahn-Teller distortions affect the UV-visible spectra of many transition metal complexes in a significant manner and sometimes it is almost impossible to find out the values of the Racah parameter and crystal splitting energy if we do not take it into account. In UV-visible absorption spectroscopy, distortion causes the splitting of bands in the spectrum due to a reduction in symmetry (O_h to D_{4h}). The complexes undergoing Jahn-Teller distortion show an increased number of bands. Consider the following examples:

1. $[Cr(H_2O)_6]^{2+}$: The Cr^{2+} ion in aqueous solutions is pale blue and its UV-visible absorption spectrum consists of a weak broad band with a maxima 14000 cm^{-1}. In normal conditions, this band is expected to provide the value of crystal field splitting value directly (from Orgel-diagram of d^4-configuration); however, the actual value of Δ_o is considerably different. The electronic configuration for high-spin Cr^{2+} in the octahedral field is $t_{2g}^3\, e_g^1$; and therefore, this complex is bound to undergo strong Jahn-Teller distortion. The energy level diagram and main transitions are shown below.

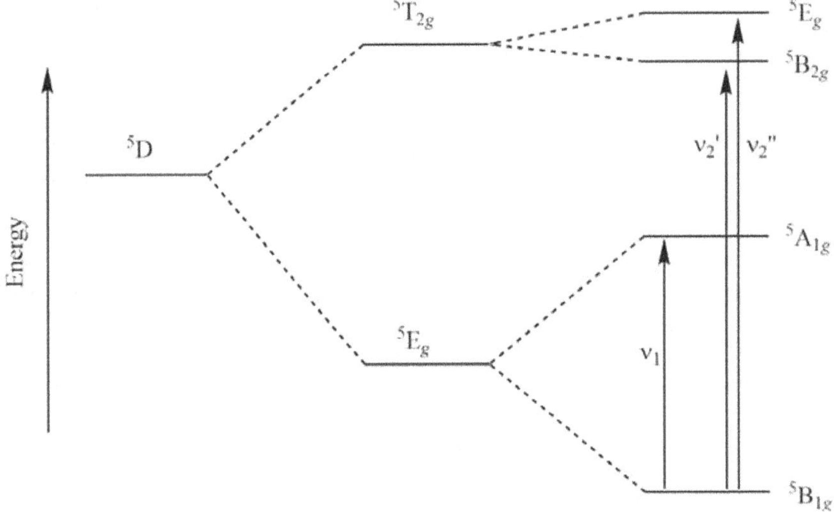

Figure 62. The schematic energy level diagram and prominent transitions in $[Cr(H_2O)_6]^{2+}$.

The ground state term for a regular octahedral $[Cr(H_2O)_6]^{2+}$ complex would be ^5E_g but the ground as well as excited electronic state will undergo a Jahn-Teller distortion. It is worthy to note that the former state shows strong while splitting of later one is weak. Hence, the Tanabe-Sugano diagram cannot be used to describe the energy levels in the complex. The gaussian deconvolution of the overall spectrum has shown that the transition between the split components of the ^5E_g level ($^5B_{1g} \rightarrow\ ^5A_{1g}$) occurs at around 10000 cm^{-1} and contributes to the low-energy tail of the broad band in aqueous solution.

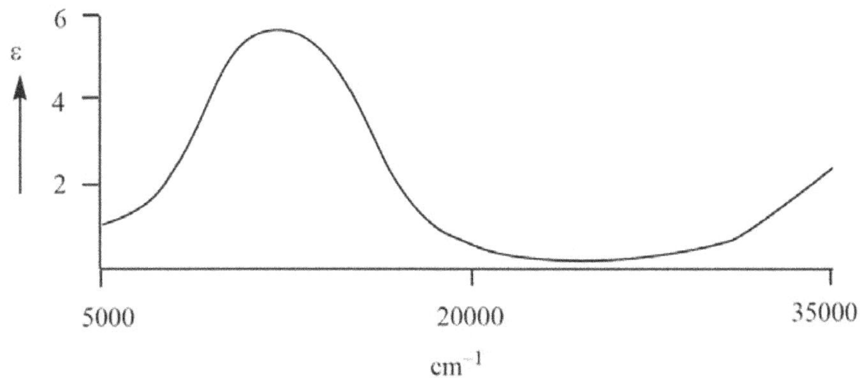

Figure 63. UV-visible spectra of $[Cr(H_2O)_6]^{2+}$.

The transitions from the $^5B_{1g}$ ground state to the components of the $^5T_{2g}$ state are unresolved and together give rise to the band at around 14000 cm^{-1}. Now, suppose that the Jahn-Teller effect produces a symmetrical splitting of the ^5E_g state, we can obtain the d-orbital splitting of the parent octahedral complex from the energy of the $^5B_{1g} \rightarrow {}^5T_{2g}$ transition minus half the energy of the $^5B_{1g} \rightarrow {}^5A_{1g}$ transition. This gives the estimated value of crystal field splitting value as follows:

$$\Delta_o = [v_2\,({}^5B_{1g} \rightarrow {}^5T_{2g})] - [v_1\,({}^5B_{1g} \rightarrow {}^5A_{1g})]/2$$

$$\Delta_o - 14000 - 10000/2 - 9000 \text{ cm}^{-1}$$

Similar results have been observed in the low-temperature spectra of other hydrate complexes with Jahn-Teller distortion.

2. $[Fe(H_2O)_6]^{2+}$: The Fe^{2+} ion in aqueous solutions is pale green and its UV-visible absorption spectrum is consisted of a weak broad band with a maxima 10000 cm^{-1}. The electronic configuration for high-spin Fe^{2+} in octahedral field is $t_{2g}^4\,e_g^2$; and therefore, this complex is bound to undergo slight Jahn-Teller distortion. The ground state term for a regular octahedral $[Fe(H_2O)_6]^{2+}$ complex would be $^5T_{2g}$ but the ground as well as excited electronic state will undergo a Jahn-Teller distortion. It is worthy to note that the former state shows strong while splitting of later one is weak.

Hence, the Orgel-diagram can be used to describe the energy levels in the complex. Moreover, as the band is obviously a doublet, we don't need gaussian deconvolution the overall spectrum. The transition from $^5T_{2g}$ to the split components of the ^5E_g level ($^5B_{1g}$ and $^5A_{1g}$) are split by an energy difference of around 2000 cm^{-1}. This contributes to the broadening of the band in aqueous solution. The energy level diagram and main transitions are shown below.

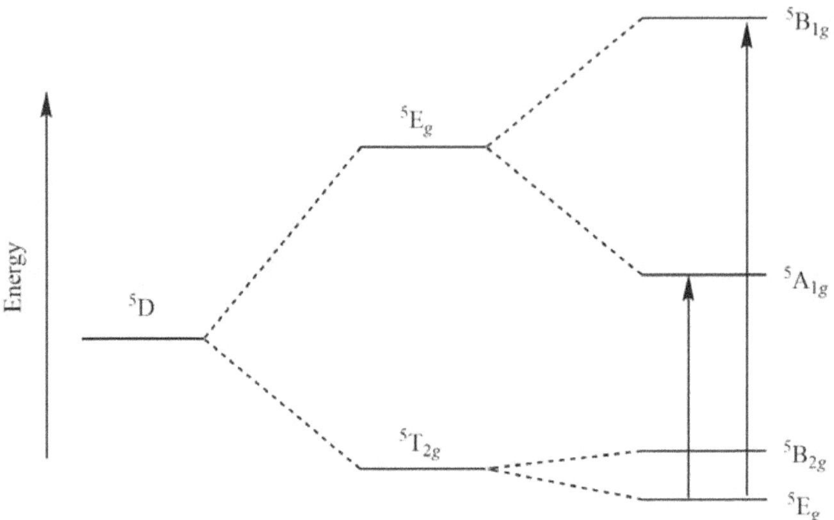

Figure 64. The schematic energy level diagram and prominent transitions in $[Fe(H_2O)_6]^{2+}$.

The corresponding spectrum is given below.

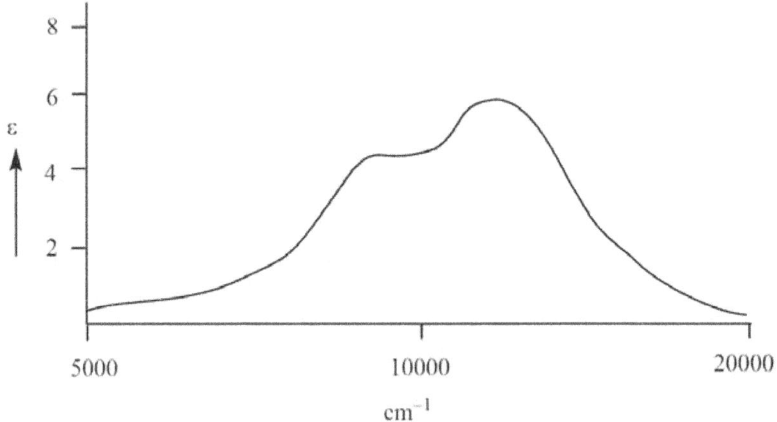

Figure 65. UV-visible spectra of $[Fe(H_2O)_6]^{2+}$.

Now, suppose that the Jahn-Teller effect produces a symmetrical splitting of the ^5E_g state, we can obtain the d-orbital splitting of the parent octahedral complex from the energy of the $^5T_{2g} \rightarrow {}^5E_g$ transition, and that value, in this case, is 10000 cm^{-1}. These results are rather similar to what we have observed in $[Ti(H_2O)_6]^{3+}$. Both complexes possess Jahn-Teller distortion associated with the electronic degeneracy of t_{2g} orbital.

> ### *Static and Dynamic Jahn-Teller Distortion*

On the basis of the observed geometry, the Jahn-Teller distortion can be classified in two types as given below.

1. Static Jahn-Teller distortion: Some molecules show tetragonal shape under all conditions i.e., in solid state and in solution state; at lower and relatively higher temperatures. This is referred to as static Jahn-Teller distortion. Hence the distortion is strong and permanent. For example, in CuF_2 lattice

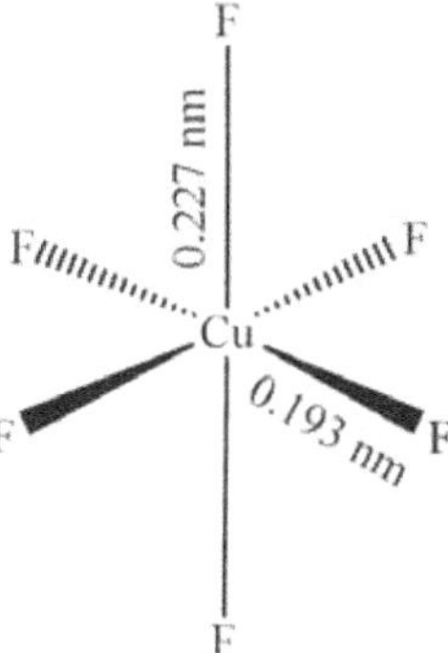

Figure 66. Static Jahn-Teller distortion in CuF_2 lattice.

2. Dynamic Jahn-Teller distortion: If the energy gap between z-out and z-in is smaller than the available thermal energy, the complex ions tend to attain both states, i.e., compressed and elongated. This is known as the "Dynamic Jahn Teller Effect". For examples, consider $K_2Pb[Cu(NO_2)_6]$ complex:

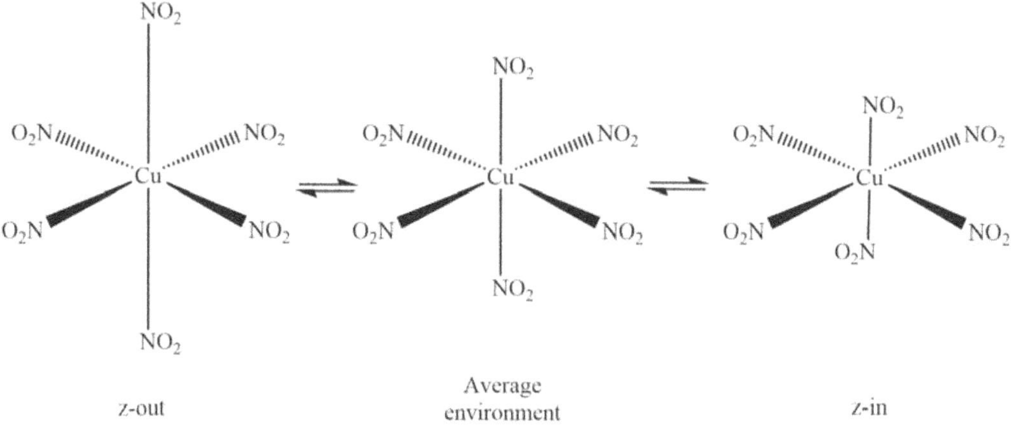

Figure 67. Dynamic Jahn-Teller distortion in $K_2Pb[Cu(NO_2)_6]$.

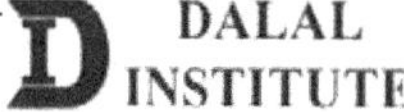

> ➢ *Consequences of Jahn-Teller Distortion*

Some of the main consequences of the Jahn-Teller effect in the field of chemical science are given below.

1. Irving-William series: Stability of metal complexes with a given ligand follows the order $Mn^{2+} < Fe^{2+} < Co^{2+} < Ni^{2+} < Cu^{2+} > Zn^{2+}$. The increase in the stability of the complexes from Mn^{2+} to Zn^{2+} is the increase effective nuclear charge. However, the exceptionally greater stability of Cu^{2+} complexes is the Jahn-Teller distortion.

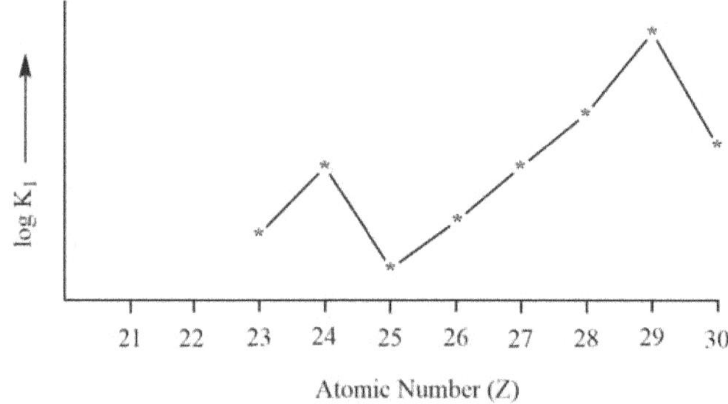

Figure 68. Variation of log K_1 for the addition of first ethylenediamine.

The sequence is generally quoted only for Mn(II) to Zn(II) as there is little or no data available for the other $3d$ series transition metal ions as their M(II) oxidation states are not very much stable. Crystal field theory is based on the idea that the interactions between the metal centre and the ligands-attached are purely ionic in nature; this suggests that the stability of the complexes should be related to the charge to radius ratio (ionic potential).

2. Disproportionation of Au^{2+} salts: Bivalent gold is less stable and undergoes disproportionation to form Au^{1+} and Au^{3+}. On the other hand, bivalent salts of Cu and Ag ions are quite common and relatively more stable. However, as far as the electronic configuration is concerned, all of the three belong to the same group and are d^9 systems.

Thus, a strong Jahn-Teller distortion is expected. The disproportion of Au^{2+} can be explained in terms of increasing Δ value down the group. Therefore, bivalent salts of gold would have the maximum magnitude of crystal field splitting, which results high destabilization associated with the filling of last electron (in $d_{x^2-y^2}$). This makes Au^{2+} to undergo either to form Au^{3+}, a d^8 system; or reduction to Au^{1+}, a d^{10} system. The d^8 system (Au^{3+}) is usually square-planar in geometry and quite stable as the electron from the $d_{x^2-y^2}$ is removed. The d^{10} system (Au^{1+}) is of linear geometry and stable due to fully filled configuration.

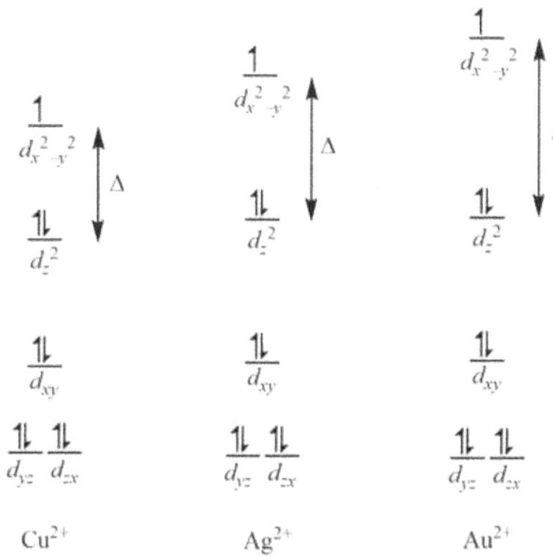

Figure 69. Variation of crystal field splitting energy for tetragonal Cu^{2+}, Ag^{2+} and Au^{2+}.

3. Stabity order of ethylenediamine complexes of Cu^{2+}: Jahn-Teller distortion is responsible for the stability order of $[Cu(en)_3]^{2+}$, cis-$[Cu(en)_2(H_2O)_2]^{2+}$, trans-$[Cu(en)_2(H_2O)_2]^{2+}$.

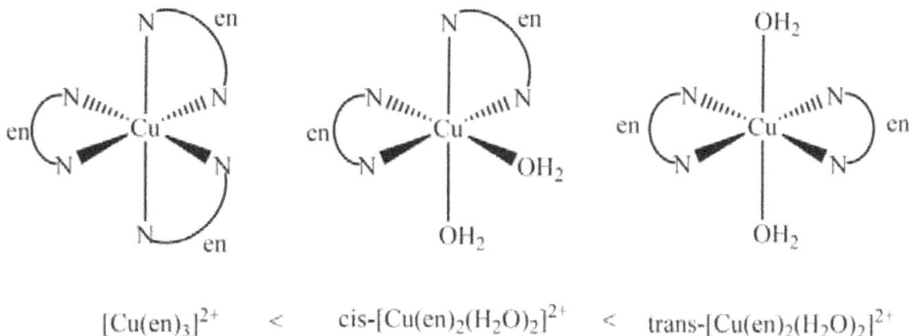

$$[Cu(en)_3]^{2+} \quad < \quad cis\text{-}[Cu(en)_2(H_2O)_2]^{2+} \quad < \quad trans\text{-}[Cu(en)_2(H_2O)_2]^{2+}$$

Figure 70. Stabity order of $[Cu(en)_3]^{2+}$, cis-$[Cu(en)_2(H_2O)_2]^{2+}$, trans-$[Cu(en)_2(H_2O)_2]^{2+}$.

Bivalent copper cannot form $[Cu(en)_3]^{2+}$ because the Jahn-Teller distortion induces strain into the ethylenediamine molecule that is added along z-axis. Therefore, the only complex that exists is $[Cu(en)_2(H_2O)_2]^{2+}$. Similarly, cis-$[Cu(en)_2(H_2O)_2]^{2+}$ is less stable in comparison than trans-$[Cu(en)_2(H_2O)_2]^{2+}$. The extra stability of trans-$[Cu(en)_2(H_2O)_2]^{2+}$ is because of the non-involvement of longer bonds in chelation.

❖ Spectrochemical and Nephelauxetic Series

➢ *The Spectrochemical Series*

The spectrochemical series is a list of ligands arranged on basis of ligand strength and a list of metal ions based on oxidation number, group and its identity.

In crystal field theory, ligands change the difference in energy between the d-orbitals (Δ) called the ligand-field splitting parameter or crystal-field splitting parameter for ligands, which is primarily reflected in differences in color of similar metal-ligand complexes. However, there are many ligands which do not form complexes with a particular metal ion and the vice-versa is also true. It clearly means that the value of crystal field splitting energy for these complexes cannot be calculated experimentally. Therefore, an empirical method must be used to find out their Δ values. In this method, two empirical parameters have been suggested for different metal ions and ligands.

Table 12. Values of parameters g and f for different metal centers and ligands.

Metal ion	g	Ligands	f
Co^{2+}	9.3	3acac	1.21
Co^{3+}	19.0	$6H_2O$	1.00
Cr^{3+}	17.0	$6CH_3COO^-$	0.96
Cu^{2+}	12.0	3en	1.28
Fe^{2+}	10.0	$6OH^-$	1.70
Mn^{2+}	8.5	$6Cl^-$	0.80
Ni^{2+}	8.9	$6F^-$	0.90

Now, on the basis of experimental and empirical results, not only the metal ions but also the ligands can be arranged in the increasing or decreasing order of Δ-values. The empirical formula to calculate the magnitude of crystal field splitting energy is given below.

$$\Delta_o = f \times g \times 10^3 \text{ cm}^{-1}$$

i) For $[Ni(H_2O)_6]^{2+}$

$$\Delta_o = 8.9 \times 1.00 \times 10^3 \text{ cm}^{-1}$$

$$\Delta_o = 8900 \text{ cm}^{-1}$$

This value is pretty close to what has been observed experimentally (8700 cm^{-1}).

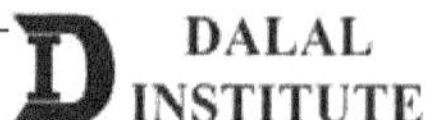

ii) For $[Co(en)_3]^{3+}$

$$\Delta_o = 19 \times 1.28 \times 10^3 \text{ cm}^{-1}$$

$$\Delta_o = 24320 \text{ cm}^{-1}$$

This experimental value of Δ for For $[Co(en)_3]^{3+}$ is 23000 cm^{-1} which is pretty much comparable.

1. Spectrochemical series of ligands: The proposal of the spectrochemical series was first given in the year of 1938 by analyzing the results of UV-visible absorption spectra of cobalt complexes. A simple spectrochemical series ordering of ligands from small Δ to large Δ is given below.

$O_2^{2-} < I^- < Br^- < S^{2-} < SCN^-$ (S-bonded) $< Cl^- < N^{3-} < F^- < NCO^- < OH^- < C_2O_4^{2-} \approx H_2O < NCS^-$ (N-bonded) $< CH_3CN < py$ (pyridine) $< NH_3 < en$ (ethylenediamine) $< bipy$ (2,2'-bipyridine) $< phen$ (1,10-phenanthroline) $< NO^{2-} < PPh_3 < CN^- < CO$

Ligands placed on the left side of this spectrochemical series are usually considered as weak-field ligands and cannot cause the electron-pairing within $3d$ shell and thus form spin-free octahedral complexes. On the other hand, ligands lying at the right end are stronger ligands and form spin-paired octahedral complexes due to forcible pairing of electrons within $3d$ level and therefore are called low spin ligands. However, it should also be noted that the spectrochemical series is definitely backward from what it should be for a reasonable prediction based on the assumptions of crystal field theory. This deviation from crystal field theory highlights the reliability of the main assumption crystal field theory's that purely ionic bonds exist between metal and ligand. The order of the spectrochemical series can be obtained from the understanding that ligands are frequently classified by their donor or acceptor abilities.

Some, like NH$_3$, are σ-bond donors only, with no orbitals of appropriate symmetry for π-bonding interactions. Bonding by these ligands to metals is relatively simple, using only the σ-bonds to create relatively weak interactions. Another example of a σ-bonding ligand would be ethylenediamine, however, ethylenediamine has a stronger effect than ammonia, generating a larger crystal field split, Δ. Ligands that have filled p-orbitals are potentially π-donors. These types of ligands tend to donate these electrons to the metal along with the σ bonding electrons, exhibiting stronger metal-ligand interactions and an effective decrease of Δ. Most halide ligands, as well as OH$-$ and H2O, are primary examples of π-donor ligands. When ligands have vacant π^* and d orbitals of suitable energy, there is the possibility of pi-backbonding, and the ligands may be π-acceptors. This addition to the bonding scheme increases Δ. Ligands that do this very effectively include CN$^-$, CO and many others.

2. Spectrochemical series of metals: Like the ligands, different metal ions can also be ordered in increasing Δ, and this order is independent of the identity of the ligand to a large extent.

$Mn^{2+} < Ni^{2+} < Co^{2+} < Fe^{2+} < V^{2+} < Cu^{2+} < Fe^{3+} < Cr^{3+} < V^{3+} < Co^{3+}$

In general, it is almost impossible to claim whether a given ligand will exert a strong or a weak field on a given metal ion. Nevertheless, if we consider the metal ion, it has been observed that the magnitude of ligand field splitting energy increases with increasing oxidation number, and also as we move down the group.

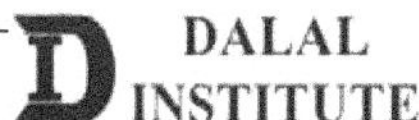

> ### ➤ *The Nephelauxetic Series*

The nephelauxetic series is a list of ligands or metal ions ordered on the basis of the strength of their nephelauxetic effect.

In the inorganic studies, the term "nephelauxetic effect" is very frequently used for transition metal complexes. This refers to a decreasing Racah parameter (B), a measure of inter-electronic repulsion, that occurs when a free transition-metal ion forms a complex with different types of ligands. The label "nephelauxetic" is for cloud-expanding in the Greek language. The presence of the nephelauxetic effect brings out the drawbacks of crystal field theory, as this suggests a somewhat covalent character in the metal-ligand bonding.

The declining value of the Racah parameter hints that in a metal complex, there is less repulsion between the two electrons in a given doubly occupied metal d-orbital than what is in the free ion counterpart, M^{n+}; which consecutively implies that the orbital size is larger after complexation. Two reasons for electron-cloud-expansion effect may be given; one is that the effective positive charge on the metal is reduced by any negative charge on the ligands, the d-orbitals can expand a slight manner; the second is the considers the overlapping with ligand orbitals and creation of covalent bonds increases the size of the orbital.

The reduction of B from its free ion value is normally reported in terms of the nephelauxetic parameter, β, as:

$$\beta = \frac{B'(\text{Complex})}{B\ (\text{Free ion})}$$

Moreover, it is also observed experimentally that the magnitude of the nephelauxetic parameter always follows a certain order with respect to the nature of the ligands attached. However, there are many ligands which do not form complexes with a particular metal ion and the vice-versa is also true. It clearly means that the value of the Racah parameter for these complexes cannot be calculated experimentally. Therefore, an empirical method must be used to find out their B values. In this method, two empirical parameters have been suggested for metal ions and ligands.

Table 13. Values of parameters k and h for different metal centers and ligands.

Metal ion	k	Ligands	h
Co^{2+}	0.24	$6CN^-$	2.0
Co^{3+}	0.35	3en	1.5
Cr^{3+}	0.21	$6H_2O$	1.0
Mn^{2+}	0.07	$6NH_3$	1.4
Ni^{2+}	0.12	$6F^-$	0.8

The empirical formula to calculate the magnitude of the Racah parameter for any metal ion in complexation is given below.

$$B' = B(1 - kh) \text{ cm}^{-1}$$

Where, B and B′ are the Racah parameters for free ion and metal center in complexation, respectively.

i) For $[Ni(H_2O)_6]^{2+}$

$$B' = 1080(1 - 0.12 \times 1.0) \text{ cm}^{-1}$$

$$B' = 950 \text{ cm}^{-1}$$

This value is pretty close to what has been observed experimentally (905 cm^{-1}).

ii) For $[Co(en)_3]^{3+}$

$$B' = 1400(1 - 0.35 \times 1.5) \text{ cm}^{-1}$$

$$B' = 665 \text{ cm}^{-1}$$

This experimental value of B for For $[Co(en)_3]^{3+}$ is 568 cm^{-1}.

The values of Racah parameter (B) for transition metal ion in the gaseous state can be noted from the table given below.

Table 14. Racah parameters for different free transition metal atoms or ions.

Metal	M^0	M^{1+}	M^{2+}	M^{3+}	M^{4+}
Ti	560	681	719	-	-
V	579	660	765	860	-
Cr	790	710	830	1030	1040
Mn	720	872	960	1140	-
Co	789	879	1117	-	-
Ni	1025	1038	1080	-	-
Cu	-	1218	1239	-	-
Fe	805	870	1059	-	1144

Now, on the basis of experimental and empirical results, not only the metal ions but also the ligands can be arranged in the increasing or decreasing order of β-values.

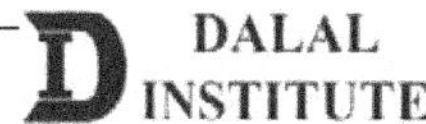

1. Nephelauxetic series of ligands: The list is shown below list commons ligands (with increasing nephelauxetic effect):

$$F^- < H_2O < NH_3 < en < NCS^- \text{ (N-bonded)} < Cl^- < CN^- < Br^- < N_3^- < I^-$$

Although parts of this series may seem quite similar to the spectrochemical series of ligands, that is not true. for instance; the fluoride, ethylenediamine and cyanide seem to occupy almost similar positions; Some other ligands such as chloride, iodide and bromide, are arranged very differently. Roughly, we can say that the ordering reflects the capability of the ligands to form batter covalent bonds with metals; means ligands at the end of the series i.e. with large nephelauxetic effect have batter tendency to for covalent bond than the ligands present at the start of the series.

2. Nephelauxetic series of metal ions: The nephelauxetic effect does not only depend upon the ligand type but also upon the central metal ion. These too can be arranged in order of increasing nephelauxetic effect as follows:

$$Mn^{2+} < Ni^{2+} \approx Co^{2+} < Mo^{2+} < Re^{4+} < Fe^{3+} < Ir^{3+} < Co^{3+} < Mn^{4+}$$

It is obvious that as the oxidation number for the metal ion increases, the nephelauxetic effect also increases.

❖ Charge Transfer Spectra

A charge transfer band may be defined as the peak arising from the transition in which an electron is transferred from one atom or group in the molecule to another one.

In other words, the transition occurs between molecular orbitals that are essentially centered on different atoms or groups. These transitions are neither Laporte nor spin-forbidden in nature; and therefore, show very intense absorption. Charge transfer transitions are primarily classified in four types as:

➢ *Ligand to Metal Charge Transfer*

The ligand to metal charge transfer (LMCT) in metal complexes arises when the electrons are transferred from a molecular orbital with a ligand-like character to those with metal-like character. This type of transfer is predominant if the following conditions are fulfilled:

i) The ligands should have lone pair of electrons with relatively high-energy such as O^{2-}, Cl^-, Br^-, S^{2-} or Se.

ii) The metal should be in a high oxidation state and must have low-lying empty orbitals.

These conditions imply that the acceptor level is available and low in energy. Moreover, the charge transfer transitions for octahedral and tetrahedral complexes are different; therefore, these two types of ligand to metal charge transfer are quite important and must be discussed in detail.

1. LMCT in octahedral complexes: Before we discuss the ligand to metal charge transfer in octahedral complexes, the molecular orbital diagram for σ and π-bonding in ML_6 geometry should be recalled. The electrons can be excited, not only from the t_{2g}^* to e_g^* but also from the bonding molecular orbitals of σ and π nature that are predominantly associated with the ligands. The latter two types of excitation modes result in

the charge transfer spectra, labeled as the ligand to metal charge transfer. This type of transition results in a formal reduction of the metal.

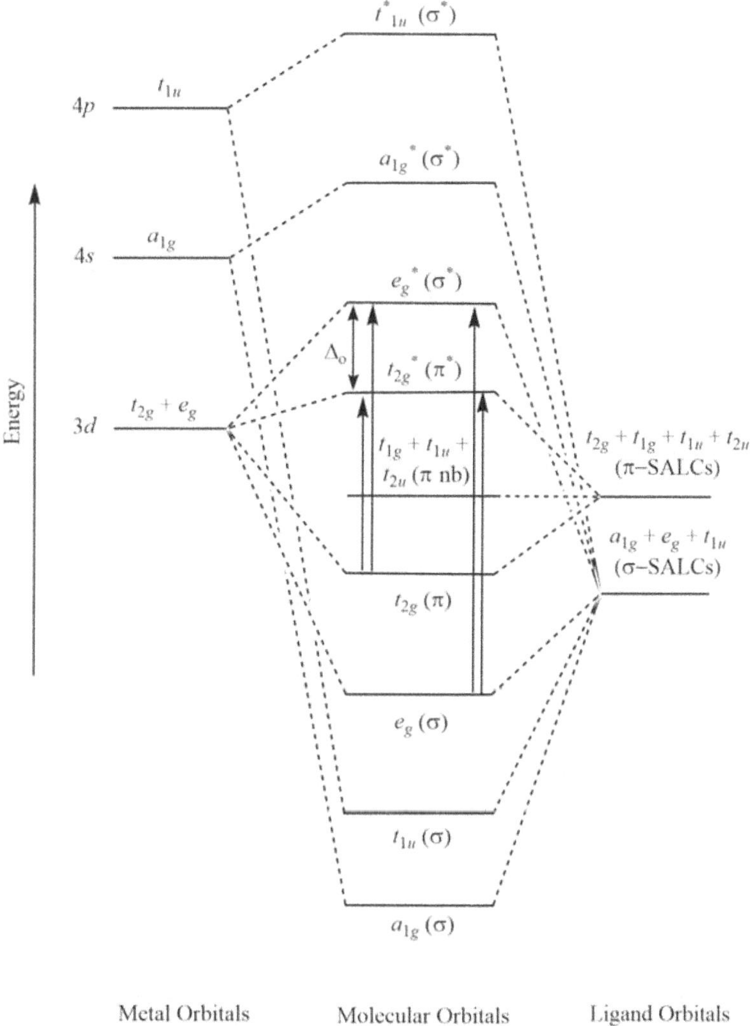

Figure 71. Ligand to metal charge transfer in octahedral (ML$_6$) complexes.

Consider a d^6 octahedral complex, such as $[\mathrm{Co(NH_3)_5(X)}]^{2+}$ (X = F⁻, Cl⁻, Br⁻ or I⁻), whose t_{2g}^* levels are filled. As a consequence, an intense absorption band in $[\mathrm{Co(NH_3)_5F}]^{2+}$ is observed above 40000 cm⁻¹; and that is corresponding to a transition from ligand σ-bonding molecular orbital to the empty e_g^* molecular orbital. However, in $[\mathrm{Co(NH_3)_5Cl}]^{2+}$ this intense band with two components is observed above 30000 cm⁻¹.

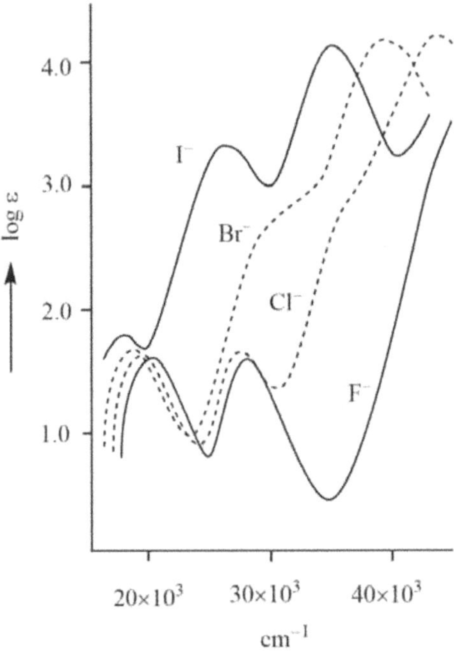

Figure 72. The UV-visible absorption spectra of $[Co(NH_3)_5(X)]^{2+}$ (X = F⁻, Cl⁻, Br⁻ or I⁻).

The charge-transfer bands appear at lower energy for $[Co(NH_3)_5Br]^{2+}$ and at still lower energy for $[Co(NH_3)_5I]^{2+}$, overlapping with ligand-field bands and masking their higher energy peaks. It must be noted that the ligand-field peaks shift slightly according to the increasing strength of the crystal field of X, but the charge-transfer bands show very large shifts. The shifts in the energies of the LMCT transitions in $[Co(NH_3)_5(X)]^{2+}$ correspond to the changes in ease of removal of the electron (oxidation) from X⁻. Though the transition corresponds to the transfer of an electron from X⁻ to Co^{3+}; and therefore, no net oxidation-reduction occurs because of the very short lifetime of the excited state. Nonetheless, this process, provide a mechanism for photochemical decomposition that occurs for many complexes stored in strong light. A similar pattern has been observed in the case of $[Cr(NH_3)_5(X)]^{2+}$ (X = F⁻, Cl⁻, Br⁻ or I⁻).

2. LMCT in tetrahedral complexes: Before we discuss the ligand to metal charge transfer in tetrahedral complexes, the molecular orbital diagram for σ and π-bonding in ML₄ geometry should be recalled. The electrons can be excited, not only from the e^* to $t_2{}^*$, but also from the π-bonding molecular orbitals of t_2-symmetry and nonbonding π-SALCs of t_1-symmetry; both of which are predominantly associated with the ligands. The latter two types of excitation modes result in the charge transfer spectra, labeled as the ligand to metal charge transfer. This type of transition results in a formal reduction of the metal.

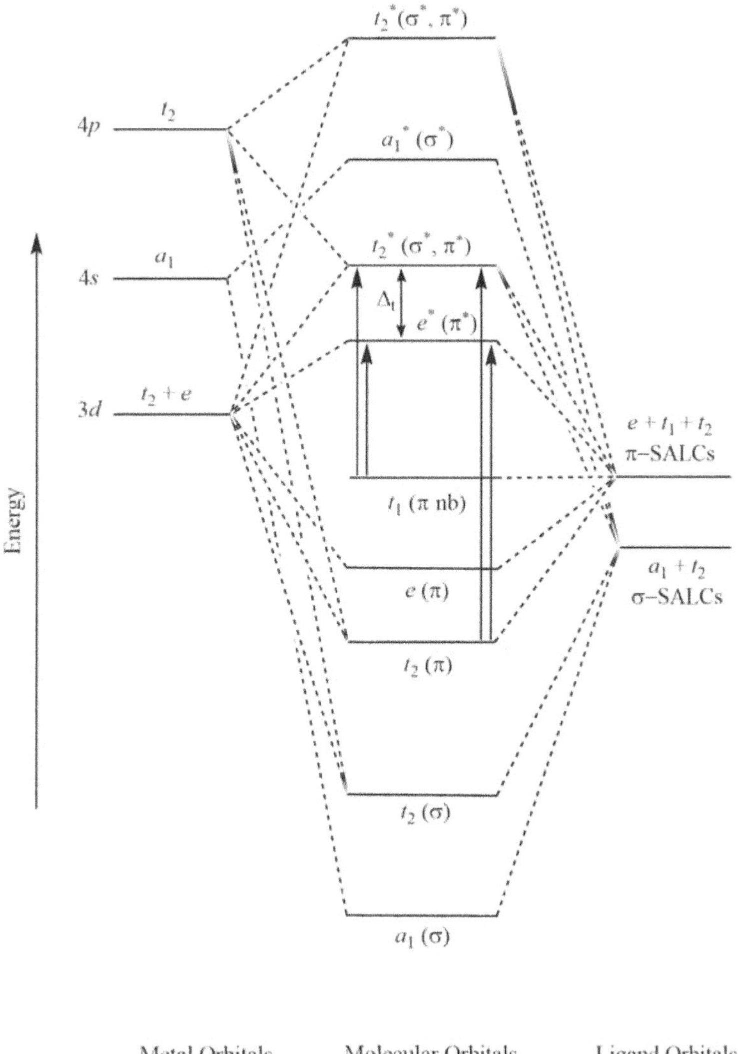

Figure 73. Ligand to metal charge transfer in tetrahedral (ML_4) complexes.

Consider a d^0 tetrahedral complex, such as MnO_4^-, whose e^* and t_2^* levels are totally empty. As a consequence, very intense absorption spectra in permanganate is obtained and all of the four ligand-to-metal-charge-transfer (LMCT) transitions are observed. However, it must be noted that three out of four peaks happen to arise in the ultra-violet region and only $t_1 \rightarrow e^*$ transition belongs to the visible range. Moreover, this particular transition is also responsible for the deep purple color of permanganate ion.

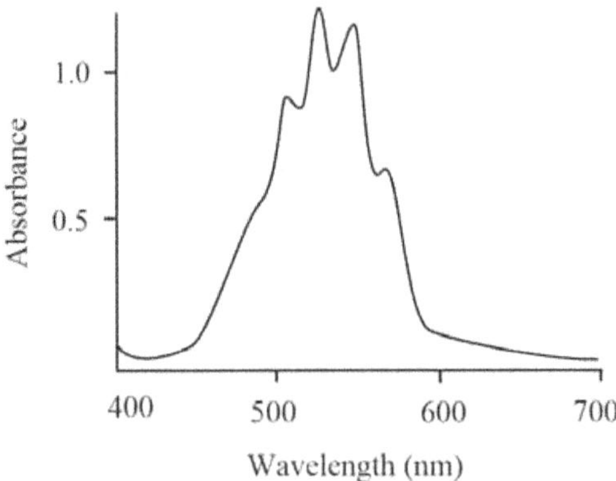

Figure 74. The UV-visible absorption spectra of MnO_4^-.

The complete profile of absorption spectra of MnO_4^- ion is given below as:

Table 15. The type of transition and their wavenumbers of the absorption peaks for MnO_4^-.

Transition type	Wavenumber (cm^{-1})
$L(t_1) \rightarrow M(e^*)$	17700
$L(t_1) \rightarrow M(t_2^*)$	29500
$L(t_2) \rightarrow M(e^*)$	30300
$L(t_2) \rightarrow M(t_2^*)$	44400

A similar pattern has been observed in the case of CrO_4^{2-}. The energies of transitions correlate with the order of the electrochemical series. Lower energy absorption is expected for the metal ions which are reduced more easily. The abovementioned trend is in accordance to the transfer of electrons from the ligand to the metal, and hence resulting in a reduction of the metal center by the ligand attached.

Oxidation state		Complexes			
+7	MnO_4^-	TcO_4^-	ReO_4^-		
+6	CrO_4^{2-}	MoO_4^{2-}	WO_4^{2-}		Blue shift
+5	VO_4^{3-}	NbO_4^{3-}	TaO_4^{3-}		

LMCT shows blue shift

> ➤ *Metal to Ligand Charge Transfer*

The metal to ligand charge transfer (MLCT) in metal complexes arises when the electrons are transferred from a molecular orbital with a metal-like character to those with a ligand-like character. This type of transfer is predominant if the following conditions are fulfilled:

i) The ligands should have high-energy empty π^* orbitals such as CO, CN^-, SCN^- or NO.

ii) The metal should be in a low oxidation state and must have high-lying filled orbitals.

These conditions imply that the empty π^* orbitals on the ligands become the acceptor orbitals on the absorption of light. The available acceptor level is relatively high in energy. However, before we discuss the ligand to metal charge transfer in transition metal complexes, the molecular orbital diagram for σ and π-bonding in ML_6 geometry should be recalled.

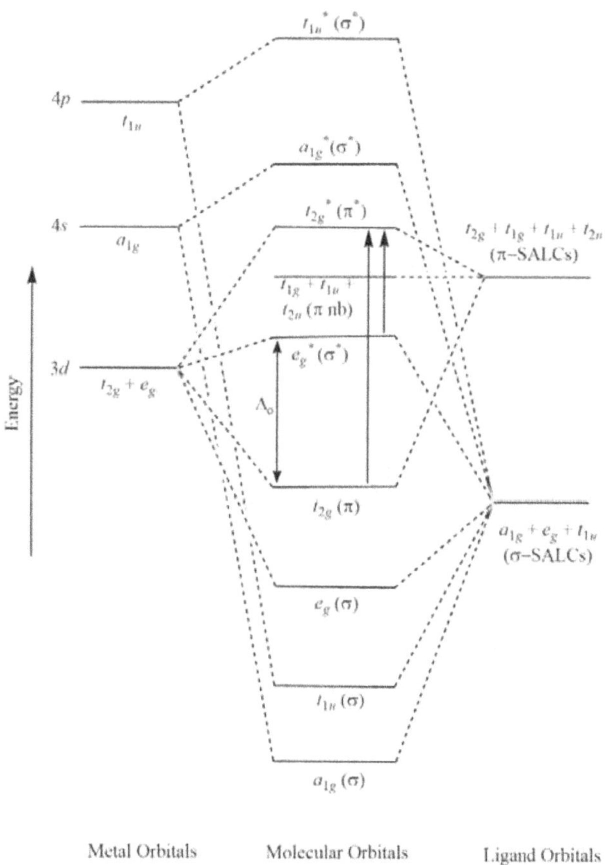

Figure 75. Metal to ligand charge transfer in octahedral (ML_6) complexes.

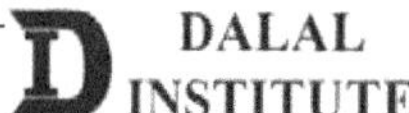

The electrons can be excited, not only from the t_{2g} to e_g^*, but also from the bonding molecular orbitals of π and antibonding molecular orbitals of σ nature (predominantly associated with the ligands) to antibonding π^*. The latter two types of excitation modes result in the charge transfer spectra, labeled as the metal to ligand charge transfer. This type of transition results in formal reduction of the metal. The common type of ligands taking part in MLCT include 2,2'-bipyridine (bipy), 1,10-phenanthroline (phen), CO, CN− and SCN−. Examples of these complexes include: Tris(2,2'-bipyridyl) ruthenium(II), W(CO)$_4$(phen), K$_4$[Fe(CN)$_6$], K$_3$[Fe(CN)$_6$], [Fe(phen)$_3$]$^{3+}$, [Fe(acac)$_3$] and Fe(CO)$_3$(bipy).

An orange-colored complex of bivalent ruthenium, [Ru(bpy)$_3$]$^{2+}$, is being analyzed because the excited electronic state that results from this charge transfer has an average life-time around microseconds and the complex can act as an adaptable photochemical reagent. The photo-reactivity of these complexes arises from the nature of the reduced ligand and oxidized metal center. Now although the states of MLCT complexes such as [Ru(bipy)$_3$]$^{2+}$ and Re(bipy)(CO)$_3$Cl were intrinsically not that much reactive, there are many MLCT complexes which are characterized by reactive MLCT states. Vogler and Kunkely proposed that a MLCT complex can be considered as an isomer of the ground state, that possesses a reduced ligand and oxidized metal. Hence, many reactions like electrophilic attack, oxidative addition at the metal ion due to the reduced ligand, the radical reactions on the reduced ligand, or the outer-sphere charge-transfer reactions can be attributed to states arising from metal-to-ligand-charge-transfer transitions. The reactivity of MLCT states usually depends on the oxidation of the metal center. The succeeding processes include exciplex formation, cleavage of metal-metal bonds, associative ligand substitution.

> ### *Metal to Metal Charge Transfer*

The metal to metal or inter-valence charge transfer may simply be defined as the excitation and subsequent transfer of an electron from a low oxidation state cation to a neighboring cation of a higher oxidation state.

The transfer is usually excited by a certain visible portion of the light of and produces a characteristic color. The electron then drops back down, giving off the extra energy as a small amount of heat. An example is corundum with the coupled substitution. The simplified chemical equation representing various oxidation states can be given as:

$$Fe^{2+} + Ti^{4+} \rightarrow Fe^{3+} + Ti^{3+}$$

This reaction absorbs red photons and gives sapphire its characteristic blue color. In hematite, the process absorbs all visible photons. Materials that exhibit this property retain their dark color regardless of how finely they are ground. Materials exhibiting metal to metal charge transfer (MMCT) are also conductors. MMCT generally shows a strong sloping spectral signature in the range 500-1000 nm. Edge-shared octahedral geometries generally exhibit MMCT in the range 700-800 nm, while face-shared octahedral complexes exhibit it is in the range of 800-900 nm. In order for metal-to-metal–charge-transfer (MMCT) to occur, orbital must overlap so electrons can flight back and forth. Examples of the systems displaying metal to metal charge transfer are:

Table 16. The type of transition and their wavenumbers of the absorption peaks for MnO_4^-.

System	Absorption maxima (nm)
$Fe^{2+} \rightarrow Fe^{3+}$ (edge-shared)	700-800
$Fe^{2+} \rightarrow Fe^{3+}$ (face-shared)	800-900
$Ti^{3+} \rightarrow Ti^{4+}$	600-800
$Mn^{2+} + Ti^{4+} \rightarrow Mn^{3+} + Ti^{3+}$	380-450

The most popular example of inter-valence charge transfer is "Prussian blue". This compound has the formula $KFe[Fe(CN)_6]$ and shows a very intense blue color owing to the transfer of an electron from Fe^{2+} to Fe^{3+}. In the crystal structure of Prussian blue, the Fe^{2+} ions are bonded with N atom while the Fe^{3+} ions are bonded with C atom of octahedrally surrounding of CN^- ligands. Therefore, the charge transfer takes place through the cyanide bridge.

> ➢ *Ligand to Ligand Charge Transfer*

The ligand to ligand or inter-ligand charge transfer may be defined as the excitation and subsequent transfer of an electron from a one ligand orbitals to a neighboring ligand orbital.

The ligand to ligand charge transfer (LLCT) or inter-ligand charge transfer transitions are quite uncommon and rarely observed. In comparison to the enormous literature on metal-to-ligand and ligand-to-metal charge transfer, very little has been published on LLCT. In most of the cases, LLCT peaks are difficult to detect in UV-visible absorption spectra; which can be attributed to the fact that these peaks may be obscured or hidden under absorption bands of different origins, or they may occur at a position very distinct from those ordinarily analyzed. The LLCT bands are of low intensity due to the poor overlap between the participating orbitals. Molecular orbitals with dominant ligand characters may have some amount of metal character also, and a transition that is labelled as ligand to ligand may in fact also involve the metal to some point. However, if the LLCT is pure, it would have transition energy that does not change considerably when the metal is changed. A more specific form of L_{red}–M–L_{ox} complexes is mixed-valence compounds (ligand-based), which possesses the same ligand in two different redox states. In this case, the interaction between ligands may yield a partial or complete electron delocalization between the oxidized and reduced form of the ligands. Therefore, the LLCT loses its charge-transfer character because now it is occurring between delocalized orbitals.

One of the recent examples of metal complexes involving ligand to ligand charge transfer is $(CuTpAsPh_3)$. The emission and UV-visible absorption spectra of $(CuTpAsPh_3)$, contain low-energy bands (with a band maximum at $16\,500$ cm^{-1} in emission and a weak shoulder at about 25000 cm^{-1} in absorption) that are not present in the corresponding spectra of the phosphine or amine complexes. The peaks are assigned to the ligand to ligand charge transfer (LLCT) may have some contribution from the metal.

❖ Electronic Spectra of Molecular Addition Compounds

The electronic absorption spectra of molecular addition compounds can be best understood by considering the example of iodine adducts. The absorption maxima for iodine (dissolved in carbon tetrachloride) appears at about 520 nm wavelength. This band, without any doubt, can successfully be assigned to a $\pi^* \rightarrow \sigma^*$ transition; and is also responsible for the violet color of free iodine. However, when certain ligands like macrocyclic ethers or any other donor molecules are added to the iodine solution, two obvious changes in the absorption spectrum occur. The characteristic band of free iodine shows a blue shift, and some intense peaks arise in the ultraviolet region. The hypsochromic shift is due to the complexation of free iodine, while the origin of additional peaks in the ultra-violet region can be ascribed to the charge transfer transitions. The later claim is also confirmed by the absence of any dominant peak in the absorption spectra of the ligand alone. Furthermore, the isosbestic point at 490 nm suggests the presence of two absorbing species in the system; and the value of the equilibrium constant K obtained over a wide range of base concentrations also demonstrates the existence of a 1:1 molecular addition compound.

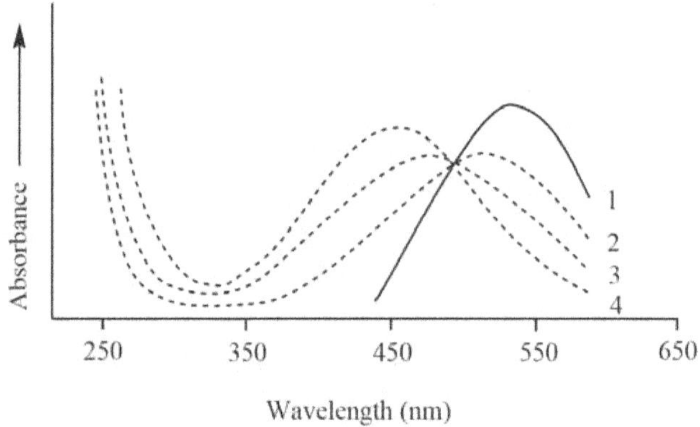

Figure 76. The UV-visible absorption spectra of iodine (1) and base-iodine solutions in CCl_4 with increasing concentration of base (2 < 3 < 4).

The observed blue shift in the free iodine band is due to the perturbation of σ^* molecular orbital of iodine by a repulsive interaction between the two components of the charge-transfer complex. Therefore, as the repulsive interactions increase, a larger blue shift of the iodine band would be expected. Hence, it is sensible to consider the amount of the hypsochromic shift in the free I_2 band as a measure of the interaction between the iodine molecule and the ligand molecule. The limiting value of the blue shift is approximately 360 nm and which is the characteristic absorption peak for I_3^- ions. The formation of triiodide ion, implicitly, would mean that the magnitude of the polarization of the iodine molecule in base-iodine complex is so large that a clear-cut charge separation has occurred. However, the confirmation of I_3^- ions in solution must be done by checking the presence of the second characteristic peak of triiodide ion at 292 nm.

The absorption spectra of I_2 in the presence of large excess of dicyclohexyl-18-crown-6 in chloroform solution at is shown in 'Figure 77'. It can clearly be seen that none of the initial reactants show any measurable absorption in 250-450 nm region, however, the addition of crown ether to the iodine solution generates in two strong absorptions bands the aforementioned wavelength range. This can be attributed to the formation of charge-transfer complexes between the dicyclohexyl-18-crown-6 and the free iodine which causes the vanishing of purple color from the solution. The spectra recorded for the electron donor-acceptor (EDA) complex between ligands and I_2 are time-dependent and the intensities of the different absorptions change significantly with time. Though the general profile of the spectra remains unaffected, after some time the purple color of solution begins to disappear slowly and the intensity of the absorption bands in the 250–600 nm region declines with time.

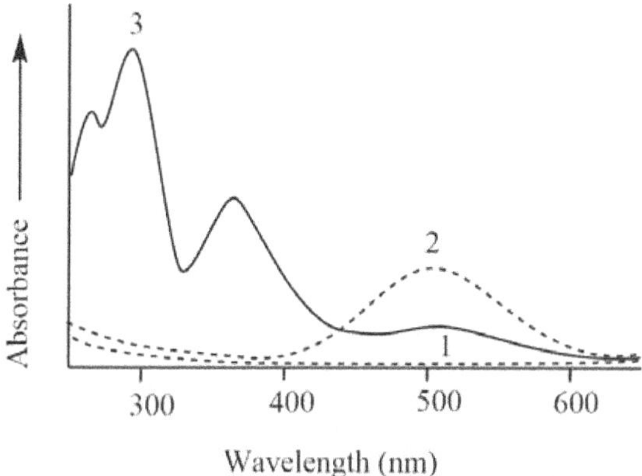

Figure 77. The UV-visible absorption spectra of I_2 in the presence of excess of dicyclohexyl-18-crown-6.

If ψ_{cov} includes contributions from covalent interactions while ψ_{ele} includes contributions from purely electrostatic forces (these are described as charge-transfer interactions); then the bonding in iodine adducts can be described by the equation:

$$\psi = a\psi_{cov} + b\psi_{ele}$$

The ground state of most adducts exists with $b > a$; however, as the charge-transfer transition occurs, the situation is actually reversed with $a > b$. This results in a band around 250 nm owing to the transition in which an electron from the ground state is promoted to an excited state. Hence, the assignment of the charge-transfer band can be approximated by a transfer of an electron from the nonbonding orbital of the base to the σ^* orbital iodine. The whole spectra (including normal $\pi^* \rightarrow \sigma^*$ transition after complexation) can be explained in terms of relative energies of the molecular orbitals of iodine and the complex. The molecular orbitals ($\sigma_{complex}$, $\sigma^*_{complex}$, and $\pi^*_{complex}$) are very much like the original base and iodine orbitals because of the weak Lewis acid-base interaction.

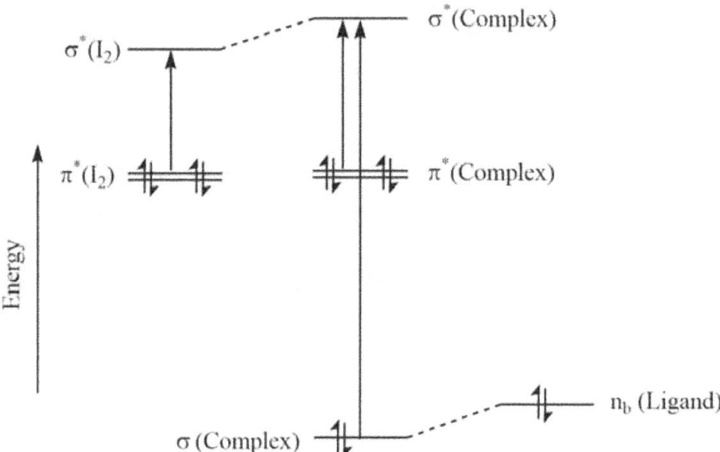

Figure 78. The partial molecular orbital diagram in base–iodine adducts.

Meanwhile, $\sigma^*_{complex}$ is slightly higher in energy than the corresponding σ^*_{iodine}, the analogs transition in complexed iodine requires slightly more energy than the corresponding transition in free and a blue shift is observed. The charge-transfer transition occurs at higher energy in the ultraviolet region. The claim that the blue shift is correlated to the magnitude of the base-iodine interaction (enthalpy of formation of adduct) would be reasonable only if the energy of $\pi^*_{complex}$ differs very little from that of π^*_{iodine} or its energy changes with the enthalpy in a linear fashion. When a wide range of different types of Lewis bases were studied with iodine solution, it has been found that the abovementioned correlation does exist roughly but is not of quantitative nature. The equation relating the charge transfer band, the ionization energy of the base, I_b, and the electron affinity of the acid, E_a, has also been reported:

$$\nu_{charge\ transfer\ band} = I_b - E_a - \Delta$$

where Δ is an empirically determined constant for a related series of bases.

Both inorganic and organic researchers have always been interested in the enthalpies for the formation of these charge-transfer complexes. This is because a lot of information about donor and acceptor interactions can be obtained which is pretty much essential to an understanding of many phenomena in inorganic systems; especially in the areas of non-aqueous solvents and coordination chemistry. It should also be noted that the thermodynamic study of these adducts is easier in the solvents like hexane or carbon tetrachloride than in polar solvents, which is due to the larger magnitude of enthalpies and entropies of solvation in the later. The characteristic results from ligand-iodine solutions with minimal solvation effects for a wide range of systems can be studied and the information that can be obtained by studying enthalpies of association in weakly basic, non-polar solvents are given below:

1. A correlation of the heat of formation of iodine adducts of a series of para-substituted benzamides with the Hammett substituent constants of the benzamides.

2. The donor properties of the π-electron systems of alkyl-substituted benzenes

3. The donor properties of sulfites, sulfoxides, sulfones, and carbonyl compounds have been studied which shows that carbon-oxygen bonding in ketones and acetates is more effective than in sulfur-oxygen bonding.

4. The ring size effect on the donor properties of cyclic ethers and sulphides showed that for $(CH_2)_n$ S, donor strength of sulfur follows order $n = 5 > 6 > 4 > 3$; and the order for the corresponding ethers is $4 > 5 > 6 > 3$.

5. The donor strength of 3°, 2°, and 1° amines have been assessed which varies with the acid studied.

Besides iodine, a lot of charge-transfer complexes of many other Lewis acids absorb light in the UV-visible region. The factors affecting the magnitude of the interaction and other useful information regarding the bonding in the molecular addition compounds of SO_2, ICl, Br_2, phenol and other acids have also been investigated.

❖ Problems

Q 1. Define term symbols and also find out the ground state term for V^{3+} ion.

Q 2. Calculate the total number of microstates for p^2 and d^7-configurations.

Q 3. Draw and discuss the pigeon-hole diagram for d^2-configuration and also comment on the energy correlation of all the free ion terms.

Q 4. How many singlet microstates do exist for a metal ion with an electronic configuration of $3d^1, 4f^1$?

Q 5. Discuss the microstate distribution in $^7F_2$ term symbol.

Q 6. Explain the energy correlation and spin-orbit coupling in Cr^{3+} and Cu^{2+}.

Q 7. What are Orgel diagrams? How do they differ from Tanabe-Sugano diagrams?

Q 8. Draw and discuss the generalized Orgel diagram for d^2. d^3, d^7 and d^8 electronic configurations.

Q 9. How can you find out the high-spin low spin nature of a metal complex using the Tanabe-Sugano diagram for d^n-systems? Explain in detail using a suitable example.

Q 10. What do you understand from the trigonal distortion of octahedral complexes? How does it affect the various d-orbital energy levels in low-spin Co^{3+} complexes?

Q 11. Write a short note on the structural evidence from the electronic spectrum of transition metal complexes.

Q 12. Define the Jahn-Teller theorem. Discuss its effect in the coordination chemistry.

Q 13. How does the Jahn-Teller distortion affect the electronic spectrum of transition metal complexes?

Q 14. Distinguish between static and dynamic Jahn-Teller distortion.

Q 15. Write a short note on the spectrochemical series.

Q 16. What is 'the nephelauxetic' effect and what is the empirical formula to calculate the Racah parameter for different metal ions in complexation?

Q 17. What is ligand to metal charge transfer? Draw and discuss in tetrahedral complexes using MnO_4^-.

Q 18. What is Prussian blue? Discuss the cause of its characteristic blue color.

Q 19. What are the molecular addition compounds? Discuss the spectra of iodine in carbon tetrachloride.

Q 20. Give any five applications of the enthalpy if adduct formation.

❖ Bibliography

[1] N. N. Greenwood, A. Earnshaw, *Chemistry of the Elements*, Butterworth-Heinemann, Oxford, Britain, 1998.

[2] B. W. Pfennig, *Principles of Inorganic Chemistry*, John Wiley & Sons, New Jersey, USA, 2015.

[3] J. E. Huheey, E. A. Keiter, R. L. Keiter, *Inorganic Chemistry: Principals of Structure and Reactivity*, HarperCollins College Publishers, New York, USA, 1993.

[4] B. R. Puri, L. R. Sharma, K. C. Kalia, *Principals of Inorganic Chemistry*, Milestone Publishers, Delhi, India, 2012.

[5] S. F. A. Kettle, *Physical Inorganic Chemistry: A Coordination Chemistry Approach*, Springer-Verlag, Berlin, Germany, 1996.

[6] B. N. Figgis, M. A. Hitchman, *Ligand Field Theory and its Applications*, Wiley-VCH, New York, USA, 2000.

[7] J. E. House, *Inorganic Chemistry*, Academic Press, California, USA, 2008.

[8] Y. Tanabe, S. Sugano, *On the Absorption Spectra of Complex Ions. I*, Journal of the Physical Society of Japan, 9, 1954, 753.

[9] Y. Tanabe, S. Sugano, *On the Absorption Spectra of Complex Ions. II*, Journal of the Physical Society of Japan, 9, 1954, 766.

[10] A. L. Tchougréeff, R. Dronskowski, *Nephelauxetic Effect Revisited,* International Journal of Quantum Chemistry, 109, 2009, 2606.

[11] C. K. Jørgensen, C. H. D. Verdier, J. Glomset, N.A. Sörensen, *Studies of Absorption Spectra IV: Some New Transition Group Bands of Low Intensity*, Acta Chem. Scand., 8, 1954, 1502.

[12] R. S. Drago, *Physical Methods for Chemists*, Surfside Scientific Publishers, Florida, USA, 1992.

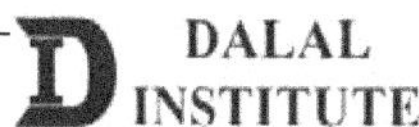

CHAPTER 9

Magnetic Properties of Transition Metal Complexes:

❖ Elementary Theory of Magneto-Chemistry

The history of magnetism starts earlier than 600 B.C., but the initiation of conceptual understanding dates back only in the twentieth century, after which the scientific community started developing technologies based on this understanding. The phenomenon of magnetism was most likely first detected in the mineral magnetite, also called "lodestone (Fe_3O_4)", which is essentially a chemical compound of iron and oxygen with inverse spinal structure. In ancient times, the Greeks were the first who used this compound and called it a magnet due to its remarkable capability to attract iron pieces or other blocks of the same material. Plato (428-348 B.C.) and Aristotle have also given some description of permanent magnets in their writings. The first record of a magnetic compass used for navigational purpose comes from a Chinese writing (1040 A.D.). The first systematic scientific investigation of the phenomenon of magnetism was carried out by a British physicist William Gilbert (1540-1603); who also discovered that the earth is also a weak magnet itself. A French military engineer and physicist Charles-Augustin de Coulomb (1736-1806) initiated the quantitative studies of magnetic phenomena in the eighteenth century. He gave the inverse square law, telling that the attraction force between two magnetic objects is directly proportional to the multiplication of their individual field strengths and inversely proportional to the square of their distance of separation. Danish physicist, H. C. Oersted (1777-1851), first proposed a link between the magnetism and electricity. French physicist Andre Marie Ampere (1775-1836) and British physicist Michael Faraday (1791-1869) carried out the experiments involving the effects of magnetic and electric fields on one another. Finally, the legendary Scotsman, James Clerk Maxwell (1831-1879), provided the theoretical basis to the physics of electromagnetism in the nineteenth century by showing that the magnetism and electricity are just the two faces of the same coin.

The modern point of view of magnetism in condensed matter originates from the work of two French physicists, Pierre Curie (1859-1906) and Pierre Weiss (1865-1940). Pierre Curie studied how the temperature affects magnetism of different materials and witnessed that magnetism vanished quickly above a certain critical temperature in materials like iron. Pierre Weiss put forward a theory about magnetism which was based upon the internal magnetic field, present at the molecular scale, which is proportional to the magnetic average that aligns the micro-magnets in magnetic substances. Today's understanding of magnetic phenomena counts on the theory of the motion and interactions of electrons in atoms, given by Ernest Ising and Werner Heisenberg. The study of the magnetic field generated by the motion of electrons and nuclei in different materials help us to rationalize various fundamental effects and phenomenon. For example, nuclear magnetic resonance is one of the most important tools to characterize organic and inorganic compounds; or the study of magnetic properties of transition metal complexes has provided a beautiful insight of stereochemistry of metal centers and the nature of metal-ligand bonding. The branch of chemistry which is especially concerned with the magnetic properties of chemical compounds is generally called as magneto-chemistry.

➢ *Basic Terminology*

Now before we start to discuss the classical and quantum mechanical aspects of magneto-chemistry, some terms, which will be used very frequently, must be defined.

1. Magnetic field strength (H): The magnetic fields produced by currents are calculated using Biot-Savart Law or Ampere's Law; and are generally measured in Tesla (T). However, when the fields so created pass through the magnetic things which can have magnetic fields induced internally; uncertainties can arise about which part of the field comes from the material considered and which part of the field comes from the external currents. Therefore, it is a common practice to distinguish the two by defining another magnetic field quantity "H" usually called as "magnetic field strength". Thus, the magnetic field strength (H) is one of two ways that can be used to express the magnetic field intensity. To be precise, a distinction is made between magnetic flux density B, measured in Newton per ampere-meter (N/mA), also called tesla (T) and magnetic field strength H, measured in amperes per meter (A/m).

2. Magnetic induction (B): The phenomenon of the rise of magnetism in a specimen of magnetic material when it is placed in an external magnetic field is called as magnetic induction. The term "magnetic induction" is sometimes also referred as "magnetic flux density" which may be defined as the total number of magnetic lines of force crossing a unit area around a point positioned inside an object placed in the magnetizing field. The generally used symbol for magnetic induction or magnetic flux density is "B"; and the relationship between total magnetic flux (φ) and magnetic flux density is B = φ/a, where a is the cross-sectional area in square meter. The SI unit for magnetic flux density is the Tesla (T) which is equal to Weber/m^2 or N/mA.

3. Magnetic permeability(μ): The magnetic permeability, or simply the permeability, may be defined as the relative decrease or increase in the total magnetic-field inside a substance compared to the magnetizing field, the given material placed within. In other words, the permeability of a material is equal to the magnetic flux density (B) created within the material by a magnetizing field divided by the intensity of magnetizing field i.e magnetic field strength (H). Therefore, magnetic permeability is defined as $\mu = B/H$. In SI units, permeability is measured in Henry per meter (H/m), or equivalently in Newton (kg m/s^2) per ampere squared (NA^{-2}).

4. Intensity of magnetization (I): The intensity of magnetization represents the extent up to which a material has been magnetized under the influence of the magnetizing field. The intensity of magnetization of a magnetic material is thus defined as the magnetic moment per unit volume of the material i.e. I = M/V, where M is the magnetic moment which is equal to the product of pole strength and the distance of separation of magnetic poles of the specimen. Like H, the intensity of magnetization is also measured in amperes per meter (A/m).

5. Magnetic susceptibility (K, χ, χ_M): The magnetic susceptibility is simply a measure of the magnetic properties of a material. The magnetic susceptibility shows whether a substance is repelled out or attracted into a magnetic field, which in turn has practical applications. Mathematically, volume susceptibility (K) is the ratio of the intensity of magnetization to the applied magnetizing field intensity i.e. $K = I/H$. Now because the units of I and H are same, volume susceptibility is a dimensionless quantity. However, volume susceptibility divided by the density of the material is called as mass susceptibility (χ) which is measured in cm^3 g^{-1}. The χ multiplied by molar mass is called as molar susceptibility (χ_M) which is measured in cm^3 mol^{-1}.

> ### *The Classical Concept of Magnetism*

The classical theory of magnetism was developed long before quantum mechanics. The Lenz's law states that when a substance is placed within a magnetic field of strength H, the field-induced within the substance (B) differs from H by $4\pi I$ i.e. the difference is proportional to the intensity of magnetization of the material. Mathematically, we can state this relationship as:

$$B = H + 4\pi I \tag{1}$$

Dividing equation (1) throughout by H, we get

$$\frac{B}{H} = 1 + 4\pi \frac{I}{H} \tag{2}$$

Now putting the value of $I/H = K$ (volume susceptibility) in equation (2), we get

$$\frac{B}{H} = 1 + 4\pi K \qquad \text{or} \qquad 4\pi K = \frac{B}{H} - 1 \tag{3}$$

For some materials, the ratio of B/H is less than one, which means the value of K is negative, these materials are labeled as diamagnetic materials. For some materials, the ratio of B/H is greater than one, which means the value of K is positive, these materials are labeled as paramagnetic materials. The mass susceptibility (χ) in cm^3 g^{-1} can be obtained as:

$$\chi = \frac{K}{d} \tag{4}$$

Or the molar susceptibility in cm^3 mol^{-1} can be calculated from equation (4) as follows:

$$\chi_M = \chi \times M \tag{5}$$

Where d and M are the density and gram molar mass of the material, respectively. Since this value includes the underlying diamagnetism of paired electrons, it is necessary to correct for the diamagnetic portion of χ_M to get a corrected paramagnetic susceptibility i.e. measured susceptibility (χ_M) = paramagnetic susceptibility (χ_M^P) + diamagnetic susceptibility (χ_M^D).

$$\chi_M^P = \chi_M - \chi_M^D \tag{6}$$

The values of these corrections are generally tabulated in the laboratory manuals and are available on-line too.

A French physicist, Pierre Curie, was investigating the effect of temperature on magnetic properties in the ending times of the nineteenth century. He discovered that, for a large number of paramagnetic substances, molar magnetic susceptibility (χ_M) varies inversely with the temperature. This observation is called as Curie law, which states that:

$$\chi_M \propto \frac{1}{T}$$

$$\chi_M = \frac{C}{T} \qquad (7)$$

Where is C is the Curie constant having different magnitude for different substances. Curie also discovered that for every ferromagnetic substance, there is a temperature T_C above which, the normal paramagnetic behavior occurs. Later work by Onnes and Perrier showed that, for many paramagnetic substances, a more precise relationship is:

$$\chi_M = \frac{C}{T + \theta} \qquad (8)$$

Where is θ is the Weiss constant and the equation (8) is popularly known as the Curie-Weiss law. The symbol "θ" used in equation (8) is sometimes replaced by T_C because in the case of ferromagnetic materials, the value of θ calculated by Curie-Weiss plot, is actually equal to the negative of their Curie temperature. That's why there is another popular form of the Curie-Weiss law as given below.

$$\chi_M = \frac{C}{T - \theta} \qquad (9)$$

The conventions shown in equation (8) are more widely accepted by the British and American academics, while the form with a negative sign is more popular in Indian and German universities. Furthermore, Louis Neel, another French physicist, observed that for every antiferromagnetic substance, there is a temperature T_N above which, the normal paramagnetic behavior occurs.

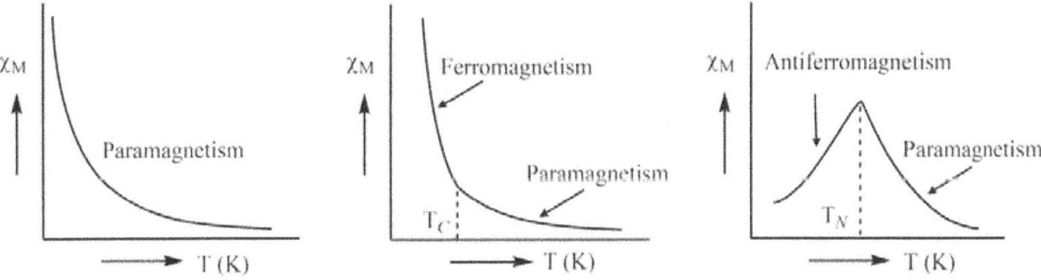

Figure 1. Plot of magnetic susceptibility vs temperature for normal paramagnetic, ferromagnetic and antiferromagnetic materials.

Normally, the reciprocal of magnetic susceptibility is plotted versus temperature ($1/\chi_M$ vs T follows a straight line equation); which makes the use of both forms of Curie-Weiss law. The symbol "θ" in equation (9) is replaced by T_C, which gives suitable form for ferromagnetic substances (magnetic moments of atoms align to produce a strong magnetic effect); while the replacement of the symbol "θ" by T_N in equation (8) gives the suitable form for antiferromagnetic materials (magnetic moments of atoms align anti-parallel to produce a strong magnetic effect).

Therefore, for ferromagnetic substances

$$\chi_M = \frac{C}{T - T_C} \tag{10}$$

For antiferromagnetic substances

$$\chi_M = \frac{C}{T + T_N} \tag{11}$$

It is also worthy to mention that the normal paramagnetic behavior of ferromagnetic or antiferromagnetic materials is observed only when $T > \theta$.

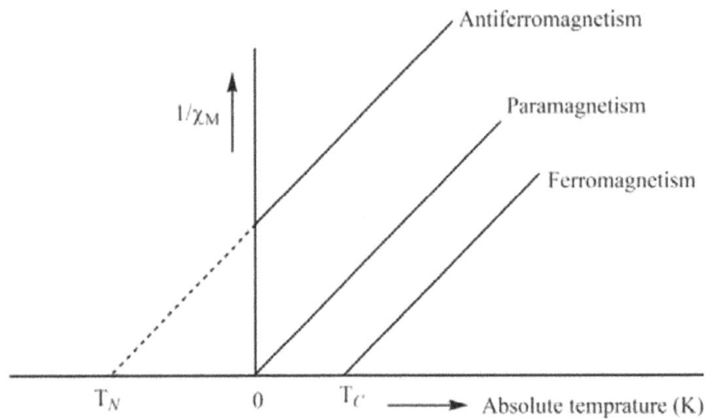

Figure 2. The plot of reciprocal of magnetic susceptibility vs temperature for normal paramagnetic, ferromagnetic and antiferromagnetic materials.

Thus, for ferromagnetic and antiferromagnetic materials, the value of θ is generally labeled T_C (Curie temperature) and T_N (Neel temperature), respectively.

> ➢ ***The Quantum Mechanical Concept of Magnetism***

　　　　The genesis of magnetic phenomena in all atoms lies in the orbital and spin motions of electrons and how these electrons interact with each other. The orbital motion of the electron gives rise to the orbital magnetic moment (μ_l), and the spin motion generates the spin magnetic moment (μ_s). The total magnetic moment of an atom is actually the resultant of the two aforementioned effects. Now, though the wave mechanical model of an atom is more precise in the rationalization of different atomic properties, the prewave mechanical model of an atom is still very much of use for understanding certain quantum mechanical effects. In the Bohr model, the electron is considered as a negatively charged hard-sphere that spins about its own axis as well as revolves around the positively charged heavy center of the atom. The pictorial representation of the rise of the magnetic moment by these two kinds of motion is shown below.

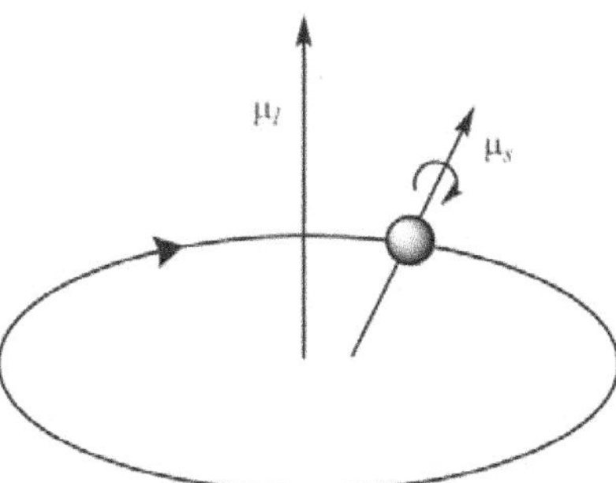

Figure 3. The generation of orbital magnetic moment (μ_l) and spin magnetic moment (μ_s) from prewave mechanical quantum theory.

Thus, we can agree on the fact that a more conceptual comprehensive understanding of the phenomena of magnetism in different chemical compounds requires us to start from the most elementary ideas of spin and orbital magnetic moments.

1. Orbital magnetic moment (μ_l): The motion of a negatively charged electron in a circular path is very much analogs to the current flowing through a ring of conducting material. Consequently, a magnetic field in a direction perpendicular to the plane of the ring or orbit is generated. The strength of the magnetic field thus produced can be obtained by multiplying the magnitude of the current flowing (i) with the surface area of that ring (A). Mathematically, the magnitude of the orbital magnetic moment (μ_l) can be given as:

$$\mu_l = iA = \left(\frac{e\omega}{2\pi c}\right)\pi r^2 \tag{12}$$

Where e is the electronic charge, ω is the angular velocity of the electron, c is the velocity of light and r is the radius of the orbit. From the quantum theory of angular momentum, we know that the magnitude of the angular momenta of an orbiting electron is given by the following relation.

$$L_{quantum\ mechanical} = \sqrt{l(l+1)}\,\frac{h}{2\pi} \tag{13}$$

Where L is magnitude angular momentum due to orbital motion and l is the quantum number for the same. Moreover, the magnitude of angular momentum from classical mechanics is given by the product of angular velocity (ω) and moment of inertia (I).

$$L_{classical} = I\omega \tag{14}$$

Putting $I = mr^2$ in equation (14) we get

$$L_{classical} = mr^2\omega \tag{15}$$

Now, it is a well-known fact that all values of the quantum domain are also present in the classical domain though the vice-versa is not true. Therefore, we can put equation (13) equal to equation (15) to find the classical analogs.

$$mr^2\omega = \sqrt{l(l+1)}\,\frac{h}{2\pi} \tag{16}$$

Or

$$\omega r^2 = \sqrt{l(l+1)}\,\frac{h}{2\pi m} \tag{17}$$

Putting the value of ωr^2 from equation (17) into equation (12), we get

$$\mu_l = \frac{e}{2c}\left(\sqrt{l(l+1)}\,\frac{h}{2\pi m}\right) \tag{18}$$

$$\mu_l = \sqrt{l(l+1)}\left(\frac{eh}{4\pi mc}\right) \tag{19}$$

$$\mu_l = \sqrt{l(l+1)}\ \text{B.M.} \tag{20}$$

Comparing equation (20) and equation (13), we can conclude that the magnitude of magnetic moment (μ_l) in the units of Bohr magneton (B.M.) is equal to orbital angular momentum (L) measured in the units of $h/2\pi$. It is also worthy to note that both the vectors (μ_l and L) are collinear but oriented in the opposite direction. In other words, if the magnetic moment is oriented upward to the orbit plane, orbital angular momentum is downward, and vice-versa.

2. Spin magnetic moment (μ_s): In 1926, two American-Dutch physicists, named Samuel Goudsmit and George Uhlenbeck, observed that the angular momentum possessed by the moving electron is actually greater than the orbital angular momentum. This excess of angular momentum was then attributed to the spinning motion of the electron. This spinning motion of a negatively charged electron about its own axis is also analogs to the current-carrying circular conductor. Accordingly, a magnetic field, in a direction along to the spinning axis, is generated. They also postulated that the ratio of the spin magnetic moment (μ_s) measured in the units of B.M. to the spin angular momentum (S) measured in the units of $h/2\pi$, must be equal to 2. This ratio is called as Lande's splitting factor or the "*g*" value.

$$\frac{\mu_s}{S} = \frac{\mu_s}{\sqrt{s(s+1)}} = g \tag{21}$$

$$\mu_s = g\sqrt{s(s+1)} = 2\sqrt{s(s+1)}\ \text{B.M.} \tag{22}$$

Where s is the quantum number defining the spin motion of the electron and $[s(s+1)]^{1/2}$ is the corresponding spin angular momentum in the units of $h/2\pi$ as discussed earlier.

Both types of magnetic moments will interact with the external magnetic field and will tend to align themselves along the direction of the field; which in turn will reinforce the magnitude of the applied field. In multi-electron systems, the spin motion of the individual electrons will interact with each other to give resultant spin motion quantum number "S"; while the orbital motion of individual electrons will interact to give resultant orbital motion quantum number "L". Now, if L and S do not interact with each other, the overall magnetic moment will just be the sum of their individual magnetic moments. However, if the resultant spin and resultant orbital motions do couple, and the overall magnetic moment will be obtained from "J" i.e. total angular momentum quantum number. The phenomena like diamagnetism, paramagnetism, or ferromagnetism arise as a result of alignments and interactions of theses micro magnates.

> ### *Classes of Magnetic Materials*

The most primitive way to classify different materials on the basis of their magnetic properties is how they respond to the externally applied magnetic field. Thereafter, we can discuss the cause or interaction responsible for such behavior. In some material, the atomic-scale magnetic moments do not interact with each other; while in some cases the strong interaction may lead to a very complex magnetic profile depending upon the structural specificity. Different magnetic materials can be classified into the following four major classes:

1. Diamagnetic materials: When some substances are placed in an external magnetic field, the number of magnetic lines of force passing through the substance is less than the number of magnetic lines of force passing through the vacuum. This eventually means that the ratio of B/H is less than one, which gives a negative value of magnetic susceptibility (K). Such substances are called as diamagnetic substances and are repelled by the external magnetic field.

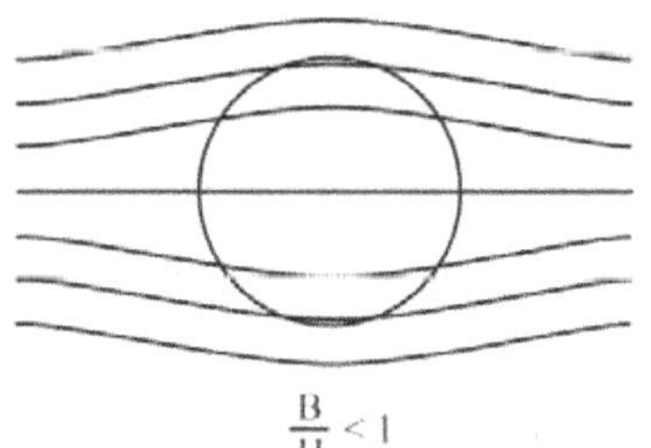

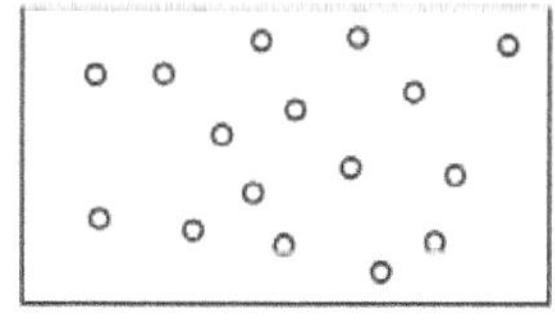

Figure 4. The behavior of a diamagnetic body in the externally applied magnetic field and corresponding magnetic domain.

Diamagnetic substances do not have unpaired electrons, and therefore the magnetic moment produced by one electron is canceled out by the other one. The phenomenon of diamagnetism is 1000 times weaker than paramagnetism, which makes it unobservable in substances with unpaired electrons. However, the measured magnetic susceptibilities must be corrected for the underlying diamagnetic effect, because most of the materials do contain paired electrons. Diamagnetic susceptibility is generally independent both of field strength and temperature.

2. Paramagnetic materials: When some substances are placed in an external magnetic field, the number of magnetic lines of force passing through the substance is more than the number of magnetic lines of force passing through the vacuum. This eventually means that the ratio of B/H is greater than one, which gives a positive value of magnetic susceptibility (K). Such substances are called as paramagnetic substances and are attracted by the external magnetic field.

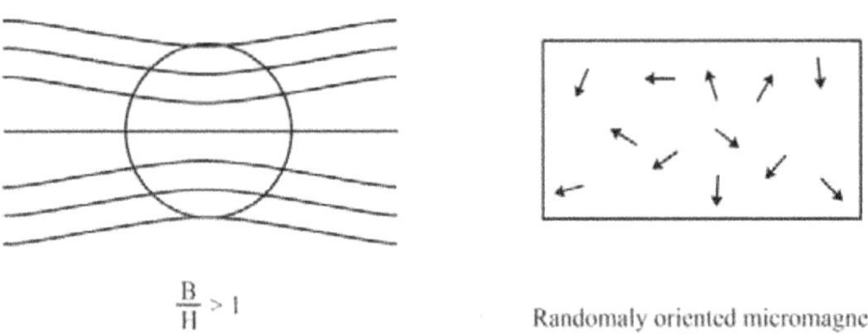

Figure 5. The behavior of a paramagnetic body in the externally applied magnetic field and corresponding magnetic domain.

Paramagnetic substances do have unpaired electrons, and therefore there is a net magnetic moment possessed by individual constituents. However, these micro-magnets are randomly oriented. The paramagnetic susceptibility of these materials decreases with the increase of temperature and follow simple Curie law. Hence, the paramagnetic susceptibility is generally independent of field strength, but markedly dependent on the temperature of the system.

3. Ferromagnetic materials: When some substances are placed in the external magnetic field, the number of magnetic lines of force passing through the substance is hugely greater than the number magnetic lines of force passing through the vacuum. This eventually means that the ratio of B/H is much greater than 1, which gives a positive value of magnetic susceptibility (K) of order as high as 10^4. Such substances are called as ferromagnetic substances and are strongly attracted by the external magnetic field.

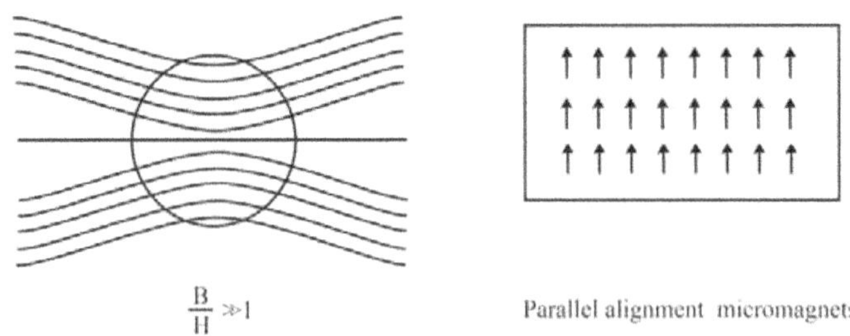

Figure 6. The behavior of a ferromagnetic body in an externally applied magnetic field and corresponding magnetic domain.

Ferromagnetic substances do have unpaired electrons, and therefore there is a net magnetic moment possessed by individual constituents. However, a special interaction of these micro-magnets makes them orient parallel to each other yielding very strong paramagnetism. The hysteresis and remanence and are characteristic features of ferromagnetic materials. Hence, the ferromagnetic susceptibility depends upon the field strength as well as the temperature of the system considered.

4. Antiferromagnetic materials: When some substances are placed in an external magnetic field, the number of magnetic lines of force passing through the substance is slightly greater than the number magnetic lines of force passing through the vacuum. This eventually means that the ratio of B/H is slightly greater than one, which gives a very small positive value of magnetic susceptibility (K). Such substances are called as antiferromagnetic substance and are weakly attracted by the external magnetic field.

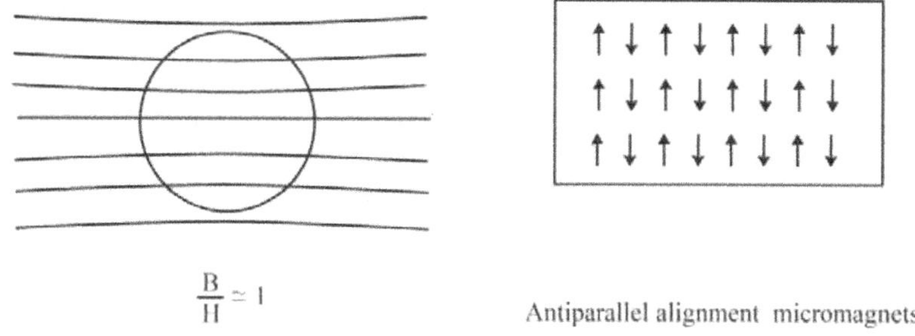

Figure 7. The behavior of an antiferromagnetic body in an externally applied magnetic field and corresponding magnetic domain.

Antiferromagnetic substances do have unpaired electrons, and therefore, are expected to show paramagnetism due to the presence of net magnetic moment possessed by individual constituents. However, a special interaction of these micro-magnets makes them orient antiparallel to each other yielding a very small value of positive magnetic susceptibility. The antiferromagnetic susceptibility usually depends on the temperature of the system only, though the dependence on field strength is also observed sometimes.

❖ Gouy's Method for Determination of Magnetic Susceptibility

The simplest method used for measuring the magnetic susceptibilities of transition metal complexes was proposed by a French physicist, named Louis Georges Gouy. In 1889, He obtained a mathematical expression revealing that the force is actually proportional to volume susceptibility (K) or the interaction of material in a uniform external magnetic field. From this derivation, Gouy suggested that the balance measurements taken for tubes of material suspended in a magnetic field could evaluate the expression for volume susceptibility. Though Gouy never tested his scientific proposal, this inexpensive and simple technique would become a blueprint of the Gouy balance; and therefore, for measuring magnetic susceptibilities.

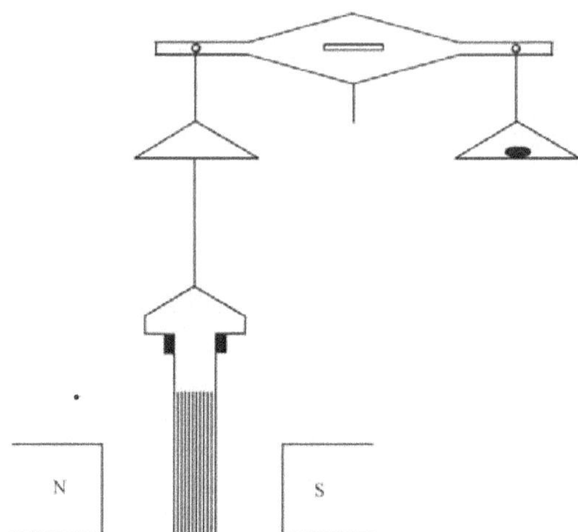

Figure 8. The schematics of Gouy balance.

The determination of a magnetic susceptibility depends on the measurement of B/H. From the classical description of magnetism, Lenz's law can be stated as:

$$\frac{B}{H} = 1 + 4\pi K \tag{23}$$

where B/H is called the magnetic permeability of the material and K is the magnetic susceptibility per unit volume (I/H). The Gouy's method includes the measurement of force on the sample by the externally applied magnetic field and depends upon the tendency of the sample to align itself in high or low magnetic field strength. The force at any given point of the sample (say dx) is given by:

$$dF = \mu^{\circ} H \, K dV \, \frac{dH}{dx} \tag{24}$$

Where μ° symbolizes the permeability of the vacuum (value is 1 when using c.g.s. system of units), dV is the volume of the sample at point dx, H is the magnitude of the magnetic field at point dx and K represents the magnetic susceptibility per unit volume. The Gouy tube, packed uniformly with the sample is placed in the magnetic field such that each end of the glass tube experiences a constant field strength. In order to achieve this situation, the Gouy's tube must be packed to a certain height like 10 or 15 cm, and the tube is then hanged between the electromagnetic poles in such a way that the bottom of the sample lies in the center of the magnetic field (an area where a uniform field strength can be easily achieved); while the top of the sample is out of the field ($H = 0$). Now, the total magnitude of the force acting on the sample can be calculated just by integrating the equation (24).

$$F = \mu^\circ A\,K\,\left[\frac{H^2}{2}\right]_0^H \tag{25}$$

$$F = \frac{\mu^\circ A\,KH^2}{2} \tag{26}$$

Where A is the area of cross-section of the sample. The force can easily be measured by the apparent change in mass when the external magnetic field is switched on.

$$F = g\Delta w \tag{27}$$

Where g is the acceleration due to gravity and Δw is the apparent deviation in mass. From equation (26) and equation (27), we get:

$$g\Delta w = \frac{\mu^\circ A\,KH^2}{2} \tag{28}$$

It is also worthy to mention that some correction must be made for the tube because it possesses its own magnetic properties due to air-filled within the tube, and the nature of its construction materials. Therefore, equation (28) takes the form:

$$g\Delta w' = \frac{\mu^\circ A\,(K - K')\,H^2}{2} \tag{29}$$

Where $\Delta w' = \Delta w + \delta$, δ is a constant allowing for the magnetic properties of the empty tube, K' is the volume susceptibility of the replaced air. This gives:

$$K = \frac{2g\Delta w'}{\mu^\circ A\,H^2} + K' \tag{30}$$

Converting from volume susceptibility (K) to mass susceptibility (χ) leads to:

$$\chi = \frac{K}{\rho} = \frac{KV}{W} = \frac{2g\Delta w'V}{\mu^\circ A\,H^2 W} + \frac{K'V}{W} \tag{31}$$

$$\chi = \frac{\beta\Delta w'}{W} + \frac{K'V}{W} \tag{32}$$

$$\chi = \frac{(\alpha + \beta\Delta w')}{W} \tag{33}$$

Where $\alpha = K'V$ is a correction constant incorporated for the air replaced by the sample, $\beta = (2gV)/(\mu^\circ AH^2)$ is also a constant which depends upon the strength of the magnetic field and W is the weight of the sample under consideration. In order to measure the mass susceptibility a sample more accurately, the predetermine of α, β and δ is necessary. Therefore, the dependence of these constants on the magnetic field strength, the amount of sample put in the tube and the tube itself emphasize on the fact that each analyst must find their value for every new configuration.

Determination of the constants can be carried out by selecting a tube and a small nichrome-wire to make an assembly which will allow the tube to be hanged from the analytical balance so that the bottom of the tube is aligned halfway between the mutually-facing poles of the electromagnets used, and sample's top is above the magnet and thus subject to a zero-field strength.

i) Calculation of δ: Adjust the zero on the Gouy's balance, then suspend the empty tube from the balance and measure its weight (W_1). Now turn on the electromagnets to desired magnetic field strength and reweigh the tube (W_2). The force on the Gouy's tube, δ, is thus $δ = W_2 - W_1$. The value will be negative because the tubes are generally diamagnetic and are pushed out of the magnetic field.

ii) Calculation of α: Fill the water in Gouy's tube to the required marking and weigh it, this will give the value of W_3. Now considering the density of water at this temperature as 1.00 g cm^{-3}, this volume of water would be equal to the volume of the sample. Hence, $V = (W_3 - W_1)/1.00$, where the changes in weight should be expressed in grams. Now, $α = K'V$ or $α = 0.029 \times (W_3 - W_1)$ in 10^{-6} c.g.s. units, where 0.029 is the volume susceptibility of the air.

ii) Calculation of β: The measurement of β requires a standard compound whose magnetic properties are already known. The most commonly used calibrants are $[Ni(en)_3]S_2O_3$ and $Hg[Co(SCN)_4]$. Now because of fact that the magnetic properties are usually temperature-dependent, the susceptibility of the calibrant must be determined at a temperature exactly similar to what is required for the sample. Record the temperature, T_1, and then fill the tube to the required height with the calibrant and weigh it with the magnetic field off (W_4) and on (W_5). For $[Ni(en)_3]S_2O_3$, use $χ = 3172/T$ in 10^{-6} c.g.s units; while for $HgCo(SCN)_4$ the $χ = 4985/(T+10)$ in 10^{-6} in c.g.s unit can be used at temperature T. Using this χ then $β = (χW - α)/Δw'$, where $Δw' = (W_5 - W_4) - δ$ in mg and $W = (W_4 - W_1)$ in grams.

❖ Calculation of Magnetic Moments

The resultant magnetic moment of any magnetic material (including free transition metal ions or their complexes) arises due to the orbital and spin motions of electrons and how these electrons interact with one another. The validation and applicability of any magnetic theory depend upon the precision of its theoretical results with the experimental one. The theoretical and experimental routes to magnetic moments are given below.

➢ *Experimental Calculation of Magnetic Moments*

It is pretty funny to say but the magnetic moment of a substance cannot be calculated from the experiment directly. The experimental determination of magnetic moment (μ) requires the measurement of magnetic susceptibility (χ) first. This experimental value of paramagnetic susceptibility is then used to find the value of the magnetic moment by using a certain quantum mechanical correlation. Now as we know that the experimentally measured molar paramagnetic susceptibility ($χ_M$) is actually the sum of paramagnetic ($χ_M^P$) and

diamagnetic ($\chi_M{}^D$) susceptibilities. Hence, the actual value of paramagnetic susceptibility is obtained by subtracting the diamagnetic susceptibility from the experimentally measured susceptibility.

$$\chi_M = \chi_M^P + \chi_M^D \tag{34}$$

$$\chi_M^P = \chi_M - \chi_M^D \tag{35}$$

The diamagnetic susceptibility is possessed by almost every material due to the presence of paired electrons in the valence or deeper shells. The root cause of diamagnetism is the motion of negatively charged particles in the applied magnetic field. This motion creates an orbital electric current and the applied magnetic field makes theses orbits precess about its direction of application. This precession is called as Larmor precession (depends upon the strength of the applied field) that generates its own internal magnetic field in a direction opposite to the external one. Consequently, the magnetic lines of force try to bypass the material which in turn imparts a negative value to magnetic susceptibility. The atomic diamagnetic susceptibility per mole of atoms ($\chi_M{}^{D\text{-}atomic}$) can be calculated using equation (36).

$$\chi_M^{D\_atomic} = -\frac{N\,e^2}{6mc^2} \sum_{i=1}^{i=n} < r_i^2 > \tag{36}$$

Where n is the number of electrons in the atom under consideration, N is the Avogadro number, m is the mass of the electron, c is the velocity of the light and $<r_i^2>$ is the mean square radius of the ith electron. Therefore, the larger the number of the electrons in an atom, the greater will be the magnitude of diamagnetic susceptibility. The total molar magnetic susceptibility of a chemical compound is the sum of contributions from diamagnetic susceptibility per mole of atoms ($\chi_M{}^{D\text{-}atomic}$) and diamagnetic susceptibility per mole of bonds ($\chi_M{}^{D\text{-}bond}$). The $\chi_M{}^{D\text{-}atomic}$ and $\chi_M{}^{D\text{-}bond}$ are generally labeled as Pascal's constants.

Table 1. The value of Pascal's constants for different atoms and bonds.

	$\chi_M{}^{D\_atomic}$ (10^{-6} cm^3 mol^{-1})			$\chi_M{}^{D\_bond}$ (10^{-6} cm^3 mol^{-1})	
H	−2.9	Fe^{2+}	−12.8	C=C	+5.5
C	−6.0	Ni^{2+}	−12.8	C≡C	+0.8
C (aromatic)	−6.2	Co^{2+}	−12.8	C=N	+8.2
N	−5.6	Cu^{2+}	−12.8	C≡N	+0.8
N (aromatic)	−4.6	Mg^{2+}	−5.0	N=N	+1.8
O	−4.6	Zn^{2+}	−15.0	N=O	+1.7
Cl	−20.1	Ca^{2+}	−10.4	C=O	+6.3

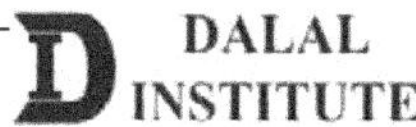

The calculation of $\chi_M{}^D$ using Pascal's constants can be exemplified by taking the case of pyridine and acetone.

i) Molar diamagnetic susceptibility ($\chi_M{}^P$) for pyridine: The contributions from diamagnetic susceptibility per mole of atoms ($\chi_M{}^{D\text{-}atomic}$) and diamagnetic susceptibility per mole of bonds ($\chi_M{}^{D\text{-}bond}$) is given below. It should also be noted that the contribution from bonds will be included in the ring values of N and C.

$$5 \text{ carbons of aromatic ring } = -31.0 \times 10^{-6} \text{ cm}^3\text{mol}^{-1}$$

$$5 \text{ hydrogens } = -14.5 \times 10^{-6} \text{ cm}^3\text{mol}^{-1}$$

$$1 \text{ nitrogen of aromatic ring } = -4.6 \times 10^{-6} \text{ cm}^3\text{mol}^{-1}$$

$$\text{Total } = -50.1 \times 10^{-6} \text{ cm}^3\text{mol}^{-1}$$

i) Molar diamagnetic susceptibility ($\chi_M{}^P$) for acetone: The contributions from diamagnetic susceptibility per mole of atoms ($\chi_M{}^{D\text{-}atomic}$) and diamagnetic susceptibility per mole of bonds ($\chi_M{}^{D\text{-}bond}$) is given below. It should also be noted that the contribution from C–C and C–H bonds is zero.

$$3 \text{ carbons } = -18.0 \times 10^{-6} \text{ cm}^3\text{mol}^{-1}$$

$$6 \text{ hydrogens } = -17.4 \times 10^{-6} \text{ cm}^3\text{mol}^{-1}$$

$$1 \text{ oxygen } = -4.6 \times 10^{-6} \text{ cm}^3\text{mol}^{-1}$$

$$1 \text{ CO double bond} = +6.3 \times 10^{-6} \text{ cm}^3\text{mol}^{-1}$$

$$\text{Total } = -33.7 \times 10^{-6} \text{ cm}^3\text{mol}^{-1}$$

It is also worthy to remember that the molar diamagnetic susceptibility is independent of temperature and is always negative.

Now once the molar diamagnetic susceptibility is known, the corrected paramagnetic susceptibility ($\chi_M{}^P$) can be obtained from equation (35). Nevertheless, $\chi_M{}^P$ should further be corrected for temperature-independent paramagnetism (TIP) for more precise and accurate results. The temperature-independent paramagnetism arises due to the mixing of the ground state and excited state under the effect of an externally applied magnetic field. From the quantum mechanical description of magnetism, the magnitude of paramagnetic susceptibility can be given as follows:

$$\chi_M^P = -\frac{N\mu_{\text{eff}}}{3kT} \tag{36}$$

Where μ_{eff} is the magnetic moment per mole of the material (in the units of B.M.), N is the Avogadro number, T is the absolute temperature and k is the Boltzmann constant. Putting the values of different constants in equation (36), we get

$$\mu_{\text{eff}} = 2.84 \sqrt{\chi_M^P\, T} \text{ B.M.} \tag{37}$$

> *Theoretical Calculation of Magnetic Moments*

The valence bond model is the most basic theoretical approach to calculate the effective magnetic moment of transition metal atoms or ions (complexed or free). Now owing to the fact that the valence bond theory does not consider the splitting of free ion term arising from inter-electronic repulsion, there are three possible scenarios for the calculation of effective magnetic moment as given below.

1. Resultant orbital motion (L) and resultant spin motion (S) contribute independently: When the J states after L-S coupling are very close to each other energetically, the thermal energy would be sufficient to populate all J-levels of ^{2S+1}L state. This situation is pretty much analogs to the case when spin and orbital motion are completely decoupled from each other. In other words, L-S would not be effective in distinguishing transition metal atoms or ions on the basis of total motion. Therefore, the effective magnetic moment will be given by the following relation:

$$\mu_{\text{eff}} = \sqrt{4S(S+1) + L(L+1)} \text{ B. M.} \tag{38}$$

Hence, the relationship shown in the equation (38) does not consider any quenching of orbital magnetic moment at all.

2. Resultant orbital motion (L) and resultant spin motion (S) couple with each other: When the J states after L-S coupling are very far to each other energetically, the thermal energy would not be sufficient to populate all J-levels of ^{2S+1}L state. This, in turn, would result in an almost hundred percent population density in the ground $^{2S+1}L_J$ state. This situation is pretty much analogs to the case when spin and orbital motion are coupled each to other. Therefore, the effective magnetic moment will be given by the following relation:

$$\mu_{\text{eff}} = g\sqrt{J(J+1)} \text{ B. M.} \tag{39}$$

Where g is the gyromagnetic ratio or Lande splitting factor whose value can be found as follows:

$$g = 1 + \frac{S(S+1) - L(L+1) + J(J+1)}{2J(J+1)} \tag{40}$$

It must be noted that resultant spin motion (S), resultant orbital motion (L) and total motion (J) quantum number used in equation (40) belong to ground state term symbol.

3. Resultant spin motion (S) contributes but the resultant orbital motion is quenched: In this case, the orbital motion contribution to the effective magnetic moment is completely quenched and μ_{eff} is calculated using spin only formula. Therefore, the effective magnetic moment should be calculated as:

$$\mu_{\text{eff}} = \sqrt{4S(S+1)} \text{ B. M.} \tag{41}$$

Equation (41) can also be written in the form given below.

$$\mu_{\text{eff}} - \sqrt{n(n+2)} \text{ B. M.} \tag{42}$$

Where n is the number of unpaired electrons.

A comparison of theoretically calculated magnetic moments (using different expressions) of various transition metal complexes along with their experimental values are listed in the following table.

Table 2. The comparison of theoretically calculated and experimental magnetic moments (B.M.).

Metal centre	Ground state	$\mu_{eff} = [4S(S+1)+L(L+1)]^{1/2}$	$\mu_{eff} = g[J(J+1)]^{1/2}$	$\mu_{eff} = [4S(S+1)]^{1/2}$	μ_{eff} (experimental)
Ti^{3+}	$^2D_{3/2}$	3.00	1.55	1.73	$1.7 - 1.8$
V^{3+}	$^3F_2$	4.47	1.65	2.85	$2.7 - 2.9$
Cr^{3+}	$^4F_{3/2}$	5.20	0.70	3.87	$3.7 - 3.9$
Mn^{3+}	$^5D_0$	5.48	0	4.90	$4.8 - 4.9$
Fe^{3+}	$^6S_{5/2}$	5.92	5.92	5.92	$5.7 - 6.0$
Fe^{2+}	$^5D_4$	5.48	6.71	4.90	$5.0 - 5.6$
Co^{2+}	$^4F_{9/2}$	5.20	6.63	3.87	$4.3 - 5.2$
Ni^{2+}	$^3F_4$	4.47	5.59	2.87	$2.9 - 3.5$
Cu^{2+}	$^2D_{5/2}$	3.0	3.55	1.73	$1.8 - 2.1$

The results clearly indicate that the experimental value of μ_{eff} is significantly different than those calculated theoretically. Nevertheless, the spin only formula seems to be the best bet out of three different approaches; which in turn implies that the orbital contribution is probably very much quenched in case metal complexes of first transition series. The plausible reason for this kind of behavior on the basis of valence bond theory is the hindrance faced by the orbital motion of metal electrons due to ligand cloud; and the extent to which orbital motion is hindered controls the contribution of the orbital magnetic moment. However, valence bond theory does not provide any sophisticated explanation for the qualitative and quantitative nature of the quenching of orbital magnetic moment. The orbital contribution to the magnetic moment and its behavior under different ligand fields will be rationalized on the basis of crystal field theory in the forthcoming sections of this chapter.

By comparing theoretically calculated magnetic moment with the experimental values, one can also comment on the outer or inner orbital nature of the complex. For example, the experimental magnetic moment for $[Fe(H_2O)_6]^{2+}$ is about 5.0 B.M.; which suggests an outer orbital configuration with four unpaired electrons (4.9 B.M.), as theoretically calculated magnetic moment for the inner orbital complex would be zero. Similarly, the zero value of experimental magnetic moment for $[Fe(CN)_6]^{4-}$ complex suggests an inner orbital configuration with zero unpaired electrons (0.0 B.M.), as theoretically calculated spin only magnetic moment for the outer orbital complex would be 4.9 B.M.

❖ Magnetic Properties of Free Ions

The resultant magnetic behavior of free ions; whether we talk about transition metals or lanthanides; arises as a result of spin-spin, orbital-orbital and spin-orbital interactions. Now stating more precisely, after considering the effect of spin-spin coupling and orbital-orbital coupling, the further degeneracy removal of the free ion terms is carried out by spin-orbital interaction (L-S coupling) that mainly leads to three different cases depending upon its relative magnitude with thermal energy. The determination of magnetic moment or magnetic susceptibility in these three situations is discussed below.

➢ *Magnitude of L-S Coupling in Ground State is Much Greater Than Thermal Energy*

If the J states after spin-orbital interaction are very far from each other energetically, the thermal energy (kT) would not be adequate to populate all J-levels of ^{2S+1}L state. This, in turn, would result in a Boltzmann distribution in which almost all of the population density is lying in the ground state $(^{2S+1}L_J)$. Consequently, the effective magnetic moment will be given by equation (43) as:

$$\mu_{\text{eff}} = g\sqrt{J(J+1)} \text{ B. M.} \qquad (43)$$

Where g is the Lande splitting factor or gyromagnetic ratio whose value can be found as follows:

$$g = 1 + \frac{S(S+1) - L(L+1) + J(J+1)}{2J(J+1)} \qquad (44)$$

In case the resultant orbital angular momentum quantum number is zero i.e. $L = 0$, The value of $J = S$ and equation (43) will become:

$$\mu_{\text{eff}} = g\sqrt{S(S+1)} \text{ B. M.} \qquad (45)$$

Now because the value of the gyromagnetic ratio or Lande splitting factor for free-electron is 2, equation (45) takes the following form.

$$\mu_{\text{eff}} = 2\sqrt{S(S+1)} \text{ B. M.} \qquad (46)$$

Owing to the relationship between resultant spin quantum number and number of unpaired electrons ($S = n/2$), equation (46) can also be written in the form given below.

$$\mu_{\text{eff}} = \sqrt{n(n+2)} \text{ B. M.} \qquad (47)$$

Where S is the resultant spin quantum number and n is the number of unpaired electrons. The equation (45 – 47) is also called as spin only formula as they do not include any contribution from the orbital motion.

This type of magnetic behavior is generally shown by lanthanide ions. A comparison of theoretically calculated magnetic moments of various lanthanide ions with their experimental observed values is presented in the following table.

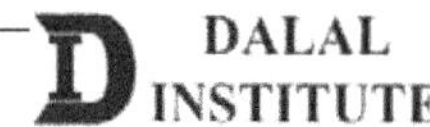

Table 3. The comparison of theoretically calculated and experimental magnetic moments (B.M.) of trivalent lanthanide ions.

Metal ion	Electronic configuration	Ground state term symbol	Gyromagnetic ratio	$\mu_{eff} = g[J(J+1)]^{1/2}$	μ_{eff} (experimental)
La^{3+}	f^0	$^1S_0$	–	0	0
Ce^{3+}	f^1	$^2F_{5/2}$	6/7	2.54	2.3 – 2.5
Pr^{3+}	f^2	$^3H_4$	4/5	3.58	3.4 – 3.6
Nd^{3+}	f^3	$^4I_{9/2}$	8/11	3.62	3.5 – 3.6
Pm^{3+}	f^4	$^5I_4$	3/5	2.68	2.7 – 2.8
Sm^{3+}	f^5	$^6H_{5/2}$	2/7	0.84	1.5 – 1.6
Eu^{3+}	f^6	$^7F_0$	–	0	3.4 – 3.6
Gd^{3+}	f^7	$^8S_{7/2}$	2	7.94	7.8 – 8.0
Tb^{3+}	f^8	$^7F_6$	3/2	9.72	9.4 – 9.6
Dy^{3+}	f^9	$^6H_{15/2}$	4/3	10.63	10.4 – 10.5
Ho^{3+}	f^{10}	$^5I_8$	5/4	10.60	10.3 – 10.5
Er^{3+}	f^{11}	$^4I_{15/2}$	6/5	9.57	9.4 – 9.6
Tm^{3+}	f^{12}	$^3H_6$	7/6	7.63	7.1 – 7.4
Yb^{3+}	f^{13}	$^2F_{7/2}$	8/7	4.50	4.4 – 4.9
Lu^{3+}	f^{14}	$^1S_0$	–	0	0

The results clearly indicate that the experimental values of μ_{eff} for most of the lanthanide ions are pretty much comparable to their theoretical counterparts. However, the magnetic moment for Eu^{3+} and Sm^{3+} calculated using equation (43) are zero, suggesting them as diamagnetic which is quite strange as they are actually having a considerable amount of paramagnetism. This is obviously due to the fact that the energy difference between their ground state term and the first excited state is comparable to thermal energy i.e. $\Delta E \approx kT$. Therefore, even at room temperature, a part of the total population of Sm^{3+} and Eu^{3+} ions would be present in their excited states, which in turn are obviously having a different value of total angular momentum quantum number (J-value) and gyromagnetic ratio (g-value).

> *Magnitude of L-S Coupling in Ground State is Comparable to Thermal Energy*

If the separation between J states after spin-orbital interaction is comparable to the thermal energy (kT) available, the resultant value of the magnetic moment is governed by a complex function of temperature. In this scenario, the total magnetic susceptibility will be having contributions from first-order Zeeman effects of involved states according to their Boltzmann weights, as well as from second-order Zeeman effects from neighboring levels. Mathematically, the magnetic susceptibility will be given as follows:

$$\chi_{\mathrm{M}} = \frac{N}{3kT} \sum g^2 \beta^2 J(J+1)(2J+1)e^{-\Delta E/kT} \sum (2J+1)e^{-\Delta E/kT} \tag{48}$$

It can clearly be seen from equation (48) that the magnitude of magnetic susceptibility does not depend directly on the reciprocal of temperature, and thus does not follow Curie-Weiss law. The theoretical calculation for Sm^{3+} ion, according to equation (48), yielded a magnetic moment of 1.38 B.M.; which pretty much comparable to the experimental one. Similarly, satisfactory results are also obtained for trivalent europium ions.

> *Magnitude of L-S Coupling in Ground State is Much Less Than Thermal Energy*

If the energy separation between J states (generated after L-S coupling) is much less than the thermal energy (kT) available, all of the J-levels of ^{2S+1}L ground state. This situation is pretty much analogs to the case when spin and orbital motion are completely decoupled from each other. In other words, L-S interaction would not be effective in distinguishing different metal ions on the basis of total motion. Therefore, the effective magnetic moment will be given by the following relation:

$$\mu_{\mathrm{eff}} = \sqrt{4S(S+1) + L(L+1)} \ \text{B. M.} \tag{49}$$

Hence, the relationship shown in the equation (49) does not consider any quenching of orbital magnetic moment at all. In this case, the resultant orbital angular momentum quantum number is zero i.e. $L = 0$, the L will be completely eliminated from equation (49) and we will get:

$$\mu_{\mathrm{eff}} = \sqrt{4S(S+1)} \ \text{B. M.} \tag{50}$$

Owing to the relationship between spin multiplicity and the number of unpaired electrons (S = $n/2$), equation (50) can also be written in the form given below.

$$\mu_{\mathrm{eff}} = \sqrt{n(n+2)} \ \text{B. M.} \tag{51}$$

Where S is the spin multiplicity and n is the number of unpaired electrons. The equations (50, 51) are also called as spin only formulas as they do not include any contribution from orbital motion. Now it is worthy to note that no free transition metal exists with very small J-separation relative to thermal energy; however, had they shown such behavior, we would have used the equation (49) to calculate their magnetic moment. In that case, a comparative analysis of the results from equation (49) and equation (51) could be used to estimate the contribution exclusively from the orbital motion.

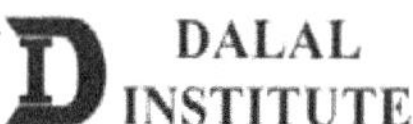

❖ Orbital Contribution: Effect of Ligand-Field

In the prewave-mechanical model of the atom, the orbital angular momentum was thought to be generated due to the circular motion of an electron in its orbit. This revolutionary motion of the negatively charged particle was supposed to contribute orbital magnetic moment to the total magnetic moment, as we have discussed in the first section of this chapter. However, as we all know that the concept of orbits is no longer valid because of its violation to the Heisenberg uncertainty principle; the physical realization of orbital angular momentum and its contribution to magnetic moment on the basis of wave mechanical model is a must. In the wave mechanical description of atom, the magnitude of orbital angular momentum along a particular axis depends upon the rotational feasibility of an atomic or molecular orbital about the same axis-line to carry itself into a degenerate and identical orbital. This can better be understood by taking the example of d-orbital set. A rotation of $45°$ about z-axis will transform $d_{x^2-y^2}$ into d_{xy} orbital, and rotation through $90°$ would give two equivalent states; resulting in a 2 units of angular momentum (unit = $h/2\pi$) about z-axis. The same can be said about d_{xy}, which is also having an angular momentum of 2 units about z-axis. Moreover, a rotation of $90°$ about z-axis will transform d_{xz} into d_{yz} orbital; and therefore, would give only one equivalent states, resulting only 1 unit of angular momentum about z-axis for both the orbitals. However, the rotation of d_{z^2} about z-axis will not be able to transform it into any other orbital; resulting in a zero-unit angular momentum about z-axis.

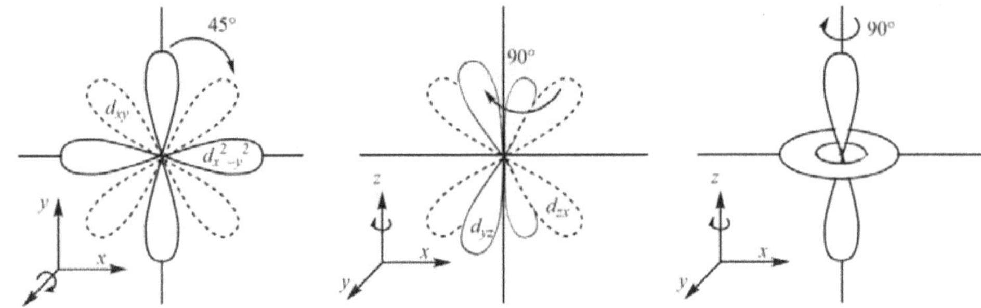

Figure 9. The circulation of electron density in a partially filled d-subshell about the z-axis.

It is also worthy to mention that the resultant orbital angular momentum would exist only if the degenerate set of interconvertible orbitals is unsymmetrically occupied. Though we already know that the generation of the orbital magnetic moment was analogs to the magnetic effect generated by the current in a solenoid; the better visualization can be made if we think of a circular wire with uniform charge density. If this circular wire is rotated by any angle about the axis perpendicular to its plane, it will be left in a physically identical state, like nothing has moved. On the other hand, if we use a solenoid, instead of a ring, the ends of the wire coil would be in a different location after the rotation. The prerequisite for a non-zero orbital motion is exactly the same. Any spherically symmetric distribution of electron density would not result in an orbital angular momentum (like a single electron in s-orbital or multi-electron cloud with S state).

Hence, we can say that a free ion present in a state other than S, will have an orbital angular momentum, and thereby orbital magnetic moment too. Nevertheless, if this orbital degeneracy is removed by complexation or by trapping it in a solid host matrix, the magnitude of orbital angular momentum would be somewhat more or less quenched. If the orbital degeneracy is slightly reduced, the contribution of the orbital magnetic moment would be quenched incompletely or partially. However, as in addition to the orbital degeneracy, the unsymmetrical filling is also a necessary condition for an orbital moment, we should work out different Mulliken states satisfying these conditions. Therefore, in order to show the orbital magnetic contribution to the total magnetic moment, there should be an unsymmetrically occupied degenerate set of orbitals that can be interconverted or transformed into each other through rotational motion.

> ### *Ligand Field Effect on the Orbital Contribution in Octahedral Complexes*

The octahedral coordination of a transition metal center removes the degeneracy of d-subshell into two main orbital-sets. The first degenerate set of d_{xy}, d_{yz}, d_{zx} is of t_{2g} symmetry; while the second degenerate set of $d_{x^2-y^2}$ and d_{z^2} is having e_g symmetry now. Thus, owing to the different energies of $d_{x^2-y^2}$ and d_{xy}, the interconversion of these orbitals through a rotation of $45°$ about z-axis is not possible anymore. This gives a zero angular momentum along z-axis for electron present in any of these two. Nevertheless, the interconversion of d_{xz} into d_{yz} orbital through a rotation of $90°$ about z-axis is still feasible because of their orbital degeneracy in t_{2g} symmetry. Hence, an unsymmetrical filling of t_{2g} orbital-set results in a non-zero orbital angular momentum; and consequently, non-zero orbital magnetism along z-axis. The unsymmetrical filling of e_g orbital-set will not result in any orbital magnetism due to lack of their interconversion.

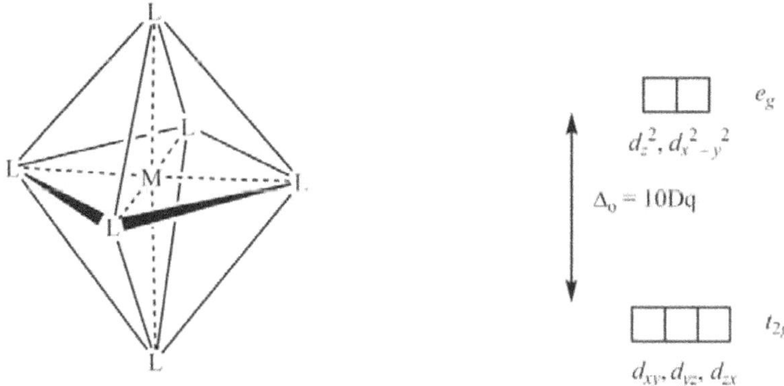

Figure 10. The octahedral coordinative environment of transition metal center and corresponding crystal field splitting pattern of d-subshell.

Now, we can find out which electronic configurations in high-spin (spin-free) or low spin (spin paired) octahedral complexes of transition metals will show an orbital contribution to the total magnetic moment, and which will have the same absent.

1. Electronic configurations with orbital magnet: Recalling the prerequisites of orbital magnetism; unsymmetrically occupied, degenerate set of interconvertible orbitals; the configurations suitable for the presence of orbital magnetic moment are given below.

Table 4. Octahedral transition metal complexes having orbital magnetic moments.

Weak-field		Strong-field	
Electronic configuration / Ground state		Electronic configuration / Ground state	
$d^1\ (t_{2g}^{\ 1})$	$^2T_{2g}$	$d^1\ (t_{2g}^{\ 1})$	$^2T_{2g}$
$d^2\ (t_{2g}^{\ 2})$	$^3T_{1g}$	$d^2\ (t_{2g}^{\ 2})$	$^3T_{1g}$
$d^6\ (t_{2g}^{\ 4}\ e_g^{\ 2})$	$^5T_{2g}$	$d^4\ (t_{2g}^{\ 4})$	$^3T_{1g}$
$d^7\ (t_{2g}^{\ 5}\ e_g^{\ 2})$	$^4T_{1g}$	$d^5\ (t_{2g}^{\ 5})$	$^2T_{2g}$

It is worthy to note that all ground Mulliken states which show an orbital contribution to the total magnetic moment are of "T" nature.

2. Electronic configurations without orbital magnet: Recalling the prerequisites of orbital magnetism; unsymmetrically occupied, degenerate set of interconvertible orbitals; the configurations suitable for the absence of orbital magnetic moment are given below.

Table 5. Octahedral transition metal complexes having no orbital magnetic moments.

Weak-field		Strong-field	
Electronic configuration / Ground state		Electronic configuration / Ground state	
$d^3\ (t_{2g}^{\ 3})$	$^4A_{2g}$	$d^3\ (t_{2g}^{\ 3})$	$^4A_{2g}$
$d^4\ (t_{2g}^{\ 3}\ e_g^{\ 1})$	^5E_g	$d^6\ (t_{2g}^{\ 6})$	$^1A_{1g}$
$d^5\ (t_{2g}^{\ 3}\ e_g^{\ 2})$	$^6A_{1g}$	$d^7\ (t_{2g}^{\ 6}\ e_g^{\ 1})$	^2E_g
$d^8\ (t_{2g}^{\ 6}\ e_g^{\ 2})$	$^3A_{2g}$	$d^8\ (t_{2g}^{\ 6}\ e_g^{\ 2})$	$^3A_{2g}$
$d^9\ (t_{2g}^{\ 6}\ e_g^{\ 3})$	^2E_g	$d^9\ (t_{2g}^{\ 6}\ e_g^{\ 3})$	^2E_g
$d^{10}\ (t_{2g}^{\ 6}\ e_g^{\ 4})$	$^1A_{1g}$	$d^{10}\ (t_{2g}^{\ 6}\ e_g^{\ 4})$	$^1A_{1g}$

It can be clearly seen that all ground Mulliken states which do not show orbital contribution are of either "A" or of "E" nature.

> *Ligand Field Effect on the Orbital Contribution in Tetrahedral Complexes*

The tetrahedral coordination of a transition metal center removes the degeneracy of d-subshell into two main orbital-sets. The first degenerate set of d_{xy}, d_{yz}, d_{zx} is of t_2 symmetry; while the second degenerate set of $d_{x^2-y^2}$ and d_{z^2} is having e symmetry now. Thus, owing to the different energies of $d_{x^2-y^2}$ and d_{xy}, the interconversion of these orbitals through a rotation of 45° about z-axis is not possible anymore. This gives a zero angular momentum along z-axis for electron present in any of these two. Nevertheless, the interconversion of d_{xz} into d_{yz} orbital through a rotation of 90° about z-axis is still feasible because of their orbital degeneracy in t_2 symmetry. Hence, an unsymmetrical filling of t_2 orbital-set results in a non-zero orbital angular momentum; and consequently, non-zero orbital magnetism along z-axis. The unsymmetrical filling of e orbital-set will not result in any orbital magnetism due to the lack of their interconversion.

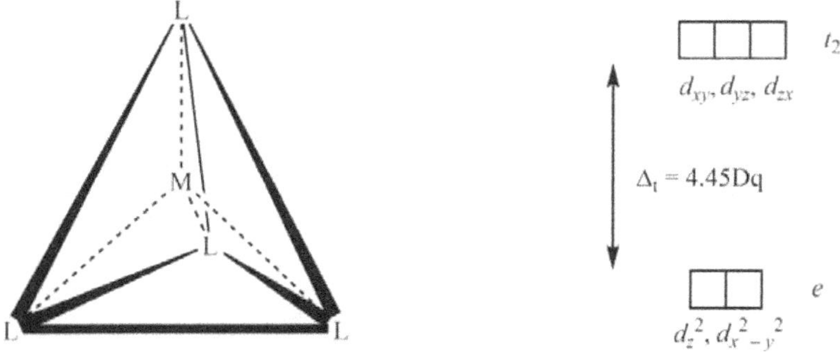

Figure 11. The tetrahedral coordinative environment of transition metal center and corresponding crystal field splitting pattern of d-subshell.

Now at this point, it is worthy to recall the electron-hole inverse relationship again, which states that the number of microstates and hence all free ion terms for d^n and d^{10-n} configuration are same. However, as the magnitude of the crystal field experienced positive electrons is same as what experienced by the negative electrons, but is of opposite sign; the splitting pattern for d^n and d^{10-n} configuration are also opposite of each other. Furthermore, owing to the hole formalism in quantum mechanics, strong field configuration of t_{2g}^4, t_{2g}^5, e_g^3 give rise to the same terms as given by the strong field configuration of t_{2g}^2, t_{2g}^1, e_g^1. However, weaker inter-electronic repulsion is considered as the perturbation over stronger V_o. Similarly, the splitting pattern of d^n tetrahedral is just the opposite of what is for d^n octahedral. However, no g or u are used in the tetrahedral case because there is no center of symmetry in a tetrahedral geometry. Hence, we can conclude that the hole-formalism and octahedral-tetrahedral inverse relationship is all term symbols including the ground state. Now, we can find out which electronic configurations in high-spin (spin-free) tetrahedral complexes of transition metals will show an orbital contribution to the total magnetic moment, and which will have the same absent. It is also important to note that here we don't need to consider the low spin cases because the smaller magnitude of crystal field splitting in tetrahedral complexes always avoids pairing of electrons.

1. Electronic configurations with orbital magnet: Recalling the prerequisites of orbital magnetism; unsymmetrically occupied, degenerate set of interconvertible orbitals; the configurations suitable for the presence of orbital magnetic moment are given below.

Table 6. Tetrahedral transition metal complexes having orbital magnetic moments.

Electronic configuration	Ground state
d^3 $(e^2\ t_2^1)$	$^4T_1$
d^4 $(e^2\ t_2^2)$	$^5T_2$
d^8 $(e^4\ t_2^4)$	$^3T_1$
d^9 $(e^4\ t_2^5)$	$^2T_2$

It can be clearly seen that all ground Mulliken states which show an orbital contribution to the total magnetic moment are of "T" nature.

2. Electronic configurations without orbital magnet: Recalling the prerequisites of orbital magnetism; unsymmetrically occupied, degenerate set of interconvertible orbitals; the configurations suitable for the absence of orbital magnetic moment are given below.

Table 7. Tetrahedral transition metal complexes with no orbital magnetic moments.

Electronic configuration	Ground state
d^1 (e^2)	^2E
d^2 (e^2)	$^3A_2$
d^5 $(e^2\ t_2^3)$	$^6A_1$
d^6 $(e^3\ t_2^3)$	^5E
d^7 $(e^4\ t_2^3)$	$^4A_2$
d^{10} $(e^4\ t_2^6)$	$^1A_1$

It can be clearly seen that all ground Mulliken states which do not show orbital contribution are of either "A" or of "E" nature.

Summarizing both the case, we can conclude that the ground Mulliken states of "T" symmetry do have an orbital magnetism; while any configuration leading to "A" or "E" state will not be able to show an orbital contribution to the total magnetic moment.

> ### *Effect of Spin-Orbital Coupling on the Orbital Contribution*

The conclusive remark after the general observation of the orbital contribution discussed so far includes that "A" and "E" terms as the ground state do not yield any orbital magnetism, while the ground state "T" terms do possess an orbital contribution to the total magnetic moment. In other words, the orbital magnetic moment is partially quenched in transition metal complexes with A or E ground Mulliken states, and the complete quenching occurs in metal complexes with ground Mulliken states of T nature. However, if a higher energy T-symmetry state with the same multiplicity lies above the "A" or "E" state, the spin-orbital coupling may mix the excited state into ground one such that a certain amount of orbital magnetic moment is introduced into the ground state also. This is because the ligand field has the ability to quench orbital angular momentum, but not spin angular momentum. Therefore, if both types of momentum are coupled by *L-S* interaction, the ligand field would not be able to separate the same multiplicity terms on the basis of their orbital angular momentum quantum number. Consequently, the ground state term is not a pure A or E anymore but does have some T character, giving significant deviation from the spin-only value of the magnetic moment.

The magnitude of orbital magnetism contribution to the total effective magnetic moment directly depends upon the value of the spin-orbital coupling parameter (λ) and is inversely proportional to the energy difference of the states to be mixed (Δ). Mathematically, the relationship between spin-orbital interaction and total effective magnetic moment can be stated as:

$$\mu_{eff} = \mu_{S.O.}\left(1 - \alpha\frac{\lambda}{\Delta}\right) \tag{52}$$

Where $\mu_{S.O.}$ is the spin-only magnetic moment, λ is the spin-orbital coupling parameter, and Δ is the energy between ground (A or E) and the excited state (T). It is also worthy to remember that the sign of the spin-orbital coupling parameter is negative for more than half field configurations while positive for less than half-filled electronic configurations. The α is a constant whose value depends upon the spectroscopic ground state term; value is 0 for ^6S ($^6A_1$), 2 for ^2D (^2E) and ^5D (^5E), and 4 for ^3F ($^3A_2$) and ^4F ($^4A_2$) ground state terms. The calculation of the effective magnetic moment for all these three cases is given below.

1. Calculation of μ_{eff} for A_1 terms: The ground Mulliken state of A_1 symmetry exists for the weak field octahedral complex with d^5 and the strong field octahedral complexes with d^6 configurations.

i) d^5 weak field ($t_{2g}^3 e_g^2$): Keeping in mind that there are no high energy terms of the same multiplicity in d^5 high-spin complexes, and putting the value of $\Delta = 10Dq$ and α as zero in equation (52), we get

$$\mu_{eff} = \mu_{S.O.}\left(1 - 0\times\frac{\lambda}{\Delta}\right) \tag{53}$$

$$\mu_{eff} = \mu_{S.O.}$$

Hence, d^5 high-spin complexes are expected to give a spin-only moment of 5.92 B.M.; which is independent of temperature.

ii) d^6 strong field (t_{2g}^6): Keeping in mind that there are no unpaired electrons in d^6 low-spin complexes and putting $\mu_{S.O.} = 0$, value of $\Delta = 10Dq$ and α as zero in equation (52), we get

$$\mu_{eff} = 0 \left(1 - 0 \times \frac{\lambda}{\Delta}\right) \tag{54}$$

$$\mu_{eff} = 0$$

Hence, d^6 low-spin complexes are expected to be diamagnetic in nature.

2. Calculation of μ_{eff} for E terms: The ground Mulliken state of E symmetry exists for the weak field octahedral complex with d^4 and d^9 configurations; and strong field octahedral complexes with d^7.

i) d^4 weak field ($t_{2g}^3 e_g^1$): Putting the value of $\Delta = 10Dq$ (the energy gap between ground ^5E_g and $^5T_{2g}$) and α as 2 in equation (52), we get

$$\mu_{eff} = \mu_{S.O.} \left(1 - 2 \times \frac{\lambda}{10Dq}\right) \tag{55}$$

Now because the sign of spin-orbital coupling parameter is positive for less than half-filled configurations, the equation (55) can also be written in the following form.

$$\mu_{eff} = \mu_{S.O.} \left(1 - 2 \times \frac{|\lambda|}{10Dq}\right) \tag{56}$$

Hence, for d^4 high-spin complexes, the effective magnetic moment is less than the spin-only magnetic moment; which is also independent of temperature.

ii) d^9 strong-field weak-field ($t_{2g}^6 e_g^3$): Putting the value of $\Delta = 10Dq$ (the energy gap between ground ^2E_g and $^2T_{2g}$) and α as 2 in equation (52), we get

$$\mu_{eff} = \mu_{S.O.} \left(1 - 2 \times \frac{\lambda}{10Dq}\right) \tag{57}$$

Now because the sign of spin-orbital coupling parameter is negative for more than half-filled configurations, the equation (57) can also be written in the following form.

$$\mu_{eff} = \mu_{S.O.} \left(1 + 2 \times \frac{|\lambda|}{10Dq}\right) \tag{58}$$

Hence, for d^9 strong-field weak-field, the effective magnetic moment is greater than the spin-only magnetic moment; which is also independent of temperature.

3. Calculation of μ_{eff} for A₂ terms: The ground Mulliken state of A₂ symmetry exists for the weak field as well as strong field octahedral complex with d^3 and d^8 configurations.

i) d^3 weak-field strong-field (t_{2g}^3): Putting the value of $\Delta = 10Dq$ (the energy gap between ground $^4A_{2g}$ and $^4T_{2g}$) and α as 4 in equation (52), we get:

$$\mu_{eff} = \mu_{S.O.} \left(1 - 4 \times \frac{\lambda}{10Dq}\right) \tag{59}$$

Now because the sign of spin-orbital coupling parameter is positive for less than half-filled configurations, the equation (59) can also be written in the following form.

$$\mu_{eff} = \mu_{S.O.} \left(1 - 4 \times \frac{|\lambda|}{10Dq}\right) \tag{60}$$

Hence, for d^3 complexes, the effective magnetic moment is less than the spin-only magnetic moment; which is also independent of temperature.

ii) d^8 strong-field weak-field ($t_{2g}{}^6\, e_g{}^2$): Putting the value of $\Delta = 10Dq$ (the energy gap between ground ^3E_g and $^3T_{2g}$) and α as 4 in equation (52), we get

$$\mu_{eff} = \mu_{S.O.} \left(1 - 4 \times \frac{\lambda}{10Dq}\right) \tag{61}$$

Now because the sign of spin-orbital coupling parameter is negative for more than half-filled configurations, the equation (61) can also be written in the following form.

$$\mu_{eff} = \mu_{S.O.} \left(1 + 4 \times \frac{|\lambda|}{10Dq}\right) \tag{62}$$

Hence, for d^8 complexes, the effective magnetic moment is greater than the spin-only magnetic moment; which is also independent of temperature.

> ➤ *Temperature Independent Paramagnetism (TIP)*

In the previous section, we concluded that owing to the A_1 symmetry of the ground state and zero unpaired electron, the strong field complexes of d^6 configuration should be diamagnetic in nature. In other words, there should be no spin-only or the orbital magnetism in spin paired complexes d^6 metal ions. However, that is not hundred-percent true because there are some complexes that have neither the spin nor the orbital degeneracy, yet show weak paramagnetism. Consider the case of MnO_4^- or $[Co(NH_3)_6]^{3+}$, which do not have unpaired electron (no spin degeneracy) and are lacking ground Mulliken state of T-symmetry (no orbital degeneracy); suggesting zero spin-only and zero orbital magnetism. Nevertheless, both of these compounds are weakly paramagnetic in nature. This exceptional behavior can be attributed to the existence of a low-lying excited state which does have the necessary orbital degeneracy and is still a singlet. When the external magnetic field is applied, some of the excited state mixes itself into the ground state resulting in a perturbed ground state that possesses some of the properties of the excited one. The experiments have shown that these complexes have small orbital paramagnetism, which at times, is enough to cancel out the inherent diamagnetism. This kind of paramagnetism is also called temperature-independent paramagnetism due to its non-reliance on the thermal population of Boltzmann levels.

❖ Application of Magneto-Chemistry in Structure Determination

The paramagnetic susceptibility of transition metal complexes can be used to obtain information about the electronic structure, stereochemistry and the nature of metal-ligand bond. The primary applications of magneto-chemistry in the structure determination of coordination compounds are given below.

➤ Determination of the Oxidation State of Transition Metal Centre

The most widely used application of paramagnetic susceptibility measurement is the determination formal oxidation number of the transition metal center. The aforementioned use relies on the finding of the number of unpaired electrons present on the central metal atom or ion. The diamagnetic allowance due to ligands attached is usually ignored; however, corrections must be made for the large orbital contribution to the magnetic moment. For instance, some of the low-spin Co^{2+} complexes have a μ_{eff} around 2.9 B.M., though they have only one unpaired electron ($t_{2g}^6\ e_g^1$). Likewise, some of the high-spin Co^{2+} complexes have a μ_{eff} around 4.9 B.M.; suggesting four unpaired electrons, though they have only three unpaired electrons ($t_{2g}^5\ e_g^2$). Therefore, in order to find the spin only magnetic moment; and thus the correct number of unpaired electrons; the temperature-dependent study of magnetic susceptibility is pretty much essential. Furthermore, the larger magnitude of spin-orbital interaction can also affect the spin-only magnetic moment to a significant extent in some cases like the low spin complexes of Os^{4+} ion. For example, the effective magnetic moments of low-spin complexes Fe^{4+} and Ru^{4+} complexes are about 2.83 B.M.; which is pretty much close to their spin-only value, suggesting two unpaired electrons, which is obviously correct. However, the μ_{eff} for $[OsCl_6]^{2-}$ complex is only 1.4 B.M.; suggesting only one unpaired electron, which is pretty far from reality. Therefore, spin-orbital coupling corrections are also necessary in some cases.

In addition to the orbital magnetism, or the spin-orbital coupling effect; there are two more difficulties in the precise measurement of the formal oxidation state of the transition metal centers. The first one is the electron delocalization and use of non-innocent ligands; for instance, the neutral [Be(bipyridyl)$_2$] complex is expected to be diamagnetic with neutral Be as the central atom. However, the compound is actually paramagnetic due to the delocalization of two electrons from the metal to two bipyridyl chelate ligands. Therefore, if the ligand attached can accommodate one or more unpaired electrons in the appropriate molecular orbitals, then a paramagnetic complex with bivalent beryllium can also be obtained. In such cases, the assignment of formal oxidation state is carried out only after X-ray and EPR (electronic paramagnetic resonance) studies. The second anomaly can arise from metal-metal interactions leading to unexpected electron pairing that depends upon the nature of the ligand used. For example, vanadium hexa-carbonyl, [V(CO)$_6$], is paramagnetic with one unpaired electron. However, one of the neutral dinuclear clusters of vanadium, [(Diarsine)(CO)$_4$V–V(CO)$_4$(Diarsine)], is diamagnetic in nature; even though the oxidation number of the vanadium is same in both. Therefore, a correct molecular weight measurement is pretty much necessary in such cases for fruitful interpretation of experimental results. Moreover, in magnetically concentrated materials, the ferromagnetic or antiferromagnetic interaction must be resolved by magnetic susceptibility measurements as a function of temperature or the field-strength.

> ### *Determination of the Electronic Configuration of Transition Metal Centre*

The energy difference between high-spin and low-spin configurations of a transition metal center in most of the complexes is pretty larger than the available thermal energy. Therefore, most of the complexes are either high-spin like $[FeF_6]^{3-}$, or low-spin like $[Fe(CN)_6]^{3-}$, over a wide range temperature; which makes is quite easy to distinguish and designate their electronic configurations. Nevertheless, there are complexes in which the energy gap between high-spin and low-spin configurations is quite comparable to thermal energy, leading to a rearrangement of electron filling as a function of temperature or a slight variation of coordinative environment. The magnetic susceptibility measurements can be used to detect such phenomena as the number of unpaired electrons varies during the course of spin-equilibria under consideration. For instance, some square-planar complexes of Ni^{2+} (d^8) are diamagnetic but can be converted into paramagnetic tetrahedral or octahedral complexes with two unpaired electrons. Similarly, penta-coordinated Schiff's-base complexes of bivalent cobalt (d^7) also show high-low spin equilibria with three and one unpaired electrons, respectively.

Moreover, the magnetic susceptibility measurements can also be used to identify the exact orbital occupied by the unpaired electron, which depends upon the formal charge and the ligand field strength. For example, the orbital occupied by the unpaired electron in potassium (K) is $4s$, while the isoelectronic Ti^{3+} ion bears the unpaired electron in its $3d$ orbital; which is obviously due to the difference in Z_{eff} arising from changes in the formal charge. The surprisingly large magnetic moment of magnetically dilute $[FeCl_4]^{2-}$ complex may also be explained in terms of mixing of $3d^6$ configuration (4 unpaired electrons) with $3d^5\,4s^1$ configuration due to change of Z_{eff}. The theoretical calculations suggest that the spin-orbital coupling can only increase the μ_{eff} of $[FeCl_4]^{2-}$ from 4.9 B.M. ($\mu_{S.O}$) to 5.1 B.M. at maximum, which is not even close to the experimentally observed magnetic moment (5.5 B.M.). The only rational explanation for such an exceptionally high magnetic moment relies on the σ-bond covalency between Fe^{2+} and Cl^- ligand which decreases the charge on the metal center to an extant where a considerable contribution from the $3d^5\,4s^1$ configuration (6.92 B.M.) to the $3d^6$ configuration is possible. Furthermore, spin pairing due to covalent bonding, [(Diarsine)(CO)$_4$V–V(CO)$_4$(Diarsine)], can also cause μ_{eff} to deviate significantly. Finally, magnetic exchange coupling (ferromagnetism and antiferromagnetism) in magnetically concentrated systems have the capacity to make everything a mess in the analysis; and therefore, requires a special treatment in such cases.

> ### *Determination of the Stereochemistry of Various Transition Metal Complexes*

One of the most valuable applications of paramagnetic susceptibility measurements is to find the stereochemical arrangements of different ligands attached to the transition metal center. The paramagnetic susceptibility measurements can be exploited in two ways to do so; the first one relies upon the determination of unpaired electrons, while the second one depends upon the determination of the exact magnitude of orbital magnetism. Both of the approaches with suitable examples are given below.

1. Stereochemistry from unpaired electrons: The possibility of stereochemistry-prediction of transition metal complexes from unpaired electrons arises from the fact the ligand coordination prefers to adapt the most symmetrical geometry unless the metals d-electron cloud is not spherically symmetrical. For instance, d^0 and d^{10} configurations are spherically symmetrical; and therefore, results in linear, triangular, perfect tetrahedral,

trigonal bipyramidal, octahedral and pentagonal bipyramidal geometries for 2, 3, 4, 5, 6 and 7 coordination number, respectively. Furthermore, if the cloud is not spherically symmetric, slightly distorted octahedral, tetragonal, or square-planar geometries can be obtained. The slightly distorted octahedron refers to an octahedron with two of the bond lengths differing by less than 0.05 Å from the other four. On the other hand, the tetragonal geometry here refers to an octahedral structure with two coaxial bonds considerably longer in comparison to the other four.

Table 8. Expected stereochemistry of transition metal complexes using the number of unpaired d-electrons from crystal field theory.

Weak-field		Strong-field	
Unpaired electrons / Stereochemistry		Unpaired electrons / Stereochemistry	
0 (d^0)	Symmetrical	0 (d^0)	Symmetrical
1 (d^1, t_{2g}^1)	Slightly distorted octahedron	1 (d^1, t_{2g}^1)	Slightly distorted octahedron
2 (d^2, e^2 or t_{2g}^2)	Tetrahedron or slightly distorted octahedron	2 (d^2 e^2 or t_{2g}^2)	Tetrahedron or slightly distorted octahedron
3 (d^3, t_{2g}^3)	Perfect octahedron	3 (d^3, t_{2g}^3)	Perfect octahedron
4 (d^4, t_{2g}^3 e_g^1)	Square-planar or tetragonal	2 or 0 (d^4, t_{2g}^4 or e^4)	Slightly distorted octahedron or tetrahedral
5 (d^5, $e^2 t_2^3$ or $t_{2g}^3 e_g^2$)	Tetrahedron or perfect octahedron	1 (d^5, t_{2g}^5)	Slightly distorted octahedron
4 (d^6, $t_{2g}^4 e_g^2$)	Slightly distorted octahedron	0 (d^6 t_{2g}^6)	Perfect octahedron
3 (d^7, $e^4 t_2^3$ or $t_{2g}^5 e_g^2$)	Tetrahedron or slightly distorted octahedron	1 (d^7, $t_{2g}^6 e_g^1$)	Square planar or tetragonal
2 (d^8, $t_{2g}^6 e_g^2$)	Perfect octahedron	0 (d^8, $t_{2g}^6 e_g^2$)	Square planar or tetragonal
1 (d^9, $t_{2g}^6 e_g^3$)	Square planar or tetragonal	1 (d^9, $t_{2g}^6 e_g^3$)	Square planar or tetragonal
0 (d^{10})	Symmetrical	0 (d^{10})	Symmetrical

Electronic configuration from d^1 to d^9 may or may not have a spherically symmetrical cloud after complexation; therefore, a distortion is expected if any asymmetry is present. The distortion of an octahedron can occur in one of three ways; the first one involves an unsymmetrical filling of t_{2g} level (t_{2g}^1, t_{2g}^2, t_{2g}^4 and t_{2g}^5) which will have little effect on the octahedron is because of the indirect encounter metal between orbitals and the ligands, leading slightly distorted perfect octahedral geometry. On the other hand, if the asymmetry occurs in e_g level, then a tetragonal or square planar geometry may be produced depending upon whether the two ligands are repelled or lost along the z-axis, respectively. However, it is also worthy to mention that the diamagnetism in d^8 complexes does not confirm a square-planar geometry as is also possible in a compatible tetragonal geometry. Metal complexes with d^4 configuration can have zero, two, or four unpaired electrons yielding tetrahedral, tetragonal or square-planar geometries. Hence, we can say that the stereochemistry-chemistry prediction on the basis of the number of unpaired electrons is not very much reliable. Furthermore, consider the example of $[Co(NO_2)_6]^{4-}$ complex in which cobalt is present in +2 oxidation state. The experimental magnetic moment for this complex is about 1.9 B.M., which unexpectedly low for an octahedral complex; because the octahedral geometries of Co(II) usually have three unpaired electrons (μ_{eff} = 4.85–5.2 B.M.), while the presence of one unpaired electron in Co(II) is generally related to a square pyramidal or square planar structure (μ_{eff} = 2.2–2.9 B.M.). The low magnetic moment and octahedral geometry in such cases are generally ascribed to the π-backbonding from the metal to the attached ligand.

2. Stereochemistry from orbital magnet: The unpaired electron approach is somewhat more or less capable of shortlisting a metal complex into one of four classes. classify not sufficient to predict the exact stereochemistry of transition metal complexes. for instance, Ni(II) complexes which contain two unpaired electrons may be perfect octahedron, square bipyramidal, or tetrahedral; while the diamagnetic ones may be square-planar, square pyramidal, trigonal bipyramidal, or six-coordinate tetragonal. Nevertheless, this information can be used in conjugation with orbital magnetism approach for more precise and accurate results. For instance, consider the spin-free complexes of Co(II), in which the magnitude of $\mu_{S.O.}$ is expected to be 3.88 B.M. Now because there is no orbital degeneracy present in Co(II) tetrahedral complexes; and of course, ignoring any spin-orbital coupling; its μ_{eff} is expected to be comparable to $\mu_{S.O}$ value. However, even after considering spin-orbital coupling, the effective magnetic moment would still be lower than 4.5 B.M. On the other hand, the presence of orbital degeneracy in spin-free octahedral complexes of Co^{2+} ($t_{2g}^5 e_g^2$) guarantees a huge orbital magnetic component, making μ_{eff} to touch a mark of 5.1 B.M. Therefore, the contribution of orbital magnetic moment can be used as a criterion for distinguishing between tetrahedral and octahedral complexes of bivalent cobalt. It should also be noted that other analytical methods must be used for final conclusion if magnetic moment is on the borderline (4.5–4.7 B.M.), or there is large distortion of the octahedron leading to splitting of the t_{2g} orbital set, or the value of crystal field splitting in tetrahedral complex is very small as it would lead to a large orbital magnetism due to spin-orbit coupling.

It must be clear again that the structure prediction discussed in this section is totally based upon the simple ligand field theory, which only considers the metals d-electron cloud-only and ignores the lattice effects or effect of ions surrounding the complex ion. The stereochemical arrangements in both spin-free and spin-paired complexes with various electron configurations are given below.

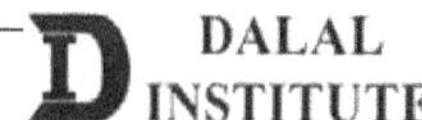

Table 9. The correlation between orbital magnetism and the stereochemistry of weak field transition metal complexes.

Stereochemistry	Supporting configuration	Unpaired electron	Examples	Expected magnetic profile considering orbital magnetism
Perfect octahedral	t_{2g}^3	3	$[Cr(H_2O)_6]^{3+}$	Spin only moment of 3.88 B.M due to absence of orbital magnet.
	$t_{2g}^3 e_g^2$	5	$[FeF_6]^{3-}$	Spin only moment of 5.9 B.M due to absence of orbital magnet.
	$t_{2g}^6 e_g^2$	2	$[Ni(H_2O)_6]^{2+}$	Spin only moment of 2.83 B.M is expected but the orbital contribution from $t_{2g}^5 e_g^3$ increases μ_{eff} due to spin-orbital coupling.
Slightly distorted octahedron	t_{2g}^1	1	$[Ti(H_2O)_6]^{3+}$	Orbital contribution is expected yielding an effective magnetic moment exceeding 1.73 B.M.
	t_{2g}^2	2	$[V(H_2O)_6]^{3+}$	Orbital contribution is expected yielding an effective magnetic moment exceeding 2.83 B.M.
	$t_{2g}^5 e_g^2$	3	$[Co(H_2O)_6]^{2+}$	The orbital contribution is expected yielding an effective magnetic moment exceeding 3.88 B.M.
Square-planar or tetragonal	$t_{2g}^3 e_g^1$	4	MnF_3	Spin only magnetic moment of 4.9 B.M., and significant orbital magnetism is expected from $t_{2g}^2 e_g^2$ states.
	$t_{2g}^6 e_g^3$	1	$[Cu(H_2O)_6]^{2+}$	Spin only magnetic moment of 1.73 B.M., and significant orbital contribution expected from $t_{2g}^5 e_g^4$.
			$[Cu(NH_3)_6]^{2+}$	Spin only magnetic moment of 1.73 B.M., and less orbital contribution expected from $t_{2g}^5 e_g^4$ states.

Table 9. Continued on the next page…

Copyright © *Mandeep Dalal*

Regular tetrahedral geometry	e^2	2	$[FeO_4]^{2-}$	Spin only magnetic moment of 2.83 B.M. is expected with no orbital contribution
	$e^2 t_2^3$	5	$[FeCl_4]^-$	Spin only magnetic moment of 5.9 B.M. with no orbital magnet.
	$e^4 t_2^3$	3	$[CoCl_4]^{2-}$	Spin only magnetic moment of 3.88 B.M. with considerable orbital magnetism possibility from $e^3 t_2^4$

Though the orbital magnetism profiling is pretty useful in the determination of the stereochemistry of various transition metal complexes, the conclusive remarks with full confidence are put after using magnetic susceptibility measurement in conjunction with other physical data like electronic absorption spectra or X-ray diffraction profile.

❖ Magnetic Exchange Coupling and Spin State Cross Over

There are two most remarkable phenomena in the magnetochemistry of transition metal complexes which are pretty much unusual too. The first one is the magnetic exchange interactions, responsible for ferromagnetism and antiferromagnetism; and the other one is the high-spin low-spin equilibria, responsible for the oxygen-carrying capacity of the blood. In this section, we will discuss the fundamental concepts and underlying mechanisms of both the phenomenon.

➤ *Magnetic Exchange Coupling*

The magnetic materials can be broadly classified into two categories; magnetically dilute and magnetically concentrated. Magnetically dilute substances simply refer to the magnetic materials in which individual paramagnetic centers cannot interact with each other due to a large distance of separation. On the other hand, the total-electron-spin of paramagnetic centers do interact through direct-exchange or superexchange interactions in magnetically concentrated substances, which is obviously due to the small inter-micromagnetic separation. In the direct-exchange, there is a coupling between nearest neighboring cations without involving any intermediary anion; while the superexchange is generally a strong magnetic exchange interaction between two nearest neighboring cations through a non-magnetic anion. Now although the orbital contribution to the magnetic moment can be of considerable magnitude, it can be neglected in magnetically concentrated materials for a simplified approach to the rationalization of overall magnetic profile. Therefore, we will consider the spin-only magnetic moment ($\mu_{S.O.}$) as an effective magnetic moment (μ_{eff}) in these cases. The energy of total-spin interactions of paramagnetic centers (ΔE) in magnetically concentrated material can be given by the following relation.

$$\Delta E = 2J(S)_i(S)_k \tag{63}$$

Where J is the exchange-coupling constant and is a measure of the magnitude of total electron spin interaction between various paramagnetic centers. The symbols $(S)_i$ and $(S)_k$ simply represent the total-electron-spin of ith and k^{th} metal centers, respectively. For ferromagnetic substances, the value of J is positive which simply implies that the total electronic spins of all paramagnetic centers are aligned parallel to each other. For antiferromagnetic substances, the value of J is negative which simply indicates that the total electronic spins of all paramagnetic centers are aligned antiparallel to each other. The important features of both of the behaviors are discussed in detail.

1. Ferromagnetic interactions: The magnetic susceptibility in ferromagnetic materials increases very rapidly with the decrease of temperature if measurements are carried out below a certain value of temperature. Moreover, the magnetic susceptibility in ferromagnetic materials also depends upon the strength of the magnetic field applied. The comparison of variation of magnetic susceptibility with temperature for ferromagnetic and paramagnetic substances is given below.

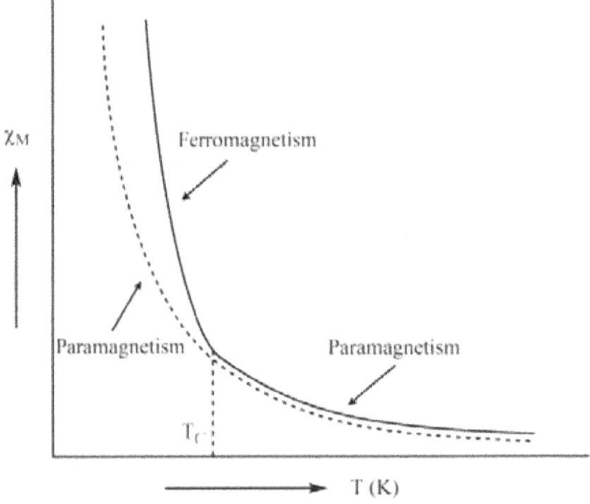

Figure 12. Plots of χ vs T for ferromagnetic and paramagnetic materials.

It can be clearly seen that the magnetic susceptibility of paramagnetic materials follows the Curie law over the complete range of temperature. On the other hand, the ferromagnetic materials obey the Curie law only up to the T_C; but if the temperature is further lowered down, we see an abrupt gain the magnetic susceptibility. Now although the ferromagnetism generally arises from direct-exchange the superexchange can also result in the same if the paramagnetic centers connected to the diamagnetic bridge are at 90° to each other. Moreover, the ferromagnetism may also arise if the two metal ions are not identical; one having an electron in e_g and the other one in t_{2g}. The corresponding modes of orbital interactions are shown below.

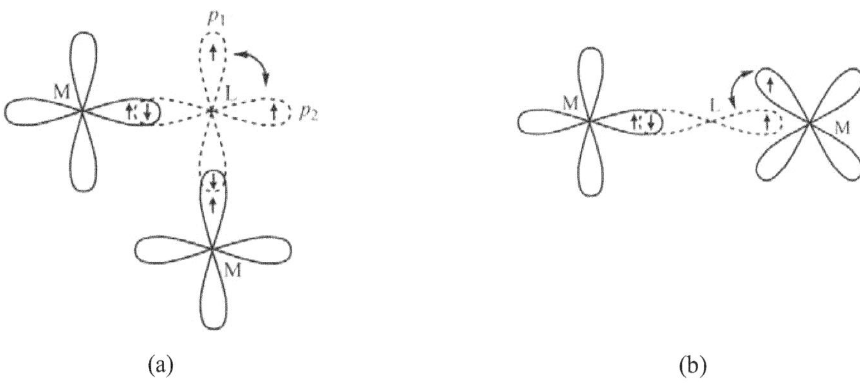

Figure 13. The ferromagnetic coupling via superexchange interactions; when (a) metal centers are at 90°, or (b) non-identical in respect of electronic configuration.

Now it's the time to discuss one of the most interesting properties of ferromagnetic materials, "the hysteresis"; which is actually responsible for the making of permanent magnets. In ferromagnetic substances, total-electron-spins of paramagnetic centers tend to align themselves parallel to each other; however, if the temperature of the system is above the Curie temperature (T_C), the thermal energy (kT) is enough to randomize the individual micro-magnets, and thus forbid them to do so. Consequently, all ferromagnetic substances behave in a paramagnetic fashion above the Curie temperature. However, if the temperature is lowered than T_C, the thermal energy would not be sufficient to decouple the individual total-electron-spins. This, in turn, would result in the creation of small magnetic domains, each of which will be having a parallel alignment of individual paramagnetic centers. Now although the magnetic moments of individual paramagnetic centers in a single "magnetic domain" are parallel to each other, the resultant magnetic moments of these "magnetic domain" are still randomly oriented; which, therefore, would not give any net magnetic moment in the absence of external magnetic field. However, when the external magnetic field is applied, the resultant magnetic moments of all magnetic domains start aligning themselves along the direction of the applied magnetic field.

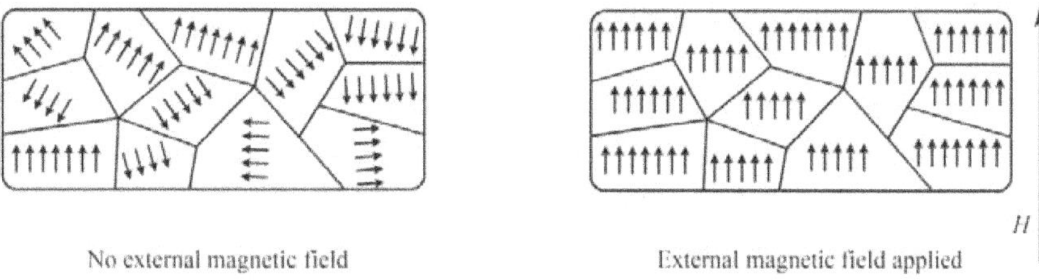

Figure 14. The alignment of magnetic domains in ferromagnetic materials when the external magnetic field is applied.

The magnitude of this domain alignment rises with the increase in the strength of the applied magnetic field, and this will continue to happen till all magnetic domains are aligned. The further increase in the magnetic field strength will not be able to enhance the magnetic susceptibility. Now though the thermal energy is continuously trying to randomize this alignment of "magnetic domains", it would not always be able to do so completely even if the external magnetic field is turned off.

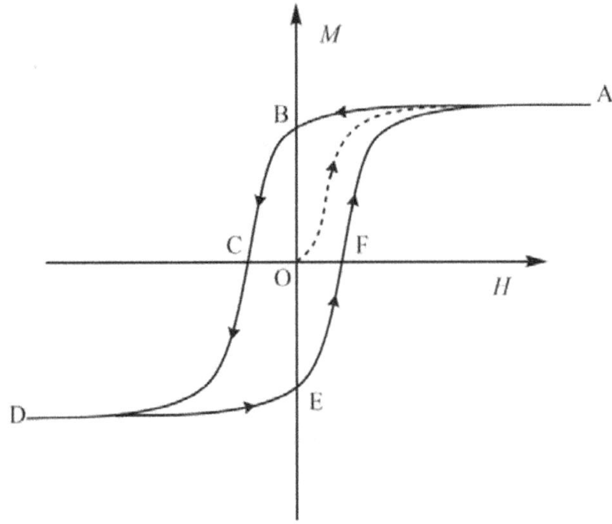

Figure 15. The magnetization of ferromagnetic material from zero-field and the hysteresis loop responsible for the retention of magnetism.

The magnetization of ferromagnetic material is governed by a non-linear curve along OA, attaining a saturation state. However, when the applied field strength decreased back to zero, the demagnetization does not follow AO but the path AB, resulting in a non-zero magnetic moment even if the applied field strength is lowered down to zero. For complete randomization of magnetic domains, the magnetic field should be applied and then increased in the opposite direction; which, consequently, would result in a decrease of magnetization along path BC. The further increase in the magnetic field strength in the opposite direction will actually induce the magnetization of the ferromagnetic material in a reversed manner with again a saturation state, represented by the path CD. Now if the reversed magnetic field is lowered down to zero, the material would again be left with non-zero magnetization (DE). For complete randomization of "magnetic domains" along EF, an increasing applied field strength in the forward direction is needed. Finally, the further increase in the magnetic field will again magnetize the material along FA, completing the hysteresis loop. This retention of "magnetic domain" alignments imparts the permanent magnetism to the ferromagnetic material and called as hysteresis, and has applications in memory-storage devices like computer hard disk drives.

2. Antiferromagnetic interactions: The magnetic susceptibility in antiferromagnetic materials decreases very rapidly with the decrease of temperature if measurements are carried below a certain value of temperature. Moreover, the magnetic susceptibility in antiferromagnetic materials also sometimes depends upon the strength of the magnetic field applied. The comparison of variation of magnetic susceptibility with temperature for antiferromagnetic and paramagnetic substances is given below.

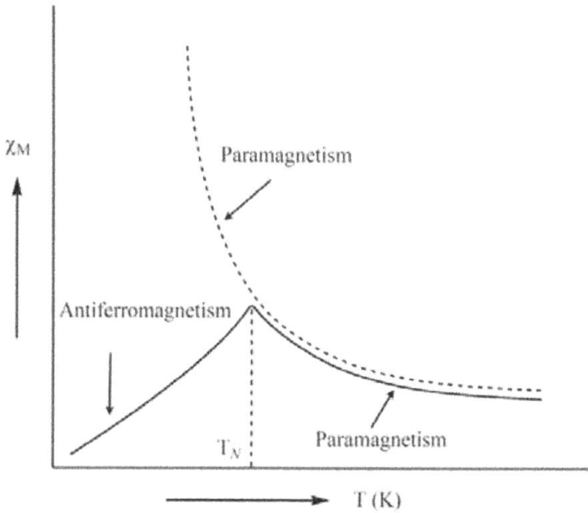

Figure 16. Plots of χ vs T for antiferromagnetic and paramagnetic materials.

It can be clearly seen that the magnetic susceptibility of paramagnetic materials follows the Curie law over the complete range of temperature. On the other hand, the antiferromagnetic materials obey the Curie law only up to the T_N; but if the temperature is further lowered down, we see an abrupt decline the magnetic susceptibility.

Now if the antiparallel alignment of total-electronic-spins of individual paramagnetic centers wants to occur without any interference of a diamagnetic center, it would be possible only via the formation of normal covalent bond in which half-filled orbitals of the participating paramagnetic centers overlap with each other. Sometimes, the pairing of electrons occurs without the formation of any covalent bond. For instance, recent studies have shown that there is no direct metal–metal bond in $Fe_2(CO)_9$, and the individual iron centers have one unpaired electron but the pairing results in a net-zero total-electron-spin. However, in most of the cases, the pairing of total-electron spins of paramagnetic centers in antiferromagnetic materials occurs via the mediation of a diamagnetic atom or ion. The second mechanism is so profound that even if the distance of separation between paramagnetic centers is very small, the pairing of total-electron-spin takes place via the diamagnetic bridge. The diamagnetic mediation is usually done by O^{2-} or halide ions. The classic example of this kind of antiferromagnetism is MnO; in which oxide ions serve in bridging.

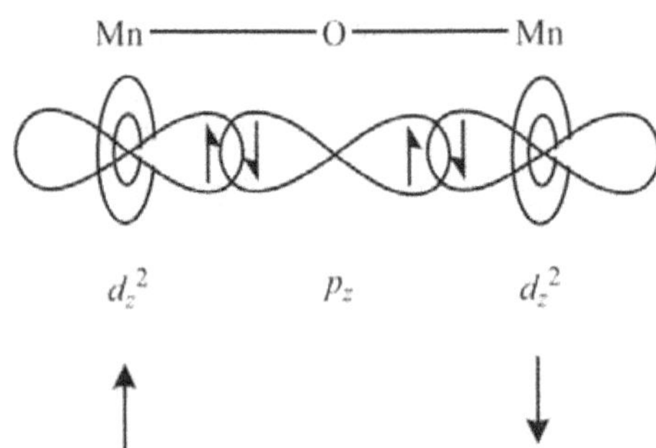

Figure 17. The antiferromagnetic coupling in MnO.

Therefore, the pairing of total-electron-spins in MnO can be understood in terms of the overlap of five half-filled d-orbital of one Mn^{2+} ion with five half-filled (opposite spin) d-orbitals from five neighboring Mn^{2+} ions via oxide ion bridging.

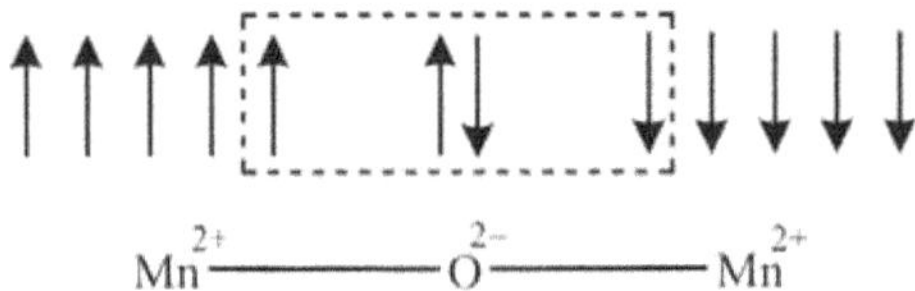

Figure 18. The total-electron-spin pairing in resulting from the antiferromagnetic coupling in MnO.

Now if we omit the oxide ion bridges for simplicity, each paramagnetic center would be surrounded by similar paramagnetic centers but with opposite total-electron-spin. This, in turn, will form an interpenetrating antiferromagnetic lattice.

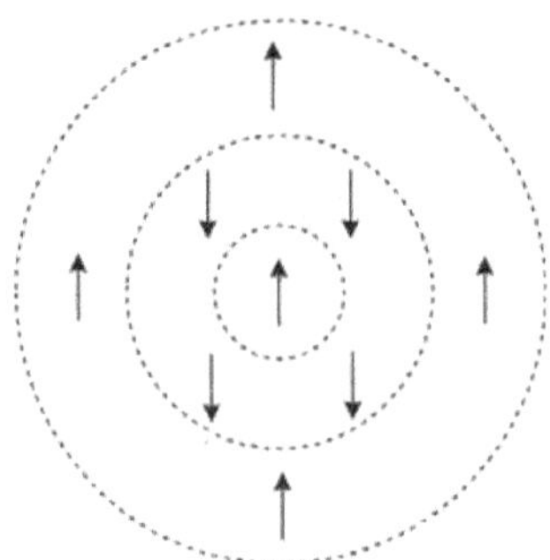

Figure 19. The interpenetrating antiferromagnetic lattice.

Hence, in antiferromagnetic substances, total-electron-spins of paramagnetic centers tend to align themselves antiparallel to each other; however, if the temperature of the system is above the Neel temperature (T_N), the thermal energy (kT) is enough to randomize the individual micro-magnets, and thus forbid them to do so. Consequently, all antiferromagnetic substances behave in a paramagnetic fashion above the Neel temperature. However, if the temperature is lowered than T_N, the thermal energy would not be sufficient to decouple the individual total-electron-spins. This, in turn, would result in the creation of antiferromagnetic lattice, in which total-electron-spins of individual paramagnetic centers are aligned antiparallel to each other. The antiferromagnetic coupling is temperature-dependent and sometimes field-dependent too. Furthermore, the phenomena of antiferromagnetism can also occur within the same molecule, and that what we call intramolecular antiferromagnetism. This kind of antiferromagnetism generally arises through a direct metal–metal bond like $[Cu_2(OAc)_4].(H_2O)_2$ and $K_4[Cl_5Ru–RuCl_5]$; however, in some compounds like $K_4[Cl_5Ru–O–RuCl_5]$, the mediation via diamagnetic center does exist.

The antiparallel alignment of total-electron-spins may also lead to a different type of material "ferrimagnetic", which have the have the magnetic profile just like ferromagnets. The magnetic susceptibility in ferrimagnetic materials increases rapidly (but at a slightly lower rate than ferromagnetic) with the decrease of temperature if measurements are carried below a certain value of temperature (T_C). Moreover, like ferromagnetic materials, the magnetic susceptibility in ferrimagnetic materials also depends upon the strength of the magnetic field applied. Now though the magnetic behavior of a ferrimagnetic material resembles ferromagnetic materials, the ferrimagnetic materials do have an antiparallel alignment of total-electron-spins like in antiferromagnetic materials. However, these paramagnetic ions are present in two different sets of lattice sites. In other words, the ferrimagnetic materials also have the antiparallel alignment of total-electronic-spins, but unlike antiferromagnetism, they do not cancel each other completely, resulting in a permanent magnetism.

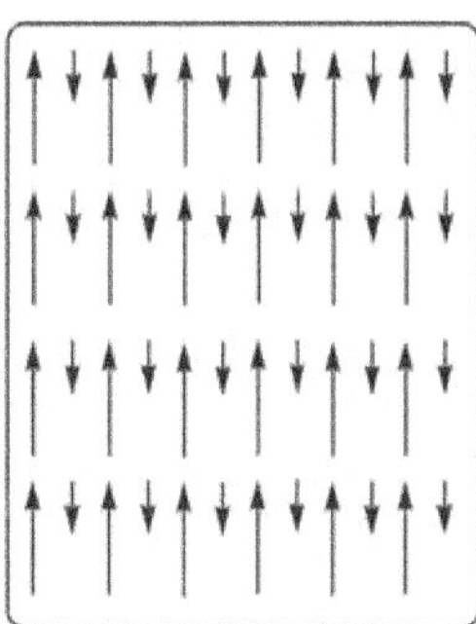

Figure 20. The ordering of total-electron-spins in Ferrimagnetic materials.

Owing to the resemblance in overall magnetic profile ferromagnetism, the oldest known magnetic substance, "loadstone" was originally classified as a ferromagnetic in nature. However, only after Neel's discovery of ferrimagnetism and antiferromagnetism in 1948, we came to know that Fe_3O_4 is actually a ferrimagnet substance.

> ### *Spin State Cross Over*

In transition metal chemistry, most of the compounds for a given stereochemistry are either high-spin or low-spin for a wide range of conditions like temperature, pressure, or the replacement of ligands, etc. However, there are complexes which show a transformation from high-spin to low-spin state with keeping its stereochemistry somewhat more or less the same. Now It is a quite well-known fact that metal complexes from d^4 to d^7 electronic configurations in octahedral crystal fields may result in a high or low-spin system depending on the strength of the surrounding crystal field. Most of the complexes with aforementioned configurations are either high-spin or low-spin over a wide range of temperature, pressure or other conditions like a slight variation in ligand field. However, in some of the complexes, the strength of the ligand field is such that the separation of high-multiplicity term and low-multiplicity terms is very small and quite comparable to the thermal energy available. Now though the high-spin–low-spin crossover can occur, and does occur, in d^4–d^7 complexes, ninety percent of the reported cases belong to Fe^{2+} (d^6) complexes. This can be better understood from the partial correlation diagram of d^6 octahedral complexes.

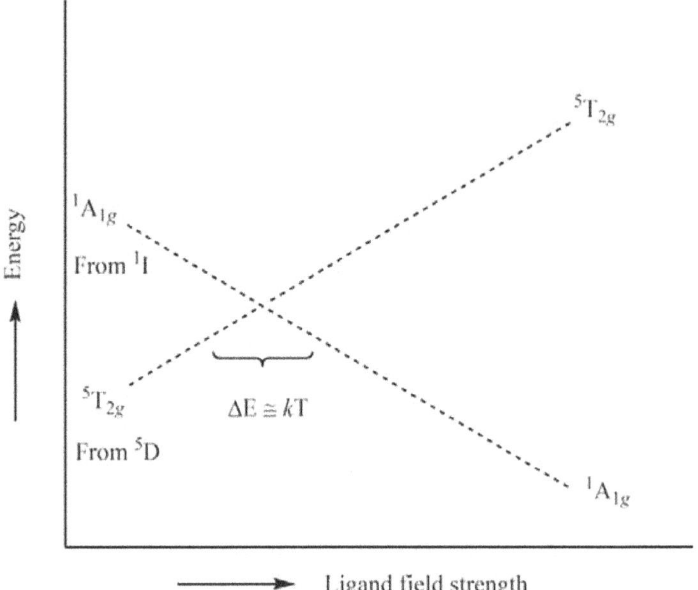

Figure 21. The partial correlation diagram of d^6 octahedral complexes showing the transformation of the ground state from $^5T_{2g}$ to $^1A_{1g}$ as the ligand-field increases.

It can be clearly seen that the energy gap between high-spin and low-spin states near the crossover point is quite small and comparable to the thermal energy; therefore, the temperature does have a great effect on the position of the high-spin–low-spin equilibria.

The study of these spin-crossover phenomena is pretty easy to do as the number of unpaired electron in the two states are different. For instance, the d^6 complexes of Fe^{2+} are diamagnetic in the low-spin state while having four unpaired electrons in the high-spin state. Thus, a variation in the magnitude of the magnetic moment is expected as we after spin crossover. Consider the example of [Fe(phen)$_2$(NCS)$_2$] and [Fe(phen)$_2$(NCSe)$_2$]; in which the high-spin state transforms into to low-spin with the decrease of temperature. Both of these complexes exist in the high-spin $^5T_{2g}$ ground state with effective magnetic moments slightly greater than the spin-only value for four unpaired electrons. However, the moment of [Fe(phen)$_2$(NCS)$_2$] at 174 K suddenly drops to below 1 B.M. Similarly, the same thing does happen in [Fe(phen)$_2$(NCSe)$_2$] but at 232 K. As the temperature is lowered than these specific temperatures, the complexes start to exist in the $^1A_{1g}$ ground state.

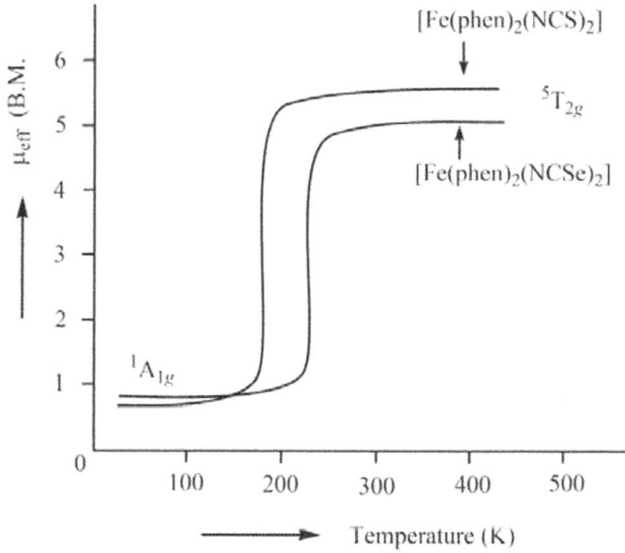

Figure 21. The partial correlation diagram of d^6 octahedral complexes.

Other examples of this include Fe^{3+} surrounded by six sulfur ligands in [Fe(S$_2$CNR$_2$)$_3$] complex. The preference of low-spin state at lower temperatures and higher pressure can be explained in terms of the volume factor. High-spin octahedral complexes have larger volumes due to the presence of electron in e_g orbitals; while low-spin octahedral complexes have electron density concentrated in t_{2g} orbitals. Now because of the lowering of temperature and increment of pressure support the lower volume, the low-spin state is also favored at the lower value of temperature. In hemoglobin, the replacement of water molecule by dioxygen also induces a spin-crossover, which in turn results in the phenomenon of cooperativity, which is essential for our life.

❖ Problems

Q 1. Discuss the classical theory of magnetochemistry in detail.

Q 2. What is Curie-Weiss law? Does it fit the phenomenon of ferromagnetism and antiferromagnetism?

Q 3. What are the different classes of magnetic materials? Give a brief comparison of their magnetic susceptibility with special emphasis upon the field strength and temperature.

Q 4. Discuss Gouy's method for the determination of magnetic susceptibility in detail.

Q 5. How would you calculate the magnetic moment of transition metal complexes experimentally?

Q 6. Discuss the relationship between spin-orbital coupling and magnetic moment.

Q 7. The experimental magnetic moment of Sm^{3+} and Eu^{3+} are different from their spin only value as well as from calculated using total angular momentum quantum number, why?

Q 8. What are the prerequisites for the orbital contribution to the magnetic moment?

Q 9. Discuss the ligand field effect on the orbital contribution in octahedral complexes.

Q 10. What is temperature independent paramagnetism?

Q 11. Discuss the effect of spin-orbital coupling upon the orbital contribution to the total magnetic moment.

Q 12. What are the applications of magnetochemistry in the structure determination of inorganic compounds?

Q 13. Define Curie temperature (T_C) and Neel temperature (T_N).

Q 14. What is magnetic exchange coupling? Explain in detail with special emphasis on ferrimagnetism.

Q 15. Explain the phenomenon of spin-state cross-over using suitable examples.

❖ Bibliography

[1] J. E. Huheey, E. A. Keiter, R. L. Keiter, *Inorganic Chemistry: Principals of Structure and Reactivity*, HarperCollins College Publishers, New York, USA, 1993.

[2] B. R. Puri, L. R. Sharma, K. C. Kalia, *Principals of Inorganic Chemistry*, Milestone Publishers, Delhi, India, 2012.

[3] A. Earnshaw, *Introduction to Magnetochemistry*, Academic Press, New York, USA, 1968.

[4] R. L. Carlin, *Magnetochemistry*, Springer-Verlag, Berlin, Germany, 1986.

[5] S. F. A. Kettle, *Physical Inorganic Chemistry: A Coordination Chemistry Approach*, Springer-Verlag, Berlin, Germany, 1996.

[6] B. N. Figgis, M. A. Hitchman, *Ligand Field Theory and its Applications*, Wiley-VCH, New York, USA, 2000.

[7] R. S. Nyholm, *Magnetism, Bonding and Structure of Coordination Compounds*, Pure and Applied Chemistry, 17, 1968,1.

[8] R. S. Nyholm, *Magnetochemistry*, Journal of Inorganic and Nuclear Chemistry, 8, 1958, 401.

[9] C. W. Bauschlicher, *On the bonding in $Fe_2(CO)_9$*, The Journal of Chemical Physics 84, 1986, 872.

[10] D. Nicholls, *Complexes and First-Row Transition Elements*, Palgrave Macmillan, Basingstoke, UK, 1974.

CHAPTER 10

Metal Clusters:

❖ Structure and Bonding in Higher Boranes

Boron hydrides or simply boranes are a class of compounds that have the generic formula B_xH_y. Though the boron does not react with hydrogen directly to form any of the boron hydrides, yet many of these compounds can easily be synthesized under special reaction conditions. However, because of the high affinity of boron for oxygen, these compounds readily oxidize on contact with air (lighter boranes even explode); and therefore, do not occur in nature. This class is derived from the borane (BH_3) itself, which exists only as a transient intermediate and dimerizes to form diborane (B_2H_6) immediately. The higher boranes are all consisted of boron clusters that are actually polyhedral in nature. Besides the neutral boranes, a large number of anionic boron hydrides also exist. The general formula for single-cluster boron hydrides is $B_nH_n^{2-}$, B_nH_{n+4}, B_nH_{n+6}, B_nH_{n+8} and B_nH_{n+10} (n is the number of boron atoms) for closo-, nido-, arachno-, hypho- and klado- type, respectively. There also exists a series of substituted neutral hypercloso-boranes that have the theoretical formulae of B_nH_n.

The naming of neutral boranes is done with a Greek prefix showing the number of boron atoms and the number of hydrogen atoms in brackets. In the naming of anions, the hydrogen count is specified first followed by the boron count and the overall charge in the bracket in the last. Furthermore, the prefix closo-nido- etc. can also be added. For example:

Formula	B_5H_9	B_4H_{10}	$B_6H_6^{2-}$
Name (IUPAC)	Pentaborane(9)	Tetraborane(10)	hexahydridohexaborate(2−)
Type	Nido-	Arachno-	Closo-

Understandably many of the compounds have abbreviated common names. The prefix like closo-, nido are actually related to geometry and the number of framework electrons of a particular cluster. Now because the boron hydrides are electron-deficient compounds, the structure and bonding in higher boranes can be understood only after rationalizing the 3-center-2-electron bond involved.

➢ *Valence Bond Treatment of Three-Centre Two-Electron Bond*

The simplest example of boron hydride containing 3-center-2-electron bond is diborane; and therefore, the first orbital based approach we can use to study the diborane structure is valence bond theory. The B_2H_6 or diborane can be considered as a dimeric unit of two BH_3 molecules. Now, as all the three half-filled sp^3 hybrid orbitals of boron are already used to bind hydrogens, the dimerization seems to be impossible. However, the valence bond theory still suggests a mode of dimerization through sp^3-hybridization. The electronic configuration of boron in its ground state and excited state are:

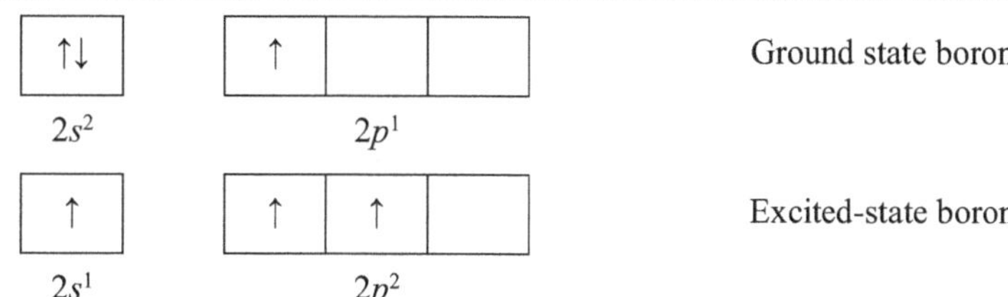

The mixing of one *s* and three *p* orbitals generate four sp^3 hybrid orbitals; three half-filled and one empty. Each boron atom in the diborane structure can use two of its sp^3 hybrid orbitals half-filled to bind two hydrogen atoms (terminal-H). The remaining two hybrid orbitals (one half-filled and one empty) are then used to bind bridging-H atoms as given below.

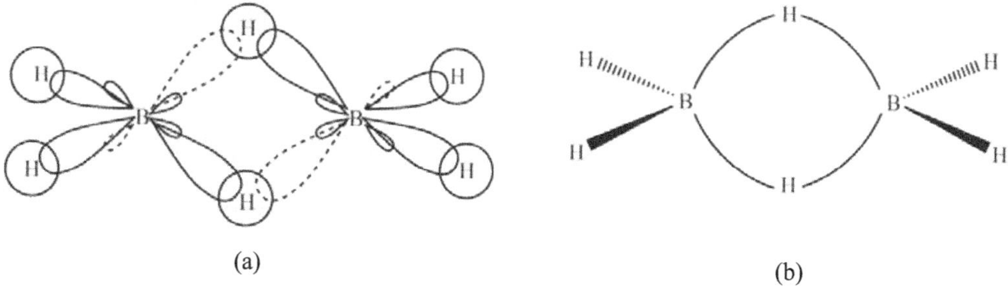

Figure 1. The (a) orbital overlap of hybrid atomic orbital of boron with hydrogens and (b) the resulting banana-shaped 3-center-2-electron bond.

It can be clearly seen that the valence electron in each bridging hydrogen must get delocalize over all three atoms (B–H–B) to bond to both of the B atoms simultaneously. This is pretty much contradictory for the valence bond theory, which likes to bound the valence electrons into regions of space that are localized between two participating nuclei only. Additionally, the molecular geometry of diborane cannot be explained using valence shell electron pair repulsion (VSEPR) theory as far as bridging hydrogens are concerned. According to VSEPR model, two bonds around each bridging hydrogen should be linear i.e. the B–H–B bond angle should be equal 180° with straight-line coordination around.

> ➤ *Molecular Orbital Treatment of Three-Centre Two-Electron Bond*

The rationalization of the 3-centre-2-electron bond is much more convincing from the molecular orbital approach, which considers the delocalization of electron density in bonding as a common phenomenon. Now because we only want to understand the nature of bonding involved in B–H–B bridges, we need to focus on just the part of the molecule. Before the formation of molecular orbitals, we need to identify the basis function required. Owing to the partial success of the valence bond theory in the rationalization of the 3-center-2-electron bond, we can make the use of four equivalent sp^3 hybrid orbitals and two *s* orbitals of hydrogen atoms to construct the symmetry adapted linear combinations (SALCs) of atomic orbitals.

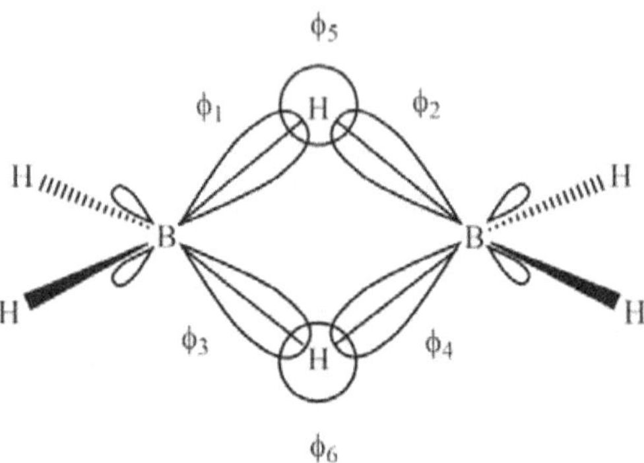

Figure 2. Basis functions for the B–H–B portion in B_2H_6.

The reducible representation based upon four hybrid orbitals of borons and two $1s$ orbitals hydrogens in the B–H–B portion of diborane can be given as:

Table 1. The reducible representation based upon four hybrid orbitals of borons and two $1s$ orbitals hydrogens in the B–H–B portion of diborane.

D_{2h}	E	C_2^z	C_2^y	C_2^x	i	σ_{xy}	σ_{xz}	σ_{yz}	Irreducible components
Γ_H	2	0	0	2	0	2	2	0	$a_{1g} + b_{3u}$
Γ_B	4	0	0	0	0	4	0	0	$a_{1g} + b_{1g} + b_{2u} + b_{3u}$

The mathematical forms of the six normalized SALCs can be deduced through the projection operator theorem and are given by the equations (1) to (6) as:

$$\psi_{a_{1g}}(\text{H}) = \frac{1}{\sqrt{2}}(\phi_5 + \phi_6) \tag{1}$$

$$\psi_{b_{3u}}(\text{H}) = \frac{1}{\sqrt{2}}(\phi_5 - \phi_6) \tag{2}$$

$$\psi_{a_{1g}}(\text{B}) = \frac{1}{2}(\phi_1 + \phi_2 + \phi_3 + \phi_4) \tag{3}$$

$$\psi_{b_{1g}}(\text{B}) = \frac{1}{2}(\phi_1 - \phi_2 + \phi_3 - \phi_4) \tag{4}$$

$$\psi_{b_{2u}}(\text{B}) = \frac{1}{2}(\phi_1 - \phi_2 - \phi_3 + \phi_4) \tag{5}$$

$$\psi_{b_{3u}}(\text{B}) = \frac{1}{2}(\phi_1 + \phi_2 - \phi_3 - \phi_4) \tag{6}$$

The single-electron wave functions and corresponding molecular orbital shapes for B–H–B portion of diborane are shown in Figure 3. For simplicity, only the largest lobes of the four sp^3 hybrid orbitals are depicted. Nevertheless, the smaller lobes on the opposite side also have a good possibility to overlap, which makes the energies two nonbonding molecular orbitals unequal. The mixing of two SALCs of a_{1g} symmetry generates two molecular orbitals; one bonding and other antibonding shown by a_{1g} and $a_{1g}{}^*$, respectively. Similarly, the mixing of two SALCs of b_{3u} symmetry yields two molecular orbitals of b_{3u} and $b_{3u}{}^*$ symmetry.

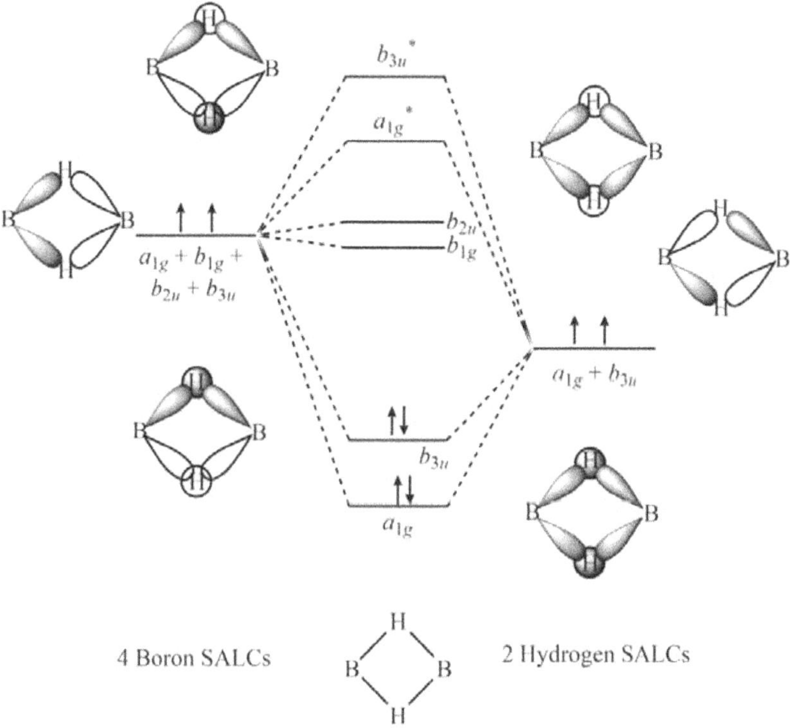

Figure 3. Partial one-electron molecular orbital energy level diagram for the B–H–B part of B_2H_6.

The shapes of molecular orbital with a_{1g} and b_{3u} symmetry evidently show that the electron density present in them is actually delocalized over all three nuclei involved in B–H–B unit of diborane. A certain amount of direct B–B bonding in diborane is also present which helps to stabilize the molecule geometry; which is attributed to small-sized H-groups.

> ➤ *Classification of Bonds Present in Higher Boranes*

The explanation of overall bonding in higher boranes needs the calculation of electrons involved in the skeletal structure of B_n cluster and these electrons are generally referred as the "framework electrons". Moreover, the B–H–B bridges are considered as part of the B_n framework while the terminal B–H are discarded. Though the electrons involved in the framework construction of B_n cluster are very much delocalized, yet the localized 3-centre 2-electron and 2-centre 2-electron orbital approach can be used to explain the main features of the boranes. Each B atom uses one electron in the normal 2-center-2-electron terminal B–H bond and the remaining two electrons are used in B_n framework. The structure of higher boranes may or may not have all of the following kinds of bonds i.e. they may possess few or all types.

1. Terminal B–H bond: This is a normal 2-center-2-electron covalent bond and is generally shown as B–H.

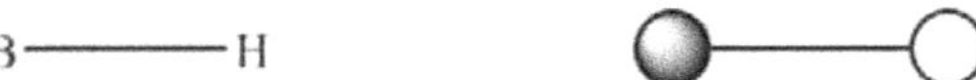

2. Direct B–B bond: This is a 2-center-2-electron bond and is able to connect to two boron atoms. These bonds are generally represented as B–B.

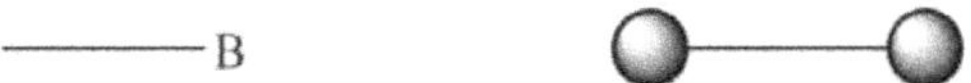

3. Bridging or open B–H–B bond: This is a 3-center-2-electron bond and is able to connect two boron and one hydrogen atoms. These bonds are generally represented as B–H–B.

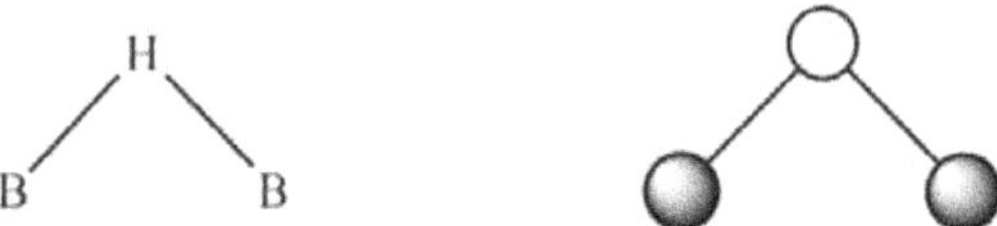

4. Closed B–B–B bond: This is a 3-center-2-electron bond and is able to connect three boron atoms that are at the corners of an equilateral triangle. These bonds are generally represented as B–B–B.

5. Open bridging B–B–B bond: This is a 3-center-2-electron bond and is able to connect three boron atoms to form a B–H–B like bridge found in diborane. These bonds are represented as B–B–B.

> *Geometry Prediction in Higher Boranes Using Lipscomb's Model Involving STYX Numbers*

The main initial approach to understand the structures of boranes and related species was Lipscomb's topological model involving STYX numbers and rules. W. N. Lipscomb was an American chemist who developed a method to find the key combinations of bonding profile that are actually possible for a specific boron hydride formula. Each boron atom in neutral boranes or hydroborate anions has at least one H attached by a normal 2-center-2-electron sigma bond; therefore, it can be assumed that one B–H bond is present per boron atom. Besides, Lipscomb also proposed that every other bond present in the cluster must belong to any of the following categories.

i) 3-centre-2-electrons B–H–B bond, labeled as "S".

ii) Closed, open, or a mixture of both 3-centre 2-electrons B–B–B bond, labeled as "T".

iii) 2-center-2-electrons B–B bond, labeled as "Y".

iv) 2-center-2-electrons B–H terminal bond (simply no. of BH_2 groups), labeled as "X".

The structure prediction of borane clusters using the Lipscomb model involves the three major components which are discussed in detail below.

1. The relationship between STYX code and valence electrons: The correlation between the boranes formula and the number and types of bonds in the cluster is given by simple equations of balance. These equations relate all kinds of sigma bonds to the number of valence electrons available.

i) Three centers orbital balance: The electron deficiency in a borane cluster can be removed only if one 3-center-2-electron bond is created by each boron atom. Thus, the sum of the number of 3-centre 2-electron B–H–B bonds and the number of 3-centre 2-electron B–B–B bonds must be equal the number of B–H units (n).

$$n = S + T \tag{7}$$

ii) The hydrogen balance: Supposing that one terminal hydrogen is attached to each boron atom, the number of hydrogen atoms leftover (m), must be distributed among bridges and additional B–H terminal bonds.

$$m = S + X \tag{8}$$

iii) The electron balance: Now, if each BH unit contributes one pair of electrons to the skeleton and each of the extra hydrogens gives one electron; the number of electron pairs contributed by additional hydrogens must be calculated by halving their number. All of these electron pairs must be participating in bonding; and therefore, the total number of bond pairs can be given as:

$$n + (m/2) = S + T + Y + X \tag{9}$$

From equation (7) and (8) we get

$$Y - \frac{1}{2}(S - X) \tag{10}$$

2. The calculation of various STYX possibilities: After knowing the correlation between various kinds of bonds with framework electron pairs in boron hydrides, we need to follow the following steps to write various STYX possibilities.

i) Write down the general formula of given borane cluster in form like B_nH_{n+m} to fix the value of n and m.

ii) Calculate the number of B–H–B bridges which are represented by S. The value of S must lie within the range of $m/2$ to m. In other words, the value of S must satisfy the following condition.

$$\frac{m}{2} \le S \ge m \tag{11}$$

The validation of the higher limit comes from the fact that m represents the total additional hydrogens out of which some are present in B–H–B bond and some are present in B–H terminal bond. Therefore, if all the additional hydrogens are present in B–H–B bond, the value of "S" at its maximum can just be equal to m. The validation of the lower limit can be derived equation (8) and equation (9) as:

$$n + (m/2) = (m - X) + T + Y + X \tag{12}$$

$$n = \left(\frac{m}{2}\right) + T + Y \tag{13}$$

From equation (7), we get

$$S + T = \left(\frac{m}{2}\right) + T + Y \tag{14}$$

$$S = \left(\frac{m}{2}\right) + Y \tag{15}$$

Thus, the value of S is always equal to or greater than $m/2$.

iii) For different values of S obtained in 2nd step, we have to calculate the equally possible solutions for the values of T, Y and X.

iv) There are many sets of STYX numbers for a given borane cluster and thus many possible topologies.

3. The shortlisting of STYX codes using empirical rules: In order to select the correct STYX code among several possibilities, empirical rules have been developed which follow as:

i) The STYX number sets with negative value are not considered as they have no physical meaning.

ii) All boranes have at least a 2-fold symmetry, so it is assumed that any new hydride probably would have at least one plane, center, or two-fold axis of symmetry. Low symmetry seems to activate the center of reactivity.

iii) Only one terminal hydrogen and no bridging hydrogen may be attached to boron that is bound to five neighboring borons. This restricts B–H–B bridges and BH_2 groups to the open edges of boron frameworks.

iv) If a boron atom is bound to four other boron, it will probably not make use of more than one B–H–B bridge.

v) A boron atom that is bound to only two other boron atoms will be involved in at least one B–H–B bridge.

The whole concept can be exemplified using the diborane structure. The general formula for diborane can be written in the form of B_2H_{2+4}; which gives the values of n and m as 2 and 4, respectively. From equation (11), we get

$$\frac{4}{2} \leq S \geq 4 \tag{16}$$

The possible values of S = 2, 3, 4. By using S = 2 in equation (7), we get

$$2 = 2 + T \tag{17}$$

$$T = 0 \tag{18}$$

Similarly, putting S = 2 in equation (8), we get

$$4 = 2 + X \tag{19}$$

$$X = 2 \tag{20}$$

Now putting the values of T and X from equation (18) and equation (20) in equation (9), we get

$$2 + (4/2) = 2 + 0 + Y + 2 \tag{21}$$

$$Y = 0 \tag{22}$$

Similarly, by taking the value of S as 3 and 4, one can get two another set of possible STYX numbers as tabulated as given below.

Table 2. Three possible sets of STYX numbers for B_2H_6.

Number of B–H–B bonds (S)	Number of B–B–B bonds (T)	Number of B–B bonds (Y)	Number of BH$_2$ groups (X)
2	0	0	2
3	−1	1	1
4	−2	2	0

It can clearly be seen that out of three sets of STYX numbers, only the first one is physically reasonable since the latter two involve negative values. Besides, this also confirmed by the experimental geometrical structure of the diborane that the STYX number of 2002 is, of course, seems to be correct; because there are two B–H–B bridges, no three-center B–B–B bonds or two-center B–B bonds and two B–H terminal bonds in addition to those already considered i.e. two BH$_2$ groups. A similar procedure can be applied to find out the topological structures of other borane clusters like B_4H_{10}, B_5H_9, $B_{10}H_{14}$.

> ➢ *Structure and Bonding Profile of Some Typical Higher Borane Clusters*

As we have already discussed the fundamentals of structure and bonding in higher borane systems, now we will implement those ideas to explain some typical examples.

1. Tetraborane-10 (B_4H_{10}): The bonding in B_4H_{10} can easily be explained by drawing the plane projections of its three-dimensional structure as shown below.

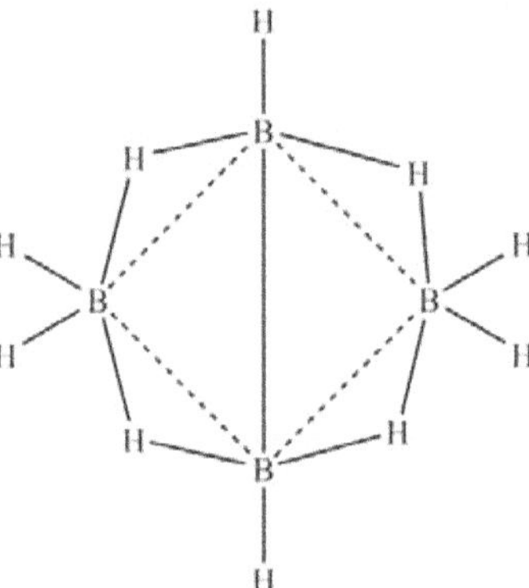

Figure 4. Planar view for the structure of B_4H_{10}.

It can be clearly seen that there are four B–H–B bridges, no closed or bridging B–B–B unit, one B–B bond and two terminal BH_2 groups. Hence, the overall STYX code for B_4H_{10} molecule is 4012. Now considering the electron requirement of various kinds of bonds present, the individual contribution of boron and hydrogen atoms can easily be tabulated.

Table 3. Nature and number of bonds (along with electrons required) present in B_4H_{10}.

Nature of the Bond	Number of Bonds	Total electron required	Contribution from 4 B atoms	Contribution from 10 H atoms
B–H–B	4	8	4	4
B–B–B	0	0	0	0
B–B	1	2	2	0
B–H	6	12	6	6

Hence, four 3-centre 2-electron and seven 2-centre 2-electron bonds require a total $4\times2 + 7\times2 = 22$ electrons. Four boron atoms have 12 valence electrons while 10 electrons are actually contributed by ten hydrogen groups that participating in both types of bonds.

2. Pentaborane-9 (B_5H_9): The pentaborane-9 (commonly called pentaborane) is an inorganic compound with the formula B_5H_9, and is different from pentaborane-11 (B_5H_{11}). It is one of the most common cluster hydrides of boron, though it is a highly reactive compound. Owing to its high reactivity toward oxygen, it was once evaluated as rocket or jet fuel. Like many of the smaller boron hydrides, pentaborane is colorless, diamagnetic, and volatile. The bonding in B_5H_9 can easily be explained by drawing the plane projections of its three-dimensional structure as shown below.

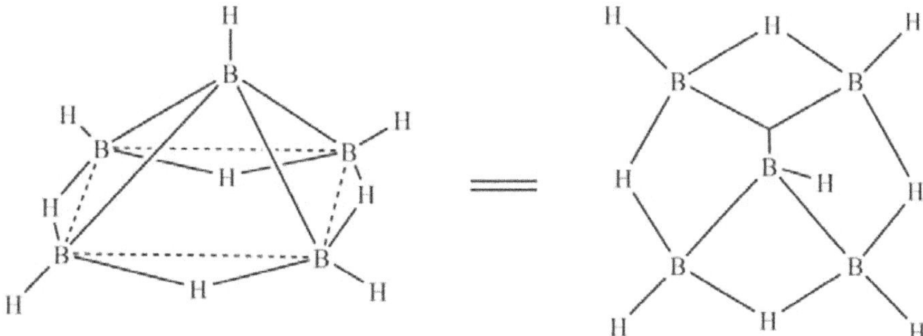

Figure 5. Structure of B_5H_9.

Its structure is that of five atoms of boron arranged in a square pyramid. Each boron has a terminal hydride ligand and four hydrides span the edges of the base of the pyramid. It can be clearly seen that there are four B–H–B bridges, one closed or triply bridged B–B–B unit, two B–B bond and no terminal BH_2 groups. Hence, the overall STYX code B_5H_9 molecule is 4120. Now considering the electron requirement of various kinds of bonds present, the individual contribution of boron and hydrogen atoms can easily be tabulated.

Table 4. Nature and number of bonds (along with electrons required) present in B_5H_9.

Nature of the Bond	Number of Bonds	Total electron required	Contribution from 5 B atoms	Contribution from 9 H atoms
B–H–B	4	8	4	4
B–B–B	1	2	2	0
B–B	2	4	4	0
B–H	5	10	5	5

Hence, five 3-centre 2-electron and seven 2-centre 2-electron bonds require a total $5{\times}2 + 7{\times}2 = 24$ electrons. Five boron atoms have 15 valence electrons while 9 electrons are actually contributed by nine hydrogen groups that are participating in both types of bonds.

3. Pentaborane-11 (B_5H_{11}): The pentaborane-11 is a compound with the general formula B_5H_{11}. It is a colorless liquid at room temperature with a boiling point of $63\,°C$. The bonding in B_5H_{11} can easily be explained by drawing the plane projections of its three-dimensional structure as shown below.

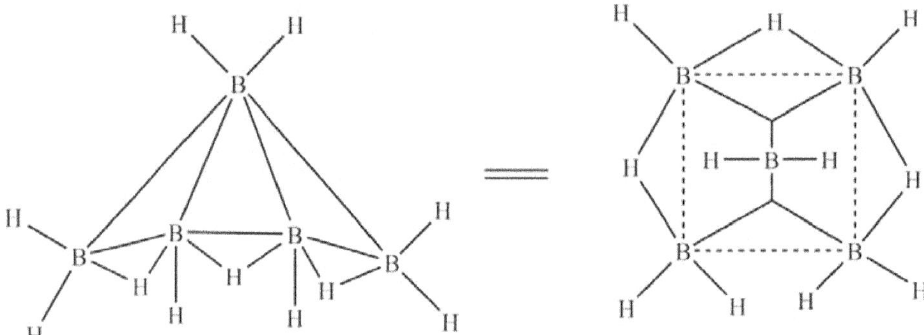

Figure 6. Structure of B_5H_{11}.

The pentaborane-11 is having unsymmetrical square-pyramidal and five boron atoms are present at the five corners of a square pyramid, just like in the case B_5H_9 molecule. Three out of five boron atoms have a terminal hydride ligand while two adjacent boron on the base of the pyramid has two hydride group each, and four hydrides span the edges of the base of the pyramid. It can be clearly seen that there are three B–H–B bridges, two closed or triply bridged B–B–B unit, zero B–B bond and three-terminal BH_2 groups. Hence, the overall STYX code B_5H_{11} molecule is 3203. Now considering the electron requirement of various kinds of bonds present, the individual contribution of boron and hydrogen atoms can easily be tabulated.

Table 5. Nature and number of bonds (along with electrons required) present in B_5H_{11}.

Nature of the Bond	Number of Bonds	Total electron required	Contribution from 5 B atoms	Contribution from 11 H atoms
B–H–B	3	6	3	3
B–B–B	2	4	4	0
B–B	0	0	0	0
B–H	8	16	8	8

Hence, five 3-centre 2-electron and eight 2-centre 2-electron bonds require a total $5×2 + 8×2 = 26$ electrons. Five boron atoms have 15 valence electrons while 11 electrons are actually contributed by eleven hydrogen groups that are participating in both types of bonds.

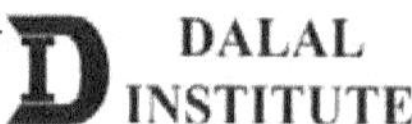

4. Hexaborane-10 (B_6H_{10}): The bonding in B_6H_{10} can easily be explained by drawing the plane projections of its three-dimensional structure as shown below.

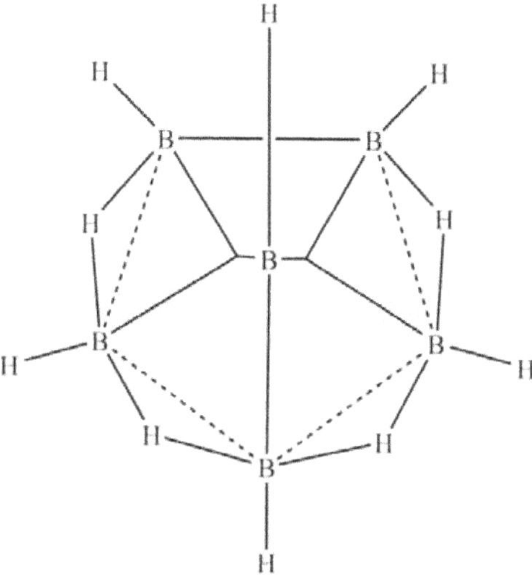

Figure 7. Structure of B_6H_{10}.

It can be clearly seen that there are four B–H–B bridges, two closed or triply bridged B–B–B unit, two B–B bond and no terminal BH_2 groups. Hence, the overall STYX code B_6H_{10} molecule is 4220. Now considering the electron requirement of various kinds of bonds present, the individual contribution of boron and hydrogen atoms can easily be tabulated.

Table 6. Nature and number of bonds (along with electrons required) present in B_6H_{10}.

Nature of the Bond	Number of Bonds	Total electron required	Contribution from 6 B atoms	Contribution from 10 H atoms
B–H–B	4	8	4	4
B–B–B	2	4	4	0
B–B	2	4	4	0
B–H	6	12	6	6

Hence, six 3-centre 2-electron and eight 2-centre 2-electron bonds require a total $6 \times 2 + 8 \times 2 = 28$ electrons. Six boron atoms have 18 valence electrons while 10 electrons are actually contributed by ten hydrogen groups that are participating in both types of bonds.

5. Decaborane-14 ($B_{10}H_{14}$): The bonding in $B_{10}H_{14}$ can easily be explained by drawing the plane projections of its three-dimensional structure as shown below.

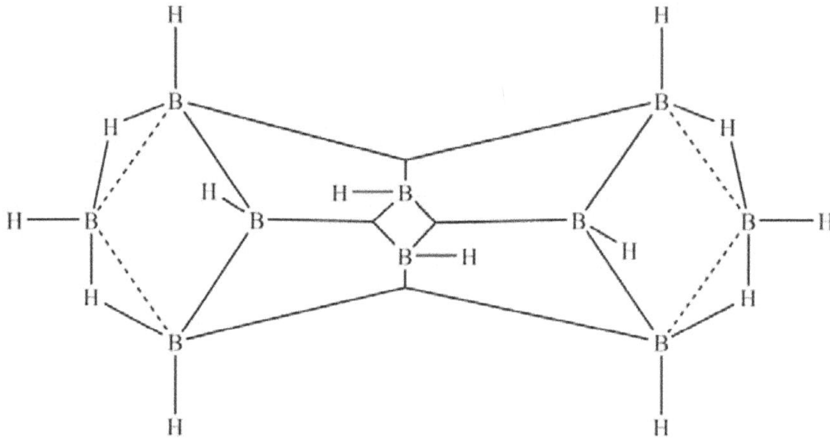

Figure 8. Structure of $B_{10}H_{14}$.

In decaborane, the B_{10} framework resembles an incomplete octadecahedron. Each boron has one radial hydride, and four boron atoms near the open part of the cluster feature extra hydrides. It can be clearly seen that there are four B–H–B bridges, six B–B–B unit (four B-B-B triple bridge bonds and two B-B-B bent bridge), two B–B bond and zero terminal BH_2 groups. Hence, the overall STYX code $B_{10}H_{14}$ molecule is 4620. Now considering the electron requirement of various kinds of bonds present, the individual contribution of boron and hydrogen atoms can easily be tabulated.

Table 7. Nature and number of bonds (along with electrons required) present in $B_{10}H_{14}$.

Nature of the Bond	Number of Bonds	Total electron required	Contribution from 10 B atoms	Contribution from 14 H atoms
B–H–B	4	8	4	4
B–B–B	6	12	12	0
B–B	2	4	4	0
B–H	10	20	10	10

Hence, ten 3-centre 2-electron and twelve 2-centre 2-electron bonds require a total $10 \times 2 + 12 \times 2 = 44$ electrons. Ten boron atoms have 30 valence electrons while 14 electrons are actually contributed by fourteen hydrogen groups which are participating in both types of bonds.

> ***Structural Relationship Between Closo, Nido and Arachno Boranes***

The structural relationship between closo, nido and arachno boranes is shown by the diagonal connecting species having the same number of skeletal electron pairs. Hydrogen atoms are omitted for clarity.

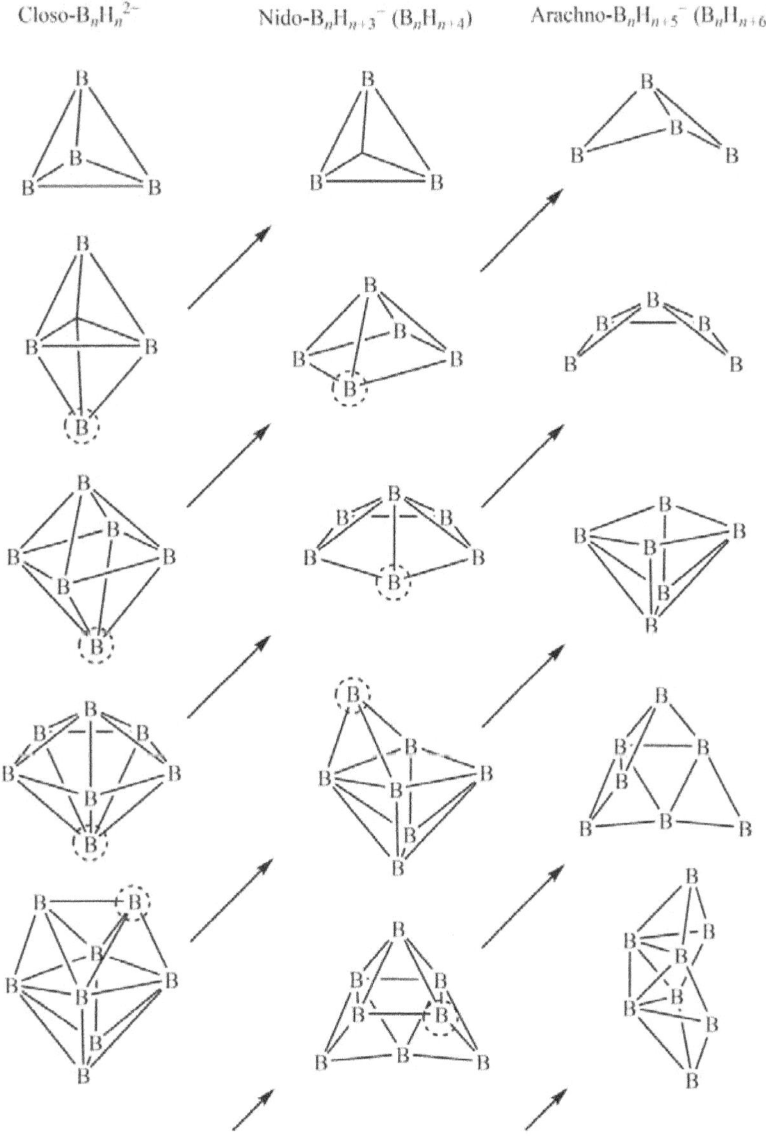

Figure 9. Continued on the next page…

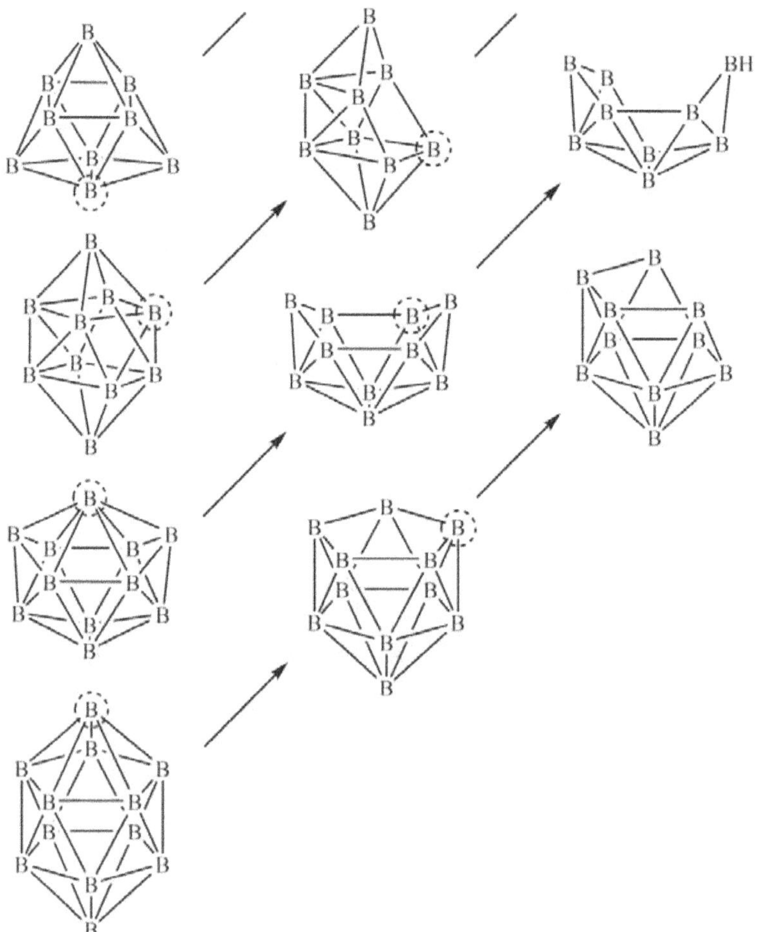

Figure 9. The structural relationship between closo, nido, and arachno boranes; the diagonal lines connect species that have the same number of skeletal electron pairs (SEP).

The structural relationships shown above explain various geometrical aspects of borane clusters; and provides a somewhat conceptual way to think about these systems. For instance, the structure of dodecahydro-closo-dodecaborate anion ($B_{12}H_{12}^{2-}$) is a regular icosahedron. Similarly, the structure of nido-$B_{10}H_{14}$ can easily be obtained just by removing a six-degree vertex from closo-decaborate ($B_{10}H_{10}^{2-}$) anion. The removal of one B–H unit and the addition of two H atoms converts nido-B_5H_9 into a butterfly-shaped arachno-B_4H_{10} borane. Nevertheless, it should also be noted that the interconversion of these structures may or may not be possible chemically; and the same statement can be made for their chemical formula.

❖ Wade's Rules

Though the Lipscomb's topological model involving STYX numbers and rules was good in the rationalization of structures of some of the boranes and related species, it was not enough to higher boranes and carbonyl clusters. Therefore, a more sophisticated and comprehensive method was needed. In 1971, a British chemist, Kenneth Wade, published a revolutionary paper in the field of cluster chemistry. The main outlines of this communication were later to be called as Wade's rules, which provided a straightforward and elegant explanation of the geometries of "electron-deficient" cluster compounds in terms of the number of skeletal electron pairs (SEPs) available. The main focus of this paper was on boranes, carboranes and low-valent transition-metal clusters; whose structures could not be explained by the normal 2-center-2-electron bond. After Wade's initial report, another British chemist, Michael Mingos, extended the principles of counting skeletal electron pairs to electron-precise and electron-rich clusters and gave a more generalized way of calculating the skeletal electron contribution of a wide range of transition-metal. Thereafter, the rules became known as the Wade-Mingos rules or, more formally, the polyhedral skeletal electron pair theory (PSEPT).

The structure prediction is based on different sets of rules ($4n$, $5n$, or $6n$), which are developed on the basis of the number of electrons present per vertex. The $4n$ rules are pretty much accurate in predicting the structures of clusters with 4 electrons per vertex, as is the case for many carboranes or boranes. The clusters following $4n$ rule are generally classified as closo-, nido-, arachno- or hypho-, depending on whether they represent a complete (closo-) deltahedron, or a deltahedron that is missing one (nido-), two (arachno-) or three (hypho-) vertices. However, if the count of electrons is near to 5 electrons per vertex, the structure changes occur and are governed by the $5n$ rules (based on 3-connected polyhedrons). Moreover, if the electron count is increased further, the clusters with $5n$ electron counts become very unstable, and the $6n$ rules find their role. The cluster compounds following $6n$ rule have structures that are dependent on rings.

➤ *4n Rule*

The basis closo-polyhedra for the $4n$ rules are composed of triangular faces. The number of vertices in the cluster prediction and the corresponding basis polyhedron is given below.

Table 8. Base polyhedrons for structure prediction using $4n$ rule.

No. of Vertex	Polyhedron	No. of Vertex	Polyhedron
4	Tetrahedron	9	Tricapped trigonal prism
5	Trigonal bipyramid	10	Bicapped square antiprism
6	Octahedron	11	Octadecahedron
7	Pentagonal bipyramid	12	Icosahedron
8	D_{2d} (trigonal) dodecahedron		

Using the electron count, the predicted structure can be found. n is the number of vertices in the cluster. The $4n$ rules are enumerated in the following table.

Table 9. Electron count and predicted structure using $4n$ rule.

Electron count	Name	Predicted structure
$4n - 2$	Bicapped closo	$n - 2$ vertex closo polyhedron with 2 capped faces
$4n$	Capped closo	$n - 1$ vertex closo polyhedron with 1 face capped
$4n + 2$	Closo	Closo polyhedron with n vertices
$4n + 4$	Nido	$n + 1$ vertex closo polyhedron with 1 missing vertex
$4n + 6$	Arachno	$n + 2$ vertex closo polyhedron with 2 missing vertex
$4n + 8$	Hypho	$n + 3$ vertex closo polyhedron with 3 missing vertex
$4n + 10$	Klado	$n + 4$ vertex closo polyhedron with 4 missing vertex

The valence electrons are enumerated when the counting of electrons for each cluster is carried out. For each transition metal atom or ion present, ten electrons are subtracted from the total number of electrons. For example, in $Rh_6(CO)_{16}$ the total number of electrons would be $6 \times 9 + 16 \times 2 - 6 \times 10 = 86 - 6 \times 10 = 26$. Therefore, the cluster is a closo polyhedron because $n = 6$, with $4n + 2 = 26$. When structure prediction is carried out, other postulates may be given as:

i) For clusters that are comprised mainly of transition metals, any main group atoms are generally counted as ligands or interstitial atoms, and not vertices.

ii) Larger atoms or more electropositive groups have the tendency to occupy vertices of higher connectivity; while the smaller or more electronegative atoms tend to fill vertices of low connectivity.

iii) In some special boron hydride clusters, every boron connected to 3 or more vertices has one terminal H, while a boron connected to two other vertices has two terminal H. Any extra hydrogens left are placed in open-face positions to even out the coordination number of different vertices.

iv) In some special cases of transition metal clusters, ligands are added to the metal centers to impart the metals a reasonable coordination numbers, and if any H atoms are present, they should be placed in bridging sites to even out the coordination numbers of different vertices.

Generally speaking, closo clusters with n vertices are n-vertex polyhedral structures. To find the nido-cluster structure, a closo cluster with $n + 1$ vertices is exploited as a starting point; if the cluster is consisted of small atoms, a vertex with high connectivity is removed; while if the cluster is consisted of large atoms, a vertex with low-connectivity is removed. To find the structure of arachno clusters, a closo polyhedron with $n+2$ vertices is exploited as the starting point, and a nido complex with $n+1$ vertex is created by the rule discussed above. However, if the cluster is of small atoms, a second vertex which adjacent to the first is removed; and if the cluster is of large atoms mainly, the second vertex removed is not adjacent to the first one.

Examples: i) P_4: Electron count = 4 × number of valence electron of P = 4 × 5 = 20 electrons. Since $n = 4$, $4n + 4 = 20$, so the cluster is a nido borane. Starting from base polyhedron of $n + 1$ vertex, one vertex has to be removed. Here, starting from trigonal bipyramid and removing one axial vertex gives the tetrahedral cluster.

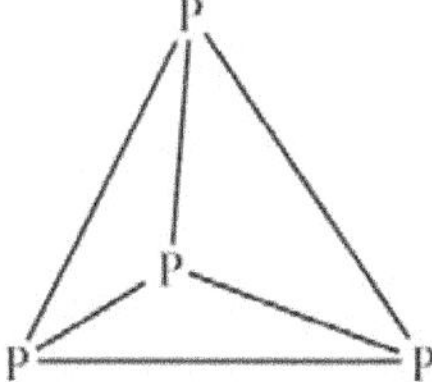

Figure 10. Structure of P_4.

ii) S_4^{2+}: Electron count = 4 × number of valence electron of S = 4 × 6 −2 (for charge) = 22 electrons. Since $n = 4$, $4n + 6 = 22$, so the cluster is an arachno borane. Starting from base polyhedron of $n + 2$ vertex, two vertexes have to be removed. In this case, it will start from octahedron and then the removal of two non-adjacent vertexes will give square cluster.

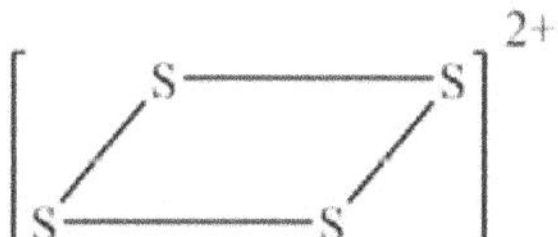

Figure 11. Structure of S_4^{2+}.

iii) $B_5H_5^{4-}$: Electron count = 5 × 3 + 1 × 5 + 4 = 24 electrons. Since $n = 5$, $4n + 4 = 24$, so the cluster is a nido borane. In this case, will start from octahedron and then the removal of one vertexes will give square pyramidal. The hydrogens have been omitted for clarity.

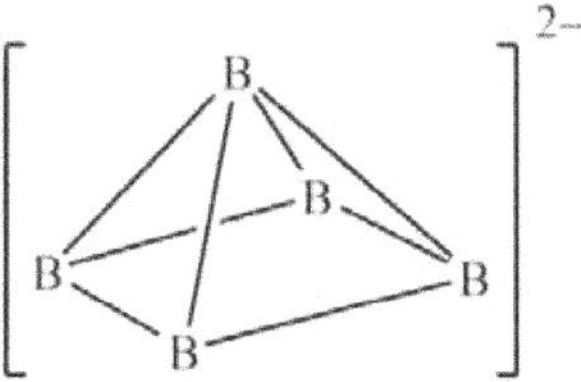

Figure 12. Structure of $B_5H_5^{4-}$.

iv) $Os_6(CO)_{18}$: Electron count: $6 \times Os + 18 \times CO - 60$ (for 6 osmium atoms) $= 6 \times 8 + 18 \times 2 - 60 = 24$. Since $n = 6$, $4n = 24$, so the cluster is capped closo. Starting from a trigonal bipyramid, a face is capped. The carbonyls have been omitted for clarity.

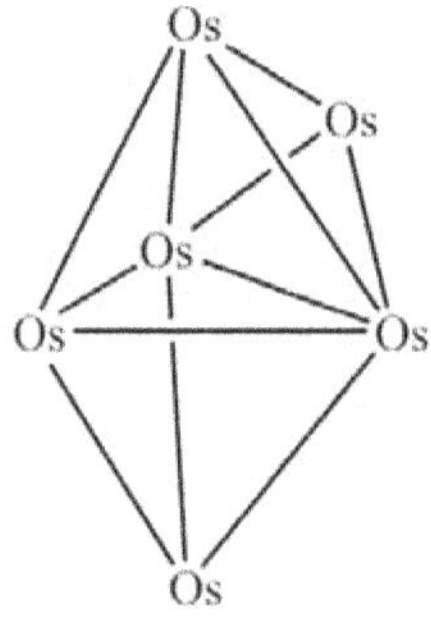

Figure 13. Structure of $Os_6(CO)_{18}$.

v) $C_2B_5H_7$: Electron count $= 2 \times C + 5 \times B + 7 \times H = 2 \times 4 + 3 \times 5 + 7 \times 1 = 30$. Since n in this case is 7, $4n + 2 = 30$, the cluster is closo. The hydrogens have been omitted for clarity.

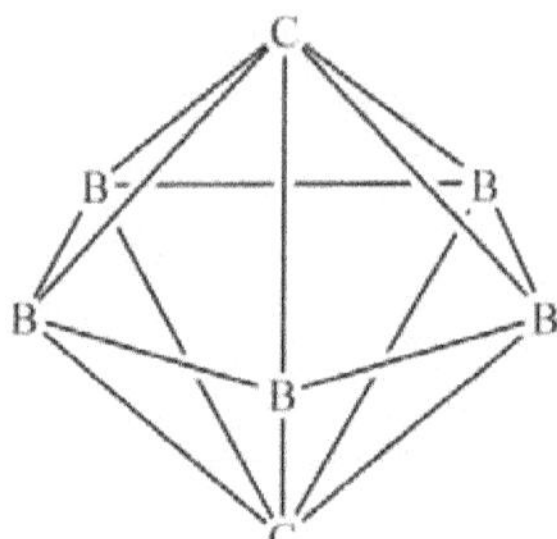

Figure 14. Structure of $C_2B_5H_7$.

The bookkeeping for deltahedral clusters is sometimes carried out by counting skeletal electrons pairs instead of the total number of electrons. The CH and BH units are considered as 3 and 2 electron donor, respectively. Additional hydrogen contributes 1 electron and charge is considered as it is. m is the sum of the number of boron and carbon atoms.

No. of e pair	$m-1$	m	$m+1$	$m+2$	$m+3$	$m+4$	$m+5$
Structure	Bicapped closo	Capped closo	Closo	Nido	Arachno	Hypho	Klado

For example, in $C_2B_7H_{13}$, m is 9, while electron pairs are $2 \times CH + 7 \times BH + 4 \times H = 6 + 14 + 4 = 24$ or 12 pairs. The number of framework electron pairs are $9 + 3 = 12$; hence the structure is arachno.

> ### *5n Rule*

If electrons per vertex are approaches 5, the *5n* rule is used based on a different series of polyhedra known as the 3-connected polyhedra, in which each vertex is connected to 3 other vertices.

Table 10. Base polyhedrons for structure prediction using *5n* rule.

No. of Vertex	Polyhedron	No. of Vertex	Polyhedron
4	Tetrahedron	14	Dual of triaugmented triangular
6	Trigonal prism	16	Square truncated trapezohedron
8	Cube	18	Dual of edge-contracted icosahedron
10	Pentagonal prism	20	Dodecahedron
12	D_{2d} pseudo-octahedron		

The *5n* rules are as follows:

Table 11. Electron count and predicted structure using *5n* rule.

Electron count	Predicted structure
$5n$	n-vertex 3-connected polyhedron
$5n + 1$	$n - 1$ vertex 3-connected polyhedron with one vertex inserted into an edge
$5n + 2$	$n - 2$ vertex 3-connected polyhedron with two vertexes inserted into an edge
$5n + k$	$n - k$ vertex 3-connected polyhedron with k vertex inserted into an edge

Examples: i) P_4S_3: Electron count = 4 × number of valence electron of P + 3 × number of valence electrons of S = 4 × 5 + 3 × 6 = 38 electrons. Since $n = 7$, $5n + 3 = 38$. Three vertices are inserted into edges.

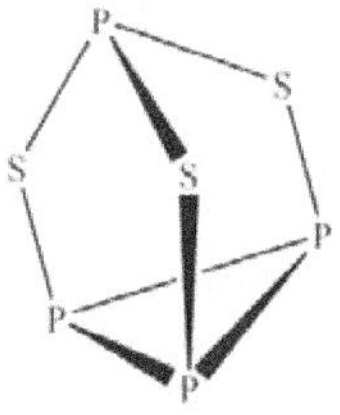

Figure 15. Structure of P_4S_3.

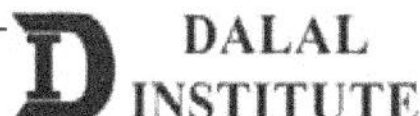

ii) P_4O_6: Electron count = 4 × number of valence electron of P + 6 × number of valence electrons of O = 4 × 5 + 6 × 6 = 56 electrons. Since $n = 10$, $5n + 6 = 56$. Six vertices are inserted into edges.

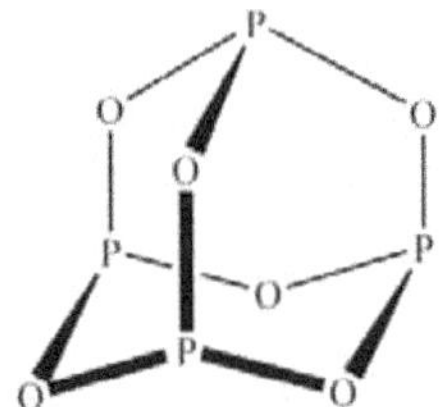

Figure 16. Structure of P_4O_6.

> ### *6n Rule*

If the electron count per vertex is about 6, a *6n* rule is used for structure prediction. This is because structures based on *4n* or *5n* rules would become unstable if the number of electron approaches 6, and clusters tend to give structures governed by the *6n* rules. The rules for the 6n structures are as follows:

Table 12. Electron count and predicted structure using *6n* rule.

Electron count	Predicted structure
$6n - k$	n-membered ring with $k/2$ transannular bonds
$6n - 4$	n-membered ring with 2 transannular bonds
$6n - 2$	n-membered ring with 1 transannular bond
$6n$	n-membered ring
$6n + 2$	n-membered chain (n-membered ring with 1 broken bond)

Examples: i) S_8: Electron count = 8 × number of valence electron of S = 48 electrons. Since $n = 8$, $6n = 48$. Therefore, the cluster is an 8-membered ring.

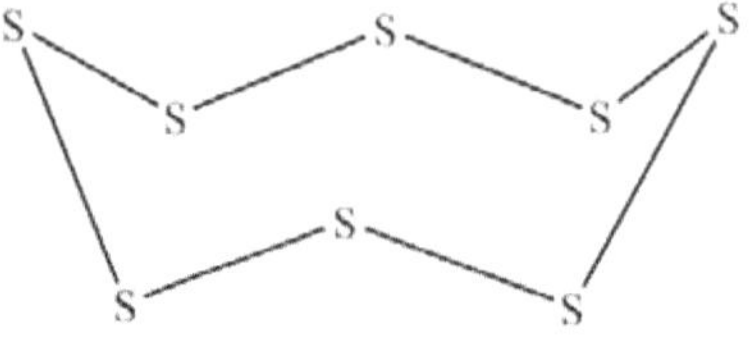

Figure 17. Structure of S_8.

ii) C_6H_{14}: Electron count = 6 × number of valence electron of C + 14 × number of valence electrons of H = 6 × 4 + 14 × 1 = 38 electrons. Since $n = 6$, $6n + 2 = 38$. Therefore, the cluster is a 6-membered chain.

Figure 18. Structure of C_6H_{14}.

It is also worth mentioning that Wade's rules not only rationalized the structures of a vast number of cluster compounds but they have also stimulated further research in cluster chemistry.

❖ Carboranes

Carboranes are the cluster composed of carbon, boron and hydrogen atoms; and just like boranes, can be classified as closo-, nido-, arachno-, hypho-, or -klado based on whether they represent a complete (closo) polyhedron, or a polyhedron that is missing one (nido-), two (arachno-), or more vertices. Carboranes are the most common examples of heteroboranes. The electronic structure of carboranes has been described by Wade-Mingos rules. Three main categories of carboranes are discussed below.

➤ *Closo-(Closed) Carboranes*

These are closed triangular polyhedral structures in which all the vertices of the triangular polyhedral geometries are occupied mainly by boron and some sites by carbon atoms. There are $n+1$ electron pairs (or $4n+2$ skeletal electrons) involved in multicentre bonding in closo-carborane; where n represents the total number of B and C atoms. Some of the common examples of closo-carboranes are:

1. $C_2B_{10}H_{12}$: In $C_2B_{10}H_{12}$, $n = 12$; according to Wade's rule, the two CH units contribute 2×3 = 6 electrons and ten BH units contribute 10×2 = 20 electrons to the bonding molecular orbitals or to the skeletal structure. Thus, there are 13 electron pairs ($n+1 = 13$) present in the multicentre bonding orbitals of $C_2B_{10}H_{12}$, confirming this as a closo kind. Three isomers (ortho-, meta- and para-) are possible.

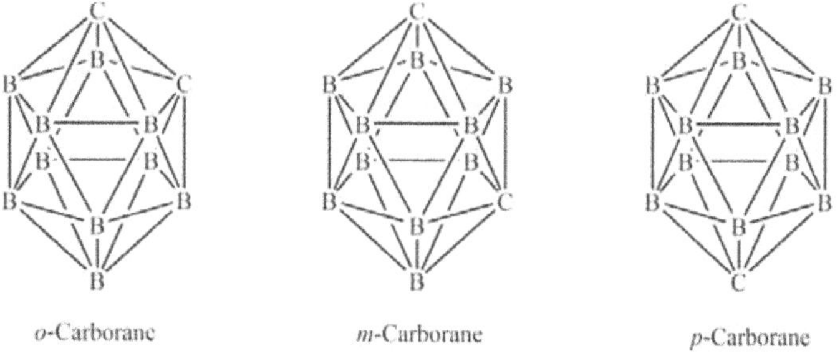

Figure 19. Structure and isomerism in $C_2B_{10}H_{12}$ (dicarba-closo-dodecaborane).

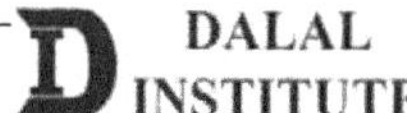

2. $C_2B_3H_5$: In $C_2B_3H_5$, $n = 5$; according to Wade's rule, the two CH units contribute $2\times3 = 6$ electrons and three BH units contribute $3\times2 = 6$ electrons to the bonding molecular orbitals or the skeletal structure. Thus there are 6 electron pairs ($n+1 = 6$) present in the multicentre bonding orbitals of $C_2B_3H_5$, confirming this as closo kind. Three isomers are possible which are given below.

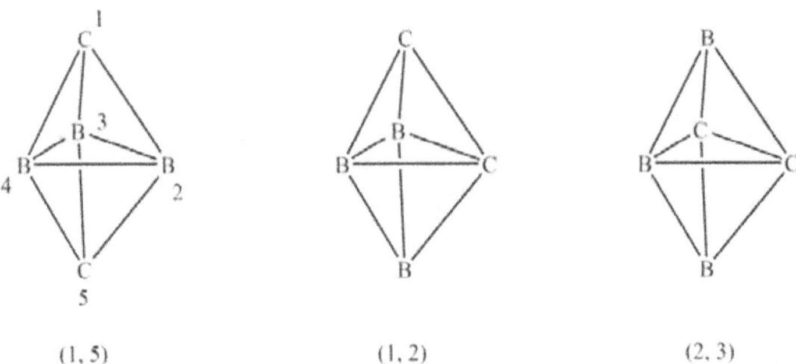

Figure 20. Structure and isomerism in $C_2B_3H_5$.

> ### Nido-(Nestlike) Carboranes

These are nest-like geometries and can be assumed as the derivatives of closed triangular polyhedral structures in which one vertex is removed. Most of the sites in these clusters are occupied by boron atoms while some sites by carbons. There are $n+2$ electron pairs (or $4n+4$ skeletal electrons) involved in multicentre bonding in nido-carboranes; where n represents the total number of B and C atoms. Common examples are:

1. $C_2B_9H_{13}$: In $C_2B_9H_{13}$, $n = 11$; and according to Wade's rule, the two CH units contribute $2\times3 = 6$ electrons, nine BH units contribute $9\times2 = 18$ electrons, and two additional hydrogens contribute $2\times1 = 2$ electrons to the bonding molecular orbitals or the skeletal structure. Thus there are total 26 electrons or 13 electron pairs ($n+2 = 13$) present in the multicentre bonding orbitals of $C_2B_9H_{13}$, confirming this as nido kind. The structure of some of the possible isomers that can be obtained experimentally are given below.

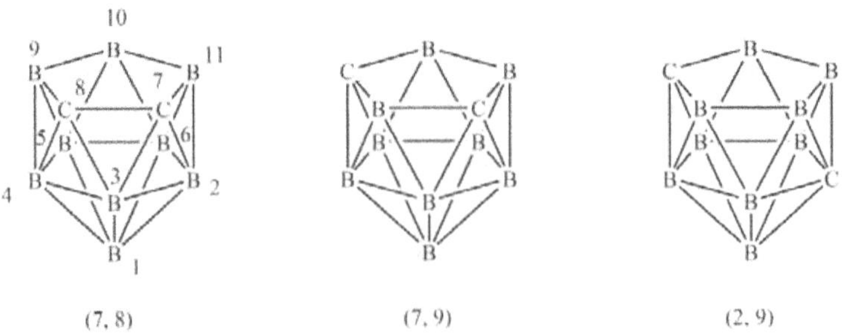

Figure 21. Structure and isomerism in $C_2B_9H_{13}$.

2. $C_2B_4H_8$: In $C_2B_4H_8$, $n = 6$; according to Wade's rule, the two CH units contribute $2\times3 = 6$ electrons and three BH units contribute $4\times2 = 8$ electrons and two additional hydrogens contribute $2\times1 = 2$ electrons to the bonding molecular orbitals or the skeletal structure. Thus there are total 16 electrons or 8 electron pairs ($n+2 = 8$) present in the multicentre bonding orbitals of $C_2B_4H_8$, confirming this as a nido kind. The structure of some of the possible isomers that can be obtained experimentally are given below.

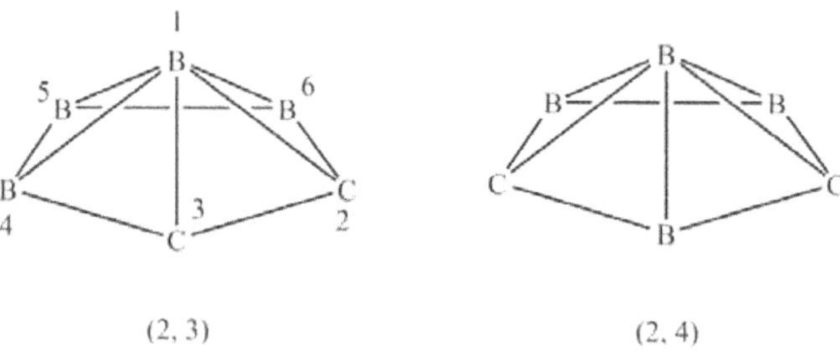

Figure 22. Structure and isomerism in $C_2B_4H_8$.

> ➢ *Arachno-(Weblike) Carboranes*

These are web-like geometries and can be assumed as the derivatives of closed triangular polyhedral structures in which two vertices are removed. Most of the sites in these clusters are occupied by boron atoms while some sites by carbons. There are $n+3$ electron pairs (or $4n+6$ skeletal electrons) involved in multicentre bonding in arachno-carboranes; where n represents the total number of B and C atoms. Some of the common examples of arachno-carboranes are:

1. $C_2B_6H_{12}$: In $C_2B_6H_{12}$, $n = 8$; according to Wade's rule, the two CH units contribute $2\times3 = 6$ electrons and six BH units contribute $6\times2 = 12$ electrons and four additional hydrogens contribute $4\times1 = 4$ electrons to the bonding molecular orbitals or the skeletal structure. Thus there are total 22 electrons or 11 electron pairs ($n+3 = 11$) present in the multicentre bonding orbitals of $C_2B_6H_{12}$, confirming this as an arachno kind. The structure of some of the possible isomers that can be obtained experimentally are given below.

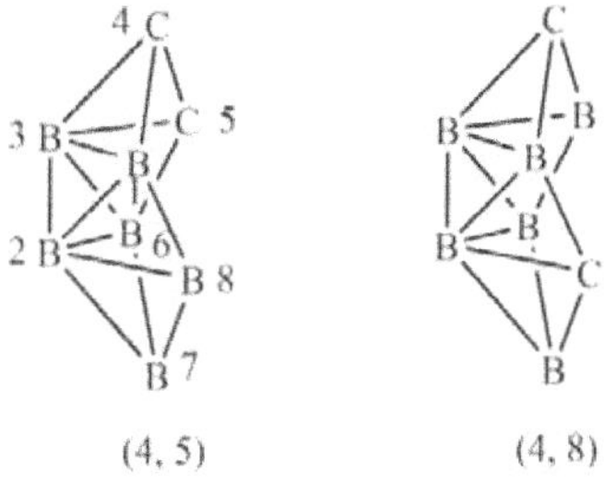

Figure 23. Structure and isomerism in $C_2B_6H_{12}$.

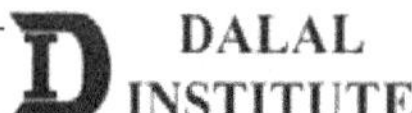

2. $C_2B_7H_{13}$: In $C_2B_7H_{13}$, $n = 9$; according to Wade's rule, the two CH units contribute $2 \times 3 = 6$ electrons and seven BH units contribute $7 \times 2 = 14$ electrons and four additional hydrogens contribute $4 \times 1 = 4$ electrons to the bonding molecular orbitals or the skeletal structure. Thus there are total 24 electrons or 12 electron pairs ($n+3 = 12$) present in the multicentre bonding orbitals of $C_2B_7H_{13}$, confirming this as an arachno-kind. The structure of some the possible isomers that can be obtained experimentally are given below.

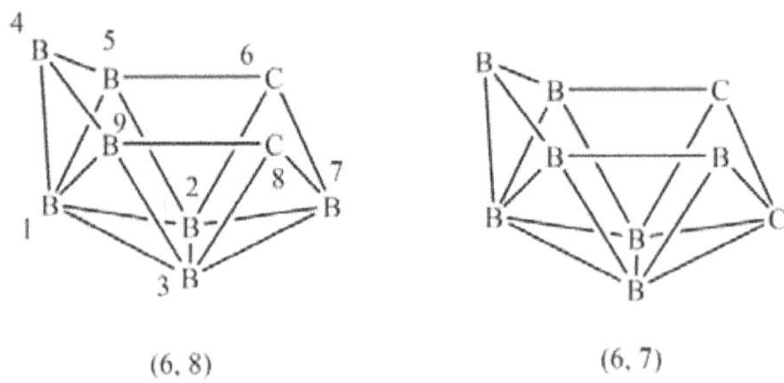

Figure 24. Structure and isomerism in $C_2B_7H_{13}$.

> ### ➢ *Structural Pattern Correlation Between Closo, Nido and Arachno Carboranes*

The structural pattern in closo, nido and arachno carboranes for different vertexes and skeletal electron pairs is very much important to to theoretical as well experimental analysis. Hs are omitted for clarity.

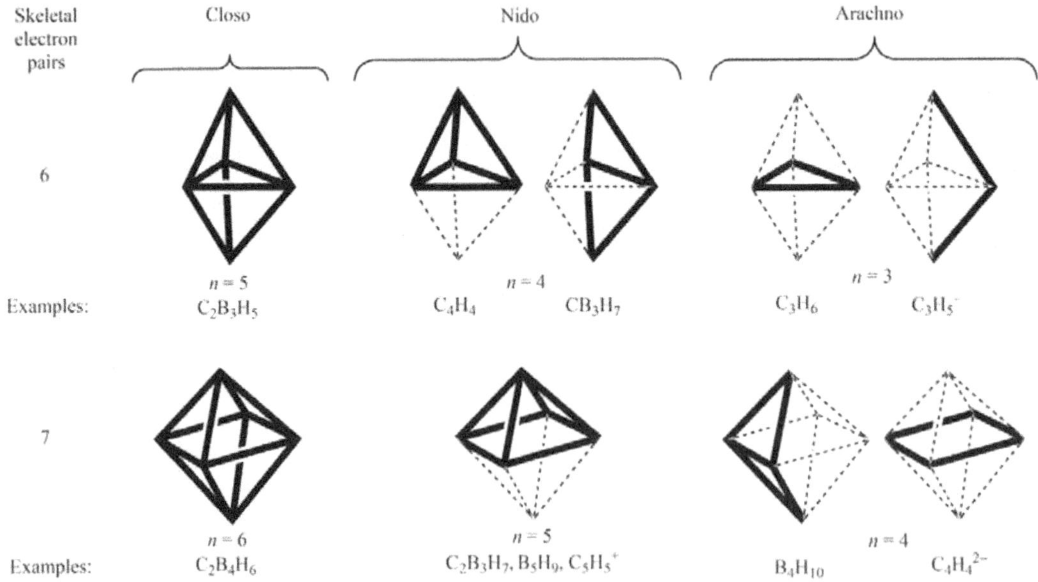

Figure 25. Continued on the next page…

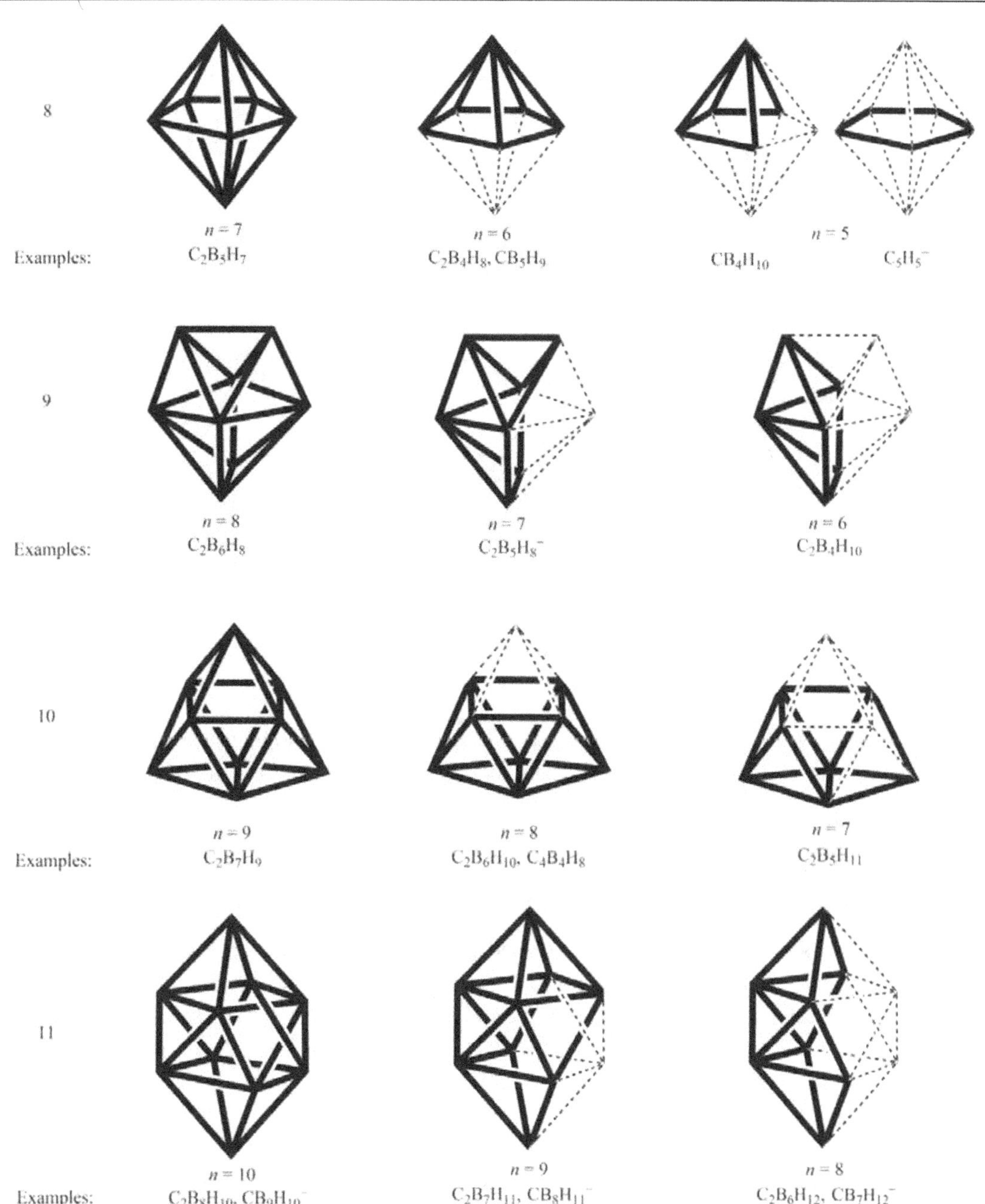

Figure 25. Continued on the next page…

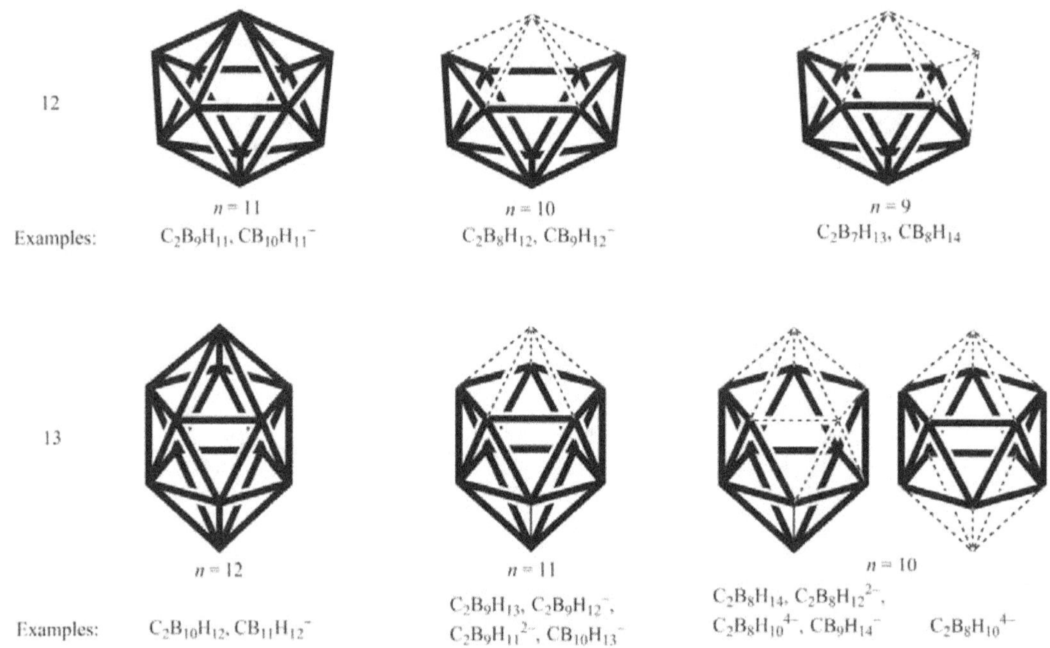

Figure 25. The structural relationship between closo, nido, and arachno carboranes.

Some hypho-carboranes ($C_2B_{n-2}H_{n+6}$) also do exist in which three vertices from the parent deltahedron are missing. Furthermore, carboranes are also formed by joining two or more preceding types; called as conjuncto-carboranes.

❖ Metal Carbonyl Clusters- Low Nuclearity Carbonyl Clusters

Metal carbonyl clusters are the metal clusters (two or more metal centers directly bonded to each other) having carbonyl groups as the ligand species. The metal centers in these cluster geometries are actually present in low oxidation state (+1, 0, −1) that can be stabilized by carbonyl ligands. Metal carbonyl clusters are mainly formed by some end-group metal (Fe, Co, Ni, Ru, Rh, Pd, Os, Ir, Pt) of the d-block elements. The primary domain of carbonyl clusters is composed of neutral carbonyls, carbonyl anions, metal carbonyl hydrides. The carbonyl hydride clusters can be obtained from neutral carbonyls by replacing one of the CO groups with two H-groups; while carbonyl anions are derived by replacing CO with one H-atom and one negative charge, or with two negative charge. Metal carbonyl clusters can be classified into two types; low nuclearity carbonyl clusters (LNCC) and high nuclearity carbonyl clusters (HNCC), depending upon the number of metal centers involved in the skeletal framework. If the number of metal centers is in the range of 2–4, they are generally labeled as low nuclearity; while on the other hand, a metal-center number of 5 and above makes them designable as high nuclearity carbonyl cluster system. Owing to the difference of electron counting scheme from high nuclearity carbonyl clusters, this section will exclusively deal with low nuclearity carbonyl clusters. The rationalization of bonding and structural profile of some important low nuclearity carbonyls clusters on the basis 18-electron scheme is discussed below.

> ➤ *Dinuclear Carbonyl Clusters:*

The structural framework of dinuclear metal carbonyl clusters is comprised of two metal centers connected by 1 metal-metal bond, and therefore, linear in geometry. The CO groups can be terminal, bridging or both. The most common examples of these are $Co_2(CO)_8$, $Fe_2(CO)_9$, Mn_2CO_{10}, Tc_2CO_{10}, and Re_2CO_{10}.

1. $Co_2(CO)_8$: This cluster is known to exist in two isomers; the first one has a D_{3d} symmetry with one metal-metal bond with zero bridging carbonyl, the second one is of C_{2v} symmetry and has two bridging CO ligands along with one metal-metal bond. The 18-electron count for $Co_2(CO)_8$ is $2×9 + 8×2 = 34$. Hence, one metal-bond (2 electrons) is needed to fulfill the requirement of two metal centers (36 electrons).

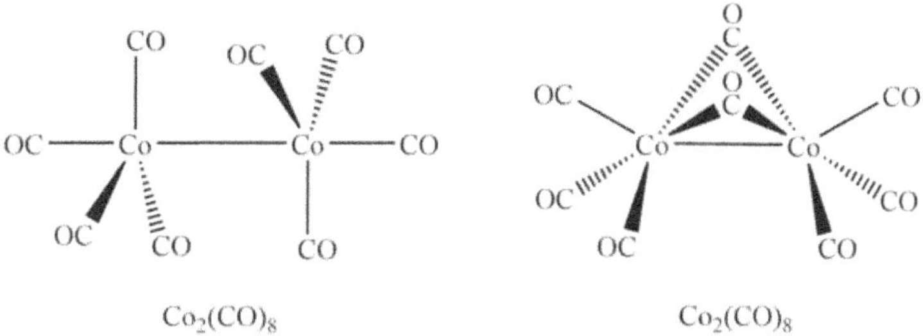

Figure 26. Structures and bonding in $Co_2(CO)_8$.

2. $Fe_2(CO)_9$: The structure of $Fe_2(CO)_9$ exist with D_{3h} symmetry, and contains three bridging CO ligands and six terminal CO groups attached. The 18-electron count for $Fe_2(CO)_9$ is $2×8 + 9×2 = 34$. Hence, one metal-bond (2 electrons) is needed to fulfill the requirement of two metal centers (36 electrons).

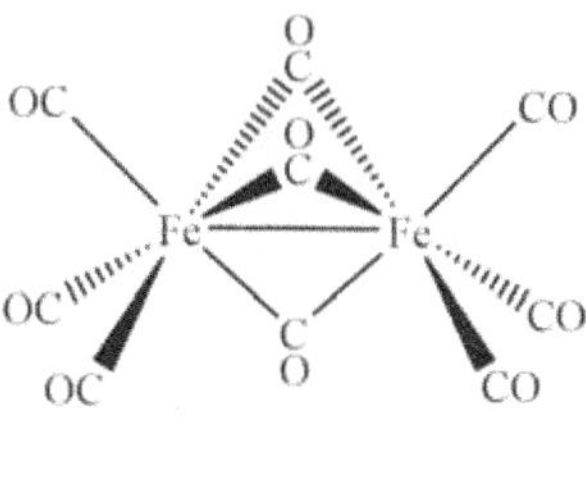

Figure 27. Structures and bonding in $Fe_2(CO)_9$.

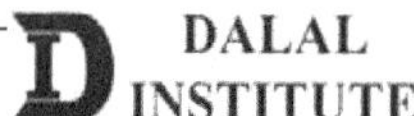

3. $M_2(CO)_{10}$: M_2CO_{10} (M = Mn, Tc, Re) exists with D_{4d} symmetry with one metal-metal bond and four CO ligands attached to each of the metal centre. The 18-electron count for M_2CO_{10} (M = Mn, Tc, Re) is $2 \times 7 + 10 \times 2 = 34$. Hence, one metal-bond (2 electrons) is needed to fulfill the requirement of two metal centers (36 electrons).

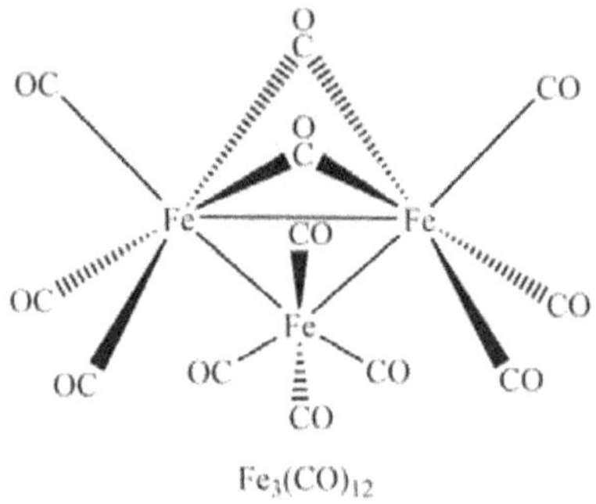

Figure 28. Structures and bonding in M_2CO_{10} (M = Mn, Tc, Re).

> ➢ *Trinuclear Metal Carbonyls*

The structural framework of trinuclear metal carbonyl clusters is comprised of three metal centers connected by three metal-metal bonds, and therefore, usually trigonal in geometry. The CO groups can be terminal, bridging or both. The most common examples of trinuclear carbonyl clusters are Fe_3CO_{12}, $Ru_3(CO)_{12}$ and $Os_3(CO)_{12}$ systems.

1. Fe_3CO_{12}: $Fe_3(CO)_{12}$ is different, with two bridging CO ligands, resulting in C_{2v} symmetry. The 18-electron count for $Fe_3(CO)_{12}$ is $3 \times 8 + 12 \times 2 = 48$. Hence, three metal-metal bonds (6 electrons) are needed to fulfill the requirement of three metal centers (54 electrons).

Figure 29. Structures and bonding in $Fe_3(CO)_{12}$.

2. $M_3(CO)_{12}$ (M = Os, Ru): For example, the $Ru_3(CO)_{12}$ cluster has D_{3h} symmetry, consisting of an equilateral triangle of Ru atoms, each of which has two axial and two equatorial CO ligands. $Os_3(CO)_{12}$ has the same structure. The 18-electron count for $M_3(CO)_{12}$ (M = Os, Ru) is $3{\times}8 + 12{\times}2 = 48$. Hence, three metal-metal bonds (6 electrons) are needed to fulfill the requirement of three metal centers (54 electrons).

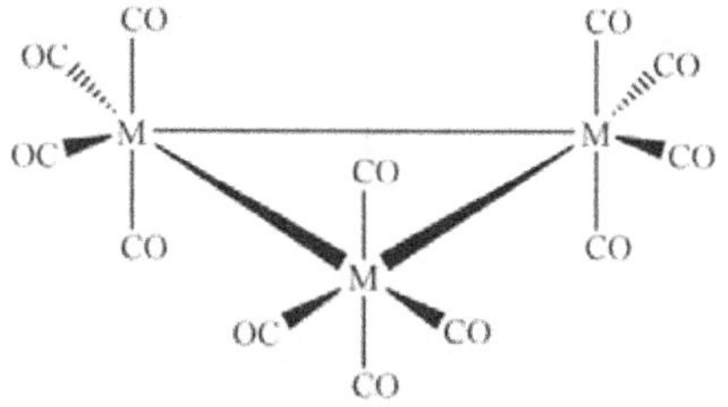

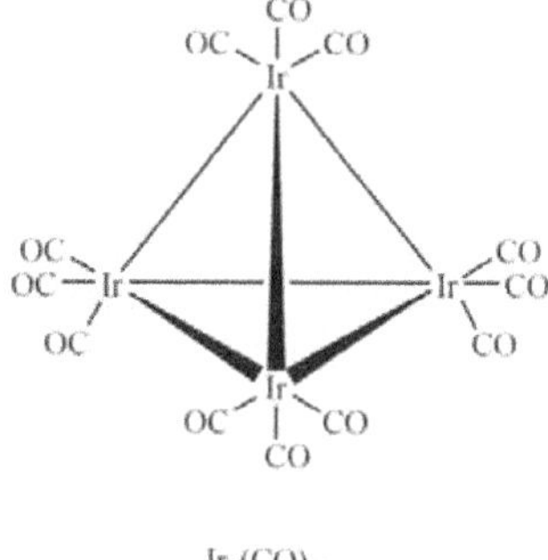

Figure 30. Structures and bonding in $M_3(CO)_{12}$ (M = Os, Ru).

> *Tetranuclear Metal Carbonyls*

The structural framework of tetranuclear metal carbonyl clusters is comprised of four metal centers connected by four to six metal-metal bonds, and therefore, usually tetrahedral in geometry. The CO groups can be terminal, bridging or both. The most common examples of tetranuclear carbonyl clusters are $Ir_4(CO)_{12}$, $Co_4(CO)_{12}$, $Rh_4(CO)_{12}$, $Re_4CO_{16}{}^{2-}$, $Ru_3(CO)_{12}$ and $Os_4(CO)_{16}$ systems.

1. $Ir_4(CO)_{12}$: The $Ir_4(CO)_{12}$ has perfect T_d symmetry with no bridging CO ligands groups. The 18-electron count for $Ir_4(CO)_{12}$ is $4{\times}9 + 12{\times}2 = 60$. Hence, six metal-metal bonds (12 electrons) are needed to fulfill the requirement of four metal centers (72 electrons).

Figure 31. Structures and bonding in $Ir_4(CO)_{12}$.

2. M₄CO₁₂ (M = Co, Rh): M_4CO_{12} (M = Co, Rh) is consisted of a tetrahedral M_4 core, but the molecular symmetry is C_{3v}. Three carbonyl ligands are bridging ligands and nine are terminal. The 18-electron count for $M_4(CO)_{12}$ (M = Co, Rh) is 4×9 + 12×2 = 60. Hence, six metal-metal bonds (12 electrons) are needed to fulfill the requirement of four metal centers (72 electrons).

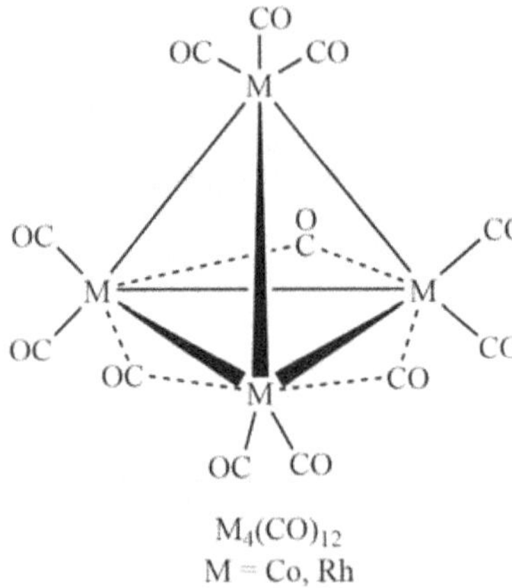

$M_4(CO)_{12}$
M = Co, Rh

Figure 32. Structures and bonding in $M_4(CO)_{12}$ (M = Co, Rh).

3. [Re₄(CO)₁₆]²⁻: $[Re_4(CO)_{16}]^{2-}$ has D_{2h} symmetry with no bridging carbonyl. The 18-electron count for $[Re_4(CO)_{16}]^{2-}$ is 4×7 + 16×2 + 2 (charge) = 62. Hence, five metal-metal bonds (10 electrons) are needed to fulfil the requirement of four metal centres (72 electrons).

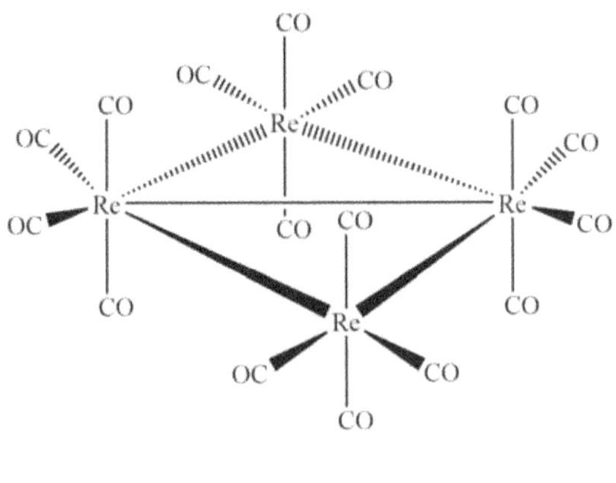

$[Re_4(CO)_{16}]^{2-}$

Figure 33. Structures and bonding in $[Re_4(CO)_{16}]^{2-}$.

4. $Os_4(CO)_{16}$: The tetranuclear $Os_4(CO)_{16}$ is analogs to the cyclobutane with a puckered structure. The X-ray diffraction analysis of $Os_4(CO)_{14}$ unveiled an irregular tetrahedral Os_4 skeleton with four weakly semi-bridging CO groups and four different Os–Os bond lengths. The 18-electron count for $Os_4(CO)_{16}$ is $4 \times 8 + 16 \times 2 = 64$. Hence, four metal-metal bonds (8 electrons) are needed to fulfill the requirement of four metal centers (72 electrons).

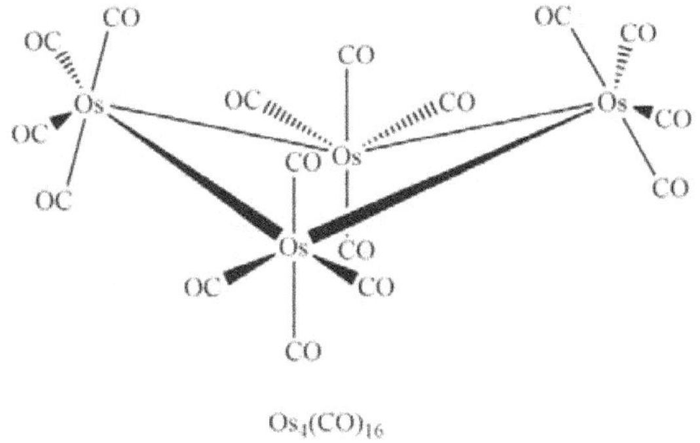

Figure 34. Structures and bonding in $Os_4(CO)_{16}$.

It is worthy to note that the electron counting scheme in low nuclearity carbonyl clusters is the same as that is used in mononuclear metal carbonyl complexes.

❖ Total Electron Count (TEC)

The simple 18-electron rule has been proven of great significance in the case of structural rationalization of low nuclearity carbonyl clusters. However, if the number of metal centers per cluster is equal or greater than five; then the conventional approach is not significant, and does not provide any satisfactory results. For example, the 18-electron count for $Rh_6(CO)_{16}$ is $6 \times 9 + 16 \times 2 = 86$; which means that eleven metal-metal bonds ($108 - 86 = 22$) are needed to fulfil the requirement of six metal centres. But the actual structure of $Rh_6(CO)_{16}$ is consisted of an octahedral Rh_6 core with twelve metal-metal bonds. Moreover, simple 18-electron treatment for high nuclearity carbonyl clusters (HNCC) does not provide any information regarding the overall geometry. The situation also becomes more and more complex if some encapsulated heteroatom like carbon is also present. Therefore, because of the lacking of any rational solution for the electronic structure of high nuclearity carbonyl clusters, most of the efforts have been devoted to find a correlation between their structure and the number of electrons available for cluster binding. It is worthy to mention that high-nuclearity carbonyl clusters are also considered as electron-deficient compounds; which is obviously due to the inadequate number of electrons to allow the assignment of all bonds as 2-centre 2-electron in nature.

> *Isolobal Analogy Between M(CO)₃ And BH Fragments*

A British chemist, Kenneth Wade, solved the unexplained problem of structure and bonding predictions in high nuclearity carbonyl clusters by developing a new scheme of electron counting, total electron count (TEC). This new proposal was actually the extension of his previously used scheme for boranes and carboranes, in which he had correlated the structure of boranes and their derivatives with the number of electrons involved in the skeletal framework. The idea behind this extension was that the $M(CO)_3$ unit is actually isolobal with the BH unit. This can be better explained by taking the example of $Ru(CO)_3$ fragment.

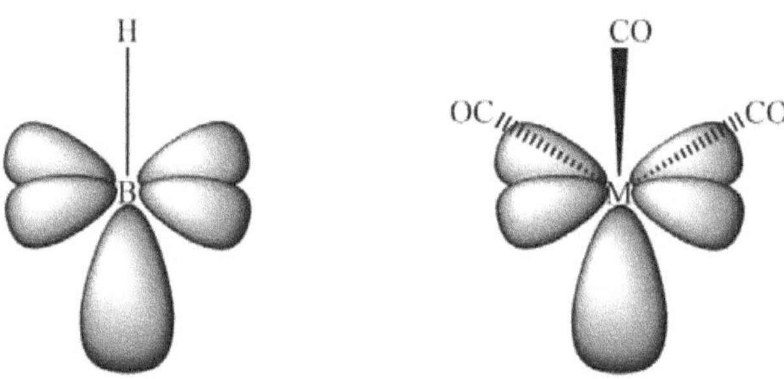

Figure 35. Structures and bonding in the isolobal pair of BH and $Ru(CO)_3$.

The BH unit uses one of four valence orbitals of boron and two electrons, one from boron and one from hydrogen. Hence, the spare valence orbitals and corresponding electrons left on boron are three and two respectively. These spare orbitals and electrons are then used in cluster bonding. Similarly, the $Ru(CO)_3$ unit uses six of nine valence orbitals of ruthenium and twelve electrons, six from ruthenium and six from three carbonyl ligands. Therefore, the spare valence orbitals and corresponding electrons left on ruthenium are three and two respectively.

Therefore, $3n$ valence orbitals and $2n$ electrons should be available in any B_nH_n cluster system. The overlap of $2n$ valence orbitals oriented towards the surface of the polyhedron generates n bonding and n antibonding molecular orbitals. The overlap of n valence orbitals oriented towards the core of the polyhedron generates 1 strongly bonding and $n-1$ weakly bonding, nonbonding or antibonding molecular orbitals. Thus we can say that a total of $n+1$ bonding molecular orbitals are generated in the process of cluster formation, and the number of electrons required to fill these bonding molecular orbitals is $2n+2$. However, in n BH units of B_nH_n cluster can supply only $2n$ electron, explain why $B_nH_n^{2-}$ or cluster is more stable than simple B_nH_n. The same argument holds true about $Ru_n(CO)_{3n}$ cluster system, i.e. $2n+2$ electrons are needed to fill all the bonding molecular orbitals. Hence, $[Ru_6(CO)_{18}]^{2-}$ anion would be more stable than $Ru_6(CO)_{18}$ cluster system.

> *Electron Counting Scheme for High Nuclearity Carbonyl Clusters*

The isolobal analogy of $M(CO)_3$ fragment with B–H (or C–H) unit inspired Kenneth Wade to explore this field further; and he of course then came with a new set of rules for electron counting in high nuclearity carbonyl cluster systems. He suggested that the total electron count of these clusters can be correlated to

skeletal electron count participating in the cluster bonding in a similar way adopted in boranes or carboranes. The total electron count can be calculated by adding the following contributions:

i) The number of valence electrons of all metal-centers.

ii) Two electrons for each carbonyl group irrespective of the fact whether it is terminal or bridged.

iii) One electron for each unit of negative charge.

iv) The number of valence electrons of each hetero or interstitial atoms like carbon or nitrogen.

This total electron count then can be used to predict the structure and bonding of carbonyl cluster systems by extracting the skeletal electron count. Now, as we know that $2n+2$, $2n+4$, $2n+6$ skeletal electrons are required for closo, nido and arachno boranes, respectively; the same is true for metal carbonyl clusters. However, in addition to the 2 valance electrons to be used in the skeletal framework, each $M(CO)_3$ also contains 12 non-skeletal electrons. The theoretical basis of this claim comes from the fact that each M–CO bond contains 4 electrons; 2 σ-electrons donated by CO ligand to the metal, and 2 π-electrons donated by the metal back to the lowest unoccupied antibonding molecular orbital of CO ligand. This gives a total electron count for closo polyhedron as $12n + 2(n+1)$. Hence, the predictions of structure and bonding in high nuclearity carbonyl cluster with n vertexes can be summed up only after considering these twelve non-skeletal electrons.

Table 9. Total electron count (TEC) and predicted structure.

Total electron count	Name	Predicted structure
$12n + 2(n - 1)$	Bicapped closo	$n - 2$ vertex closo polyhedron with 2 capped faces
$12n + 2n$	Capped closo	$n - 1$ vertex closo polyhedron with 1 face capped
$12n + 2(n + 1)$	Closo	Closo polyhedron with n vertices
$12n + 2(n + 2)$	Nido	$n + 1$ vertex closo polyhedron with 1 missing vertex
$12n + 2(n + 3)$	Arachno	$n + 2$ vertex closo polyhedron with 2 missing vertex
$12n + 2(n + 4)$	Hypho	$n + 3$ vertex closo polyhedron with 3 missing vertex
$12n + 2(n + 5)$	Klado	$n + 4$ vertex closo polyhedron with 4 missing vertex

It should also be noted that Wade's predictions based on total electron count are also applicable to low nuclearity carbonyl cluster systems in which the number of metal centers present is three or four. For example, the total electron count for $Ir_4(CO)_{12}$ is $4 \times Ir + 12 \times CO = 4 \times 9 + 12 \times 2 = 60$. Since $n = 4$, $12n + 2(n + 2) = 60$, so the cluster is nido. Starting from a trigonal bipyramid, a vertex is removed. Similarly, the total electron count for $Fe_3(CO)_{12}$ is $3 \times Fe + 12 \times CO = 3 \times 8 + 12 \times 2 = 48$. Since $n = 3$, $12n + 2(n + 3) = 48$, so the cluster is arachno. Therefore, the structure prediction Starts from a trigonal bipyramid, and then two axially opposite vertexes are removed.

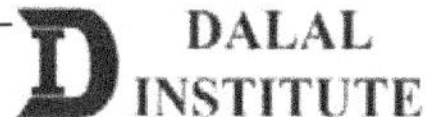

> ➤ *Structure and Bonding Profile of Some Typical High Nuclearity Carbonyl Clusters Using TEC*

As we have already discussed the fundamentals of structure and bonding in high nuclearity carbonyl clusters systems, now we will implement those ideas to explain some typical examples.

1. $M_6(CO)_{16}$ (M = Co, Rh): The total electron count for $Rh_6(CO)_{16}$ is $6 \times Rh + 16 \times CO = 6 \times 9 + 16 \times 2 = 86$. Since $n = 6$, $12n + 2(n + 1) = 86$, so the cluster is closo. Therefore, the structure prediction starts from an idealized octahedron. The predicted structure for $Rh_6(CO)_{16}$ cluster system is shown below.

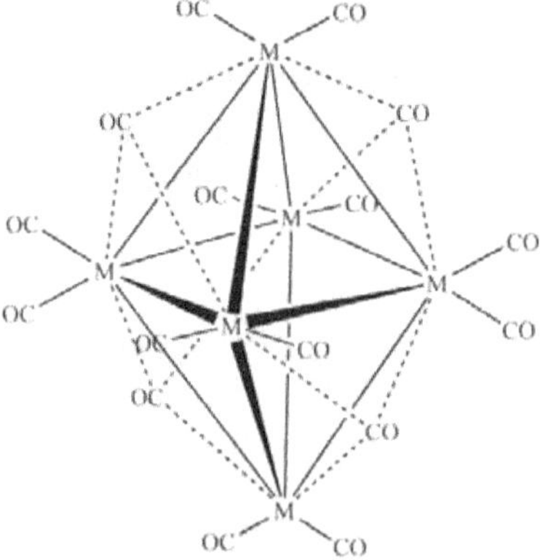

Figure 36. Structure of $M_6(CO)_{16}$ (M = Co, Rh).

2. $Os_5(CO)_{16}$: The total electron count for $Os_5(CO)_{16}$ is $5 \times Os + 16 \times CO = 5 \times 8 + 16 \times 2 = 72$. Since $n = 5$, $12n + 2(n + 1) = 72$, so the cluster is closo. Therefore, the structure prediction starts from an idealized trigonal bipyramid. The predicted structure for $Os_5(CO)_{16}$ cluster system is shown below.

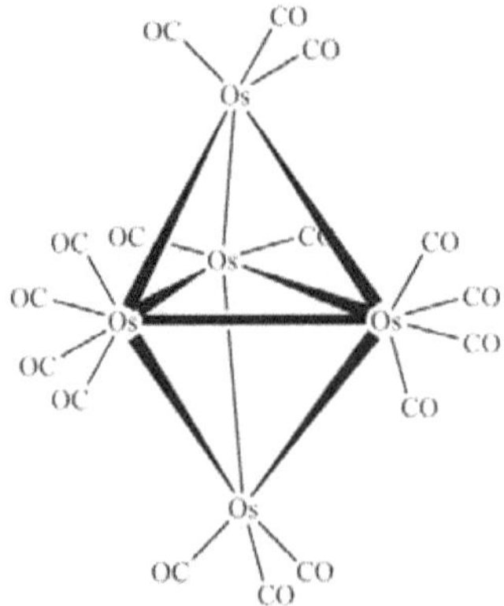

Figure 37. Structure of $Os_5(CO)_{16}$.

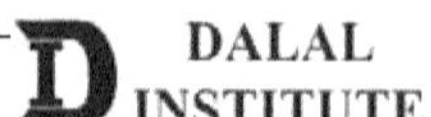

3. $[Os_6(CO)_{18}]^{2-}$: The total electron count for $[Os_6(CO)_{18}]^{2-}$ is $6\times Os + 18\times CO + 2$ (for negative charge) $= 6\times 8 + 18\times 2 + 2 = 86$. Since $n = 6$, $12n + 2(n + 1) = 86$, so the cluster is closo. Therefore, the structure prediction starts from an idealized octahedron. The predicted structure for $[Os_6(CO)_{18}]^{2-}$ cluster system is shown below.

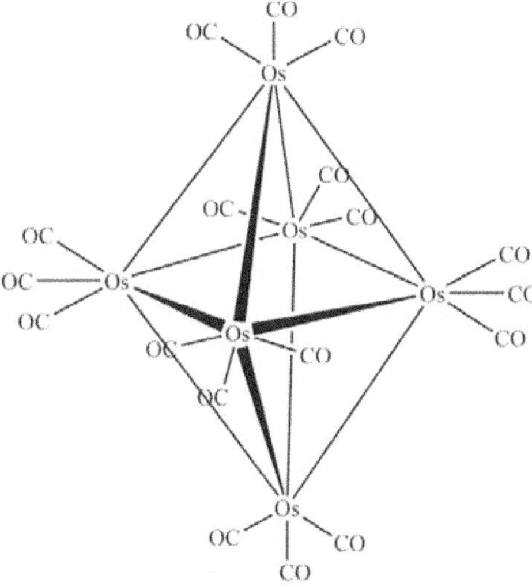

Figure 38. Structure of $[Os_6(CO)_{18}]^{2-}$.

4. $Os_5C(CO)_{15}$: The total electron count for $Os_5C(CO)_{15}$ is $5\times Os + 15\times CO + 1\times C = 5\times 8 + 15\times 2 + 4\times 1 = 74$. Since $n = 5$, $12n + 2(n + 2) = 74$, so the cluster is nido. Therefore, the structure prediction starts from an idealized octahedron, and then one vertex is removed. The predicted structure for $Os_5C(CO)_{15}$ cluster system is shown below.

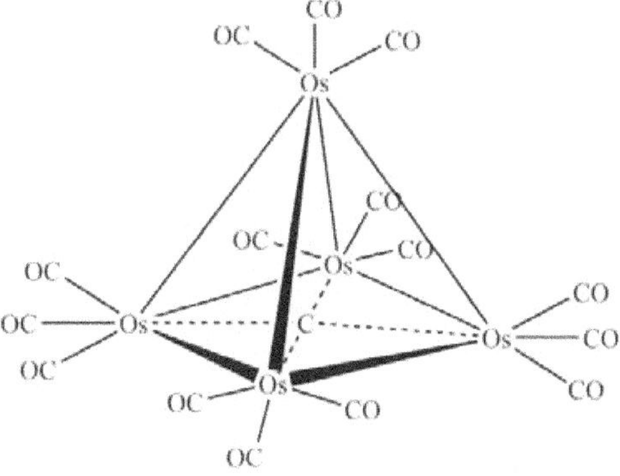

Figure 39. Structure of $Os_5C(CO)_{15}$.

5. $[Ru_6N(CO)_{16}]^-$: The total electron count for $[Ru_6N(CO)_{16}]^-$ is $6\times Ru + 16\times CO + 1\times N + 1$ (for negative charge) $= 6\times8 + 16\times2 + 1\times5 + 1 = 86$. Since $n = 6$, $12n + 2(n + 1) = 86$, so the cluster is closo. Therefore, the structure prediction starts from an idealized octahedron. The predicted structure for $[Ru_6N(CO)_{16}]^-$ cluster system is shown below.

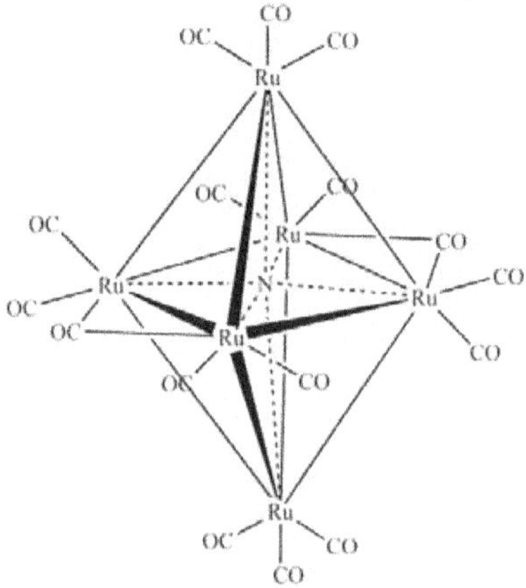

Figure 40. Structure of $[Ru_6N(CO)_{16}]^-$.

6. $[Fe_4N(CO)_{12}]^-$: The total electron count for $[Fe_4N(CO)_{12}]^-$ is $4\times Fe + 12\times CO + 1\times N + 1$ (for negative charge) $= 4\times8 + 12\times2 + 1\times5 + 1 = 62$. Since $n = 4$, $12n + 2(n + 3) = 62$, so the cluster is arachno. Therefore, the structure prediction starts from an idealized octahedron, then two adjacent vertexes are removed. The predicted structure for $[Fe_4N(CO)_{12}]^-$ cluster system is shown below.

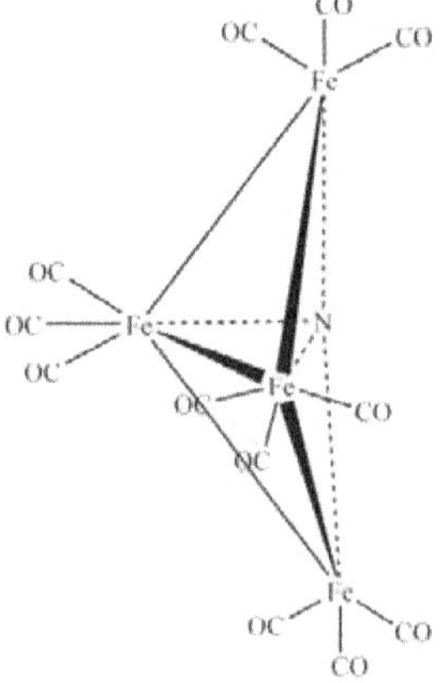

Figure 41. Structure of $[Fe_4N(CO)_{12}]^-$.

> ➤ *Correlation Between TEC and SEP*

The implementation of the rules listed in 'Table 9' is more often carried out in reverse gear, i.e. 12 electrons for each metal center are subtracted from the total electron count. The number of electrons thus obtained is then divided by 2 to get skeletal electron pairs. The number of skeletal electron pairs (SEP) can be calculated using equation (23) as described below:

$$\text{SEP} = \frac{1}{2}(\text{TEC} - 12 \text{ electrons per metal center}) \tag{23}$$

Now the structure prediction of some common high nuclearity carbonyl cluster systems on the correlative basis of total electron count (TEC) and skeletal electron pairs (SEP) can be summed up as:

Table 10. Structure prediction of some common clusters on the correlative basis of TEC and SEP.

Carbonyl cluster	Total electron count (TEC)	Skeletal electron pairs (SEP)	Vertex in parent polyhedron	Predicted structure
$Rh_6(CO)_{16}$	$(6 \times 9) + (16 \times 2) = 86$	$0.5 \times [86 - (6 \times 12)] = 7$	6	Closo
$Os_5(CO)_{16}$	$(5 \times 8) + (16 \times 2) = 72$	$0.5 \times [72 - (5 \times 12)] = 6$	5	Closo
$Os_5C(CO)_{15}$	$(5 \times 8) + (15 \times 2) + (4) = 74$	$0.5 \times [74 - (5 \times 12)] = 7$	5	Nido
$[Fe_4C(CO)_{12}]^{2-}$	$(4 \times 8) + (12 \times 2) + (4) + (2) = 62$	$0.5 \times [62 - (4 \times 12)] = 7$	6	Arachno
$[H_3Ru_4(CO)_{12}]^-$	$(4 \times 8) + (12 \times 2) + (3) + (1) = 60$	$0.5 \times [60 - (4 \times 12)] = 6$	5	Nido
$[Ru_5N(CO)_{14}]^-$	$(5 \times 8) + (14 \times 2) + (5) + (1) = 74$	$0.5 \times [74 - (5 \times 12)] = 7$	7	Nido
$[Fe_4N(CO)_{12}]^-$	$(4 \times 8) + (12 \times 2) + (5) + (1) = 62$	$0.5 \times [62 - (4 \times 12)] = 7$	6	Arachno

Finally, it should also be noted that the total electron counting scheme is applicable to most of the high nuclearity carbonyl cluster systems, yet the exceptions like $[Ni_5(CO)_{12}]^{2-}$ do exist.

❖ Problems

Q 1. Discuss the molecular orbital treatment of three-center two-electron bond in detail.

Q 2. What are the STYX numbers? How would you use the Lipscomb's model to find out the STYX code for B_4H_{10} cluster system?

Q 3. Draw and discuss the structure and bonding in B_5H_9 and B_5H_{11}.

Q 4. How many B–H–B and B–B–B bonds are present in $B_{10}H_{14}$ cluster?

Q 5. Discuss the structural relationship between closo, nido and arachno type boranes.

Q 6. What are Wade's rules? How can we use these rules to predict the structures of $B_5H_5^{4-}$ and P_4 clusters?

Q 7. What are carboranes? Explain with suitable examples.

Q 8. Draw and discuss the structural pattern correlation between closo, nido and arachno type carboranes.

Q 9. What are the low nuclearity carbonyl clusters? How do they differ from the high nuclearity ones?

Q 10. How would you calculate the number of metal-metal bonds in $Fe_3(CO)_{12}$ and $Ir_4(CO)_{12}$?

Q 11. Define total electron count (TEC).

Q 12. Explain the electron counting scheme for high nuclearity carbonyl clusters in detail.

Q 13. How would you explain the isolobal Analogy Between $Ru(CO)_3$ And BH units?

Q 14. Using total electron count (TEC), explain the structure of $Rh_6(CO)_{16}$ and $Os_5C(CO)_{15}$.

Q 15. Describe the correlation between total electron count (TEC) and skeletal electron pairs (SEP), and use same to predict the structures of $[Fe_4N(CO)_{12}]^-$ and $[H_3Ru_4(CO)_{12}]^-$.

❖ Bibliography

[1] B. W. Pfennig, *Principles of Inorganic Chemistry*, John Wiley & Sons, New Jersey, USA, 2015.

[2] G. Raj, *Advanced Inorganic Chemistry Vol-1*, Krishna Prakashan Media, Uttar Pradesh, India, 2008.

[3] A. J. Welch, *The significance and impact of Wade's rules*, Chem. Commun., 49, 2013, 3615.

[4] K. Wade, *The Structural Significance of the Number of Skeletal Bonding Electron-pairs in Carboranes, the Higher Boranes and Borane Anions, and Various Transition-metal Carbonyl Cluster Compounds*, Chem. Commun., 1971, 792.

[5] D. M. P. Mingos, *A General Theory for Cluster and Ring Compounds of the Main Group and Transition Elements*, Nature Physical Science, 236, 1972, 99.

[6] R. N. Grimes, *Carboranes*, Academic Press, London, UK, 2011.

[7] B. R. Puri, L. R. Sharma, K. C. Kalia, *Principals of Inorganic Chemistry*, Milestone Publishers, Delhi, India, 2012.

[8] F. A. Cotton, G. Wilkinson, C. A. Murillo, M. Bochmann, *Advanced Inorganic Chemistry*, John Wiley & Sons, New Jersey, USA, 1999.

[9] M. McPartlin, C. R. Eady, B. F. G. Johnson, J. Lewis, *X-Ray structures of the hexanuclear cluster complexes $[Os_6(CO)_{18}]^{2-}$, $[HOs_6(CO)_{18}]^-$, and $[H_2Os_6(CO)_{18}]$*, Chem. Commun., 1976, 884.

[10] C. R. Eady, B. F. G. Johnson, J. Lewis, B. E. Reichert, G. M. Sheldrick, *$[Os_5(CO)_{16}]$: X-Ray Crystal and Molecular Structure*, Chem. Commun., 1976, 271.

[11] A. Taheri, L. A. Berben, *Making C-H Bonds with CO_2: Production of Formate by Molecular Electrocatalysts*, Chem. Commun., 52, 2015, 1768.

[12] B. M. Gimarc, J. J. Ott, *Isomers of $C_2B_3H_5$ and the Diamond-Square-Diamond Rearrangement Mechanism*, Inorg. Chem., 25, 1986, 83.

[13] A. A. A. Attia, A. Lupan, R. B. King, *Tetracarbaboranes: Nido Structures Without Bridging Hydrogens*, Dalton Trans, 45, 2016, 18541.

[14] B. D. Gupta, A. J. Elias, *Basic Organometallic Chemistry: Concepts, Synthesises and Applications*, Universities Press (India) Private Limited, Andhra Pradesh, India, 2010.

[15] J. E. Huheey, E. A. Keiter, R. L. Keiter, *Inorganic Chemistry: Principals of Structure and Reactivity*, HarperCollins College Publishers, New York, USA, 1993.

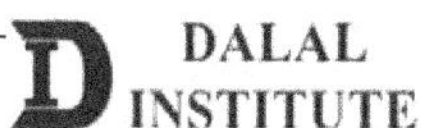

CHAPTER 11

Metal-Π Complexes:

❖ Metal Carbonyls: Structure and Bonding

The bonding in coordination compounds is usually visualized as the donation of ligand electron pair to the metal center only. However, there are some ligands which not only have filled atomic orbitals (donor orbitals) but also have some empty orbitals (acceptor orbitals) of appropriate symmetry and energy to accept electron density from a central metal atom or ion. This interaction is called π-backbonding or π-backdonation; and is generally shown by CO, NO, PR_3 and alkene-alkyne type ligands. Two of the most common examples where π-backbonding occurs include $Ni(CO)_4$ and Zeise's salt.

Furthermore, metal carbonyls are one of the most widely studied types of metal-π complexes, that can simply be defined as the coordination compounds of transition metals with carbon monoxide as a ligand. Metal carbonyls are very useful in synthetic organic chemistry and in homogeneous catalysis, like the process of hydroformylation. In the Mond process, nickel carbonyl is used to produce pure nickel. In organometallic chemistry, metal carbonyls act as precursors for the synthesis of many organometallic compounds. Metal carbonyls are toxic by inhalation, skin contact, or ingestion, in part due to their ability to attach to the iron of hemoglobin to give carboxyhemoglobin, which inhibits the binding of dioxygen. Metal carbonyls can be classified on the basis of the number of metal centers; mononuclear carbonyls have only one metal atom or ion such as $Fe(CO)_5$, while polynuclear carbonyls contain more than one metal center like homonuclear $Fe_2(CO)_9$ and heteronuclear $MnRe(CO)_{10}$. One more categorization basis of metal carbonyls is the bonding profile of carbonyl ligand; non-bridging carbonyls and bridging carbonyls. Non-bridging carbonyls may or may not contain metal-metal bonds. For instance, $Ru(CO)_5$ and $Mn_2(CO)_{10}$ both have only terminal carbonyl groups but $Mn_2(CO)_{10}$ has one metal-metal bond also. On the other hand, bridging metal carbonyls like $Fe_3(CO)_{12}$, in addition to terminal CO groups, do have CO groups bridged to more than one metal center.

> ### *General Methods of Preparation*

1. By direct reaction: Some of the mononuclear carbonyls can be prepared by the direct reaction of carbon monoxide with metal powder.

$$Ni + 4CO \xrightarrow[\text{1 atm}]{25°C} Ni(CO)_4$$

$$Fe + 5CO \xrightarrow[\text{200 atm}]{200°C} Fe(CO)_5$$

$$2Co + 8CO \xrightarrow[\text{35 atm}]{150°C} Co_2(CO)_8$$

2. By reduction: One of the most widely used methods to synthesize metal carbonyls is the reduction of corresponding metal salts in the presence of carbon monoxide.

$$CrCl_3 + Al + 6CO \xrightarrow[\text{benzene}]{AlCl_3} Cr(CO)_6 + AlCl_3$$

$$VCl_3 + 4Na + 6CO \xrightarrow[\text{high pressure}]{\text{diglyme}, 100°C} [(diglyme)_2Na][V(CO)_6] + 3NaCl$$

$$2CoI_2 + 4Cu + 8CO \xrightarrow[\text{200 atm}]{200°C} Co_2(CO)_8 + 4CuI$$

$$2CoCO_3 + 2H_2 + 8CO \xrightarrow[250 - 300 \text{ atm}]{120 - 200°C} Co_2(CO)_8 + 2H_2O + 2CO_2$$

$$Re_2O_7 + 17CO \xrightarrow[350 \text{ atm}]{250°C} Re_2(CO)_{10} + 7CO_2$$

In the last reaction, carbon monoxide is the reducing agent on its own.

3. From mononuclear carbonyls: Iron pentacarbonyl is sensitive to light and air and can be used to synthesize $Fe_2(CO)_9$ by direct photolysis.

$$2Fe(CO)_5 \xrightarrow{h\nu} Fe_2(CO)_9 + CO$$

Similarly

$$2Os(CO)_5 \xrightarrow{h\nu} Os_2(CO)_9 + CO$$

$$2Ru(CO)_5 \xrightarrow{h\nu} Ru_2(CO)_9 + CO$$

4. From iron pentacarbonyl: Carbon monoxide ligands in $Fe(CO)_5$ are labile and therefore can be used to synthesize other metal carbonyls.

$$MoCl_6 + 3Fe(CO)_5 \xrightarrow[\text{ether}]{110°C} Mo(CO)_6 + 3FeCl_2 + 9CO$$

$$WCl_6 + 3Fe(CO)_5 \xrightarrow[\text{ether}]{110°C} W(CO)_6 + 3FeCl_2 + 9CO$$

5. From metathesis reaction: Mixed-metal carbonyls can successfully be prepared via a metathesis reaction route as:

$$KCo(CO)_4 + [Ru(CO)_3Cl_2]_2 \longrightarrow 2RuCo_2(CO)_{11} + 4KCl$$

> ➤ *Structures of Metal Carbonyls*

The structure of metal carbonyls can mainly be classified into three categories; first, as the mononuclear systems that contain only one metal atom, the second one as binuclear systems that may or may not contain bridging carbonyls, and the last one as the polynuclear systems which contain more than two metal centers with all terminal, all bridging, or a mixture of two types of carbonyl groups.

1. Mononuclear metal carbonyls: The structure of mononuclear metal carbonyls is pretty simple and easy to visualize. This is definitely due to the presence of only one metal center. The general examples of mononuclear metal carbonyls include tetrahedral $Ni(CO)_4$ and $Pd(CO)_4$; the trigonal bipyramidal case of $Fe(CO)_5$, $Ru(CO)_5$ and $Os(CO)_5$; and the octahedral geometries of $V(CO)_6$, $Cr(CO)_6$, $Mo(CO)_5$ and $W(CO)_6$.

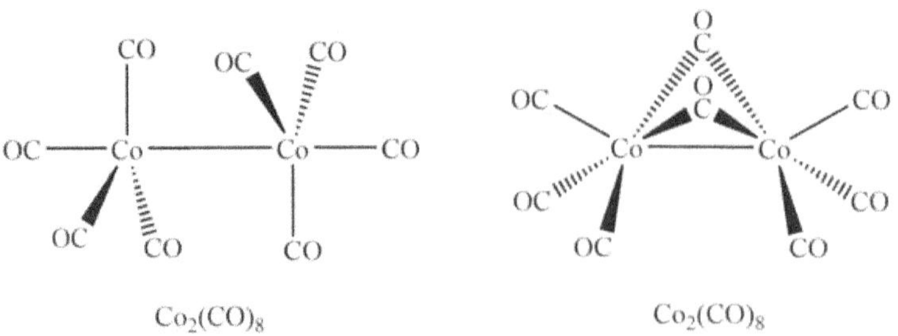

Figure 1. The structures of some mononuclear metal carbonyls.

2. Binuclear metal carbonyls: The structure of the binuclear metal carbonyls comprises of two metal centers and involve either metal-metal bonds or bridging CO groups or both. For example, the $Co_2(CO)_8$ is known to exist in two isomers. The first one has a D_{3d} symmetry with one metal-metal bond with zero bridging carbonyls; the second one is of C_{2v} symmetry and has two bridging CO ligands along with one metal-metal bond. The structure of $Fe_2(CO)_9$ exist with D_{3h} symmetry and contains three bridging CO ligands and six terminal CO groups attached. Furthermore, M_2CO_{10} (M = Mn, Tc, Re) exists with D_{4d} symmetry with one metal-metal bond and four CO ligands attached to each of the metal centers.

Figure 2. Continued on the next page...

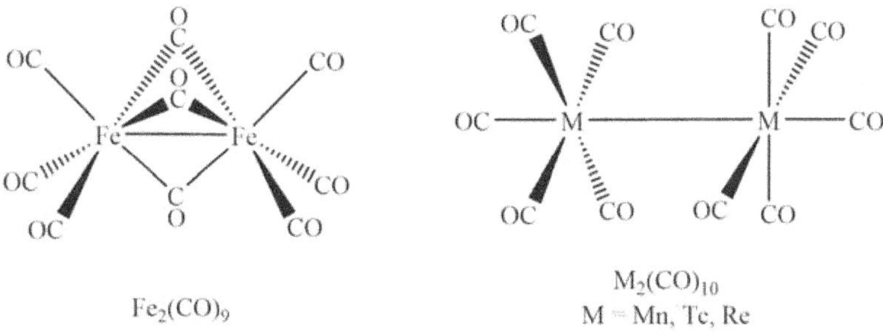

Figure 2. The structures of some binuclear metal carbonyls.

3. Polynuclear metal carbonyls: The structures of the polynuclear metal carbonyls comprises of three or more metal centers and involve all bridging, all terminal, or a mixture of two types of CO groups. For example, the $Ru_3(CO)_{12}$ cluster has D_{3h} symmetry, comprises of an equilateral triangle of Ru centers, each of which has two axial and two equatorial CO ligands. $Os_3(CO)_{12}$ has the same structure, whereas $Fe_3(CO)_{12}$ is different, with two bridging CO ligands, resulting in C_{2v} symmetry. M_4CO_{12} (M = Co, Rh) is consisted of a tetrahedral M_4 core, but the molecular symmetry is C_{3v}. Three carbonyl ligands are bridging ligands and nine are terminal. However, $Ir_4(CO)_{12}$ has perfect T_d symmetry with no bridging CO ligands groups. The Rh_4 and Ir_4 clusters are more thermally robust than that of the Co_4 compound, reflecting the usual trend in the strengths of metal-metal bond for second and third-row metals vs those for the first row metals. Furthermore, $[Re_4(CO)_{16}]^{2-}$ has D_{2h} symmetry with no bridging carbonyl. Furthermore, the structure of $Os_4(CO)_{16}$, $Os_4(CO)_{15}$ and $Os_4(CO)_{14}$ are somewhat more complex because of non-rigidity. The tetranuclear $Os_4(CO)_{16}$ is analogs to the cyclobutane with a puckered structure. The X-ray diffraction analysis of $Os_4(CO)_{14}$ unveiled an irregular tetrahedral Os_4 skeleton with four weakly semi-bridging CO groups and four different Os–Os bond lengths. The experimental structure of $Os_4(CO)_{15}$ was determined to have a planar butterfly-like geometry consisting of two triangles sharing an edge. The hexanuclear $M_6(CO)_{16}$ (M = Rh, Co) exists with an octahedral core with alternate faces participating in the bridging; i.e. with four triply bridged and twelve terminal carbonyls.

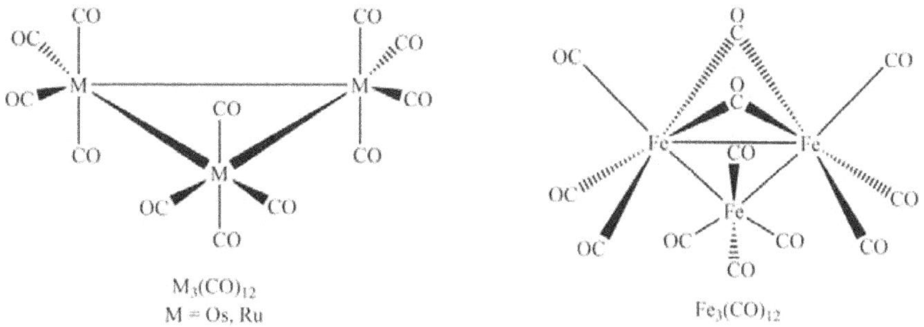

Figure 3. Continued on the next page...

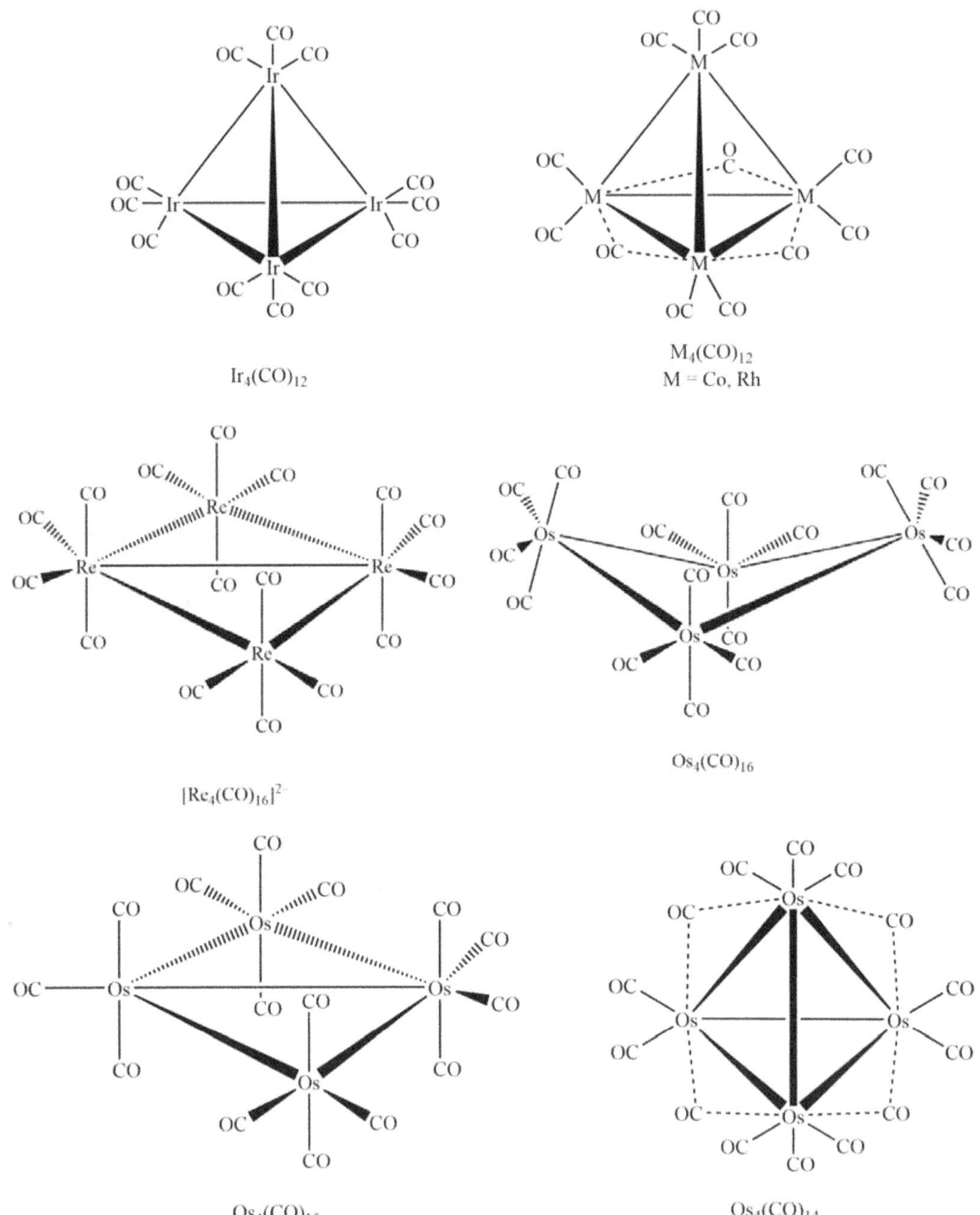

Figure 3. Continued on the next page…

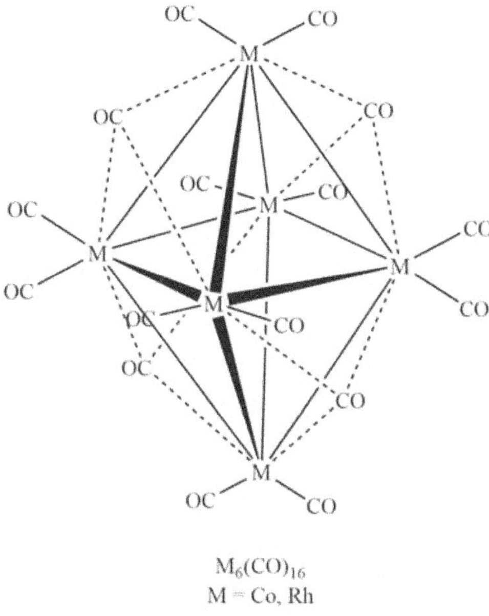

$M_6(CO)_{16}$
M = Co, Rh

Figure 3. The structures of some polynuclear metal carbonyls.

> ### ➤ *Bonding in Metal Carbonyls*

In order to rationalize the nature of the bonding between the metal center and the carbonyl ligand, we must understand the bonding within the carbonyl ligand itself first. It is a quite well-known fact that the CO group acts as a good sigma donor as well as a good π-acceptor ligand. Two popular approaches to study the bonding in carbon monoxide, as well as metal carbonyls, are discussed below.

1. Valence bond theory: According to this model, the bonding within the CO molecule can be best shown as:

$$O\equiv\!\!\!\rightarrow C$$

The carbon and oxygen atoms in CO are *sp*-hybridized with the following electronic configurations.

$$C \text{ (ground state)} = 1s^2, 2s^2, 2p_x^1, 2p_y^1, 2p_z^0$$

$$C \text{ (hybridized state)} = 1s^2, (sp_x)^2, (sp_x)^1, 2p_y^1, 2p_z^0$$

Similarly,

$$O \text{ (ground state)} = 1s^2, 2s^2, 2p_x^1, 2p_y^1, 2p_z^2$$

$$O \text{ (hybridized state)} = 1s^2, (sp_x)^2, (sp_x)^1, 2p_y^1, 2p_z^2$$

Now, the one half-filled sp_x-hybridized orbital of carbon atom overlap with half-filled sp_x-hybridized orbital of the oxygen atom to form a σ bond, while sp_x-hybridized lone pairs on both atoms remain non-bonding in nature. Moreover, two π bonds are formed as a result of the sidewise overlap; one between half-filled $2p_y$ orbitals, and the second one as dative or coordinative interaction of fully filled $2p_z$ orbital of oxygen with empty $2p_z$ orbital of carbon.

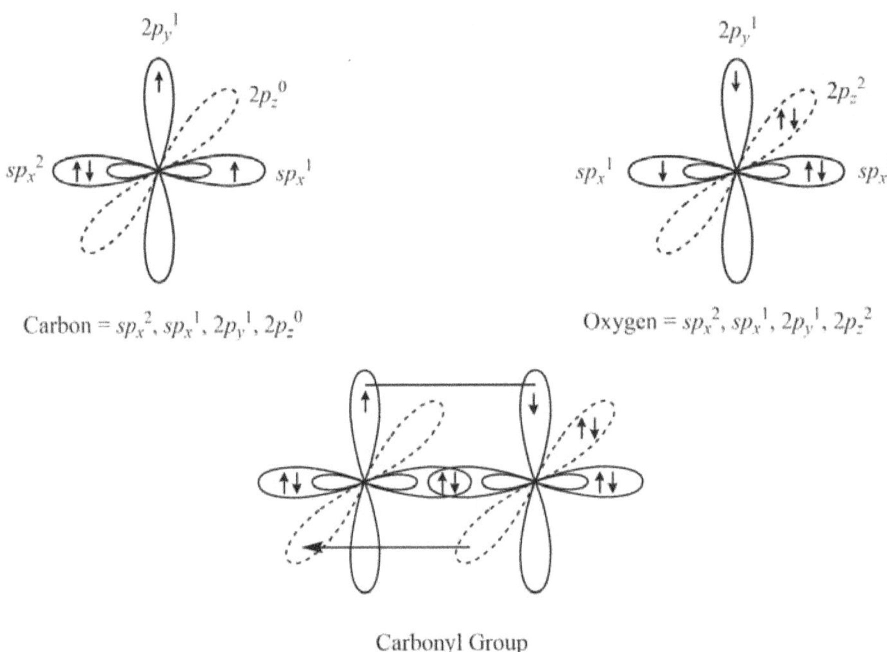

Figure 4. Bonding behavior in CO ligand according to the valence bond method.

Furthermore, the valence bond theory treats the bonding mode of the carbonyl with the metal center in terms of hybridization and resonance phenomena. The central metal atom or ion provides the required number of empty hybrid orbitals with proper orientation to accept the electron pair from surrounding ligands. For instance, in $Cr(CO)_6$ the chromium atom undergoes a d^2sp^3 hybridization to generate six empty hybrid orbital of equivalent shapes and the same energy. When one of the carbonyl ligands approaches this metal ion with its internuclear axis along x-axis, the filled hybrid lone pair of electron on carbon atom overlap with one of the two empty hybrid orbitals orientated oppositely in x-direction. The metal-carbon multiple bonds is explained in terms of various resonating structures which consequently reduces the bond strength of the carbon-oxygen bond. It should also be noted that, though there are two hybrid lone pairs (one on carbon and the other on the oxygen); the bonding of carbonyl group with metal takes place via a donation through carbon end always. This can be explained in terms of the higher energy of hybrid lone pair carbon than oxygen.

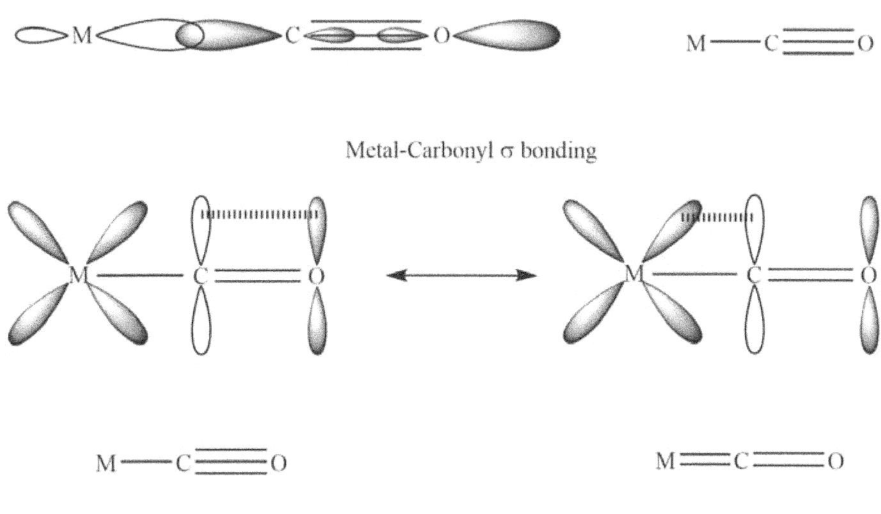

Metal-Carbonyl σ bonding

Metal-Carbonyl π bonding

Figure 5. The valence bond nature of σ and π overlap in metal carbonyls.

2. Molecular orbital theory: This is the best model to explain the bonding within the CO ligand as well as in metal carbonyl complexes. There are total three molecular diagrams for carbonyl ligand which were proposed from time to time. Though, all three molecular orbital (MO) diagrams are able to explain the nature of metal-carbonyl π-bonding; the initial treatment was not so effective to explain the σ donation, the second one does also suffer from some minor anomalies. The third molecular orbital diagram is most widely accepted in the scientific community as it gives a logical explanation to what had been a mystery in metal carbonyl chemistry. We will study these MO diagrams in the order they were proposed.

i) The first molecular orbital diagram of carbon monoxide assumes that the atomic orbitals of carbon and oxygen interact with each other to create molecular orbitals. The electronic configurations of C and O are:

$$\text{Carbon} = 1s^2, 2s^2, 2p^2$$

$$\text{Oxygen} = 1s^2, 2s^2, 2p^4$$

The number of outer electrons in carbon and oxygen are four and six, respectively. Thus, a total of 10 electrons are to be filled in the molecular orbitals of the carbon monoxide molecule. The higher energy of corresponding atomic orbitals of carbon is due to its lower electronegativity, which makes the bonding and antibonding molecular orbitals to receive different contributions from atomic orbitals of carbon and oxygen. The bonding molecular orbitals will be rich in atomic orbitals of oxygen while antibonding molecular orbitals, that are closer to carbon in energy, would be rich in atomic orbitals of carbon. The bonding molecular orbitals will have more characteristics of atomic orbitals of Oxygen and antibonding Molecular orbitals would have more characteristics of carbon. The electronic configuration of CO molecule will be $\sigma 2s^2$, σ^*2s^2, $\sigma 2p_z^2$, $\pi 2p_x^2$, $\pi 2p_y^2$ which gives a bond order three i.e. triple bond between carbon and oxygen. The molecular orbital diagram and expected bonding mode of the carbonyl ligand are given below.

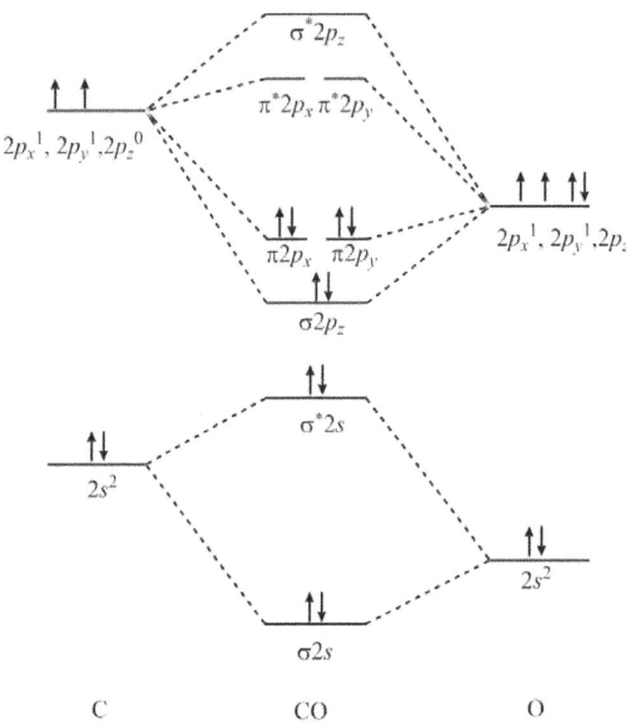

Figure 6. The first generation molecular orbital diagram of carbonyl ligand.

The formation of molecular orbitals given above is actually an oversimplification of a ticklish problem. This statement is made on the basis of two facts. The first one is that when one electron removed from CO to form CO^+, the bond order actually increases, which is actually the opposite of what is expected if the electron is lost from the highest occupied molecular orbital (HOMO) of bonding nature. Its bond order should be decreased from the removal of an electron from $\pi 2p_x^2$ or $\pi 2p_y^2$, which suggests that the HOMO of carbon monoxide should be of antibonding nature rather bonding. The second anomaly also arises from the MO diagram of CO ligand which clearly shows that in order to donate electron density from π-bonding molecular orbital, the carbonyl ligand must approach the metal center with its carbon-oxygen internuclear axis perpendicular to x, y or z-axis assigned to the central atom. This explains how the empty π^*2p_x and π^*2p_y could be used to accept electron density from filled d-orbitals of central metal atom or ion. However, in actual practice, the carbonyl ligand binds to the metal center in linear fashion via carbon end only.

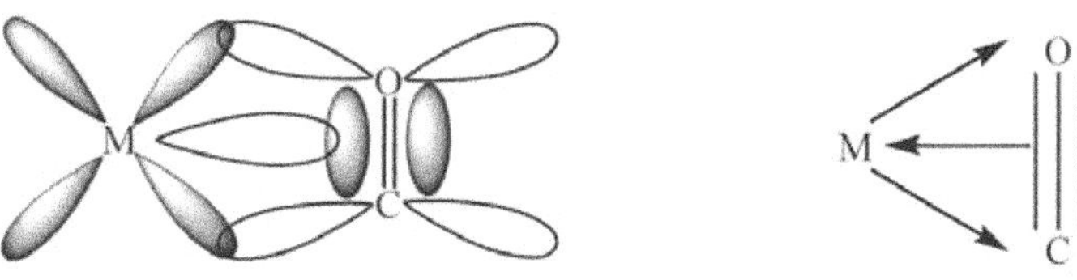

Metal-Carbonyl σ donation and π acceptance

Figure 7. The expected nature of σ and π overlap in metal carbonyls from the first MO of CO.

ii) The second molecular orbital diagram of carbon monoxide was suggested by Coulson which assumes that the first molecular orbital diagram of CO is not correct. According to Coulson, $2s$ and $2p_x$ atomic orbitals of both carbon and oxygen undergo hybridization before they create molecular orbitals. The carbon and oxygen atoms in carbon monoxide are sp-hybridized with the following electronic configurations.

$$C \text{ (hybridized state)} = 1s^2, (sp_x)^2, (sp_x)^1, 2p_y^1, 2p_z^0$$

$$O \text{ (hybridized state)} = 1s^2, (sp_x)^2, (sp_x)^1, 2p_y^1, 2p_z^2$$

The total number of valence electrons in carbon and oxygen are four and six, respectively; and thus, ten electrons are to be filled in the molecular orbitals of CO molecule. The half-filled sp_x hybrid orbitals of carbon and oxygen interact to form σ and σ* molecular orbitals; while the fully-filled sp_x hybrid lone pair orbitals of carbon and oxygen remain non-bonding. Moreover, doubly degenerate sets of π-bonding and π-antibonding molecular orbitals are also formed due to the sidewise overlap of $2p_y$ orbitals and $2p_z$ orbitals.

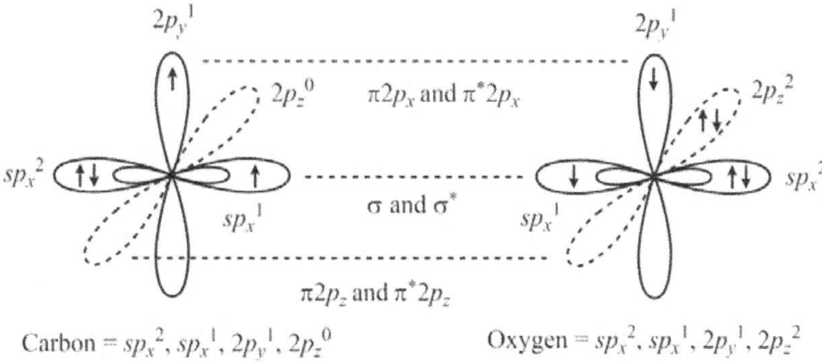

Figure 8. The nature of σ and π in carbonyl ligand.

The bonding molecular orbitals will be rich in atomic orbitals of oxygen while antibonding molecular orbitals, that are closer to carbon in energy, would be rich in atomic orbitals of carbon. The bonding molecular orbitals

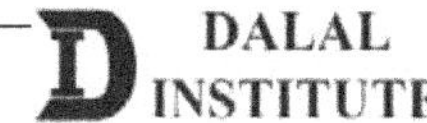

will have more characteristics of atomic orbitals of oxygen and antibonding molecular orbitals would have more characteristics of carbon. The molecular orbital diagram carbon monoxide proposed by Coulson is given below.

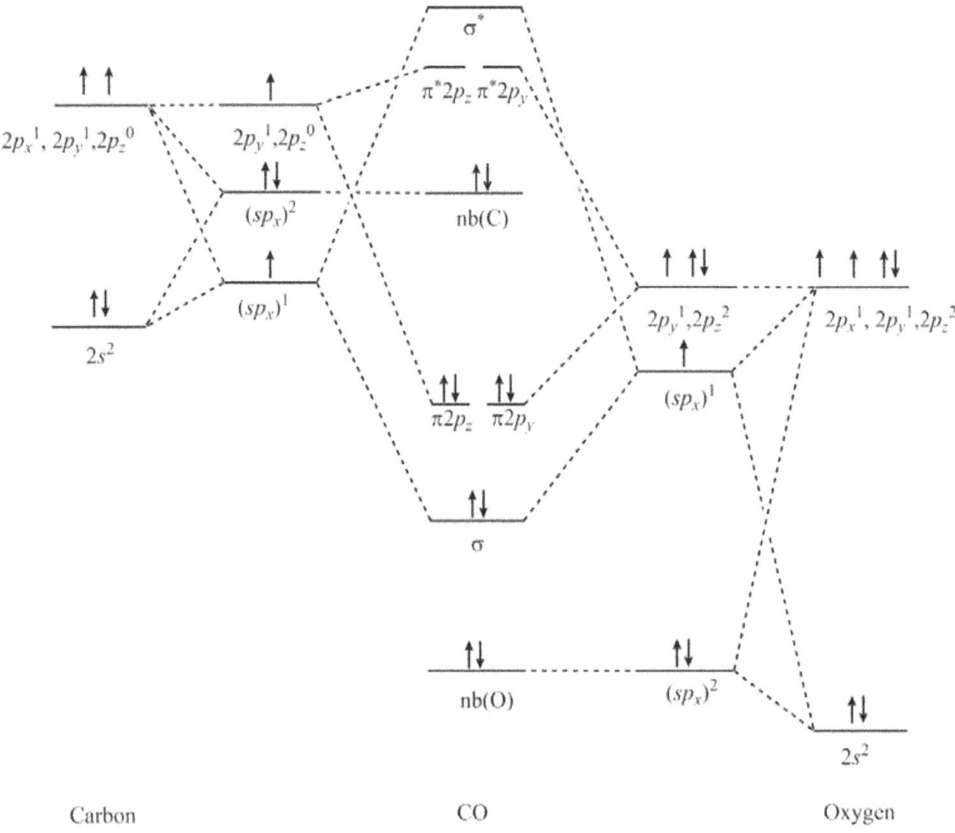

Figure 9. The second molecular orbital diagram of carbonyl ligand.

The MO diagram shown above is very useful in explaining the bonding between the metal center and carbonyl ligand. This diagram eliminates the possibility of sigma donation through bonding molecular orbital and perpendicular orientation CO ligand as the HOMO is now non-bonding hybrid lone pair rather π-bonding. This also explains why the carbonyl group prefers to bond via carbon end in a linear manner. This also explains how the lowest unoccupied molecular orbital (LUMO) π^*2p_z and π^*2p_y could be used to accept electron density from filled d-orbitals of central metal atom or ion. Moreover, the reduced CO stretching frequency of metal coordinated carbonyl can be attributed to the reduced bond order due to the transfer of d-electron density from metal to π^* orbital carbonyl ligand. However, the increase in bond order when one electron is removed from

CO to form CO^+ is still a mystery because the electron is lost from the highest occupied molecular orbital (HOMO) of nonbonding bonding nature, and the bond order should have remained the same.

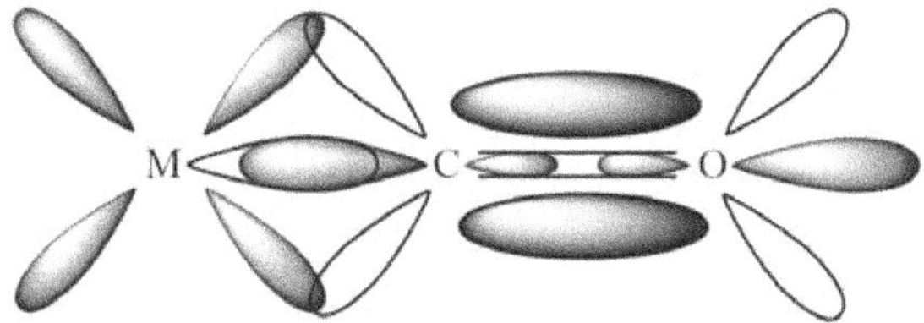

Figure 10. The nature of σ and π overlap in metal carbonyls.

iii) This molecular orbital diagram of carbon monoxide is most widely accepted to rationalize its σ-donor and π-acceptor strength. The total number of valence electrons in carbon and oxygen are four and six, respectively; and thus, ten electrons are to be filled in the molecular orbitals of CO molecule. A total of four singly degenerate σ- molecular orbitals and two doubly degenerate sets of π- molecular orbitals are formed. One doubly degenerate set of π molecular orbitals will be bonding while the other one will be antibonding in nature. The nature of σ molecular orbitals is more complex as three out of four are of bonding character. Initially, the σ_5 was thought to be of antibonding to justify the higher bond order of CO^+. However, the σ_5 is slightly bonding in nature because there is some mixing with the p atomic orbitals of the right symmetry. Out of four σ-molecular orbitals, only σ_6 possesses the antibonding character, while σ_5 goes with expected bonding characteristics. The σ_5 is essentially non-bonding and almost centered on the oxygen atom. Moreover, doubly degenerate sets of π-bonding and π-antibonding molecular orbitals are also formed due to the sidewise overlap of $2p_y$ orbitals and $2p_z$ orbitals. The π-bonding molecular orbitals set will be rich in atomic orbitals of oxygen while antibonding molecular orbitals, that are closer to carbon in energy, will be rich in atomic orbitals of carbon atom.

However, the problem that why does the bond order increases when an electron is removed from CO still persists. Because we are removing the electron from a bonding molecular orbital, its bond order must be decreased. The possible explanation for the shortening of the bond after ionization is that the ionization induces a shift of the electron-polarization in CO ligand. In other words, the ionization occurs as the loss of an electron from a σ-HOMO orbital which is mostly carbon-centered; and since the HOMO-σ orbital is only slightly bonding in nature, the loss of bonding character is quite small and could easily be compensated by the advantage in covalent character; i.e. the formation of a positive partial charge on the carbon atom increases the strength of the covalence of the bond and thus decreases the bond length. This enhanced covalent character can also be visualized in terms of better interaction of two atomic orbitals if their energies are comparable. In carbon monoxide molecule the atomic orbitals of oxygen lie energetically a lot below then the atomic orbitals of carbon; But when the CO is oxidized to CO^+, the partial positive charge on carbon shifts the atomic orbitals of carbon down in energy, and thereby makes the energies closer to the related atomic orbitals of oxygen, which leads to a stronger interaction when bonds are made.

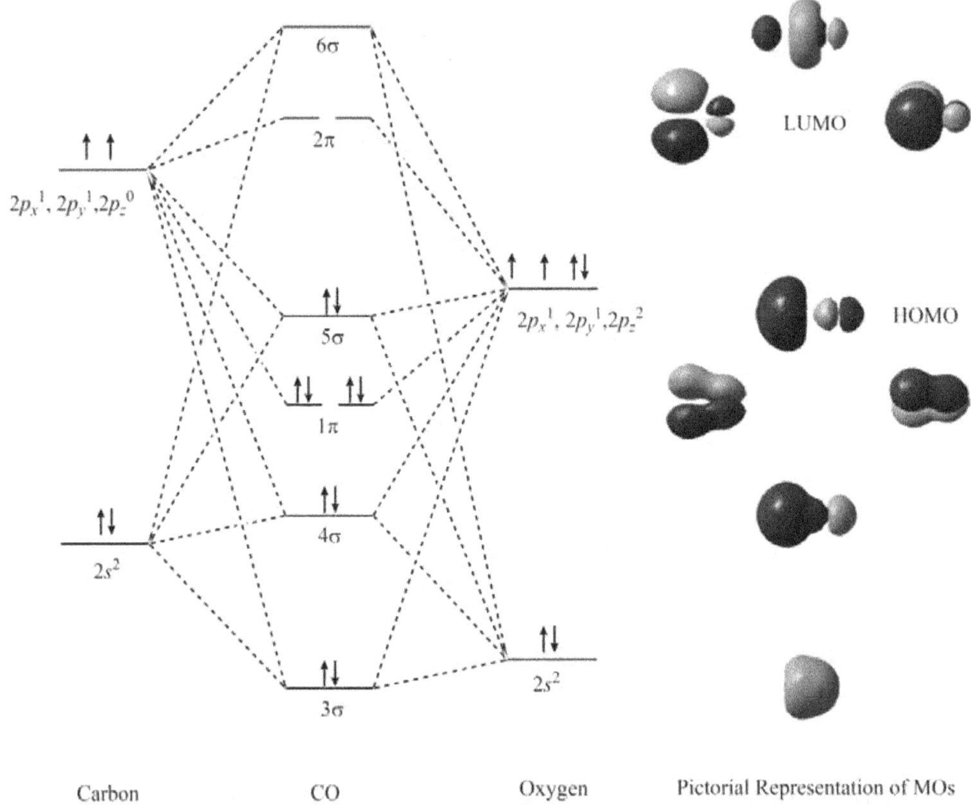

Figure 11. Third molecular orbital diagram of carbonyl ligand.

The MO diagram shown above is very useful in explaining the bonding between the metal center and carbonyl ligand. The carbonyl ligand uses its HOMO for sigma donation while simultaneously accepts electron density from filled metal d-orbital to its π^* LUMO.

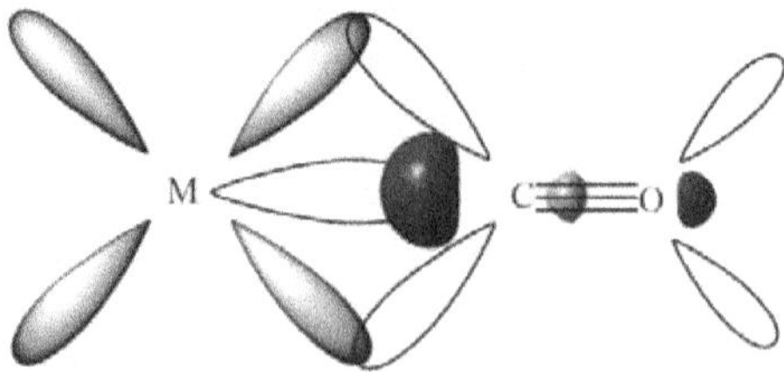

Figure 12. The nature of σ and π overlap in metal carbonyls using third MO diagram of CO.

❖ Vibrational Spectra of Metal Carbonyls for Bonding and Structure Elucidation

Vibrational spectroscopy is one of the most important methods used for the characterization of metal carbonyls. This technique provides very useful information not only about the structural prototype of different metal carbonyl compositions but also rationalizes the nature of bonding in them. Now, it is quite a well-known fact that the C–O vibration for free carbonyl group (CO gas) is typically denoted as v_{CO}, and absorbs at 2143 cm^{-1}. However, this C–O absorption shifts downward (sometimes upward) to cover a very wide range of wavenumber as the carbonyl ligand gets attached to a metal center. This is obviously due to the fact that the energies of the v_{CO} band for the metal carbonyls directly correlate with the strength of the carbon-oxygen bond, and are inversely correlated with the strength of the π-backbonding between the metal and the carbon. In other words, the molecular orbital diagram of carbonyl group suggests that the highest occupied molecular orbital, used for σ-donation is weakly bonding; but the lowest unoccupied molecular orbital, used for accepting d-electron density from metal center is strongly antibonding; therefore, the σ-donation does not affect the CO bond order very much but the acceptance of electron density in π^* orbital decreases the bond order and consequently the bond strength in a significant way. This effect reduces the force constant of C–O bond, while the magnitude of force constant for M–C will be increased by this backbonding. As a result, the enhancement of backbonding shifts the metal-carbon and carbon-oxygen stretching to higher and lower values, respectively. The main features about the bonding and structure of metal carbonyls which can be obtained from the vibrational spectra of metal carbonyls are discussed below:

> ### *1. π-Basicity of the Metal Centre*

The π-basicity of the metal center (and thus the C–O stretching frequency) depends upon a lot of factors like the nature and magnitude of the charge on metal center, and the π-accepting tendency of ligands attached other than the carbonyl. A negative charge on the metal center, or ligands with greater σ-donation and weaker π-acceptor strength, are expected to decrease the CO stretching frequency; while an accumulation of positive charge on metal center, or ligands with weaker σ-donation stronger π-accepter strength are bound to increase the CO stretching frequency. For example, in the isoelectronic series of Ti→Fe, the hexacarbonyls show decreasing π-backbonding as one increases (makes more positive) the charge on the metal.

Compound	$[Ti(CO)_6]^{2-}$	$[V(CO)_6]^{1-}$	$[Cr(CO)_6]$	$[Mn(CO)_6]^{1+}$	$[Fe(CO)_6]^{2+}$
v_{CO} (cm^{-1})	1748	1859	2000	2095	2204

Compound	$[Hf(CO)_6]^{2-}$	$[Ta(CO)_6]^{1-}$	$[W(CO)_6]$	$[Re(CO)_6]^{1+}$	$[Os(CO)_6]^{2+}$
v_{CO} (cm^{-1})	1757	1850	1977	2085	2190

Hence, π-basic ligands increase π-electron density at the metal, and improved backbonding reduces v_{CO}.

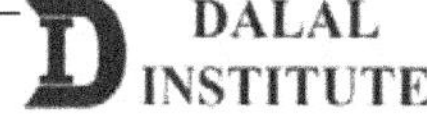

➢ *2. Toleman Electronic Parameter*

The electron-donating or withdrawing ability of a ligand is calculated in terms of the Tolman electronic parameter (TEP), named after the inventor, Chadwick A. Tolman. It is determined by measuring the frequency of the A_1 vibrational mode of the carbonyl group in complex, $Ni(CO)_3L$ by infrared spectroscopy, where L is the ligand being studied. $Ni(CO)_3L$ was chosen as the model compound because such complexes can easily be synthesized from $Ni(CO)_4$.

The CO band is pretty much unique and is rarely affected by other bands in the analyte's IR spectra. Moreover, owing to the small size of carbonyl ligand, steric factors do not muddle the analysis. The coordination of CO to a metal typically decreases v_{CO} value from 2143 cm^{-1} (free CO). This can be rationalized by π-backbonding, resulting from the sidewise overlap of metal orbitals of π-symmetry with the empty anti-bonding π^* orbitals on CO ligand. This increases the strength of the metal-carbon bond, but also weakens the C–O bond. If some other ligand enhances the π-electrons-density on the metal, the CO bond strength decreases and so the v_{CO} value. On the other hand, if other ligands present compete with CO for π-backbonding, v_{CO} increases. TEPs for selected phosphines are given below.

L	P(t-bu)$_3$	P(Me)$_3$	P(Ph)$_3$	P(OEt)$_3$	PCl$_3$	PF$_3$
v_{CO} (cm^{-1})	2056.1	2064.1	2068.9	2076.3	2097.0	2110.8

The Tolman electronic parameter has been widely used to characterize the electronic properties of phosphine based ligands.

➢ *3. Structural Prototype of Metal Carbonyls*

The symmetry behavior of different metal carbonyls can be used to determine the number of infrared active vibrational modes, which in turn enable us to comment on structural prototypes. In other words, it is well known from the group theory that only those vibrational modes will be observed which transform as the electric-dipole moment operator; therefore, the number of observable infrared transitions for a particular geometry can be predicted theoretically. These predictions are then matched with the experimentally observed infrared spectra to shortlist the various possible structural prototypes. For instance, consider the case of five-coordinated homoleptic metal carbonyl complexes. The two possible geometries are square-pyramidal and trigonal bipyramidal.

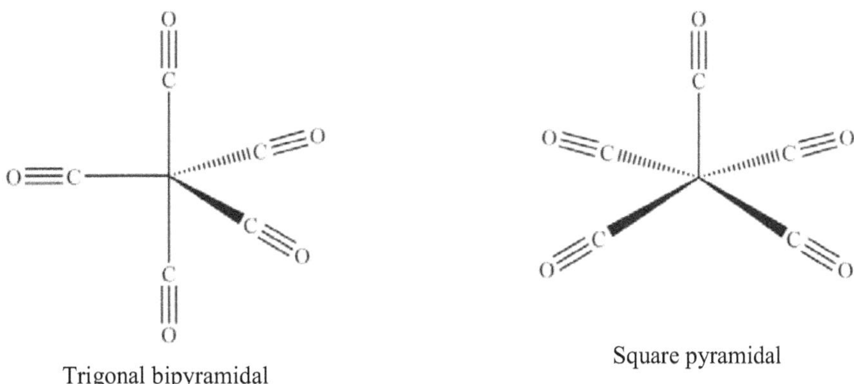

Trigonal bipyramidal   Square pyramidal

Figure 13. Two possible geometries of Iron pentacarbonyl i.e. Fe(CO)₅ complex.

For trigonal bipyramidal geometry, a reducible representation based on five C–O bonds is:

D_{3h}	E	$2C_3$	$3C_2$	σ_h	$2S_3$	$3\sigma_d$	Irreducible components
Γ_π	5	2	1	3	0	3	$2A_1' + A_2'' + E'$

Out of four irreducible representations, A_2'' transforms as z-component while E' transforms with x- and y-components of the dipole moment. Now, owing to two doubly degenerate vibrational modes (E'), only two peaks are expected in the experimental infrared spectrum. However, in the Raman infrared spectrum, three peaks are expected as only $2A_1'$ and E' irreducible components transform alongside the polarizability tensors.

For square pyramidal geometry, a reducible representation based on five C–O bonds is:

C_{4v}	E	$2C_4$	C_2	$2\sigma_v$	$2\sigma_d$	Irreducible components
Γ_π	5	2	1	3	3	$2A_1 + B_1 + E$

Out of four irreducible representations, $2A_1$ transforms as z-component while E transforms with x- and y-components of the dipole moment. Now, owing to two doubly degenerate vibrational modes (E), only three peaks are expected in the experimental infrared spectrum. However, in the Raman infrared spectrum, four peaks are expected as all of the irreducible components transform alongside the polarizability tensors.

Similarly, the CO ligands of octahedral complexes, e.g. Cr(CO)₆, transform as A_{1g}, E_g, and T_{1u}, but only the T_{1u} mode (anti-symmetric stretch of the apical carbonyl ligands) is infrared-active; and therefore, only a single carbonyl stretching is observed in the IR-spectra of the octahedral metal hexacarbonyls. Spectra for complexes of lower symmetry are more complex. For example, the IR spectrum of Fe₂(CO)₉ displays CO bands at 2082, 2019, 1829 cm⁻¹. The number of observable infrared-active vibrational modes for some metal carbonyls are listed in the following below.

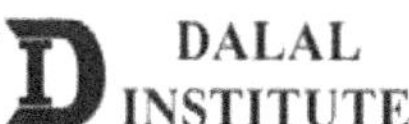

Table 1. The number of IR-active vibrational modes of several prototypical metal carbonyl complexes.

No. of carbonyls	Structural prototype (No. of IR-peaks)	Structural prototype (No. of IR-peaks)
Three	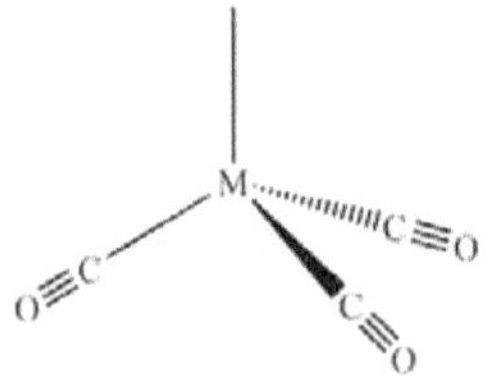 (2)	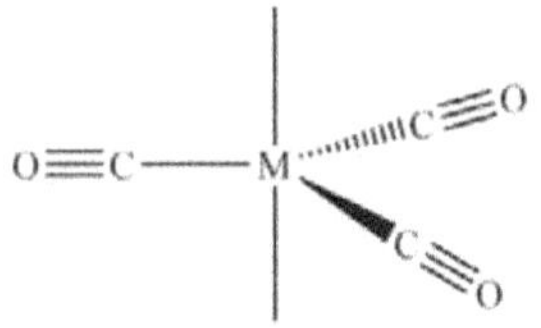 (1)
	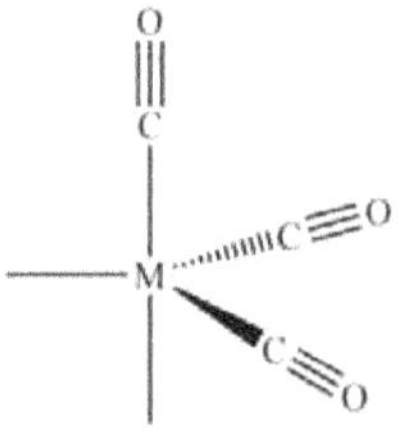 (3)	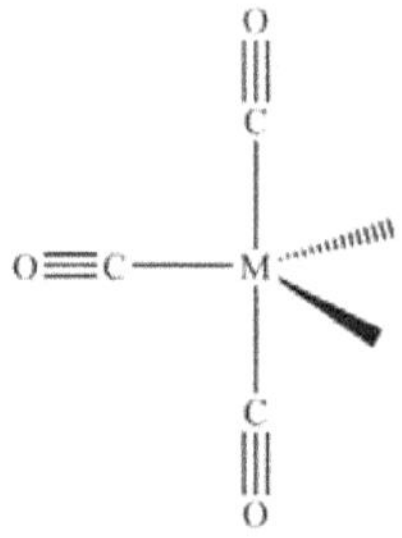 (3)
	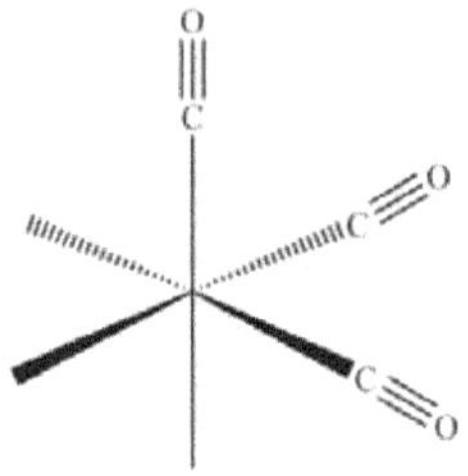 (2)	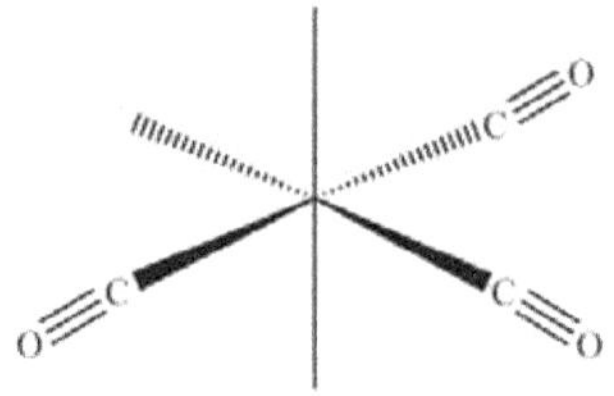 (3)
Four	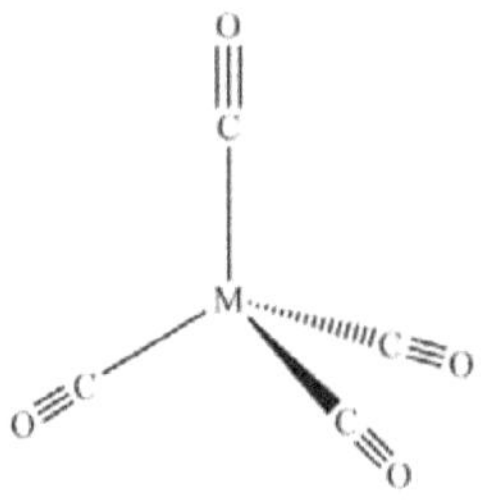 (1)	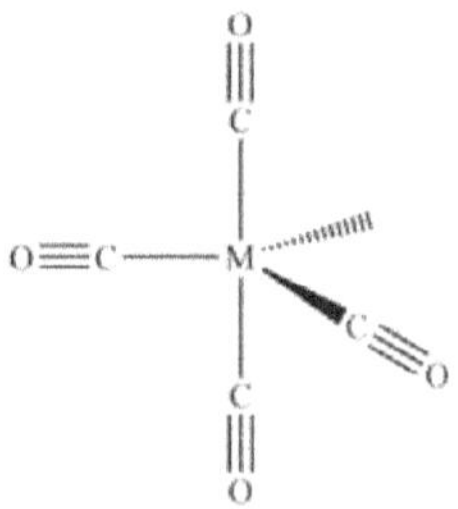 (4)

Table 1. Continued on the next page…

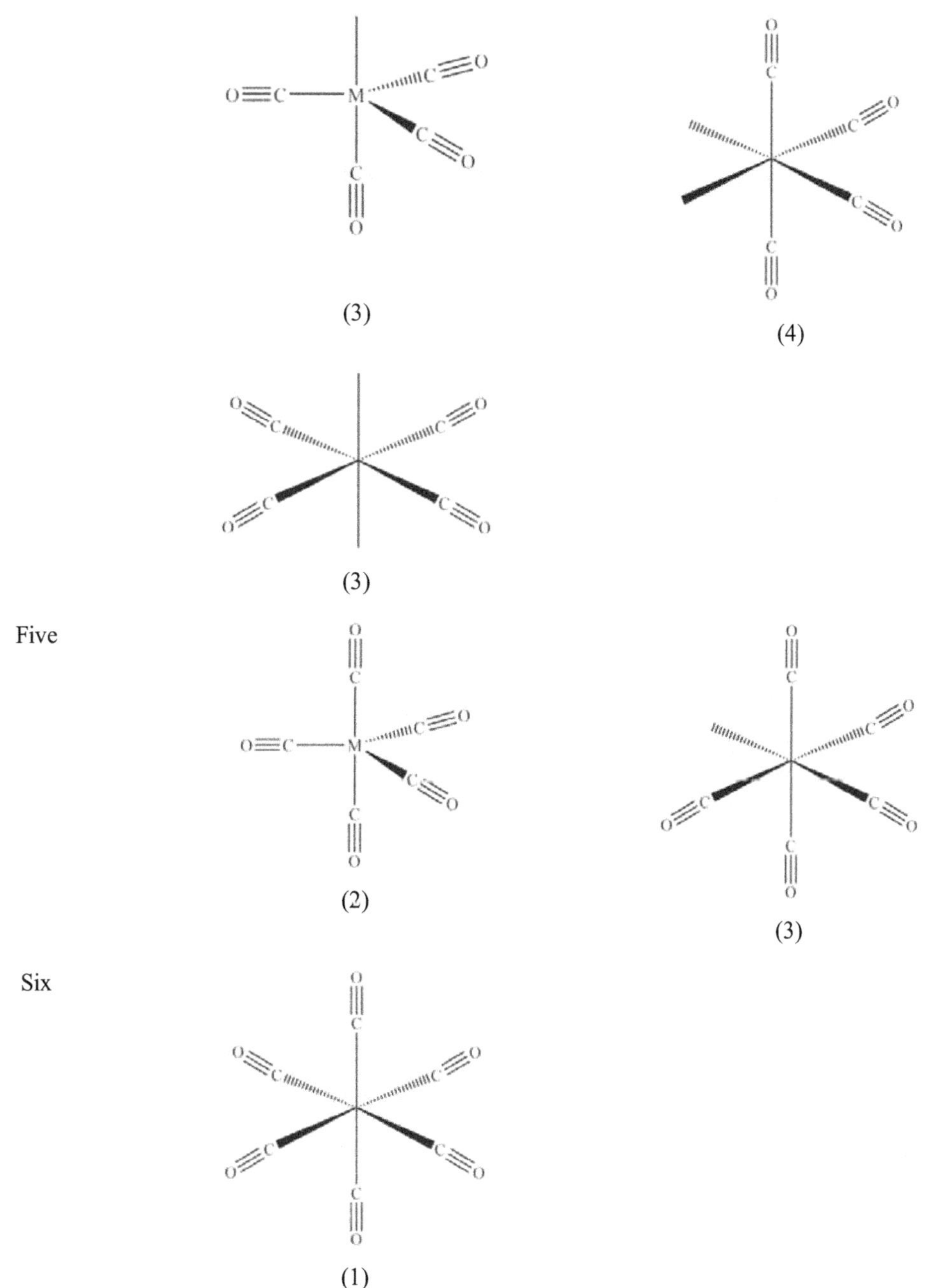

(3)

(4)

(3)

Five

(2)

(3)

Six

(1)

It is worthy to note that these rules apply to metal carbonyls in solution or the gas phase.

> ### ➤ *4. Differentiation of Terminal and Bridging Carbonyl Groups*

The mode of attachment of the carbonyl group to the metal center can also be determined by observed CO stretching frequencies. Terminal carbonyls absorb at the higher wavenumber in comparison to the bridging ones, which is obviously due to the fact that the extant of backbonding increases with the number of metal centers. Three main modes of attachment of the carbonyl group with the metal center are:

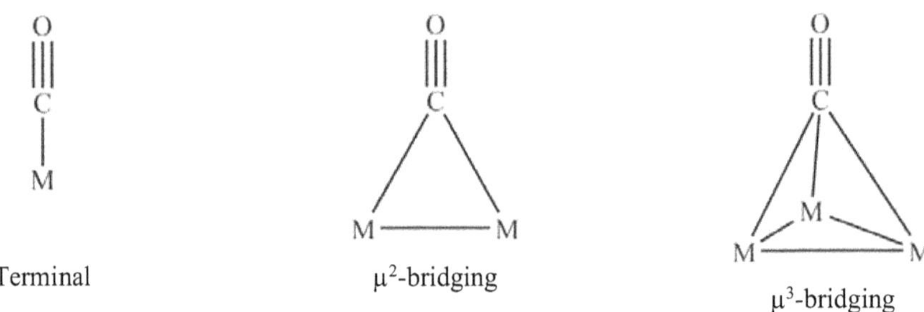

Terminal μ^2-bridging μ^3-bridging

The ν_{CO} follows the order: Terminal CO > μ^2-bridging > μ^3-bridging. The IR-range absorption for various types of carbonyls groups is listed below.

Table 2. The CO stretching frequencies in different metal carbonyl complexes.

Bonding mode of CO	ν_{CO} (cm^{-1})
Free CO	2143
Free CO$^+$	2184
Terminal CO	2120 - 1850
Symmetric μ^2-CO	1860 - 1750
Symmetric μ^3-CO	1730 - 1600

For instance, consider the rhodium carbonyl complexes:

Compound	μ^1-CO, ν_{CO} (cm^{-1})	μ^2-CO, ν_{CO} (cm^{-1})	μ^3-CO, ν_{CO} (cm^{-1})
$Rh_2(CO)_8$	2060, 2084	1846, 1862	
$Rh_4(CO)_{12}$	2044, 2070, 2074	1886	
$Rh_6(CO)_{16}$	2045, 2075		1819

There is also a semi-bridging mode that lies in between bridging and terminal bonding profile and is usually labeled as asymmetric bridging carbonyls.

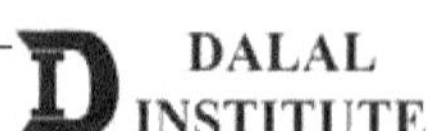

> ### 5. Calculation of CO Bond Order

Owing to the well-known fact that the highest occupied molecular orbital (HOMO) set in carbonyl ligand is of antibonding nature, a close correlation between the magnitude of backbonding and carbon-oxygen bond order can easily be established. The vibrational frequency of a bond depends upon the strength of the bond (which is measured by its force constant), and is given as:

$$v_{CO}(cm^{-1}) = \frac{1}{2\pi c}\sqrt{\frac{k}{\mu}} \tag{1}$$

Where k is the force constant and μ is the reduced mass. The physical significance of k is implied in the fact that it is proportional to the strength of the bond involved. Therefore, any factor that increases the electron density on CO group will decrease its bond order and its force constant; and the vice-versa is also true. In other words, the decrease in v_{CO} indicates a decrease in CO bond order while the increase in carbonyl stretching is associated with increasing CO bond order. For instance, the infrared absorption of free CO occurs at 2143 cm^{-1} while the metal coordinated CO absorbs generally in the range of 2120-1800 cm^{-1}. This clearly establishes the fact that metal to ligand back donation does occur which in turn reduces its bond order. Hence, the accumulation of positive charge or the deficiency of electron density will impart a larger CO bond order than neutral or ionic carbonyl complexes. Consider the following trend in isoelectronic-isostructural carbonyls.

Compound	$[Ni(CO)_4]$	$[Co(CO)_4]^{1-}$	$[Fe(CO)_4]^{2-}$
v_{CO} (cm^{-1})	2046	1890	1730
M–C Bond order	1.33	1.89	2.16
C–O Bond order	2.64	2.14	1.85
M–C + C–O Bond order	3.97	4.03	4.01

A similar trend is observed in the case of mixed carbonyls i.e. metal carbonyls having some other ligands alongside CO group. If the other ligand is electron-withdrawing in nature, it would attract the electron density from the metal center, which in turn would oppose the metal-carbonyl back bonding, yielding somewhat higher carbon-oxygen bond order and higher carbonyl stretching frequency. On the other hand, If the other ligand is electron-donating in nature, it would donate the electron density to the metal center, which in turn would support the metal-carbonyl back bonding, yielding somewhat lower carbon-oxygen bond order and lower carbonyl stretching frequency.

Compound	$[Ni(CO)_4]$	$[Ni(PF_3)(CO)_3]$	$[Ni(PMe_3)(CO)_3]$
v_{CO} (cm^{-1})	2046	1990	1980

Hence, the bond order of CO in $[Ni(PMe_3)(CO)_3]$ is definitely lower than in $[Ni(PF_3)(CO)_3]$ because PMe$_3$ is a weaker π-acceptor than PF$_3$ ligand.

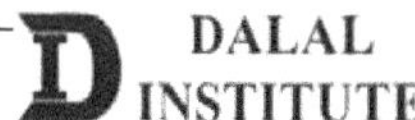

> ### *6. Study of Reaction Kinetics in Metal Carbonyls*

Infrared spectroscopy is a very useful tool in the determination of reaction kinetics of metal carbonyl complexes. From Beer's Lambert Law, we know that the absorbance (A) is related to concentration (c) and path length (l) of the sample as:

$$A = \varepsilon cl = log\frac{1}{T} \tag{1}$$

Where ε is the molar extinction coefficient and T is the transmittance of the solution. Therefore, the rate of disappearance or appearance of a characteristic infrared absorption band can be used to estimate the rate of change of concentration for the corresponding reactant or product. In other words, the rate of intensity decrease in IR peaks of reactant, or the rate of intensity increase in IR peaks of product metal carbonyl is proportional to the rate of change of concentration; which in turn enable us to record various kinetic parameters from reaction order to activation energy or rate constants.

❖ Important Reactions of Metal Carbonyls

Metal carbonyls are important precursors for the synthesis of a mixed carbonyl or some important organometallic complexes. Some of the main reactions shown by metal carbonyls are discussed below.

> ### *1. Ligand Displacement Reactions*

The displacement or substitution of CO ligands can be induced photochemically or thermally by some other donor ligands. The ligand-domain is quite wide and comprises of cyanide (CN^-), phosphines, nitrogen donors, and ethers also. Olefins are very effective ligands that can afford synthetically useful derivatives. The displacement reaction in 18-electron complexes usually follows a dissociative pathway, via a 16-electron intermediate complex. The ligand-displacement-rate in 18-electron complexes is catalyzed by catalytic amounts of oxidants through the electron-transfer phenomena. The displacement in 17-electron complexes proceeds via an associative route with a 19-electron intermediate complex. It is worthy to note that the replacement by bidentate ligands like o-phenanthroline(o-phen) and o-phenylene-bis(dimethyl arsine) (diars) occurs in the multiple of two for carbonyl groups. For example:

$$Ni(CO)_4 + o-phen \rightarrow Ni(CO)_2(o-phen) + 2CO$$

$$Ni(CO)_4 + 4CNR \rightarrow Ni(CNR)_4 + 4CO$$

$$Ni(CO)_4 + 4PF_3 \rightarrow Ni(PF_3)_4 + 4CO$$

$$Mo(CO)_6 + 3Py \rightarrow Mo(CO)_3(Py_3) + 3CO$$

$$Fe(CO)_5 + 2CNR \rightarrow Fe(CO)_3(CNR)_2 + 2CO$$

$$Mn_2(CO)_{10} + 2PR_3 \rightarrow 2Mn(CO)_4(PR_3) + 2CO$$

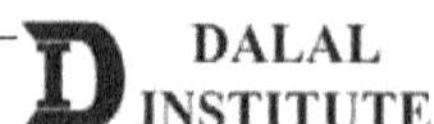

The displacement reactions in metal carbonyls also point toward the differences in the bonding nature of the attached ligands. For instance, radiochemical tracer studies for $Mn(CO)_5Br$ have unveiled that only four carbonyl groups undergo exchange with ^{14}CO.

$$Mn(CO)_5Br + 4\,^{14}CO \rightarrow Mn\big(^{14}CO\big)_4(CO)(Br) + 4CO$$

In the structure of $Mn(CO)_5Br$, four carbonyl groups undergoing exchange phenomena are present in the same plane; which means that the CO group trans to bromido is bound more firmly because the Br group is not competition for π-backbonding. On the other hand, four in-plane CO groups compete with each other all of them are good acceptors, causing a labilization of each other.

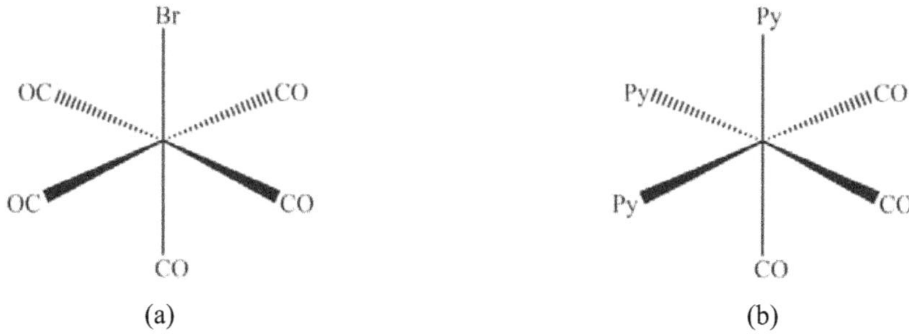

Figure 14. The structures of (a) $Mn(CO)_5Br$ and (b) $Mn(CO)_3(Py)_3$.

Furthermore, the reaction of $Mn(CO)_6$ with py always results in the formation of a facial isomer of $Mn(CO)_3(py)_3$ i.e. all the three carbonyl groups are located trans to the py ligands. This is obviously due to the fact that the extent of π-donation to the three CO groups is maximum in facial configuration, owing to the difference in the ability of CO and py for π-backbonding. When entering ligands are very good π-acceptors, all the carbonyl groups may be displaced, just in $Ni(CO)_4$ where PF_3 ligands lead to the formation of $Ni(PF_3)_4$.

> ### ➤ *2. Formation of Carbonylate Anions*

It is quite a well-known fact that many carbonylate anions like $[Fe(CO)_4]^{2-}$, $[Co(CO)_4]^-$, $[Mn(CO)_5]^-$ and $[V(CO)_6]^-$ follow the effective atomic number (EAN) rule. These ions are generally synthesized either by the reaction of metal carbonyl with a strong reducing agent or by the reaction with strong bases.

$$Cr(CO)_6 + 2Na \rightarrow Na_2[Cr(CO)_5] + Co$$

$$Mn_2(CO)_{10} + 2Na \rightarrow 2Na[Mn(CO)_5]$$

$$Co_2(CO)_8 + 2Na \rightarrow 2Na[Co(CO)_4]$$

$$Fe(CO)_5 + 3NaOH \rightarrow Na[HFe(CO)_4] + Na_2CO_3 + H_2O$$

$$Fe_2(CO)_9 + 4OH^- \rightarrow [Fe_2(CO)_8]^{2-} + CO_3^{2-} + 2H_2O$$

> ## 3. Formation of Carbonylate Cations

Carbonylate cations are not as common as carbonylate anions and can be synthesized either by the protonation of metal carbonyl in strong acids or, by the reaction with carbon monoxide and some Lewis acids.

$$Fe(CO)_5 + HCl + BCl_3 \rightarrow [HFe(CO)_5]^+[BCl_4]^-$$

$$[Mn(CO)_5Cl] + CO + AlCl_3 \rightarrow [Mn(CO)_6]^+[AlCl_4]^-$$

The use of strong acids succeeded in preparing gold carbonyl cations such as $[Au(CO)_2]^+$, which is used as a catalyst for the carbonylation of olefins. The cationic platinum carbonyl complex $[Pt(CO)_4]^+$ can be prepared by working in so-called super acids such as antimony pentafluoride.

> ## 4. Synthesis of Metal Carbonyl Hydrides

Carbonyl hydrides are generally synthesized either by the acidification of the solutions containing the corresponding carbonylate anion, or by the reactions of metal carbonyls with hydrogen. For instance:

$$Co_2(CO)_8 + H_2 \rightarrow 2[HCo(CO)_4)]$$

$$Mn_2(CO)_{10} + H_2 \rightarrow 2[HMn(CO)_5)]$$

$$[Co(CO)_4]^- + H^+(aq) \rightarrow [HCo(CO)_4]$$

$$Na[HFe(CO)_4] + H^+(aq) \rightarrow [H_2Fe(CO)_4] + Na^+(aq)$$

Some of the metal carbonyl hydrides can be prepared by direct reaction of the metal with CO and H_2. For example:

$$2Co + 8CO + H_2 \xrightarrow[\text{50 atm}]{\text{150°C}} 2[HCo(CO)_4)]$$

> ## 5. Synthesis of Metal Carbonyl Halides or Metal Halides

The metal carbonyl complexes are relatively unreactive toward many electrophiles; however, most metal carbonyls do undergo halogenation. For example, $Fe(CO)_5$ forms ferrous carbonyl halides. In some reactions, metal-metal bonds are also broken by halogens. On the basis of the electron-counting scheme used, this can be considered as oxidation of the metal center.

$$Fe(CO)_5 + X_2 \rightarrow [Fe(CO)_4X_2] + CO$$

$$Mn_2(CO)_{10} + Cl_2 \rightarrow 2[Mn(CO)_5Cl]$$

Furthermore, some metal carbonyls also get decomposed into metal halides when treated with halogens as:

$$Ni(CO)_4 + Br_2 \rightarrow NiBr_2 + 4CO$$

$$Co_2(CO)_8 + 2X_2 \rightarrow 2CoX_2 + 8CO$$

> *6. Formation of Metal Carbonyl Nitrosyls or Metal Nitrosyls*

The metal carbonyl complexes react with nitrosyl ligand to form either metal carbonyl nitrosyls or metal nitrosyl complexes. For example:

$$Fe(CO)_5 + 2NO \rightarrow [Fe(CO)_2(NO)_2] + 3CO$$

$$Co_2(CO)_8 + 2NO \rightarrow 2[Co(CO)_3NO] + 2CO$$

$$Fe_3(CO)_{12} + 6NO \rightarrow 3[Fe(CO)_2(NO)_2] + 6CO$$

$$Ni(CO)_4 + 4NO \rightarrow [Ni(NO)(NO)_2] + 4CO + N_2O$$

The displacement of carbonyl by the nitrosyl cation may be achieved using $[NO][BF_4]$ i.e nitrosyl tetrafluoroborate. On applying to the hexacarbonyls of tungsten and molybdenum, the NO binds to the metal. Some indirect methods involve the use of NO group from some other species, usually accompanied by oxidation and reduction processes. For instance, the brown ring test in which the nitric oxide ligand is actually supplied by the nitrate ion.

> *7. Disproportionation Reactions*

Many metal carbonyls show disproportionation reactions when exposed to some other coordinating ligands. For instance, $Fe(CO)_5$ reacts with amines to produce hexaaminoiron(II) tetracarbonylferrate(−II); or $Co_2(CO)_8$ reacts with amine to form Hexaamminecobalt(II) bis-tetracarbonylcobaltate(−I) as:

$$2Fe(CO)_5 + 6NH_3 \rightarrow [Fe(NH_3)_6]^{2+}[Fe(CO)_4]^{2-} + 6CO$$

$$Co_2(CO)_8 + 6NH_3 \rightarrow [Co(NH_3)_6][Co(CO)_4]_2$$

The driving force for the above two reactions can be understood in terms of the ease of formation of the carbonylate ions and the favorable coordination number for iron(II) involved. The disproportionation in both cases generates a positive metal center and a metal center with a negative oxidation state. The carbonyl ligand, being a soft base, prefers to bind to the softer acids, i.e., metal center with negative charge; on the other hand, the ligands with nitrogen as donor site are hard Lewis bases, and therefore, prefer to bind to the harder acid i.e. metal center with more positive charge on it.

A variety of carbonylate complexes can be synthesized via these disproportionation reactions which makes this type quite important as far as the practical applications are concerned. For example, the $[Ni_2(CO)_6]^{2-}$ anion and $[Co(CNR)_5][Co(CO)_4]$ can be synthesized by the following reactions.:

$$3Ni(CO)_4 + 3phen \rightarrow [Ni(phen)_3][Ni_2(CO)_6] + 6CO$$

$$Co_2(CO)_8 + 5RNC \rightarrow [Co(CNR)_5][Co(CO)_4] + 4CO$$

It is also worthy to note that the range of coordinating agents that will cause disproportionation is rather wide.

❖ Preparation, Bonding, Structure and Important Reactions of Transition Metal Nitrosyl, Dinitrogen and Dioxygen Complexes

Besides carbon monoxide, there are many other important ligands which form π-complexes with transition metal center; some of them are NO, N_2 and O_2 molecules. The methods of preparations, nature of bonding, structure and their important reactions are discussed in detail below.

➢ *1. Metal Nitrosyl Complexes*

Metal nitrosyls are the complexes that contain nitric oxide (NO) bonded to a metal center (usually transition element). There are various kinds of nitrosyl complexes known so far, which vary in respect of coligand, structure and nature of bonding. Metal complexes having nitrosyl ligands only are labeled as isoleptic nitrosyls. They are very rare, one of the special members of this class is $Cr(NO)_4$. On one hand, polycarbonyl complexes are very much common, even trinitrosyl complexes are rare on the other.

Preparation: i) Metal nitrosyl complexes can be prepared via many routes, but direct formation from nitric oxide gas is much more common. For example:

$$Co_2(CO)_8 + 2NO \rightarrow 2[Co(CO)_3NO] + 2CO$$

$$Ni(Ph_3P)_2(CO)_2 + 2NO \rightarrow Ni(Ph_3P)_2(NO)_2 + 2CO$$

$$Cr(CO)_6 + 4NO \rightarrow Cr(NO)_4 + 6CO$$

$$[(\eta^5\text{-}C_5H_5)Fe(CO)_2]_2 + 2NO \rightarrow [(\eta^5\text{-}C_5H_5)Fe(CO)(NO)]_2 + 2CO$$

ii) From nitrosonium salts:

$$[Ir(CO)(PPh_3)_2Cl] + (NO)BF_4 \rightarrow [Ir(CO)(NO)(PPh_3)_2Cl]BF_4$$

iii) From nitrosyl halides:

$$Et_4N[HB(3,5\text{-}Me_2\text{-}pz)_3Mo(CO)_3] + ClNO$$
$$\rightarrow [HB(3,5\text{-}Me_2\text{-}pz)_3Mo(CO)_2(NO)] + CO + [Et_4N]Cl$$

iv) From N-nitrosamides:

$$HMn(CO)_5 + p\text{-}tolylSO_2N(Me)(NO) \rightarrow Mn(CO)_4(NO) + CO + p\text{-}tolylSO_2NHMe$$

v) From nitrite salts:

$$Na[Co(CO)_4] + Na(NO_2) + 2HOAc \rightarrow [Co(CO)_3(NO)] + CO + 2NaOAc + H_2O$$

vi) From nitronium (NO_2^+) salts:

$$(\eta^5\text{-}C_5H_5)Re(CO)_3 + [NO_2]PF_6 \rightarrow [(\eta^5\text{-}C_5H_5)Re(CO)_2(NO)]PF_6 + CO_2$$

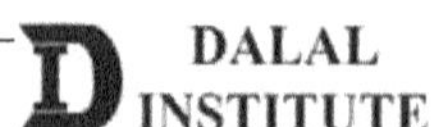

Bonding: Most of the metal nitrosyl complexes can be viewed as derivatives of the nitrosyl cation (NO^+) or anion (NO^-). The nitrosyl cation (with a bond order of 3) is isoelectronic with carbon monoxide, thus the bonding between a nitrosyl ligand and a metal follows the same principles as the bonding in carbonyl complexes. However, the bond order of neutral nitrosyl and anion are 2.5 and 2, respectively; and therefore, in order to rationalize the nature of the bonding between metal center and the nitrosyl ligand, we must understand molecular orbital diagram for the nitrosyl ligand first.

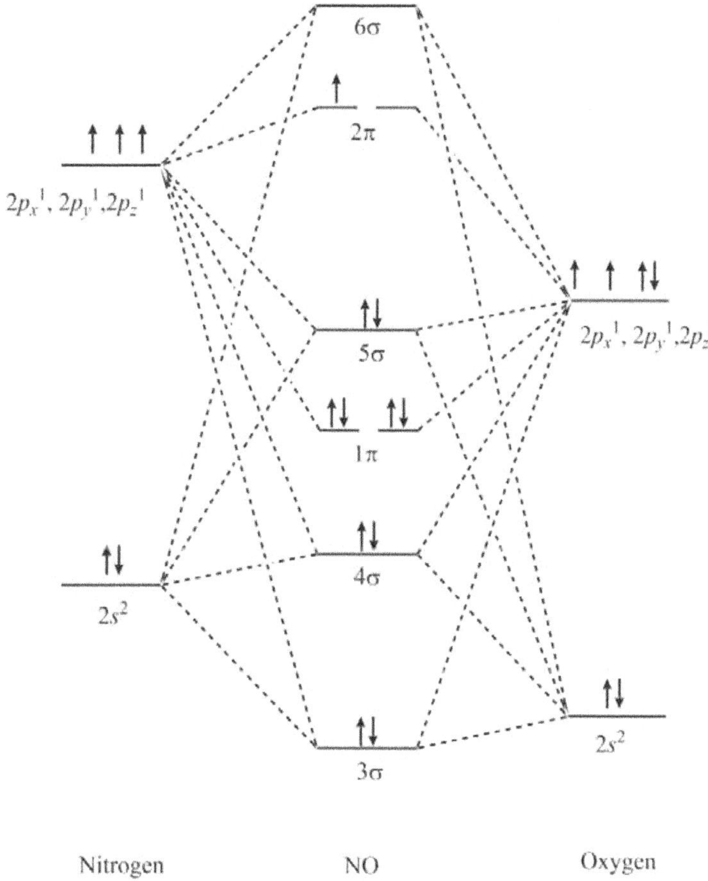

Figure 15. The molecular orbital diagram of nitric oxide (NO).

After looking at the molecular orbital diagram for NO, one can immediately recognize the difference from the carbonyl ligand that there is one extra electron in the π^*-orbital. This suggests that NO ligand can be one, two or three electron donors, depending upon the type of orbital used in bonding. If the electron present in π^*-orbital resides on NO, i.e., not transferred to the metal center, nitrosyl ligand would behave as a two-electron donor, and the nature of metal complex should be paramagnetic. There are some nitrosyl complexes of iron and cobalt such as $[Fe(NO)_2(CO)_2]$ and $[Co(NO)(CO)_3]$, which were thought to be derived from neutral NO ligand but these compounds are diamagnetic in nature; and therefore do not contain unpaired electrons. This suggests that the nitrosyl ligand is not neutral in these complexes. This assumption is also supported by the fact that the displacement of a previously attached ligand by any other neutral ligand in metal carbonyl nitrosyl complexes is always accompanied by the release of CO group. Moreover, there are also many complexes like $[Cr(NO)(CN)_5]^{3-}$ and $[Mn(NO)(CN)_5]^{2-}$, which are actually paramagnetic and were also thought to be having neutral nitrosyl ligand. However, in the later years, it was found that the unpaired electron of the nitrosyl group is actually transferred to the metal center making NO as NO^+ ligand. Thus, we can say that the coordination of neutral nitrosyl is highly unlikely.

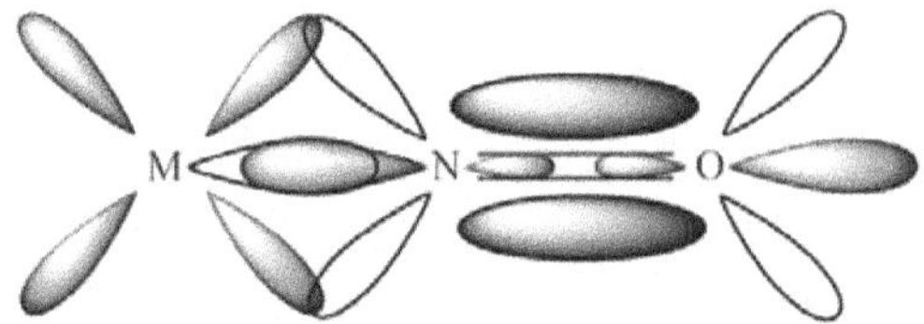

Figure 16. The nature of σ and π overlap in metal nitrosyl complexes.

i) Complexes containing NO^+: Nitric oxide molecule can easily release the odd electron from its antibonding molecular orbital to form a stable nitrosonium ion (NO^+). This also very obvious from the comparison of infrared absorption wavenumbers of free NO (1880 cm^{-1}) with the nitrosonium salts (2200 – 2300 cm^{-1}); and can be explained in terms of increased bond order from 2.5 to 3, and consequently the force constant value. Actually, most of the metal nitrosyls exist with nitrosyl ligand as three electron donor. For instance, the effective atomic numbers (EAN) for $[Mn(CO)(NO)_3]$ and $[Fe(NO)_2(CO)_2]$ complexes are 36 for each, which is possible only if the NO ligand act as NO^+ i.e. three electron donor.

$$M + NO \longrightarrow M^- + :N\!\!\equiv\!\!\overset{+}{O}: \longrightarrow \overset{2-}{M} \leftarrow \overset{+}{N}\!\!\equiv\!\!\overset{+}{O} \underset{\text{bonding}}{\overset{\pi\text{-back}}{\longleftrightarrow}} \overset{-}{M}\rightleftharpoons \overset{+}{N}\!\!\equiv\!\!O$$

In the initial step, the π^* electron of the NO transfers to the metal center, reducing M to M^- and itself forming NO^+ ion. Then NO^+ donates a lone pair of electron via N just like the carbon in metal carbonyls. However, the total number of electrons donated by NO, in this case, would be three while carbonyl can donate only two. The back donation of electron charge from filled d-orbital of metal to π^*-orbital of NO would result in a considerable decrease in the nitrogen-oxygen bond order. The infrared absorption peak of NO^+ in metal nitrosonium complexes lies in the range of 1900−1600 cm^{-1} which is far less than what has been observed in nitrosonium ionic salts. Moreover, the magnitude of decrease in carbonyl stretching frequency is less than the magnitude of decrease in nitrosonium stretching frequency as we go from their corresponding free unit to metal-coordinated unit. This, therefore, confirms the better π-acceptor strength of NO^+ ligand; which is further increased by an accumulation of negative charge on nitrosonium complexes.

ii) Complexes containing NO⁻: Nitric oxide molecule can also accept an electron from the metal center to its antibonding molecular orbital forming a NO^- ion. The metal center in this process would get oxidized from M^{n+} to $M^{(n+1)+}$ ion. Then the NO^- ion donates a lone pair of electron via N just like the carbon in metal carbonyls. However, the total number of electrons donated by NO, in this case, would be one while the carbonyl can donate only two. The infrared absorption wavenumbers of NO^- in metal nitrosyl complexes are found in the range of $1100-1200$ cm^{-1} (much lower than NO), which can be explained in terms of complete transfer of one electron from d-π orbital of metal center to antibonding molecular orbital on nitrosyl ligand. For example, during the formation of $[Co(CN)_5(NO)]^{3-}$ and $[Co(NH_3)_5(NO)]^{2+}$ complexes (passing NO through amine and cyanide salts of Co^{2+}), the Co(II) gets converted into low spin Co(III) with t_{2g}^6 configuration. Both of these complexes, in respect of charge and magnetic moment, resemble $[Co(CN)_5Br]^{3-}$ and $[Co(NH_3)_5Cl]^{2+}$, respectively.

In nitrosyl complexes, the M−N−O unit is generally linear, or no more than 15° from linear. However, in some complexes, especially where back-bonding is not that much important, the M−N−O angle can largely deviate from 180°. The linear and bent NO ligands can be differentiated using infrared spectroscopy. The linear M−N−O groups absorb in the range 1650–1900 cm^{-1} (close to metal coordinated NO^+); whereas the bent nitrosyls absorb in the range 1525–1690 cm^{-1} (close to metal coordinated NO^-). The difference of vibrational frequencies reflects the difference in N−O bond orders for linear (triple bond) and bent NO (double bond). The bent NO ligand is sometimes described as the anion, NO^-. Prototypes for such compounds are the organic nitroso compounds, such as nitrosobenzene. A complex with a bent NO ligand is trans-$[Co(en)_2(NO)Cl]^+$. The adoption of linear vs bent bonding can be analyzed with the Enemark-Feltham notation. In their framework, the factor that determines the bent vs linear NO ligands is the sum of electrons of π-symmetry.

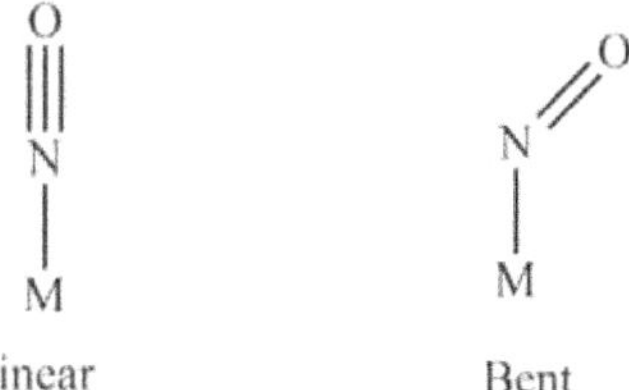

Linear   Bent

Complexes with "π-electrons" in excess of 6 tend to have bent NO ligands. Thus, $[Co(en)_2(NO)Cl]^+$, with seven electrons of π-symmetry (six in t_{2g} orbitals and one on NO), adopts a bent NO ligand, whereas $[Fe(CN)_5(NO)]^{3-}$, with six electrons of pi-symmetry, adopts a linear nitrosyl. In a further illustration, the M−N−O d-electron count of the $[Cr(CN)_5NO]^{3-}$ anion is shown. In this example, the cyanide ligands are "innocent", i.e., they have a charge of −1 each, −5 total. To balance the fragment's overall charge, the charge on Cr−N−O is thus +2 (−3 = −5 + 2). Using the neutral electron counting scheme, Cr has 6 d-electrons and NO has one odd electron for a total of 7. Two electrons are subtracted to take into account that fragment's overall charge of +2, to give 5. Written in the Enemark-Feltham notation, the d electron count in Cr−N−O unit is five. The results are the same if the nitrosyl ligand were considered NO^+ or NO^-.

Structure: The structure of metal nitrosyls can mainly be classified into three categories; first as the metal-complex systems that contain NO as a terminal ligand only, the second one as having only bridging nitrosyl group, and the third one with nitrosyl groups with the terminal as well as bridging profile.

i) Metal complexes with terminal nitrosyl: The NO group as a monodentate ligand in metal complexes acts either as a 12-electron unit (when NO^- forms a single bond with the metal center and M−N−O unit is linear), or as a 10-electron unit (when NO^+ forms a multiple bonds with the metal center and M−N−O unit is bent).

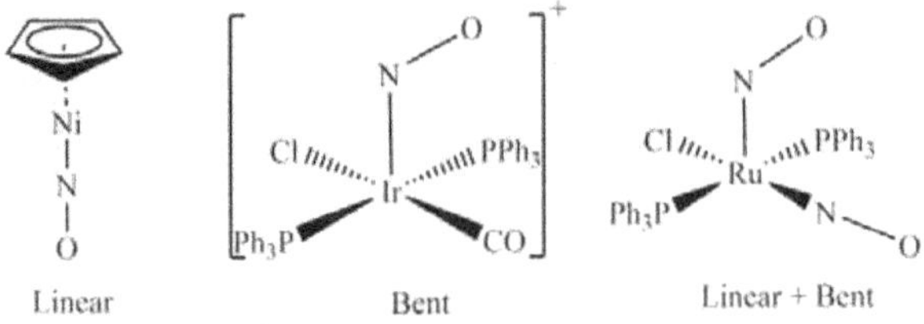

Figure 17. Metal complexes with terminal nitrosyls.

ii) Metal complexes with bridging nitrosyl: In some of the metal nitrosyl complexes, all NO groups are present in bridging mode. One of the common examples is given below.

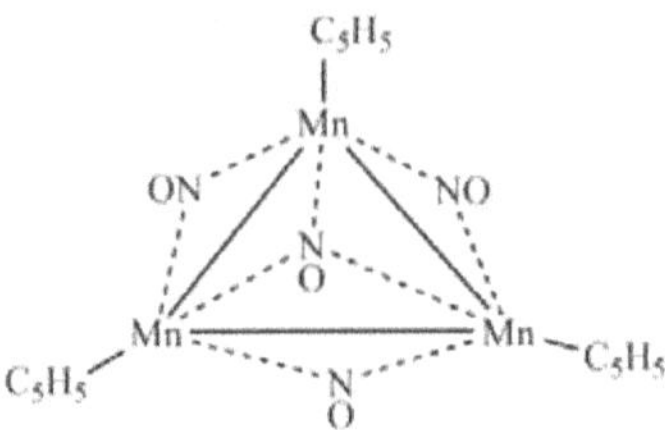

Figure 18. Metal complexes with terminal nitrosyls.

iii) Metal complexes with the terminal as well as bridging nitrosyl: In some of the metal nitrosyl complexes, NO groups are present in the terminal as well as in bridging mode. Some of the common examples are:

Figure 19. Metal complexes with terminal nitrosyls.

Reactions: i) The nucleophilic attack:

$$[Fe(CN)_5NO]^{2-} + 2OH^- \rightleftharpoons [Fe(CN)_5(NO_2)]^{4-} + H_2O$$

$$[Fe(CN)_5NO]^{2-} + 2NH_3 \rightleftharpoons [Fe(CN)_5(NH_3)]^{3-} + N_2 + H_3O^+$$

ii) Reduction of metal nitrosyls:

$$[Fe(CN)_5NO]^{2-} + e \rightleftharpoons [Fe(CN)_5NO]^{3-}$$

$$[Ru(NH_3)_5NO]^{3+} + e \rightleftharpoons [Ru(NH_3)_5NO]^{2+}$$

iii) Reactions of nitrosyls with electrophiles:

$$2[Co(en)_2NO]^{2+} + 2CH_3CN + O_2 \rightleftharpoons 2[Co(CH_3CN)(en)_2(NO_2)]^{2+}$$

$$2[Ir(PPh_3)(NO)_2]^+ + O_2 \rightleftharpoons 2[Ir(PPh_3)(NO_2)(NO)]$$

iv) Formation of carbon-nitrogen bonds:

$$[Co(CH_3)_2(PMe_3)_2(NO)] \rightarrow [Co(CH_3)(PMe_3)_2(CH_3NO)]$$

$$2[(C_5H_5)Mo(CO)_2(NO)] + 3PPh_3$$
$$\rightarrow [(C_5H_5)Mo(PPh_3)(CO)(NO)] + CO$$
$$+ [(C_5H_5)Mo(CO)(PPh_3)(NCO)] + PH_3PO$$

> ### 2. Metal Dinitrogen Complexes

Metal dinitrogen complexes are the coordination compounds that contain the dinitrogen ligand (N_2) attached to a metal center. The first complex of dinitrogen, $[Ru(NH_3)_5(N_2)]^{2+}$ was reported by Allen and Senoff in 1965, which is consisted of a 16e-$[Ru(NH_3)_5]^{2+}$ center attached to one end of N_2. The interest in such complexes arises because N_2 comprises the majority of the atmosphere and there are many useful compounds containing nitrogen atoms.

Preparation: i) Metal dinitrogen complexes can be prepared via many routes, but direct formation from dinitrogen is very common. For example:

$$[CoH_2(PPh_3)_3] + N_2 \rightarrow [Co(PPh_3)_3(N_2)] + H_2$$

$$[RuH_4(PPh_3)_3] + N_2 \rightarrow [RuH_2(PPh_3)_3(N_2)] + H_2$$

$$[CoH_3(PPh_3)_3] + N_2 \rightarrow [CoH(PPh_3)_3(N_2)] + H_2$$

$$[FeH_4(PEtPh_2)_3] + N_2 \rightarrow [FeH_2(PEtPh_2)_3(N_2)] + H_2$$

ii) From compounds containing chains of nitrogen atoms:

$$[Ru(NH_3)_5L]^{n+} + N_3^- \rightarrow [Ru(NH_3)_5(N_2)]^{2+} + L$$

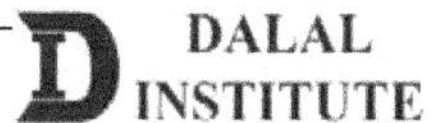

$$trans\text{-}[Ir(Cl)(CO)(PR_3)_2] + RCON_3 + EtOH$$
$$\rightarrow trans\text{-}[Ir(Cl)(N_2)(PR_3)_2] + RCONHCO_2Et$$

$$trans\text{-}[Rh(Cl)(CO)(PR_3)_2] + RCON_3 + EtOH$$
$$\rightarrow trans\text{-}[Rh(Cl)(N_2)(PR_3)_2] + RCONHCO_2Et$$

$$[OsH_4(PR_3)_3] + CH_3C_6H_4SO_2N_3 \rightarrow [OsH_2(N_2)(PR_3)_3]$$

iii) Preparations in which two nitrogen atoms are combined to give a dinitrogen group:

$$[Fe(CN)_5NO]^{2-} + N_2H_4 \rightarrow [Fe(CN)_5(N_2)]^{2-}$$

$$RuCl_3 + Zn + NH_3 \rightarrow [Ru(NH_3)_6]Cl_2 + [Ru(NH_3)(N_2)]Cl_2$$

$$[Os(NH_3)_5(CO)]^{2+} + HNO_2 \rightarrow cis\text{-}[Os(NH_3)_4(N_2)(CO)]^{2+} + 2H_2O$$

$$[Ru(NH_3)_6]^{3+} + NO + OH^- \rightarrow [Ru(NH_3)_5(N_2)]^{2+} + 2H_2O$$

Bonding: In order to rationalize the nature of the bonding between the metal center and the N_2, we must understand the bonding within the dinitrogen ligand first. The molecular orbital diagram for N_2 is given below.

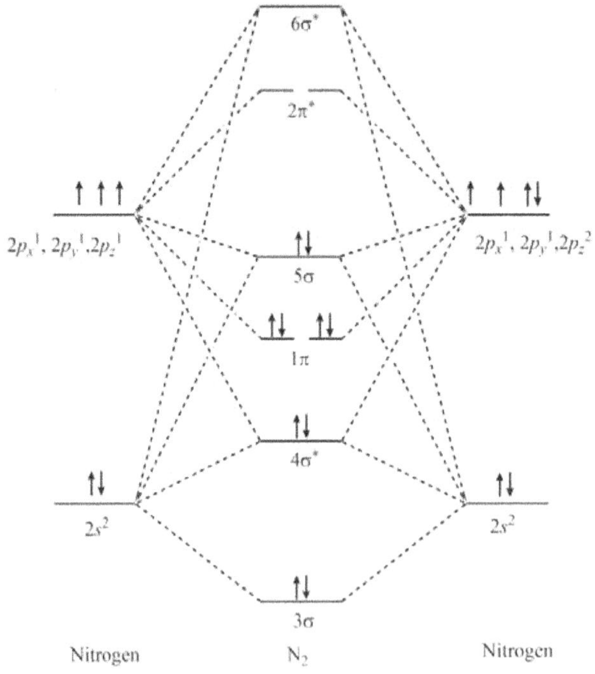

Figure 20. The molecular orbital diagram of dinitrogen molecule.

Though the N_2 molecule is isoelectronic with CO and NO^+, it does not form a large number of metal complexes like the two. This is obviously due to the fact that it is a poor ligand and cannot act as a strong π-acceptor due to the lack of polarity. In other words, N_2 ligand is neither a good σ-donor nor good π-acceptor as there is no polarity in N−O bond. The different bridging modes of dinitrogen ligand are given below.

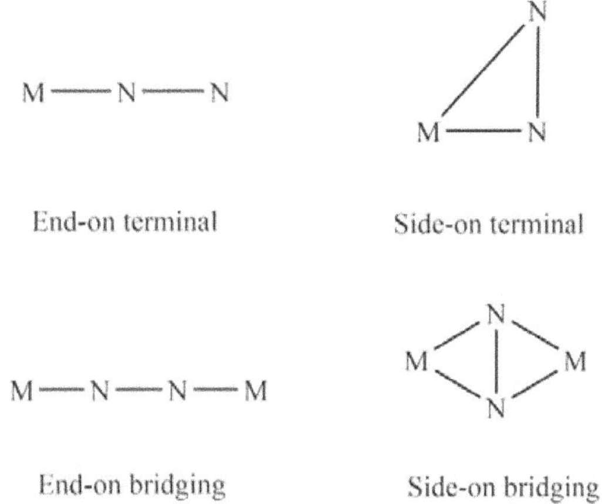

Figure 21. Bonding modes of N_2 ligand in metal-dinitrogen complexes.

The most common binding mode of dinitrogen ligand with transition metal center is end-on i.e. just like in the case of isocyanides, carbon monoxide and nitric oxide. These observations are also supported by the theoretical treatment of some dinitrogen complexes which indicated that end-on bonding is more beneficial than side on as far as the stability of the complex is concerned. The end-on bonding involves the donation of the lone pair of N_2 to the empty metal orbital and, the back-donation from filled metal d-orbital to the empty π^* orbitals of N_2 ligand. In contrast, the side-on bonding comprises of electron donation from the π and σ bonding molecular orbital of the dinitrogen to the empty orbitals of the metal and the back-donation of electron density from filled orbitals of the metal to the π^* molecular orbital of the N_2 ligand. Though the side-on bonding mode is quite common in metal-acetylenes and metal-olefin complexes, there are very few reports of side-on bonded dinitrogen complexes. Consider the example of $[Ru(NH_3)_5(N_2)]^{2+}$ complex, Ru−N bond length in Ru−N−N unit is shorter than Ru−N bond length in Ru−NH_3 unit. This shows that there is some extent of backbonding from filled d-orbital the metal center to the empty π^* molecular orbital of dinitrogen. This is also very obvious from the vibrational Raman stretching frequency of free N_2 (2331 cm^{-1}) and infrared active stretching frequency of metal coordinated ligand (2105 cm^{-1}) in $[Ru(NH_3)_5(N_2)]Cl_2$ complex. However, it is also worthy to note that metal-carbon bond in carbonyl complexes is stronger than the metal-nitrogen bond in dinitrogen complexes, which shows that CO is definitely a stronger σ-donor a better π-acceptor as highest occupied molecular orbital (HOMO) is predominantly concentrated on carbon due to high polarity.

Structure: The structures of metal-dinitrogen complexes can mainly be classified into two categories; metal-complex that contains N_2 as an end-on ligand and with N_2 group as a side-on ligand.

i) Metal complexes with end-on dinitrogen: As a ligand, N_2 usually binds to metals as an end-on ligand, as illustrated by Allen and Senoff's complex. Such complexes are usually analogous to related CO derivatives.

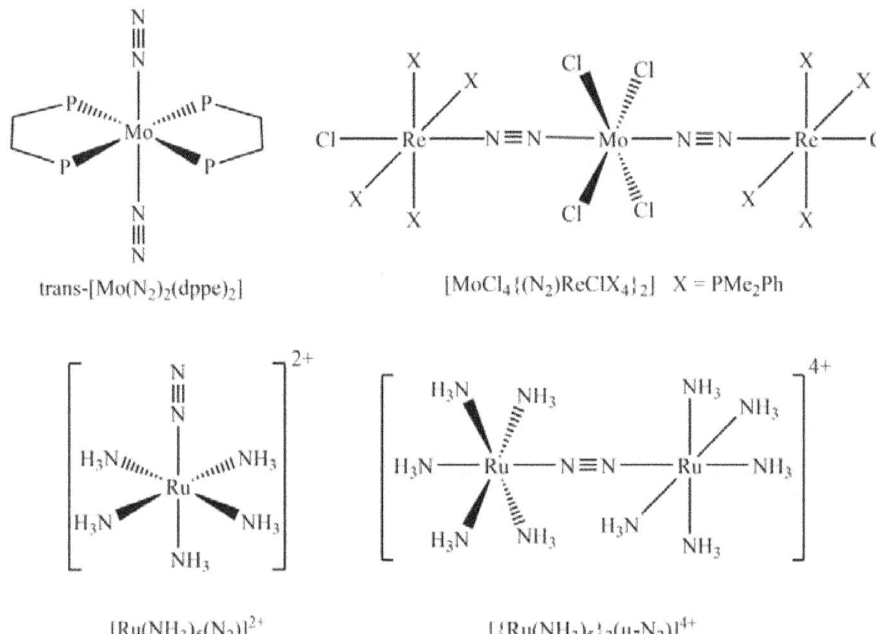

Figure 22. Metal-dinitrogen complexes with end-on N_2 ligand.

ii) Metal complexes with side-on dinitrogen: In some of the metal-dinitrogen complexes, the N−N vector is perpendicular to the M−M vector. Some of the most common examples are given below.

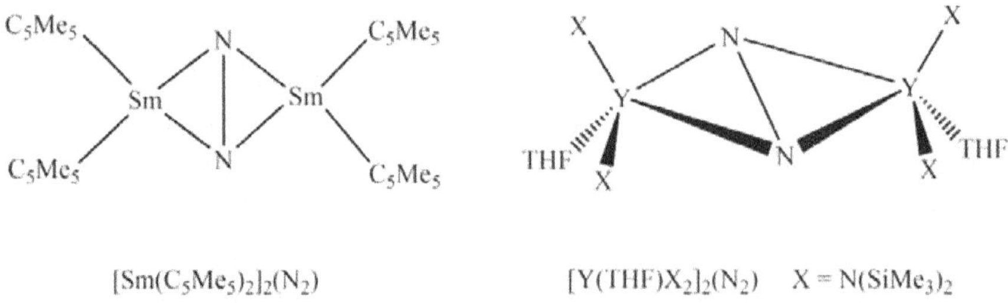

Figure 23. Metal-dinitrogen complexes with side-on N_2 ligand.

Reactions: i) The displacement of dinitrogen ligand by some other groups:

$$trans\text{-}[Mo(N_2)_2(dppe)_2] + 2C_2H_4 \;\rightarrow\; trans\text{-}[Mo(C_2H_4)_2(dppe)_2] + 2N_2$$

$$trans\text{-}[Mo(N_2)_2(dppe)_2] + 2RNC \rightarrow trans\text{-}[Mo(RNC)_2(dppe)_2] + 2N_2$$

$$trans\text{-}[Mo(N_2)_2(dppe)_2] + 2CO \;\rightarrow\; trans\text{-}[Mo(CO)_2(dppe)_2] + 2N_2$$

ii) Reactions of Ligating N_2 with Lewis Acids:

$$[ReCl(N_2)(PMe_2Ph)_4] + [Mo(Cl_4)(THF)_2]$$
$$\xrightarrow{\;CH_2Cl_2,\,MeOH\;} [(PMe_2Ph)_4ClReN_2MoCl_4(OMe)]$$

$$[ReCl(N_2)(PMe_2Ph)_4] + [MoCl_4(PPh_3)_2] \rightarrow [MoCl_4\{(N_2)ReCl(PMe_2Ph)_4\}_2]$$

iii) Formation of metal-hydrazido complxes from ligating N_2:

$$trans\text{-}[Mo(N_2)_2(dppe)_2] + 2HCl \rightarrow [MoCl(NNH_2)(dppe)_2]Cl + N_2$$

$$trans\text{-}[W(N_2)_2(dppe)_2] + 2HBF_4 \xrightarrow{\;THF\;} [WF(NNH_2)(dppe)_2][BF_4]$$
$$+ BF_3.THF + N_2$$

iv) Formation of carbon-nitrogen bonds:

$$[W(N_2)_2(dppe)_2] + RCOCl + HCl \rightarrow [WCl(N_2HCOR)(dppe)_2]Cl$$

$$[Mo(N_2)(RCN)(dppe)_2] + PhCOCl \rightarrow [MoCl(NNCOPh)(dppe)_2]$$

$$[Mo(N_2)_2(dppe)_2] + RCOCl + HCl \rightarrow [MoCl(N_2HCOR)(dppe)_2]Cl$$

> ### 3. Metal Dioxygen Complexes

    Metal dioxygen complexes are the coordination compounds which contain O_2 ligand attached to a metal center. The principal driving force behind the analysis of these compounds are oxygen-carrying proteins such as myoglobin, hemoglobin, hemocyanin, and hemerythrin. Many transition metals form complexes with O_2, and many of these complexes form reversibly. The binding of O_2 is the first step in many important phenomena, such as cellular corrosion, respiration and in industrial chemistry. The first synthetic oxygen complex was demonstrated in 1938 with Co^{2+} complex reversibly bound O_2. Most of the organometallic metal-dioxygen complexes are synthesized by the reaction of gaseous molecular oxygen with complexes (having d^7, d^8, or d^{10} electronic configuration) solution.

Preparation: i) Formation of a mononuclear dioxygen adduct with or without displacement of ligands

$$[IrCl(CO)(PPh_3)_2] + O_2 \rightleftharpoons [IrCl(CO)(PPh_3)_2O_2)]$$

$$[Co(DMG)_2] + O_2 + Pyridine \rightarrow [Co(DMG)_2(O_2)(Pyridine)]$$

ii) Formation of a dimer or a binuclear dioxygen adduct:

$$2[Rh(PPh_3)_2Cl] + 2O_2 \rightarrow [Rh(PPh_3)_2(O_2)Cl]_2$$

$$2[Co(histidine)_2] + O_2 \rightarrow [Co_2(histidine)_4(O_2)]$$

iii) Oxidation of ligands with the oxidized ligand remaining coordinated:

$$[Ru(PPh_3)_2(CO)_2(SO_2)] + O_2 \rightarrow [Ru(PPh_3)_2(CO)_2(SO_4)]$$

iv) Displacement of free oxidized ligand:

$$[Pt(PPh_3)_4] + 2O_2 \rightarrow [Pt(PPh_3)_2(O_2)] + 2Ph_3PO$$

Bonding: The nature of bonding in metal-dioxygen complexes is usually evaluated by single-crystal X-ray crystallography, focusing both on the overall geometry as well as the O–O distances, which reveals the bond order of the O_2 ligand. However, in order to rationalize the initial idea of the metal-ligand bonding, we must understand the bonding within the dioxygen ligand first. The molecular orbital diagram for O_2 is given below.

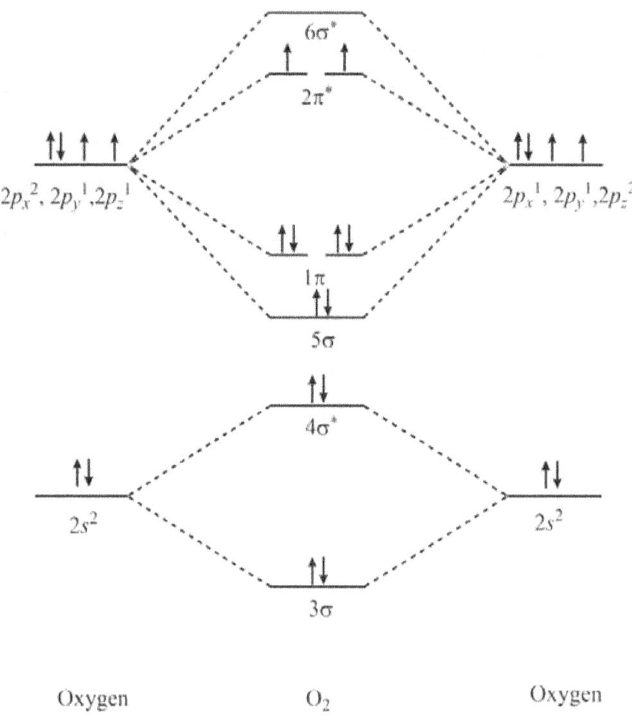

Figure 24. The molecular orbital diagram of the dioxygen molecule.

The experimental results show that the bent structure of the dioxygen unit in oxyhemoglobin is actually diamagnetic in nature; which suggests that π^*2p_x and π^*2p_y orbitals are not completely degenerate in the complex as they are in free dioxygen. A linear M−O−O structure would result in a degenerate set of π^*2p_x and π^*2p_y orbitals with a triplet state. In addition to the σ-bonding, there are two types of interactions between the d-orbitals of the metal and the π^* orbitals of the dioxygen. One is through an overlap of the $3d_{xz}$ of the metal with the π^* orbitals perpendicular to the Fe−O−O plane; while the other is through the overlap of the $3d$ orbital of the metal with the π_y^* orbital in the Fe−O−O plane. The different bridging modes of dinitrogen ligand are given below.

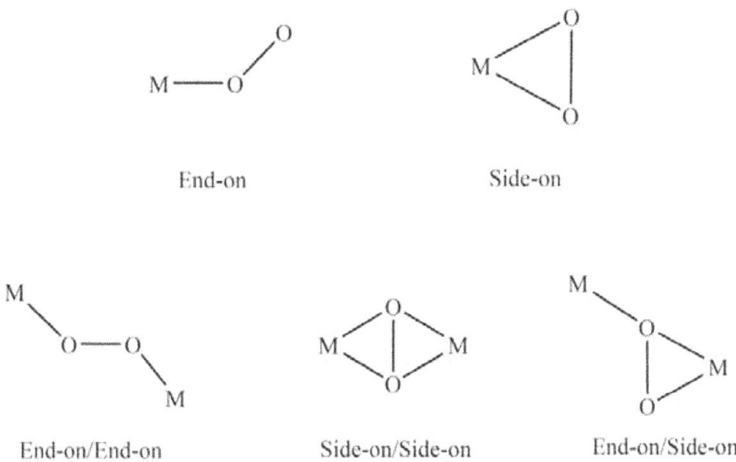

Figure 25. Bonding modes of O_2 ligand in metal-dioxygen complexes.

The O_2 ligand binds to a single metal center either end-on (η^1-) or in a side-on (η^2-) manner. Dioxygen adducts derived from Co^{2+} and Fe^{2+} complexes of porphyrin (and related anionic macrocyclic ligands) exhibit this bonding mode. The hemoglobin and myoglobin are two famous examples, and many other synthetic analogs have been reported which behave in a similar manner. Binding of O_2 is usually described as proceeding by electron transfer from the metal(II) center to give superoxide (O^{-2}) complexes of metal(III) centers. The η^2-bonding is the most common motif seen in the coordination chemistry of dioxygen. Since O_2 has a triplet ground state and Vaska's complex is a singlet, the reaction is slower than when singlet oxygen is used. Complexes containing η^2-O_2 ligands are fairly common, but most are generated using hydrogen peroxide, not O_2. The O_2 can bind to one metal of a bimetallic unit via the same modes discussed above for mononuclear complexes. A well-known example in nature is hemerythrin, which features a diiron carboxylate that binds O_2 at one Fe center. The dinuclear complexes can also cooperate in the binding, although the initial attack of O_2 probably occurs at a single metal. These binding modes include μ^2-η^2, η^2-, μ^2-η^1, η^1-, and μ^2-η^1, η^2-. Depending on the degree of electron-transfer from the dimetal unit, these O_2 ligands can again be described as peroxo or superoxo. In nature, such dinuclear dioxygen complexes often feature copper.

Structure: The structures of metal-dioxygen complexes can mainly be classified into two categories; the first one as the metal-complex that contains dioxygen (O_2) ligand attached to one metal center and the second one with O_2 group connected to two or more metal centers. A brief discussion on both of the categories is given below with suitable examples.

i) Mononuclear metal-dioxygen complexes: As a ligand, O_2 can bind to a single metal-center either as an end-on ligand or as a side-on ligand. The structures of some representative metal-dioxygen complexes of such type are given below.

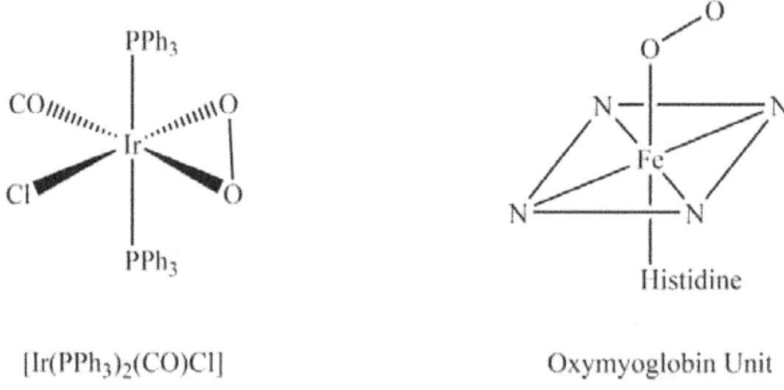

Figure 26. Mononuclear metal-O_2 complexes.

ii) Binuclear metal-dioxygen complexes: In binuclear metal-dioxygen complexes, the O_2 molecule can bind either to one or both metals center as an end-on/end-on, end-on/side-on and side-on/side-on ligand. The structures of some representative metal-dioxygen complexes of such type are given below.

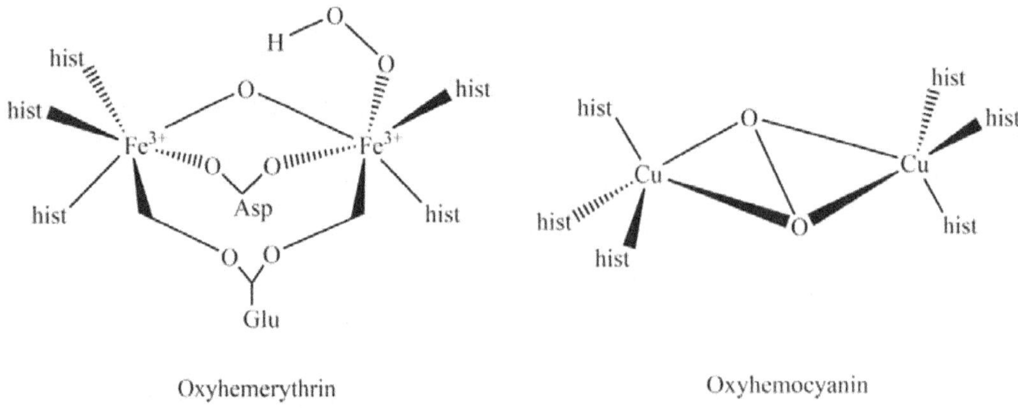

Figure 27. Metal-dioxygen complexes with side-on O_2 ligand.

Reactions: i) The displacement of molecular oxygen by some other ligand:

$$[Ni(O_2)(t\text{-}BuNC)_2] + 4t\text{-}BuNC \rightleftharpoons [Ni(t\text{-}BuNC)_4] + 2t\text{-}BuNCO$$

$$[Ni(O_2)(t\text{-}BuNC)_2] + 4PPh_3 \rightleftharpoons [Ni(t\text{-}BuNC)_2(PPh_3)_2] + 2Ph_3PO$$

$$[Pt(PPh_3)_2(O_2)] + 3PPh_3 \rightleftharpoons [Pt(PPh_3)_3] + 2Ph_3PO$$

ii) The formation of complexes of oxidized substrates:

$$[Ir(PPh_3)_2(CO)Cl(O_2)] + SO_2 \rightarrow [Ir(PPh_3)_2(CO)Cl(SO_4)]$$

$$[Pt(PPh_3)_2(O_2)] + SO_2 \rightarrow [Pt(PPh_3)_2(SO_2)]$$

$$[Pd(PPh_3)_2(O_2)] + NO_2 \rightarrow [Pd(PPh_3)_2(NO_3)_2]$$

❖ Tertiary Phosphine as Ligand

The compound phosphine (PH_3) is extremely important in coordination chemistry due to its large number of derivatives which can be used as L-type ligands (2 electron donor neutral ligands) for many metal complexes. These phosphine based ligands are the compounds with the formula PR_nH_{3-n}, and are often classified according to the value of n; the values 1, 2, 3 correspond to primary, secondary and tertiary phosphines, respectively. For instance, the most popular phosphine ligand used is organometallic chemistry is triphenylphosphine, which is obviously a tertiary phosphine. All of these phosphine ligands adopt pyramidal structures that are quite favorable for the metal center to be approached. Unlike most metal ammine complexes, metal phosphine complexes tend to be lipophilic, showing very good solubility in organic solvents. They also are compatible with metals in multiple oxidation states. Therefore, owing to these two features, metal phosphine complexes are quite useful in homogeneous catalysis.

➢ *General Methods of Preparation and Reactivity*

The first phosphine complexes were *cis*- and *trans*-[PtCl$_2$(PEt$_3$)$_2$], reported by Cahours and Gal in 1870. Being a L-type ligand, the phosphines do not change the overall charge of the metal complex. These complexes may simply be prepared by the addition of phosphines to a coordinatively unsaturated metal precursor, or by ligand displacement of another L-type complex, such as a solvent molecule acting as a ligand. However, it should also be noted that the replacement of X-type ligand (1-electron donor neutral ligand) can also yield such complexes.

$$[PdCl_2]_n + 2nPPh_3 \rightarrow n[PdCl_2(PPh_3)_2]$$

$$[Cr(CO)_6] + PPh_3 \rightarrow [Cr(CO)_5(PPh_3)] + CO$$

$$K_4[Ni(CN)_4] + 4PPh_3 \rightarrow [Ni(PPh_3)_4] + 4KCN$$

As far as the reactivity is concerned, the nature of phosphine ligands is a spectator rather than the actor. Coordinated phosphines generally do not participate in chemical reactions, except their dissociation from the metal center. However, in some high-temperature hydroformylation processes, the scission of phosphorus–carbon bonds is also observed. The thermal stability of phosphine ligands is enhanced when they are incorporated into pincer complexes. Some complexes dissociate in solutions to yield a product with a lower coordination number.

$$[Rh(PPh_3)_3Cl] \rightarrow [Rh(PPh_3)_2Cl] + PPh_3$$

$$[Pt(PPh_3)_3] + MeI \rightarrow [Pt(I)(Me)(PPh_3)_2] + PPh_3$$

> ### Structure and Bonding in of Tertiary Phosphine Complexes

The metal-phosphine complexes are not considered as the organometallic compounds because of the lack of metal-carbon bonds. However, many good inorganic textbooks discuss them in their organometallic section due to their excellent bonding similarities with the carbonyl ligand. For instance, phosphines can act as very good σ-donors as well as respectable π acceptors just like CO ligand. The σ-donation occurs via a hybrid lone pair on phosphorus and the π-acceptance in orbitals that are a mixture of empty *d*-orbital on phosphorus and σ* of the phosphorus-carbon bond. For a long time in coordination chemistry, the pure empty *d*-orbitals of phosphorus were thought to be used for the acceptance of π-electron density; and this theory was also affirmed by the fact that as the R groups attached to phosphorus become more electronegative, phosphine becomes stronger π-acceptor and vice-versa. The concept of mixing of σ* of the phosphorus-carbon bond into empty *d*-orbitals of phosphorus was developed later on.

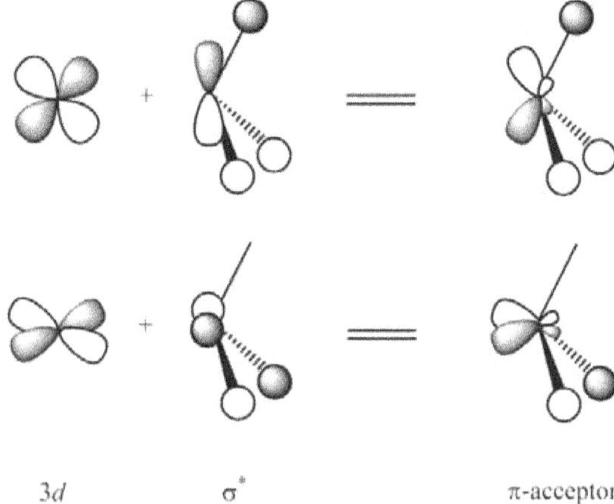

Figure 28. π-acceptor orbitals of phosphines in metal-phosphine complexes.

Therefore, the nature of R groups in tertiary phosphines governs the relative donor-acceptor strength of the corresponding ligand. For instance, PMe_3 acts as a stronger σ donor than PF_3 due to the large electron-donating effect of the three methyl groups. However, it is a relatively weaker π-acceptor in comparison to PF_3, which easily be explained in terms of a larger electron-withdrawing effect of 3 fluorines. Hence, the σ-donor and π-acceptor strength of tertiary phosphines can be fine-tuned just by changing the R groups.

> ### Steric and Electronic Properties

The most valuable thing about tertiary phosphines as ligands is that their steric and electronic properties can easily be manipulated just by changes in one or more of the three organic substituents. This fine-tuning of the electronic and steric profile of phosphine ligands is of great importance in the manipulation of catalytic properties transition metals. The ligand cone angle and Tolman electronic parameter are used to categorize various phosphines on the basis of their steric and electronic behavior, respectively. However, these parameters are not just limited to phosphines but have been extended to other ligand types also.

1. Ligand cone angle: The ligand cone angle is a measure of the size of a ligand. It is defined as the solid angle formed with the metal at the vertex and the other atoms at the perimeter of the cone. Tertiary phosphine ligands are commonly classified using this parameter, but the method can be applied to any ligand. The term cone angle was introduced by the American chemist, Chadwick A. Tolman. Originally applied to phosphines (called Tolman cone angle), the cone angles were originally determined by taking measurements from accurate physical models of them. The concept of the cone angle is most easily visualized with symmetrical ligands, e.g. PR_3. Nevertheless, this approach has been refined to include less symmetrical ligands of the type PRR'R″ as well as diphosphines by using equation (1).

$$\theta = \frac{2}{3} \sum_i \frac{\theta_l}{2} \tag{1}$$

Where $\theta_i/2$ are the half of individual angles made by R, R′ and R″ on M–P bond. In the case of diphosphines, the $\theta_i/2$ of the backbone is estimated as half of the bite angle of chelate, taking a bite angle of 90°, 85° and 74°, for propylene, ethylene and diphosphines with methylene backbones, respectively.

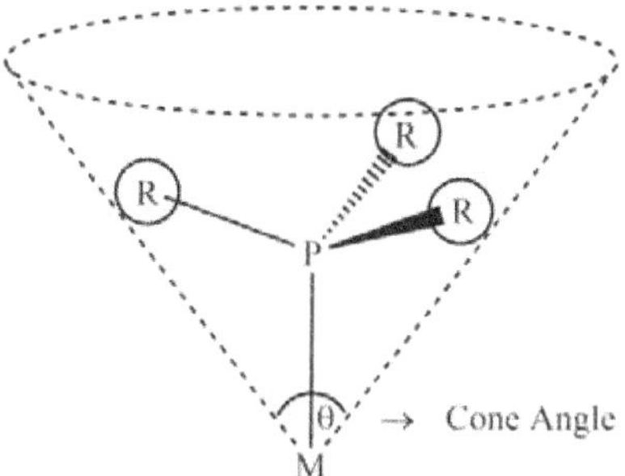

Figure 29. The ligand cone angle in metal-phosphine complexes.

The ligand-cone-angle method accepts empirical bond data and considers the perimeter as the maximum possible circumscription of a substituent which spins freely. The concept of solid-angle derives the bond length and the perimeter from the empirical crystal structure profile obtained from X-ray diffraction of solid sample. Each system has its own advantages. The exact cone angle (θ) can be found if the geometry of a ligand is known from either crystallography or computational studies. Unlike the Tolman route, no assumptions regarding the geometry are made. The cone-angle-concept is practically important in homogeneous catalysis due to the affected reactivity of the metal center from the ligand size. In a famous example, the selectivity of hydroformylation catalysts is strongly affected by the size of the coligands attached. Now though phosphines are monovalent, some are large enough to fill more than half of the coordination sphere around the metal center. The values of the cone angles of selected ligands are given below.

Table 3. The cone angle values of selected ligands.

Ligand	Cone angle	Ligand	Cone angle
PH_3	87°	$P(Me)(PPh)_2$	136°
PF_3	104°	$P(CF_3)_3$	137°
$P(OMe)_3$	107°	$P(O\text{-}o\text{-}C_6H_4CH_3)_3$	141°
$P(OEt)_3$	109°	PPh_3	145°
PMe_3	118°	$P(cyclo\text{-}C_6H_{11})_3$	170°
PCl_3	124°	$P(t\text{-}C_4H_9)_3$	182°
$P(Me)_3Ph$	127°	$P(C_6F_5)_3$	184°
PBr_3	131°	$P(o\text{-}C_6H_4CH_3)_3$	194°
PEt_3	132°	$P(mesityl)_3$	212°

The presence of bulky ligands (ligands with large cone angles) is found to enhance the rate of dissociation; which can be explained in terms of increased steric repulsion due to crowding around the metal center.

2. Tolman electronic parameter: The Tolman electronic parameter (TEP) is named after its inventor, Chadwick A. Tolman, and is a measure of the electron-donating or withdrawing strength of a ligand. It is determined by measuring the frequency of the A_1 vibrational mode of carbonyl stretching frequency of $Ni(CO)_3L$ complex by infrared spectroscopy, where L is the ligand being studied. The $Ni(CO)_3L$ is chosen as the model compound because such complexes can easily be prepared from $Ni(CO)_4$. Moreover, the carbonyl band is quite distinctive, and is rarely obscured by other bands in the observed infrared spectrum; and being a small ligand, steric factors are unable to complicate the analysis.

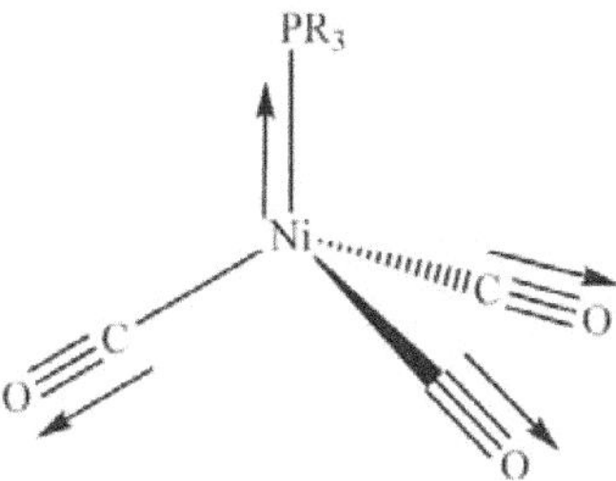

Figure 30. The A₁ stretch of Ni(CO)₃L used to determine the TEP in metal-phosphine complexes.

Upon coordination to a metal, $\nu(CO)$ typically decreases from 2143 cm^{-1} of free CO. This can be explained by π backbonding i.e. the metal is able to form a π-bond with the carbonyl ligand by donating electrons through its d-orbitals into the empty antibonding π^* orbitals on CO. This increases the strength of metal–carbon bond, but also weakens the bond between carbon and oxygen. Moreover, the C−O bond is further weakened if other ligands present increase the π-electrons-density on the metal center, and ν_{CO} decreases. However, if other ligands compete with CO for π backbonding, ν_{CO} would increase. Tolman analyzed a series of seventy ligands in reference to P(t-Bu)₃, because three t-butyl groups make it as the strongest donor ligand. Consequently, the A₁ absorption band of carbonyl ligands in Ni(CO)₃[P(t-Bu)₃] complex appears at the lowest wavenumber (2056.1 cm^{-1}), which is obviously due to the largest magnitude of back-bonding among all phosphines. Based on these observations, Tolman proposed a parameter χ_i to show the effect of individual substituents (R, R′, R″) on this band as accorded by equation (2):

$$\nu_{CO} = 2056.1 \text{ cm}^{-1} + \sum_{i=1}^{3} \chi_i \text{ cm}^{-1} \tag{2}$$

Now the calculation of the value of χ_i for each individual substituent in symmetrical phosphines can be done just by putting their observed spectral band. Let's say, we want to calculate the Tolman electronic parameter for PCl₃. The infrared absorption of A₁ symmetry for Ni(CO)₃(PF₃) is observed at 2097.0 cm^{-1}. On putting this value for ν_{CO} in the equation (2), we get:

$$2097.0 \text{ cm}^{-1} = 2056.1 \text{ cm}^{-1} + \sum_{i=1}^{3} \chi_i \text{ cm}^{-1} \tag{3}$$

$$2097.0 \text{ cm}^{-1} = 2056.1 \text{ cm}^{-1} + 3 \times \chi_i(\text{Cl}) \text{ cm}^{-1} \tag{4}$$

$$\chi_i = 14.8 \text{ cm}^{-1} \tag{5}$$

The values of χ for different substituents are listed in 'Table 4'; which in turn can easily be used to calculate the theoretical position of A₁ absorption band in case of tertiary phosphine ligands in which R, R′, and R″ are not the same anymore.

Table 4. Substituent χ Factors for Phosphine and Related Ligands.

Substituent	χ_i (cm^{-1})	Substituent	χ_i (cm^{-1})
$-t$Bu	0.0	$-$OEt	6.8
$-$Cyclohexyl	0.1	$-$OMe	7.7
$-i$Pr	1.0	$-$H	8.3
$-$Et	1.8	$-$OPh	9.7
$-$Me	2.6	$-$C$_6$H$_5$	11.2
$-$Ph	4.3	$-$Cl	14.8
$-p$-C$_6$H$_4$F	5.0	$-$F	18.2
$-m$-C$_6$H$_4$F	6.0	$-$CF$_3$	19.6

The Tolman electronic parameter (TEP) and the ligand cone angle are used to characterize the electronic and steric profile of phosphines, which are very popular ligands for catalysts.

❖ Problems

Q 1. Give the principal routes to synthesize metal carbonyl complexes.

Q 2. Discuss the nature of bonding for transition metal complexes in the framework of the molecular orbital theory.

Q 3. Draw and discuss the structure of $Fe_3(CO)_{12}$ and $Ir_4(CO)_{16}$.

Q 4. Write a short note on the structure of osmium-carbonyl complexes with special emphasis on $Os_4(CO)_{14}$, $Os_4(CO)_{15}$ and $Os_4(CO)_{16}$.

Q 5. How many metal-metal bonds and terminal carbonyls are present in $Rh_6(CO)_{16}$?

Q 6. How would you distinguish between square-pyramidal and trigonal bipyramidal geometries of $Fe(CO)_5$ using their vibration spectra?

Q 7. Discuss the effect of backbonding on the carbonyl stretching frequency in hexacarbonyl complexes of the first transition series.

Q 8. How would you calculate the reaction rate for metal carbonyls using infrared spectroscopy?

Q 9. Discuss the points of difference between linear and bent mode of bonding for nitrosyl ligand.

Q 10. Explain the structure and bonding of metal-dioxygen complexes in detail.

Q 11. How does the end-on bonding mode differ from side-on bonding mode in dinitrogen complexes of transition metals?

Q 12. Give important reactions of coordinated ligands with reference to N_2, O_2 and NO complexes.

Q 13. Why is tertiary phosphine so important in organometallic chemistry?

Q 14. What kind of orbitals do the tertiary phosphine use for backbonding? Explain in detail.

Q 15. Define the ligand cone angle and the Tolman electronic parameter.

❖ Bibliography

[1] J. E. House, *Inorganic Chemistry*, Academic Press, California, USA, 2008.

[2] B. R. Puri, L. R. Sharma, K. C. Kalia, *Principals of Inorganic Chemistry*, Milestone Publishers, Delhi, India, 2012.

[3] J. E. Huheey, E. A. Keiter, R. L. Keiter, *Inorganic Chemistry: Principals of Structure and Reactivity*, HarperCollins College Publishers, New York, USA, 1993.

[4] B. Xu, Q. S. Li, Y. Xie, R. B. King, H. F. Schaefer, *Homoleptic Tetranuclear Osmium Carbonyls: From the Rhombus via the Butterfly to the Tetrahedron*, Dalton Trans., 2008, 1366.

[5] T. T. Shi, Q. S. Li, Y. Xie, R. B. King, H. F. Schaefer, *Neutral Homoleptic Tetranuclear Iron Carbonyls: Why Haven't They Been Synthesized as Stable Molecules?*, New J. Chem., 34, 2010, 208.

[6] V. J. Johnston, F. W. B. Einstein, R. K. Pomeroy, *Structures of the New Binary Metal Carbonyl $OS_4(CO)_{15}$ and $(\eta^5\text{-}C_5Me_5)(OC)IrOs_3(CO)_{11})$. Clusters with Three-Center-Two-Electron Metal-Metal Bonds?*, J. Am. Chem. Soc., 109, 1987, 7220.

[7] P. S. Braterman, *Reactions of Coordinated Ligands: Volume 2*, Plenum Press, New York, USA, 1989.

[8] L. L. Ingraham, D. L. Meyer, *Biochemistry of Dioxygen*, Plenum Press, New York, USA, 1985.

[9] M. Merce, R. H. Crabtree, R. L. Richar, *A μ-Dinitrogen Complex with a Long N-N Bond. X-Ray Crystal Structure of [(PMe₂Ph)₄CReN₂MoCl₄(OMe)]*, J.C.S. Chem. Comm., 1973, 808.

[10] P. L. Holland, *Metal-Dioxygen and Metal-Dinitrogen Complexes: Where Are the Electrons?*, Dalton Trans., 39, 2010, 5415.

[11] A. D. Allen, R. O. Harris, B. R. Loescher, J. R. Stevens, R. N. Whiteley, *Dinitrogen Complexes of the Transition Metals*, Chemical Reviews, 73, 1973, 11.

[12] J. S. Valentine, *The Dioxygen Ligand in Mononuclear Group VIII Transition Metal Complexes*, Chemical Reviews, 73, 1973, 235.

[13] G. O. Spessard, G. L. Miessler, *Organometallic Chemistry*, Oxford University Press, New York, USA, 2010.

[14] N. N. Greenwood, A. Earnshaw, *Chemistry of the Elements*, Butterworth-Heinemann, Oxford, Britain, 1998.

[15] B. W. Pfennig, *Principles of Inorganic Chemistry*, John Wiley & Sons, New Jersey, USA, 2015.

INDEX

DALAL
INSTITUTE

DALAL
INSTITUTE

I

J

K

L

M

DALAL
INSTITUTE

DALAL
INSTITUTE

W

www.ingramcontent.com/pod-product-compliance
Lightning Source LLC
LaVergne TN
LVHW080420200726
843507LV00004B/675